5th International Conference on Mechanical, Manufacturing, Modeling and Mechatronics (IC4M 2020) and 5th International Conference on Design Engineering and Science (ICDES 2020)

IOP Conference Series: Materials Science and Engineering
Volume 825

Shenzhen, China
27 - 29 February 2020

ISBN: 978-1-7138-3609-4
ISSN: 1757-8981

Printed from e-media with permission by:

Curran Associates, Inc.
57 Morehouse Lane
Red Hook, NY 12571

Some format issues inherent in the e-media version may also appear in this print version.

Institute of Physics
Dirac House, Temple Back
Bristol BS1 6BE UK

Phone: 44 1 17 929 7481
Fax: 44 1 17 920 0979

techtracking@iop.org

Additional copies of this publication are available from:

Curran Associates, Inc.
57 Morehouse Lane
Red Hook, NY 12571 USA
Phone: 845-758-0400
Fax: 845-758-2633
Email: curran@proceedings.com
Web: www.proceedings.com

TABLE OF CONTENTS

DESIGN AND ANALYSIS

MATERIAL PROCESSING AND SURFACE TREATMENT

TESTING AND CONTROL

PREFACE

2020 5th International Conference on Mechanical, Manufacturing, Modeling and Mechatronics (IC4M 2020) and 2020 5th International Conference on Design Engineering and Science (ICDES 2020) were held in Shenzhen, China, February 27-29, 2020. IC4M 2020 & ICDES 2020 was organized by Hong Kong Society of Mechanical Engineers(HKSME). This conference aims to provide a platform for researchers and engineers to share their ideas, recent developments and successful practices in Mechanical, Manufacturing, Modeling and Mechatronics. The participants of the conference were from almost every part of the world, with background of either academia or industry, even well-known enterprise. The success and prosperity of the conference is reflected high level of the papers received.

The proceedings are a compilation of the accepted papers and represent an interesting outcome of the conference. This book covers 3 chapters: 1. Design and Analysis; 2. Material Processing and Surface Treatment; 3. Testing and Control.

We would like to acknowledge all of those who supported IC4M 2020 & ICDES 2020. Each individual and institutional help were very important for the success of this conference. Especially we would like to thank the organizing committee for their valuable advices in the organization and helpful peer review of the papers.

We sincerely hope that IC4M 2020 & ICDES 2020 will be a forum for excellent discussions that will put forward new ideas and promote collaborative researches. We are sure that the proceedings will serve as an important research source of references and the knowledge, which will lead to not only scientific and engineering progress but also other new products and processes.

Prof. Ruxu Du
Conference Chair

IOP Conf. Series: Materials Science and Engineering **825** (2020) 011002 doi:10.1088/1757-899X/825/1/011002

COMMITTEES

Conference Co-Chairs
Ruxu Du, South China University of Technology, China
Cees de Bont, Loughborough University, UK

Program Committee
Yong Zeng, Concordia University, Canada
Yan Wang, Georgia Institute of Technology, USA

Organizing Committee
Mohd Hamdi Bin Abd Shukor, University of Malaya, Malaysia
James Andrew Self, Ulsan National Institute of Science and Technology (UNIST), South Korea
Junjie Zhang, Harbin Institute of Technology, China
Stefan Dimo, University of Birmingham, UK

Scientific Committee
Jan Detand, Ghent University, Belgium
Kai Cheng, Brunel University, UK

International Technical Committee
Abraham Mansouri, American University in Dubai, UAE
Aidy Ali, National Defence University of Malaysia, Malaysia
Alain Bernard, IRCCyN Ecole Centrale Nantes, France
Amar Faiz Bin Zainal Abidin, Universiti Teknologi MARA, Malaysia
Amaresh Chakrabarti, Indian Institute of Science, India
Andrew Rae, University of the Highlands and Islands, UK
ANG Marcelo Jr Huibonhoa, National University of Singapore, Singapore
Chi Kit Au, The University of Waikato, New Zealand
Gopinath Chintala, Alliance University, India
Ismail Fidan, Tennessee Tech University, USA
Jian Jun Zhu, Sr. CAE in vehicle dynamics and simulation at General Motors, USA
Jos Gunsing, Avans University of Applied Science, Netherlands
Katz Zvi, University of Pretoria, South Africa
Kim da Costa Carneiro, Technical University of Denmark, Denmark
Lydia Anggraini, President University, Indonesia
Megat Mohamad Hamdan bin Megat Ahmad, National Defence University of Malaysia, Malaysia
Mohd Nasir Tamin, Universiti Teknologi Malaysia, Malaysia
Mojtaba Ahmadi, Carelton University, Canada
Moussaoui Kamel, Altran Company, France
N.ETHIRAJ, Dr. M.G.R Educational and Research Institute, India
Redouane Zitoune, Clement Ader Institute of Toulouse University, France
S.C. Johnson Lim, Universiti Tun Hussein Onn Malaysia, Malaysia
Samad Taghizadeh, Islamic Azad University-Maragheh branch, Iran
Sheng Li, Wright State University, USA
Sim Kok Swee, Multimedia University, Malaysia

IOP Conf. Series: Materials Science and Engineering **825** (2020) 011002 doi:10.1088/1757-899X/825/1/011002

Peer review statement

All papers published in this volume of *IOP Conference Series: Materials Science and Engineering* have been peer reviewed through processes administered by the proceedings Editors. Reviews were conducted by expert referees to the professional and scientific standards expected of a proceedings journal published by IOP Publishing.

IOP Conf. Series: Materials Science and Engineering **825** (2020) 012001 doi:10.1088/1757-899X/825/1/012001

Workspace Analysis and Dynamics Simulation of Manipulator based on MATLAB

Niu Guoxian[1,*] and Li Liang[2]

[1] Beijing Institute of Mechanical Equipment, Beijing, China
[2] School of Mechanical Science and Engineering, Huazhong University of Science and Technology, Wuhan, China

[*]E-mail: niuguoxian@126.com;

Abstract. Workspace analysis and joint moment calculation are the main contents of manipulator's design and calculation. According to the joint layout characteristics of the light cooperative manipulator, the workspace of the manipulator is divided into two parts: the arm workspace and the wrist workspace. The geometric method and Monte Carlo method are used to calculate the workspace of the manipulator and to guide the determination of the structural parameters of the manipulator. Based on Simulink/Simmechanics library, the dynamics model of six-axis serial manipulator is built and the joint rotational moment is calculated. The simulation results are in good agreement with the results of Newton Euler method. The simulation model can be used to replace Newton Euler method in the design process of the manipulator to improve the design efficiency.

1. Introduction

The six-axis manipulator has high degree of freedom and versatility, and is suitable for various industrial occasions.[1] In the design of the manipulator, firstly, according to the application scenario of the manipulator, the workspace and load requirements of the manipulator should be determined. Then, the structural dimensions of the manipulator should be determined. Finally, the reducers and servo motors of each axis should be selected.

2. Workspace Analysis of Manipulator

Generally, the six-axis manipulator is divided into four parts: base, upper arm, lower arm and wrist. A common joint layout is that the base of the manipulator is connected with the upper arm through the first and second joints, the upper arm is connected with the lower arm through the third joint, and the end of the lower arm is the fourth, fifth and sixth joints connected sequentially. At present, most of the light cooperative manipulators in the market adopt this joint layout, such as UR series cooperative manipulators.

When designing a manipulator, the reachable range of the center point of the flange at the end of the manipulator is often defined as the workspace of the manipulator. However, when the center point of the flange of the manipulator is in the edge area of the workspace, the flexibility of the manipulator is limited, thus, not all the workspace is the effective workspace. Figure 1 shows the structure and workspace of UR5 manipulator.

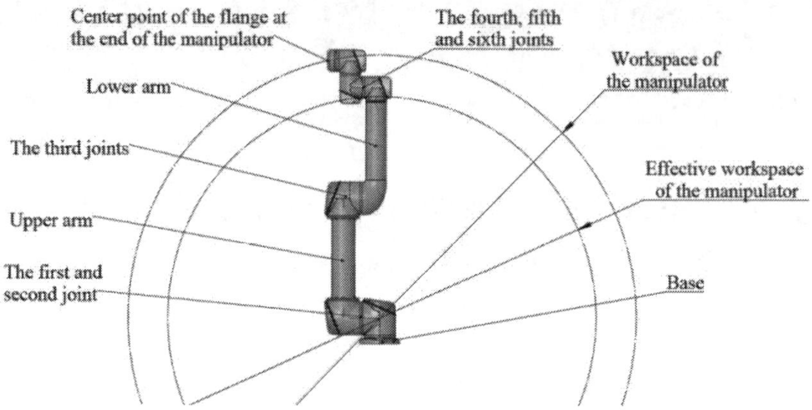

Figure 1. The structure and workspace of UR5 manipulator.

The joint layout is very similar to that of the human arm: the length of upper arm is similar to that of lower arm, and the range of motion of the end of the lower arm is approximately spherical, pushing the wrist to the area where it needs to work; the fourth, fifth and sixth joints of the lower arm are similar to human wrist, with high flexibility and can rotate to different posture to complete the work.

In order to describe the reachable range and flexibility of the manipulator better, the workspace of the manipulator is divided into two parts: arm workspace and wrist workspace. The arm workspace is used to measure the effective operation range of the manipulator, and the wrist workspace is used to measure the flexibility of the wrist at the end of the manipulator. Figure 2 is a schematic illustration of workspace partitioning using the UR5 manipulator as an example.

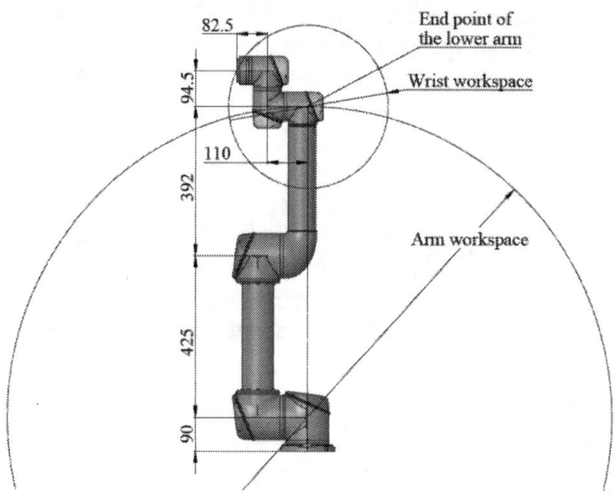

Figure 2. Schematic illustration of workspace partitioning of the UR5 manipulator.

The axes of the second, third and fourth joints of UR5 manipulator are parallel to each other, so the division of workspace is based on the intersection of the axes of the first joint and the axes of the fourth joint as show in figure 2. At this time, the shape of the arm workspace is a solid sphere.

The coordinate system is established with the intersection point of the first and second joint axes as the origin point. The range of motion of the arm workspace, i.e. the range of end point P (x, y, z) of the lower arm, can be written out directly.

IOP Conf. Series: Materials Science and Engineering **825** (2020) 012001 doi:10.1088/1757-899X/825/1/012001

$$\begin{cases} x^2+y^2+z^2 \le 817 \\ x^2+y^2+z^2 \ge 33 \end{cases} \tag{1}$$

Monte Carlo method is used to calculate the wrist workspace because of that shape of the wrist workspace is irregular. The steps of using Monte Carlo method to calculate the workspace of manipulator are as follows: establishing coordinate system, selecting the position of each joint randomly, finding the position of the end point of manipulator relative to the base center in different positions and postures, drawing the point in the coordinate system, repeating this process, when the number of end points reaches sufficient quantity, the workspace of manipulator can be obtained.[2]

The coordinate system of the manipulator from the end of the lower arm to the center of the flange at the end of manipulator is established. As shown in Figure 3, the DH parameters of each coordinate system are as shown in Table 1.

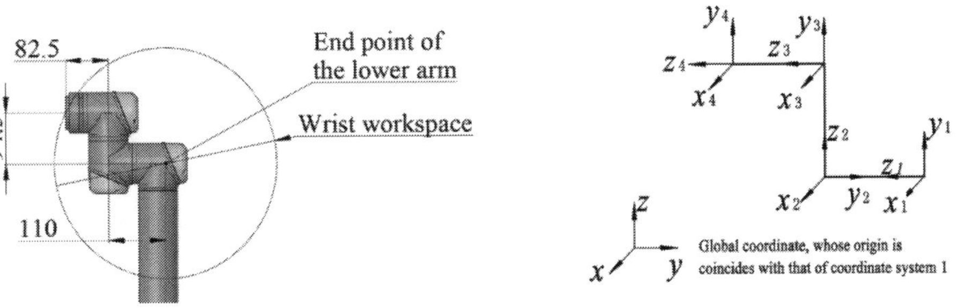

Figure 3. Coordinate system of the wrist workspace.

Table 1. DH parameter of wrist workspace coordinates.

Ordinate order i	Connecting rod offset d_i (mm)	Twist angle α_{i-1} (°)	Joint angle θ_i (°)
1	0	90	0+θ1
2	110	-90	0+θ2
3	94.5	90	0+θ3
4	82.5	0	0

Coordinate system of the wrist workspace is non-standard DH coordinate system, adjacent coordinate system transformation according to the following formula:

$$T_i = Trans(z,d_i)Rotate(x,\alpha_{i-1})Rotate(z,\theta_i) \tag{2}$$

$$= \begin{bmatrix} 1 & 0 & 0 & 0 \\ 0 & 1 & 0 & 0 \\ 0 & 0 & 1 & d_i \\ 0 & 0 & 0 & 1 \end{bmatrix} \begin{bmatrix} 1 & 0 & 0 & 0 \\ 0 & \cos\alpha_{i-1} & -\sin\alpha_{i-1} & 0 \\ 0 & \sin\alpha_{i-1} & \cos\alpha_{i-1} & 0 \\ 0 & 0 & 0 & 1 \end{bmatrix} \begin{bmatrix} \cos\theta_i & -\sin\theta_i & 0 & 0 \\ \sin\theta_i & \cos\theta_i & 0 & 0 \\ 0 & 0 & 1 & 0 \\ 0 & 0 & 0 & 1 \end{bmatrix}$$

$$= \begin{bmatrix} \cos\theta_i & -\sin\theta_i & 0 & 0 \\ \cos\alpha_{i-1}\sin\theta_i & \cos\alpha_{i-1}\cos\theta_i & -\sin\alpha_{i-1} & 0 \\ \sin\alpha_{i-1}\sin\theta_i & \sin\alpha_{i-1}\cos\theta_i & \cos\alpha_{i-1} & d_i \\ 0 & 0 & 0 & 1 \end{bmatrix}$$

The transformation matrix from coordinate system 4 to global coordinate system is:

$$T_w = T_1T_2T_3T_4 = \begin{bmatrix} & R & & P \\ 0 & 0 & 0 & 1 \end{bmatrix} \tag{3}$$

Among them, R is 3x3 matrix, P is 3x1 matrix and P is the origin of coordinate system 4 expressed in the global coordinate system. Given the rotation angles of the 4th, 5th and 6th joints with random functions, the origin of coordinate system 4 (the center point of the end flange) relative to the global coordinate system is obtained by the above coordinate transformation, and the points are repeatedly taken out in the coordinate system. Finally, the UR5 wrist workspace point map is as Figure 4:

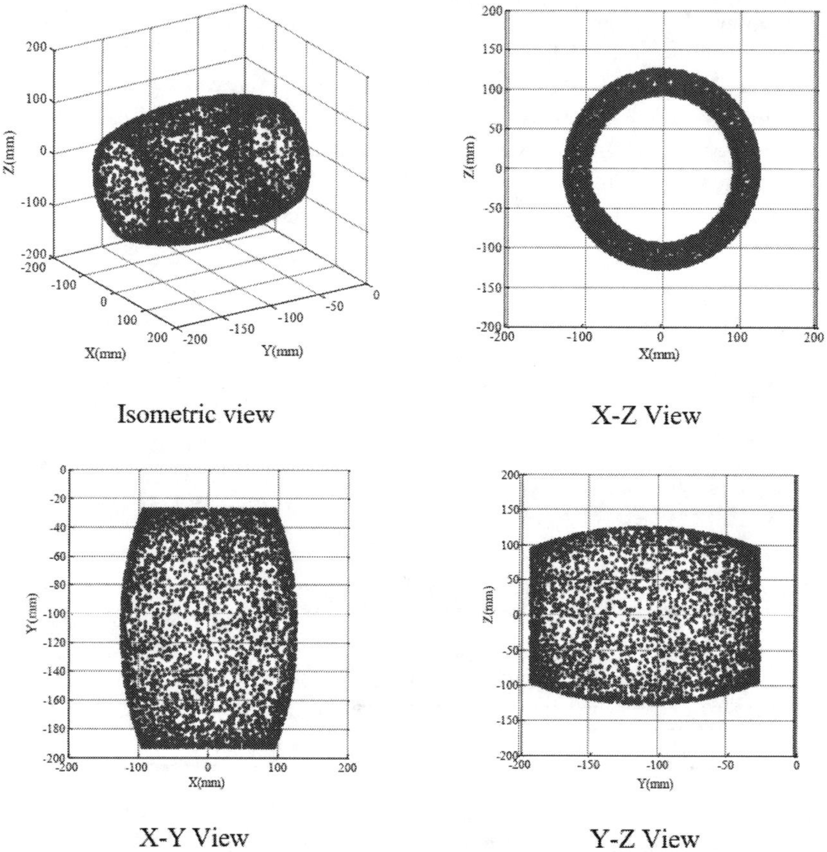

Figure 4. Cloud chart of the wrist workspace.

From the sketch of the wrist workspace, it can be found that the wrist workspace is roughly hollow olive-shaped with flattened ends, and the scale of each direction is as follows:

$$\begin{cases} -125.5 \le x \le 125.5 \\ -125.5 \le y \le 125.5 \\ -192.5 \le y \le -27.5 \end{cases} \tag{4}$$

When the manipulator moves, the center of the wrist workspace moves within the scope of the arm workspace. All space swept by the wrist workspace are all positions that can be reached by the center point of the end flange. This method can accurately measure the effective operating space and the flexibility of the end joints of the manipulator. It has certain significance to guide the structural design of the manipulator.

3. Dynamics Simulation of Manipulator

After the workspace design of the manipulator is completed, it is necessary to calculate the required moment of each joint in the course of movement, and select the servo motors and reducers. The joint moment of universal six-axis manipulator can be calculated by Newton-Euler method. Craig gives the formula of Newton-Euler method for serial manipulator. The Newton-Euler method is simple and reliable, but the calculation is slightly cumbersome, and the formulas need to be deduced again when the structure of the manipulator changes.

Simmechanics, a multi-body dynamics simulation toolbox, is provided in MATLAB. The dynamics model of the manipulator based on Simmechanics is established as follows.[3] The coordinate system of the UR5 manipulator is established by DH parameter method,[4] as shown in Fig. 5. The coordinate transformation parameters are shown in Table 2.

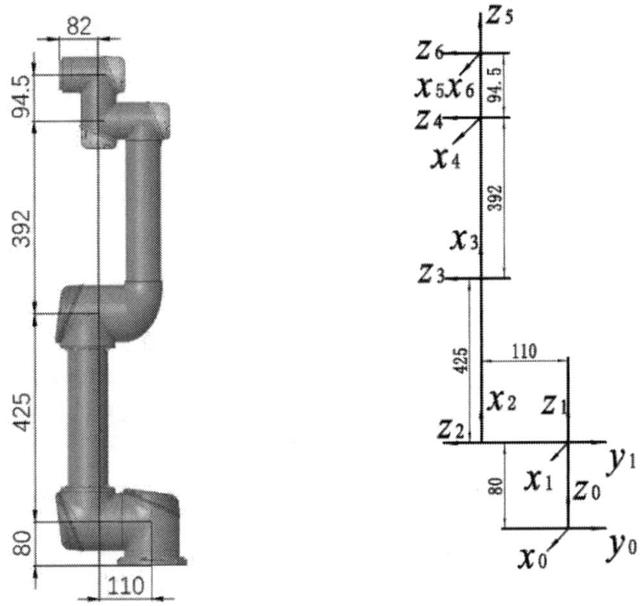

Figure 5. Coordinate system of the UR5 manipulator.

Table 2. DH parameters of wrist coordinates for UR5 manipulator.

Ordinate order i	Joint angle θ_i (°)	Connecting rod length a_{i-1}(mm)	Twist angle α_{i-1} (°)	Connecting rod offset d_i (mm)
1	$0+\theta_1$	0	0	80
2	$90+\theta_2$	0	90	110
3	$0+\theta_3$	425	0	0
4	$-90+\theta_4$	392	0	0
5	$0+\theta_5$	0	-90	94.5
6	$0+\theta_6$	0	90	0

The coordinate transformation matrix of adjacent connecting rods is:

$$T_i = Rotate(x,\alpha_{i-1})Trans(x,a_{i-1})Rotate(z,\theta_i)Trans(z,d_i)$$
$$= \begin{bmatrix} 1 & 0 & 0 & 0 \\ 0 & \cos\alpha_{i-1} & -\sin\alpha_{i-1} & 0 \\ 0 & \sin\alpha_{i-1} & \cos\alpha_{i-1} & 0 \\ 0 & 0 & 0 & 1 \end{bmatrix} \begin{bmatrix} 1 & 0 & 0 & a_{i-1} \\ 0 & 1 & 0 & 0 \\ 0 & 0 & 1 & 0 \\ 0 & 0 & 0 & 1 \end{bmatrix} \begin{bmatrix} \cos\theta_i & -\sin\theta_i & 0 & 0 \\ \sin\theta_i & \cos\theta_i & 0 & 0 \\ 0 & 0 & 1 & 0 \\ 0 & 0 & 0 & 1 \end{bmatrix} \begin{bmatrix} 1 & 0 & 0 & 0 \\ 0 & 1 & 0 & 0 \\ 0 & 0 & 1 & d_i \\ 0 & 0 & 0 & 1 \end{bmatrix}$$
$$= \begin{bmatrix} \cos\theta_i & -\sin\theta_i & 0 & a_{i-1} \\ \sin\theta_i\cos\alpha_{i-1} & \cos\theta_i\cos\alpha_{i-1} & -\sin\alpha_{i-1} & -\sin\alpha_{i-1}d_i \\ \sin\theta_i\sin\alpha_{i-1} & \cos\theta_i\sin\alpha_{i-1} & \cos\alpha_{i-1} & \cos\alpha_{i-1}d_i \\ 0 & 0 & 0 & 1 \end{bmatrix} \quad (5)$$

The three-dimensional model of the manipulator is established by using CAD software, and the DH coordinate system is established in the CAD software. The inertia tensor and centre of mass of the links relative to the coordinate system is calculated. The physical model of the manipulator is established in Simulink according to the connection relationship of the links of the manipulator, and the inertia attributes of the links are set up. Finally, the manipulator's dynamics model is as figure 6.

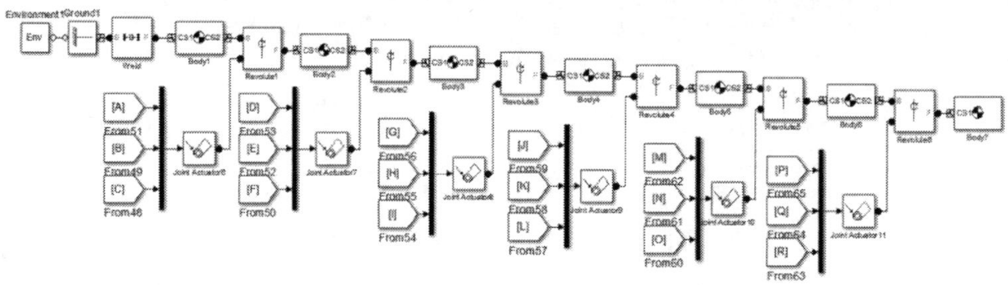

Figure 6. Illustration of the manipulator's dynamics model.

During the working process of the manipulator, the second joint receives the greatest load. Assuming that the other joints are fixed in accordance with the position shown in Figure 3, the second joint moves from - 90 to 90 degrees, which takes 50 seconds. The position, velocity and acceleration of the second joint are generated by the trajectory planning function "jtraj". The motion data generated by trajectory planning function are input into the second joint actuator, and the reaction moment curve of the second joint varying with time is obtained, as shown in Figure 7.

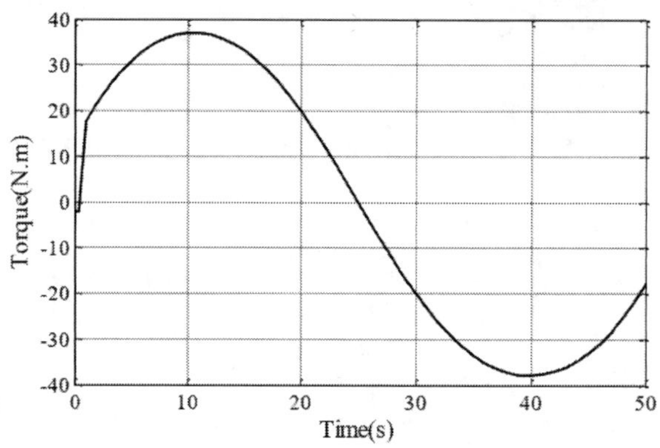

Figure 7. The reaction moment curve of the second joint varying with time.

The Newton Euler method uses the knowledge of rigid body mechanics to derive the recursive formula of kinematics, and then derives the mathematical model of robot dynamics. The Newton Euler method calculates the torque of each joint as follows. [5]

The motion of each link is first recurved according to the position, velocity and acceleration of each joint:

$$\begin{aligned}
{}^{i+1}\omega_{i+1} &= {}^{i+1}_{i}R\,{}^{i}\omega_i + \dot{\theta}_{i+1}\,{}^{i+1}Z_{i+1} \\
{}^{i+1}\dot{\omega}_{i+1} &= {}^{i+1}_{i}R\,{}^{i}\dot{\omega}_i + \ddot{\theta}_{i+1}\,{}^{i+1}Z_{i+1} + {}^{i+1}_{i}R\,{}^{i}\omega_i \times \dot{\theta}_{i+1}\,{}^{i+1}Z_{i+1} \\
{}^{i+1}v_{i+1} &= {}^{i+1}_{i}R\left({}^{i}v_i + {}^{i}\omega_i \times {}^{i}P_{i+1}\right) \\
{}^{i+1}\dot{v}_{i+1} &= {}^{i+1}_{i}R\left({}^{i}\dot{v}_i + {}^{i}\dot{\omega}_i \times {}^{i}P_{i+1} + {}^{i}\omega_i \times \left({}^{i}\omega_i \times {}^{i}P_{i+1}\right)\right) \\
{}^{i}\dot{v}_{ci} &= {}^{i}\dot{v}_i + {}^{i}\dot{\omega}_i \times {}^{i}P_{ci} + {}^{i}\omega_i \times \left({}^{i}\omega_i \times {}^{i}P_{ci}\right)
\end{aligned} \tag{6}$$

According to the Euler equation of motion, the resultant force and resultant moment of each connecting rod are obtained:

$$\begin{aligned} {}^{i}F_{i} &= m_{i}\,{}^{i}\dot{v}_{C_{i}} \\ {}^{i}N_{i} &= I_{C_{i}}\,{}^{i}\dot{\omega}_{i} + {}^{i}\omega_{i} \times I_{C_{i}}\,{}^{i}\omega_{i} \end{aligned} \tag{7}$$

f_i is used to express the force for connecting rod applied by previous rod, and n_i is used to express the moment connecting rod receives.

$$\begin{aligned} {}^{i}f_{i} &= {}^{i}F_{i} + {}_{i+1}^{\ i}R\,{}^{i+1}f_{i+1} \\ {}^{i}n_{i} &= {}^{i}N_{i} + {}_{i+1}^{\ i}R\,{}^{i+1}n_{i+1} + {}^{i}P_{C_{i}} \times {}^{i}F_{i} + {}^{i}P_{i+1} \times {}_{i+1}^{\ i}R\,{}^{i+1}f_{i+1} \end{aligned} \tag{8}$$

In Simulink, Newton Euler method is used to calculate the moment change process of the second joint in the process, and the results of Newton Euler method are subtracted from the simulation results. The difference changes with time as shown in Figure 8.

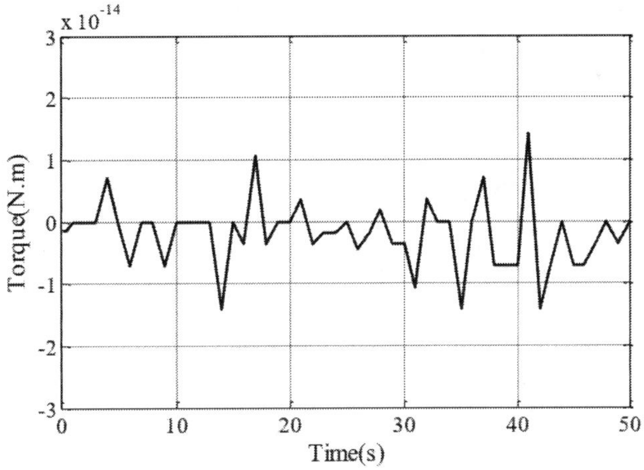

Figure 8. Deviation between simulation model and Newton Euler method.

As can be seen from the figure, the simulation results are in good agreement with the results of Newton Euler method, and the simulation model can be used instead of Newton Euler method.

4. Conclusion

In order to describe the reachable range and flexibility of the manipulator, the workspace of the UR5 manipulator is divided and calculated. The Monte Carlo method is a random sampling method used in the calculation of the wrist workspace. Compared with the actual workspace, the workspace obtained by Monte Carlo method still has some errors, but by increasing the number of sampling points, we can get a higher accuracy, so as to approximate the actual workspace until the demand is met. Because the shape of the arm workspace is regular, the analytic formula of the arm workspace is written directly, while the shape of the wrist space is irregular, the Monte Carlo method is used for analysis. When the joint layout of manipulator adopts other forms, the division and analysis method of workspace should be selected according to the situation.

The dynamics model of the manipulator is established based on Simmechanics library, and the joint moment is calculated. The results are compared with those of Newton Euler method. The simulation model is in good agreement with those of Newton Euler method. The simulation model is more intuitive and visual, and the model can be established or changed more quickly, which is more suitable for the rapid development environment. Because the working environment of the second joint is the worst, this paper only simulates the motion of the second joint alone. If necessary, it can simulate the force moment of all joints when joints move together.

Reference

[1] Xu Guangsha. Research on Robot Control System Based on POWERLINK Bus 2015 *D. Shandong University*.
[2] Wang yang, Guo Linfu. Kinematics and Workspace Analysis of Light Cooperative Manipulator 2017 *J.Mechanical Engineering and Automation*. (1) 44-46.
[3] Ru Feifei. Research on the Application of Double Closed-loop Control Method in Two-wheel Self-balancing Car 2013 *D. Henan University*.
[4] Li Xianhua, Zhang Leigang, Guo Shuai, et al. A Rapid Construction Method of Open Manipulator 3D Virtual Simulation Platform 2018 *J. Mechanical science and technology*. 37 (04): 599-606.
[5] John J. Craig. Introduction to Robotics Mechanics and Control 2006 *M. China Machine Press*. 137-139.

IOP Conf. Series: Materials Science and Engineering **825** (2020) 012002 doi:10.1088/1757-899X/825/1/012002

Structure Design and Analysis of a Novel Forward-folding Rotor used in a Downwind Horizontal-axis Turbine

Haoran Meng, Pan Zeng and Liping Lei*

Key Laboratory for Advanced Materials Processing Technology of MOE, Department of Mechanical Engineering, Tsinghua University, Beijing 100084, China

* E-mail address: leilp@tsinghua.edu.cn.

Abstract. To alleviate the stiffness constraints of the conventional blade and thus reduce the rotor mass, a novel forward-folding rotor (Downwind Forward-folding Rotor, DFFR) used in a downwind horizontal-axis turbine is presented. This novel rotor is designed to align the combination of gravitational, centrifugal, and thrust forces along the blade path, resulting in primarily tensile loads instead of cantilever loads on the blades. The DFFR blades fold forward at a power-limited condition, which induces the change of the blade pitch angle and cone angle and thus maintains a constant power output. To quantify the mass savings, a 5-MW DFFR was designed based on the NREL 5-MW reference rotor. According to the results calculated by an improved BEM method, the rotor power and torque of the 5-MW DFFR have a slight increase compared with those of the NREL-5MW reference rotor, while the rotor thrust of the 5-MW DFFR is smaller than that of the NREL-5MW reference rotor. Furthermore, based on the finite element analysis, the blade of the 5-MW DFFR had an 18.9% mass saving and an 8.4% peak stress reduction compared with the blade of the NREL-5MW reference rotor, over a range of operating wind speeds and azimuthal angles.

1. Introduction

Wind energy, a kind of renewable energy, has been widely utilized over the past decades in the world. The average wind turbine rated power has increased 60-fold over the past 40 years and is expected to reach 10-20 MW in the future [1, 2]. One of the most significant design issues associated with the large-scale wind turbines is limiting the blade mass. It is difficult to reduce the blade mass for conventional large-scale wind turbines since the blade mass is fundamentally related to the stiffness required to prevent the blade-tower strike and resist the peak loads. In this regard, downwind configuration was considered to be used in the large-scale wind turbines since the blade-tower collisions are avoided in operation, thus the blade can be designed lighter and more flexible to reduce the manufacture and installation cost [3-5].

Some innovative downwind rotor concepts, such as a downwind coned rotor (present by P M Jamieson et al. [6]) and a two-bladed 15 kW flexible downwind turbine (presented by F Rasmussen et al. [7]), were proposed to relax stiffness constraints of the conventional blade. However, these rotor concepts were mainly used in the small or medium scale turbine. Therefore, a morphing downwind-aligned rotor (MoDaR) [8] and a downwind pre-aligned rotor (DPAR) [9] applied in extreme-scale wind turbines (≥10 MW) were proposed by B Ichter et al. and E Loth et al., respectively, to achieve blade mass saving and peak structural stress reduction. Nevertheless, the additional morphing mechanism in the MoDaR may negate these mass savings [8], while the rotor cone angle cannot be adjusted in the DPAR [9].

Based on the previous work, a novel rotor concept with blade forward-folding along the oblique axis used in the large-scale downwind turbine was previously introduced by H Meng et al. [10] and defined as a downwind forward-folding rotor (DFFR). In the DFFR, the rotor cone angle and the blade pitch angle can be adjusted coupledly by only one mechanism (an oblique folding axis), to achieve peak structural stress reduction and power output control simultaneously. In this paper, the structural design and mass saving analysis of a 5-MW DFFR were presented. The structural analysis results of the 5-MW DFFR was also compared with that of the NREL 5-MW reference rotor to validate the peak stress reduction by blade forward-folding method.

2. Concept of the downwind forward-folding rotor

As shown in Figure 1, the blade of the DFFR can fold along a near-hub oblique axis instead of rotating around the pitch axis, to maintain a constant power output under the power-limited condition. The incline angle of the folding axis (γ), is defined as the angle between the folding axis and the axis that is perpendicular to the blade length direction, and is in the range from -90° to 90°. The blade fold angle (δ) is defined as the blade rotation angle along the folding axis, as shown in Figure 1. When the blade folds along the folding axis, both the blade coning (control the load alignment) and the pitching (control the power output) is generated.

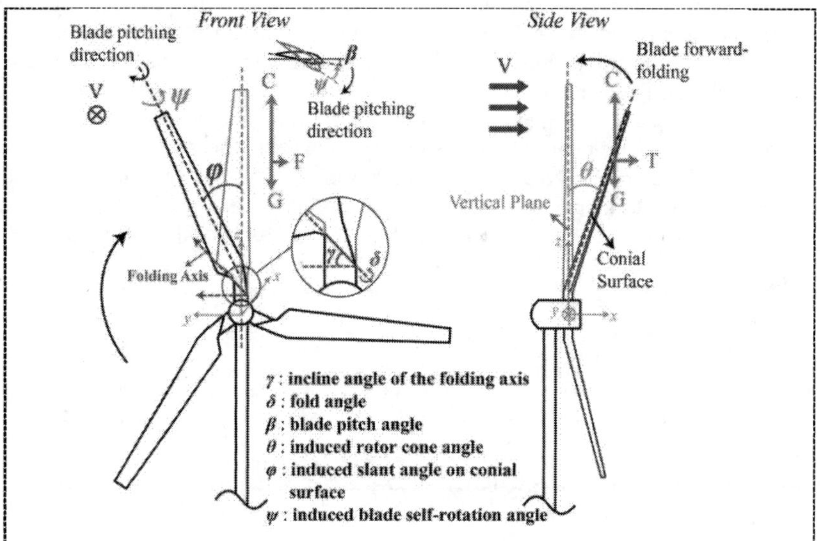

Figure 1. Schematic diagram of the DFFR.

Inspired by the load alignment concept [8], the DFFR folds at a certain angle (defined as initial fold angle (δ_0)) at the sub-rated and rated wind speeds ($V_{cut\text{-}in} \leq V \leq V_{rated}$) to align the combination of the gravitational, centrifugal, and thrust forces along the blade path (in Figure 1), resulting in primarily tensile loads. When the wind speed increases above the rated wind speed, the blades fold forward, leading to the increase of the blade pitch angle (β) to maintain constant power output, as well as the decrease of the induced rotor cone angle (θ) to achieve the load alignment (because of the reduced downwind thrust). As shown in Figure 1, the fold angle (δ) can be decomposed into three components: induced rotor cone angle (θ), induced slant angle on conical surface (φ), and induced blade self-rotation angle (ψ). These three angles can be shown as follows:

$$\theta = \arctan \frac{\cos\gamma \sin\delta}{\sin^2\gamma + \cos^2\gamma \cos\delta} \tag{1}$$

$$\varphi = \arcsin\left(\sin\gamma\cos\gamma\left(1-\cos\delta\right)\right) \quad (2)$$

$$\psi = \arctan\frac{\sin\gamma\sin\delta}{\cos^2\gamma+\sin^2\gamma\cos\delta} \quad (3)$$

Figure 2 shows the relationship of the fold angle (δ), the blade angle of attack (α), and the power output (P) with respect to the wind speed. It can be seen that under (a) sub-rated and (b) rated conditions, the DFFR operates with an initial fold angle (δ_0). The blades fold forward when the wind speed increases above the rated wind speed, and the blade angle of attack (α) of the DFFR decreases, which can degrade the blade aerodynamic performance and thus maintain a constant power output.

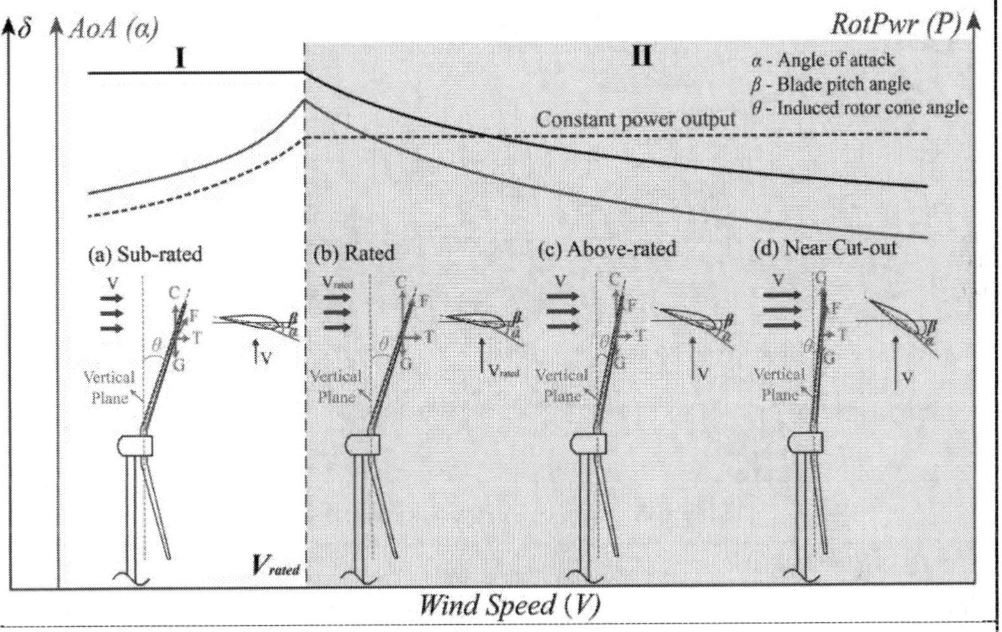

Figure 2. Relationship of the fold angle, the blade angle of attack, and the rotor power output with respect to the wind speed under: (a) sub-rated, (b) rated, (c) above-rated and (d) near-cut-out conditions.

3. Blade design for a 5-MW DFFR

To quantify the mass saving benefits of the DFFR, the blade model of a 5-MW DFFR was established based on the NREL 5-MW reference wind turbine rotor [11] (defined as 5-MW ConvR). The incline angle of the folding axis (γ) and the initial fold angle (δ_0) are designed as 60° and 35°, respectively, and the location of the folding axis was defined as 11.3% blade span. The 5-MW DFFR has the same airfoil, blade size, and chord length distribution as the 5-MW ConvR, while the blade twist angle distribution of the DFFR need to be re-designed based on the ConvR to eliminate the change of blade pitch angle induced by the initial fold angle (δ_0). Based on the improved blade element momentum (BEM) method proposed by W Xie et al. [12], the blade angles of attack of the DFFR were ensured to be the same as those of the ConvR, by adding different offset angles to the twist angle of each blade element all along the blade span of the ConvR. The twist angle and the angle of attack distributions of the ConvR and the DFFR along blade span were shown in Figure 3.

The blade internal structure and material of the ConvR and the DFFR were referred from the technical report presented by B R Resor [13]. Different parts (leading edge, spar cap, trailing edge, reinforcement, and shear web) of the entire blade are composed by several layers of different materials. Material properties used in the blade model are summarized in Table 1.

To provide a comparison to the DFFR concept, a conventional 5-MW turbine blade was first created according to the literature [13], the internal structure of the conventional blade can be seen in Figure 4(a). For the 5-MW DFFR blade, compared with the 5-MW ConvR, the numbers of material layers of the spar cap (UD Carbon), leading edge (FOAM) and the trailing edge (FOAM) were reduced, while the number of material layers of the blade root (Triax) was slightly increased, as shown in Figure 5. Furthermore, the shear webs thickness of the DFFR blade were reduced from 50 mm to 30 mm, and a third shear web was added to the trailing edge. The internal structure of the DFFR blade can be seen in Figure 4(b). The blade mass of the DFFR was calculated as 10,302 kg after mass reduction design, compared to the counterpart of the ConvR blade with a mass of 12,703 kg, having an 18.9% mass reduction.

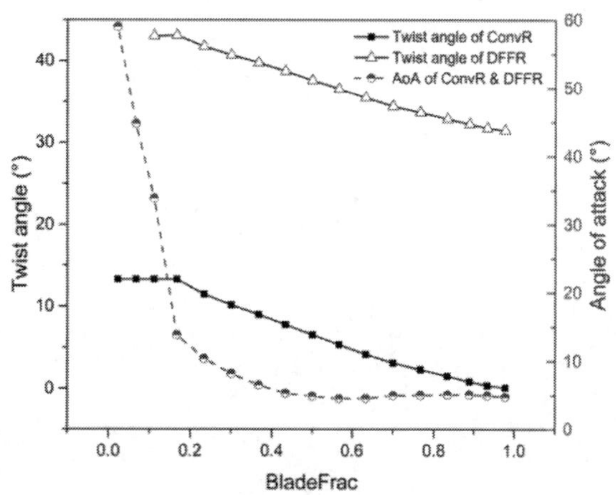

Figure 3. Twist angle and angle of attack (AoA) distributions of the ConvR and the DFFR along blade span station.

Table 1. Summary of material properties [13].

	Layer Thickness (mm)	Ex (MPa)	Ey (MPa)	Gxy (MPa)	prxy	Dens. (kg/m^3)	Main material for which part
Gelcoat	0.05	3440		1380	0.3	1235	Blade surface
E-LT-5500 (UD)	0.47	41,800	14,000	2630	0.28	1920	Trailing edge reinforcement
SNL (Triax)	0.94	27,700	13,650	7200	0.39	1850	Blade root
Saertex (DB)	1	13,600	13,300	11,800	0.49	1780	Shear web
FOAM	1	256	256	22	0.3	200	Shear web, Leading Edge, Trailing Edge
Carbon (UD)	0.47	114,500	8390	5990	0.27	1220	Spar cap

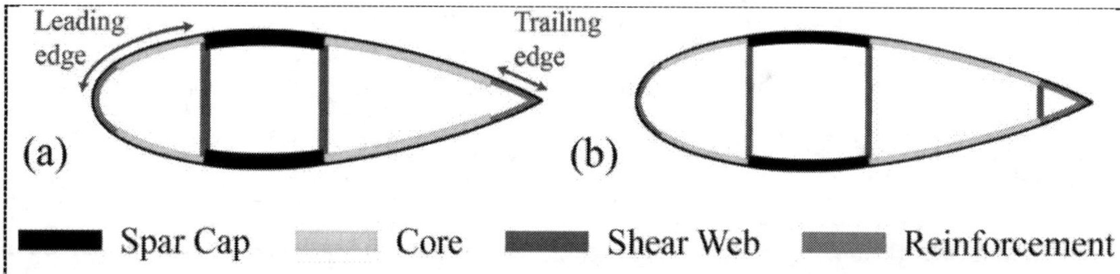

Figure 4. Cross section of internal structure for the (a) ConvR blade and (b) DFFR blade.

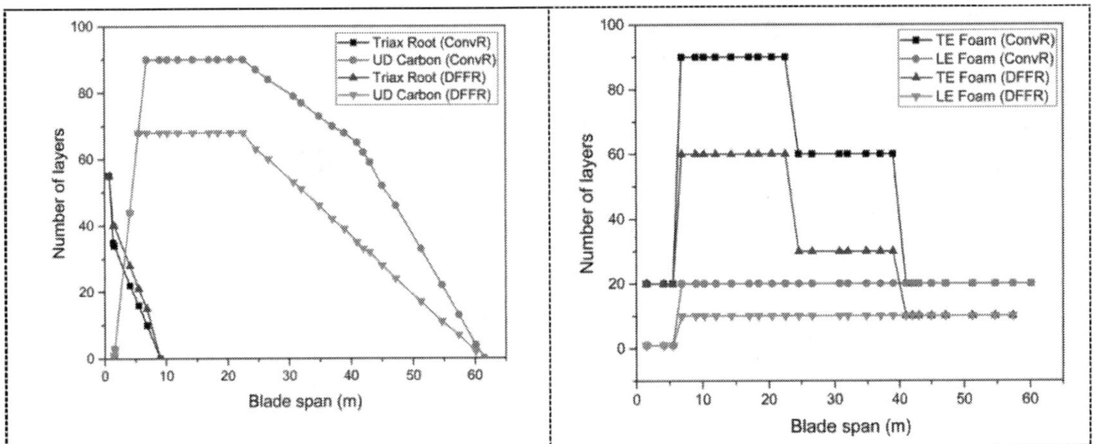

(a) (b)

Figure 5. Numbers of layers comparison for the (a) triax root and UD carbon and (b) trailing edge (TE) foam and leading edge (LE) foam

4. Aerodynamic and structural analysis for the conventional and DFFR blades

Before conducting the structural analysis of the two blades, the aerodynamic forces of the two rotors should be determined first. Blade Element Momentum (BEM) method, a basic theory in the field of rotor aerodynamics, was applied to calculate the rotor torque, thrust and power of the 5-MW ConvR under various operating wind speeds. The results of calculation were quite coincident with the values in the NREL 5-MW reference turbine report [11]. For the 5-MW DFFR, an improved BEM method [12] was used to calculate the aerodynamic forces. The calculated results of the two rotors were shown in Figure 6, it can be seen that the rotor power and torque of the 5-MW DFFR have a slight increase compared with those of the 5-MW ConvR, while the rotor thrust of the 5-MW DFFR is smaller than that of the 5-MW ConvR.

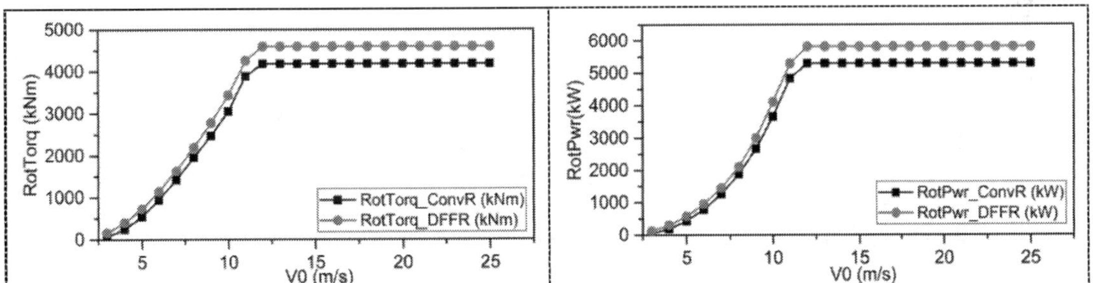

(a) (b)

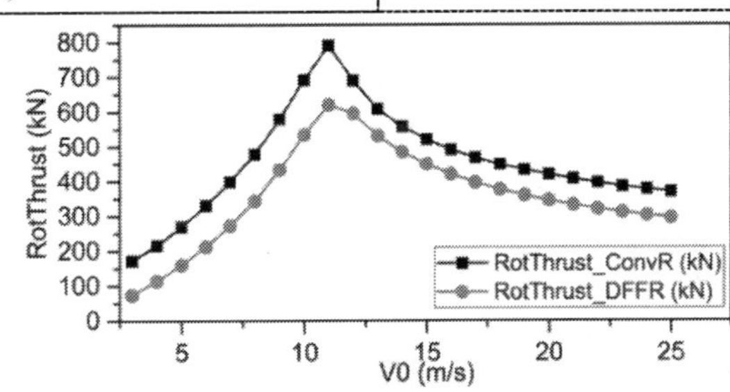

(c)

IOP Conf. Series: Materials Science and Engineering **825** (2020) 012002 doi:10.1088/1757-899X/825/1/012002

Figure 6. Comparison of the (a) rotor torque, (b) rotor power, and (c) rotor thrust between the ConvR and the DFFR

To compare the peak stress of the two blades, a ConvR blade model and a DFFR blade model were analysed using Finite Element Analysis (FEA) with ANSYS. The SHELL181, which is suitable for thin applications, was used in ANSYS, and the element size is 0.08 m for the entire blade model. The ConvR blade was simulated using an 103,992 element mesh, and the DFFR blade was simulated using an 102,553 element mesh. The aerodynamic forces calculated by improved BEM method were assigned to each node averagely by ANSYS commands, while the gravitational and centrifugal forces were applied via the ANSYS built-in inertial loading function. In addition, a fixed constraint was applied at the end of blade root for the ConvR, and the location where the folding axis is attached to blade for the DFFR, respectively.

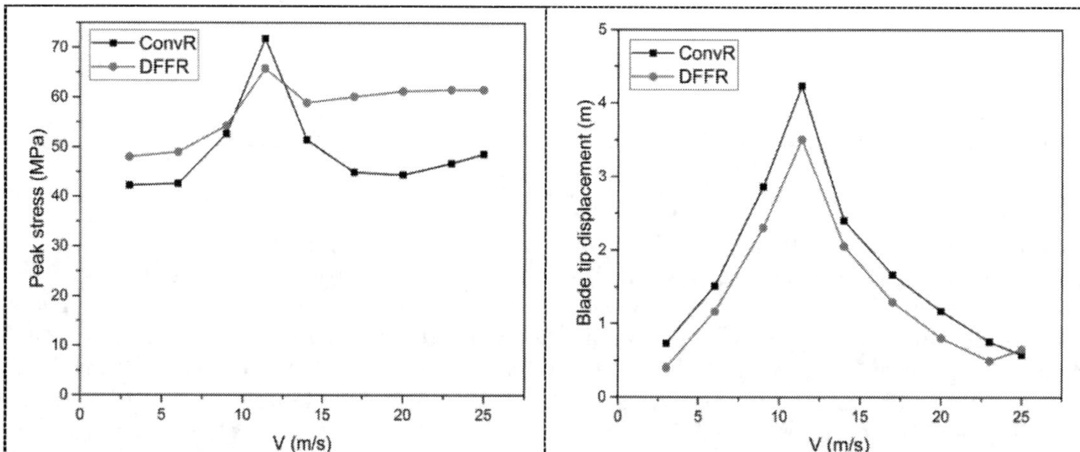

(a) (b)

Figure 7. (a) Peak Von Mises Stresses and (b) blade tip displacements of the ConvR and the DFFR blade as a function of operating wind speeds

Both the ConvR blade and the DFFR blade were simulated over a full range of operating wind speeds and azimuthal angles. Under a certain wind speed, the peak stress was selecting from the maximum Von Mises stresses over various azimuthal angles. The variation in peak Von Mises stresses of the ConvR blade and the DFFR blade is shown in Figure 7(a). It can be seen that both the ConvR blade and the DFFR blade has a peak stress at rated conditions; and the DFFR blade peaks with a stress of 65.7 MPa, having an 8.4% reduction compared with the ConvR blade (peaks with a stress of 71.8 MPa). As shown in Figure 7(b), the blade tip displacement of the DFFR blade is also smaller than that of the ConvR blade. Some results of the two rotors were listed in Table 2.

Table 2. Comparisons between the DFFR and the ConvR.

	DFFR	ConvR	Reduction ratio
Blade peak stress (MPa)	65.7	71.8	8.4%
Shear web peak stress (MPa)	48.1	57.0	15.6%
Blade tip displacement (m)	3.50	4.23	17.2%
Blade mass (kg)	10,302	12,703	18.9%

5. Conclusion

The DFFR concept may be able to alleviate the blade stiffness constraints and thus reduce the rotor mass, since it is designed to align the combination of gravitational, centrifugal, and thrust forces along the blade path, resulting in primarily tensile loads instead of cantilever loads on the blades. To quantify the mass savings of the DFFR concept, a Finite Element Analysis was conducted for a 5-MW DFFR, which is designed based on the NREL 5-MW wind turbine rotor. The blade mass of the DFFR (10,302 kg) has an 18.9% reduction compared with that of the ConvR (12,703 kg), by reducing the numbers of material layers of the blade. The aerodynamic performances of the DFFR and the ConvR were investigated based on the improved BEM method. The results show that the rotor power and torque of the DFFR have a slight increase compared with those of the ConvR, while the rotor thrust of the DFFR is smaller than that of the ConvR. At last, according to the FEA results, the peak stress of the DFFR (65.7 MPa) is lower than that of the ConvR (71.8 MPa) by 8.4%, which means the blade mass saving and the peak stress reduction can be achieved by DFFR concept. However, the rotor concept feasibility requires the more elaborate aerodynamic analysis and structure design, including the design of the folding mechanism and associated control strategy.

Acknowledgments

This work is supported by the National Natural Science Foundation of China (No. 51575296 and No. 51875305).

References

[1] Fichaux N, Beurskens J, Jensen P H and Wilkes J 2011 Design Limits and Solutions for Very Large Wind Turbines—A 20 MW Turbine Is Feasible. European Wind Energy Association)
[2] Ashwill T 2009 Materials and innovations for large blade structures: research opportunities in wind energy technology. In: *50th AIAA/ASME/ASCE/AHS/ASC Structures, Structural Dynamics, and Materials Conference,* p 2407
[3] Koh J and Ng E 2016 Downwind offshore wind turbines: Opportunities, trends and technical challenges *Renewable Sustainable Energy Reviews* **54** 797-808
[4] Kress C, Chokani N and Abhari R S 2016 Design Considerations of rotor cone angle for downwind wind turbines *Journal of Engineering for Gas Turbines Power* **138** 052602
[5] Wang Z, Tian W and Hu H 2018 A Comparative study on the aeromechanic performances of upwind and downwind horizontal-axis wind turbines *Energy Conversion and Management* **163** 100-10
[6] Jamieson P M and Jaffrey A 1995 Advanced Wind Turbine Design *Journal of Solar Energy Engineering* **119** 315-20
[7] Rasmussen F, Thirstrup J T, Vølund P, Leconte P, Szechenyi E and Westergaard C 1998 Soft Rotor Design for Flexible Turbines. Risø National Laboratory, Roskilde, Denmark)
[8] Ichter B, Steele A, Loth E, Moriarty P and Selig M 2016 A morphing downwind-aligned rotor concept based on a 13-MW wind turbine *Wind Energy* **19** 625-37
[9] Loth E, Steele A, Chao Q, Ichter B, Selig M S and Moriarty P 2017 Downwind pre-aligned rotors for extreme-scale wind turbines *Wind Energy* **20** 1241-59
[10] Meng H, Ma Z, Dou B, Zeng P and Lei L 2020 Investigation on the performance of a novel forward-folding rotor used in a downwind horizontal-axis turbine *Energy* **190** 116384
[11] Jonkman J, Butterfield S, Musial W and Scott G 2009 Definition of a 5-MW reference wind turbine for offshore system development. National Renewable Energy Lab., Golden, CO, United States)
[12] Xie W, Zeng P and Lei L 2017 Wind tunnel testing and improved blade element momentum method for umbrella-type rotor of horizontal axis wind turbine *Energy* **119** 334-50
[13] Resor B R 2013 Definition of a 5MW/61.5 m wind turbine blade reference model. Sandia National Laboratories, Albuquerque, New Mexico, USA)

IOP Conf. Series: Materials Science and Engineering **825** (2020) 012003 doi:10.1088/1757-899X/825/1/012003

The Aero-elastic-wake Coupling Behavior for a Two-wind-turbines Case with Power Control Process

Zhe Ma, LiPing Lei[*] and Pan Zeng

Department of Mechanical Engineering, Tsinghua University, Beijing 100084, China

[*] E-mail: leilp@mail.tsinghua.edu.cn

Abstract. This study focuses on the aero-elastic-wake coupling behavior of wind turbines. A newly developed code called FALM (Flexible Actuator Line Model), which combines the wake simulation ability of the actuator line mothed (ALM) and the structural simulation ability of the flexible multibody dynamics method, was employed to achieve these simulations. The power output and thrust, the natural frequency and deformation and the wake characteristics of a single NREL 5MW wind turbine were studied to validate this code. It shows that nonlinear effects such as spin softening and stress stiffening were fully considered by FALM and it can also guarantee a reliable prediction of power and thrust of wind turbines. Furthermore, a case of two-wind-turbines with the inlet wind speed of 14m/s (exceeding the rated wind speed) were carried out to study the aero-elastic-wake coupling behavior in a wind farm. It shows that FALM is able to simulate multiple wind turbines with power control system involved. The pitch control process of the upstream wind turbine was predicted and the dynamic loads of the downstream wind turbine caused by the wake effect were studied. These results will contribute to the study of reducing the fatigue load caused by wake effect.

Key words: Actuator line method; flexible multibody dynamics; aero-elastic-wake coupling behavior; wind turbine simulation

1. Introduction

With the development of wind industry, wind turbines are heading to larger scale and offshore. A conceptual 5MW wind turbine with rotor diameter of 126m was published by NREL in 2009[1] and the diameter of the largest wind turbine nowadays, which is built by Vestas with rated power of 9.5MW, has reached 164m. However, larger wind turbines are more unsteady and more susceptible to large wind loads and large deflections. The wake effect is also important since it will dramatically affect the power output and the structural load of the downstream wind turbines in a wind farm[2]. As the main reference in the design phase, wind turbine simulation tools are facing new challenges.

Blade element momentum (BEM) theory is the most widely used method in wind turbine analysis for its low computational cost and its accuracy in power prediction. However, BEM theory can hardly simulate wake characteristics of wind turbines. Actuator line method (ALM) was proposed in 2002 by Sørensen and Shen[3] to overcome this disadvantage of BEM. By combining the computational fluids dynamics and the blade element theory, ALM can simulate the power and wake characteristics of wind turbines simultaneously. Nowadays this method has become the most potential method in wind farm simulation due to its wake simulation ability and low computational cost.

The wind turbine structure is a typical flexible multi-body system, which is composed of blades, nacelle and tower. The deformation of wind turbines is significant due to the slender structure of blades and tower. Furthermore, the coupling behavior of the rigid motion and the deformation is significant due to the wind turbine rotation. In this study, flexible multibody dynamics was chosen as the structural analysis method, because of its accuracy in the motion-deformation coupling analysis[4].

In this paper, a new method called FALM (Flexible Actuator Line Method) based on ALM and flexible multibody dynamics was developed to study the aero-elastic-wake coupling behavior of wind turbines. Simulations of a NREL 5MW wind turbine with uniform inlet of 8m/s and 14m/s were carried out to validate this new method. The two-wind-turbines case with atmospheric boundary layer (ABL) inlet of 14m/s (which is over the rated wind speed) was studied. The power control process was considered to simulate the real dynamic response of the downstream wind turbine.

2. Approach

The FALM introduced in this paper is mainly composed of two parts: the aerodynamic part which is based on the actuator line method, and the structural part which is based on the flexible multibody dynamics. Figure 1 shows the diagram of the FALM approach. The details of the realization of FALM will be introduced in this section.

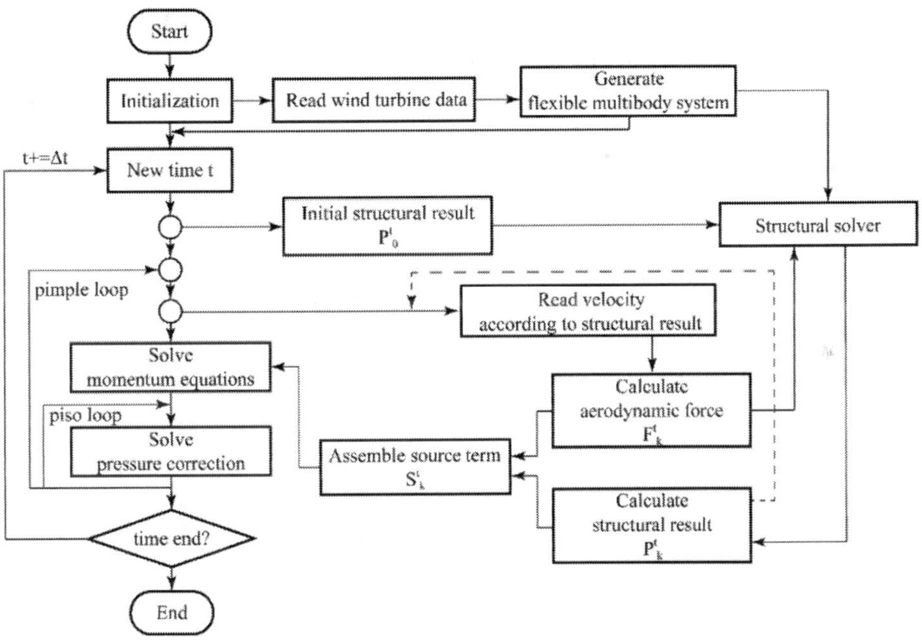

Figure 1. The diagram of the FALM approach.

2.1. Actuator line method

The actuator line method is a computational fluid dynamics (CFD) based approach. Regardless of the resolved shape, actuator line method represents wind turbine blades by source term. The governing equations used in FALM is shown as Eq. (1) and Eq. (2)

$$\frac{\partial \overline{u_i}}{\partial x_i} = 0 \tag{1}$$

$$\rho\frac{\partial \overline{u}_i}{\partial t} + \rho\frac{\partial(\overline{u}_i\overline{u}_j)}{\partial x_j} = -\frac{\partial p}{\partial x_i} + \frac{\partial}{\partial x_j}\left[\mu\left(\frac{\partial \overline{u}_i}{\partial x_j} + \frac{\partial \overline{u}_j}{\partial x_i}\right)\right] + \frac{\partial \tau^s_{ij}}{\partial x_j} + f \tag{2}$$

where $\overline{u}$ is the filtered velocity vector field, p is the scalar field of pressure, μ is a scalar represent the kinematic viscosity, $\tau_{ij}^{s} = -\rho(\overline{u_i u_j} - \overline{u}_i\,\overline{u}_j)$ is called the subgrid-scale (SGS) Reynolds stress, and the standard Smagorinsky SGS model is employed in this study.

f is the source term which represents the wind turbine blade in ALM and it can be calculated based on the blade element theory, as shown in Eq. (3). Here, v, c and L are the inlet velocity, chord length and element length, respectively, of each blade element, C_l and C_d are the lift and drag coefficient of the airfoil and ρ is the air density.

$$f_{element} = (dL, dD) = (\frac{1}{2}\rho v^2 c C_l L, \frac{1}{2}\rho v^2 c C_d L) \quad (3)$$

The forces calculated by Eq. (3) are point forces and a regularization kernel must be employed to avoid a numerical singular, as shown Eq. (4) and Eq. (5). Here, ϵ is Gaussian width which adjusts the strength of this regularization kernel.

$$f = \sum f_{element} \otimes \eta_\epsilon \quad (4)$$

$$\eta_\epsilon = \frac{1}{\epsilon\pi^{3/2}} e^{-\left(\frac{r}{\epsilon}\right)^2} \quad (5)$$

2.2. Flexible multibody dynamics

In this study, flexible multibody dynamics method with floating frame of reference formulation were employed as the structural model. The position of a point in the flexible body is represented by eq.(6)

$$P = r + A(u_0 + Sq) \quad (6)$$

Where r is the vector from the origin of the inertial frame to the floating frame, A is the transformation matrix from the floating frame to inertial frame, u_0 is the initial position before the deformation, S is the shape function and q is the node deformation.

Therefore, the kenetic energy and potential energy could be calculated using Eq. (6) and the motion of the multibody system could be depicted according to the Hamilton's theory, shown as Eq. (7). The details of the derivation could be found in Shabana's book[5].

$$\begin{bmatrix} I & -A\tilde{u}G & AS \\ & -G^T\tilde{u}^T\tilde{u}G & -G^T\tilde{u}^T S \\ sym & & S^T S \end{bmatrix} \begin{bmatrix} \ddot{r} \\ \ddot{\theta} \\ \ddot{q} \end{bmatrix} + \begin{bmatrix} 0 & 0 & 0 \\ 0 & 0 & 0 \\ 0 & 0 & K_{ff} + K_s \end{bmatrix} \begin{bmatrix} r \\ \theta \\ q \end{bmatrix} + C_q^T \lambda = Q_e + Q_v \quad (7)$$

Here, K_{ff} is the stiffness matrix, Q_e and Q_v are generallized external forces and quadratic velocity vector, respectively. C represents the constraint of the multibody system and C_q is the jacobian matrix of constraint equations.

K_s is the stress stiffening term and it must be taken into consideration because it will significantly affect the natural frequency of the wind turbine structure[6]. This increament of stiffness is related to the tightening stress which is mainly caused by the centrifugal force of the blade rotation in wind turbine analysis.

2.3. Wind turbine model and control system

The power control system of NREL 5MW wind turbine was modeled in FALM. This control system is composed of three main parts: the low-pass filter, the generator controller, and the pitch controller. The control process of NREL 5MW wind turbine is as follow:

(1) Read the generator speed and apply a low-pass filter to obtain the input value for controllers.
(2) If the input is lower than the rated speed, enable the generator controller to control the generator torque.
(3) If the input is higher than the rated speed, enable the generator-torque controller and the blade-pitch controller to control the generator torque and the pitch angle, respectively.

All the parameters used in the control system is derived according to the definition of NREL 5MW[1].

3. Results

3.1. Validation

Table 1 compared the aerodynamic performance of a NREL 5MW wind turbine predicted by FALM and previous studies. In this study, two typical inlet velocity, 8m/s and 14m/s, were chosen. It shows that the rotor power calculated by FALM agrees well with previous studies. The rotor thrust calculated by FAST is obviously higher but results of other studies including FALM agrees well with each other. In summary, FALM provides reliable simulations of the aerodynamic performance of NREL 5MW wind turbine.

Table 1. Contrast of power and thrust results between previous results and FALM

Inlet velocity		FAST[1]	HAWC2	Li[7]	Jeong[8]	Ponta[9]	FALM
8m/s	Power(MW)	1.856	1.928	1.865	2.242	1.817	1.840
	Thrust(kN)	466	391	389	410	340	388
14m/s	Power(MW)	5.296	No data	No data	5.249	5.088	5.259
	Thrust(kN)	541	No data	No data	433	408	491

Table 2. compared the simulation results of tip deflections. Oopdefl represents the out of plane deflection and IPDefl represents the in-plane deflection. The deflection result of FALM with and without nonlinear effect are both listed to make a fully comparison. It's clear that the simulation results of FALM agree with FAST and Li's result when the nonlinear effect are neglected and agree with Yu's study, which also simulated the nonlinear effect of wind turbine blade, when the nonlinear effect are involved. It could be concluded that FALM provides reliable simulations of the wind turbine deflection.

Table 2. Contrast of tip deflection results between previous results and FALM

Inlet velocity(m/s)		FAST[1]	Li[7]	Yu[10]	FALM linear	FALM nonlinear
8	OoPDefl(m)	3.22	3.592	2.958	3.637	2.938
	IPDefl(m)	-0.350	-0.345	No data	-0.257	-0.255
14	OoPDefl(m)	3.10	No data	3.195	4.275	3.01
	IPDefl(m)	-0.739	No data	No data	-0.793	-0.696

Figure 2. shows the results of natural frequency vs. rotating speed of the wind turbine rotor. It should be noticed that the difference of the natural frequency of two 2nd flapwise waving mode is calculated by FALM and this agrees with the results of ADAMS when rotating speed is zero. It's clear that FALM provides a better simulation than FAST of this frequency difference. Furthermore, the nonlinear effect such as the spin softening and stress stiffening is fully considered by FALM. The natural frequency calculated by FALM varies with the rotating speed because of the nonlinear effects. The stress stiffening mainly affects the flapwise mode and there is a dramatic increment of the natural frequency when the rotating speed goes up. However, the natural frequency of rotor torsion and tower waving mode will decrease when the rotating speed goes up. It's clear that neglecting of the nonlinear

effect will cause a big simulation error for a rotating wind turbine and FALM is suitable for these simulations.

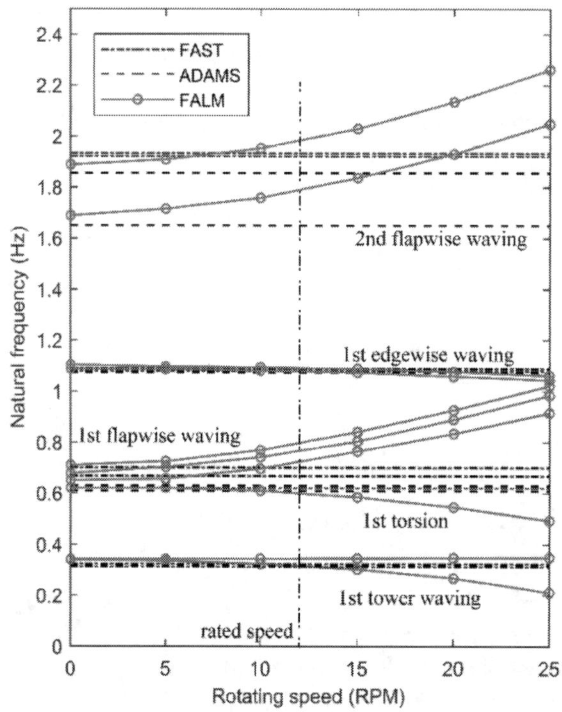

Figure 2. The natural frequency vs. rotating speed of the wind turbine rotor calculated by FAST, ADAMS and FALM. Results of the first 10 modes were plotted.

3.2. Two-wind-turbines study

A two-wind-turbines case were studied using FALM. Figure 3 shows the calculation domain of the simulation. D represents the diameter of the NREL 5MW wind turbine, which is 126m. the gap between the upstream wind turbine and the downstream wind turbine is 6D. The inlet wind speed is 14m/s which is over the rated wind speed of NREL 5MW. The atmospheric boundary layer (ABL) inlet condition and the power control process were considered to simulate the real dynamic responses of the downstream wind turbine. The results (Figure 4) showed that:

1. The pitch angle of the downstream wind turbine will drop and reach zero with oscillation. This is caused by the power control process and the pitch motion will also cause an oscillation to the aerodynamic power and blade tip deflection.

2. The rotor thrust of the downstream wind turbine is much higher than the upstream wind turbine because the pitch angle not only reduces the rotor power but also reduces the rotor thrust.

3. The rotor power of the downstream wind turbine is significantly lower than the Jensen model's prediction, which is a widely used wake model of wind turbines. This is mainly caused by the difference of velocity distribution between the FALM simulation and the Jensen model.

4. The atmospheric boundary layer (ABL) inlet will cause an oscillation on the tip deflection of wind turbines. However, this unbalance of blade tip deflection caused by ABL inlet are receded due to the velocity recovery in the wake region of the upstream wind turbine.

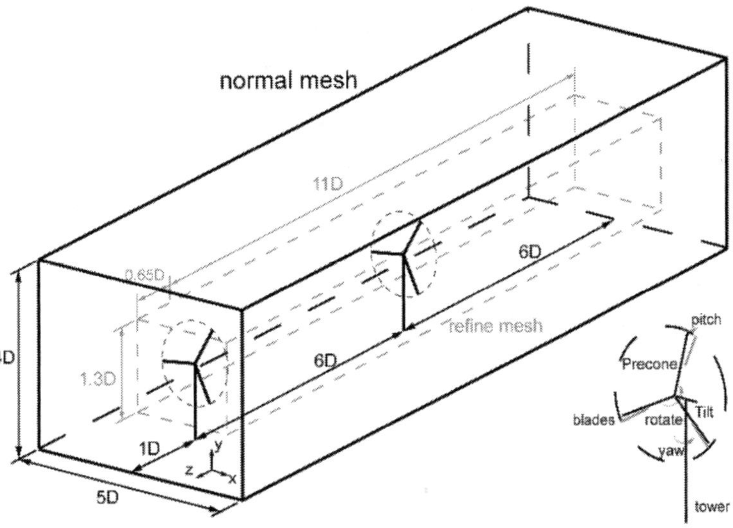

Figure 3. calculation domain of the two-wind-turbines simulation

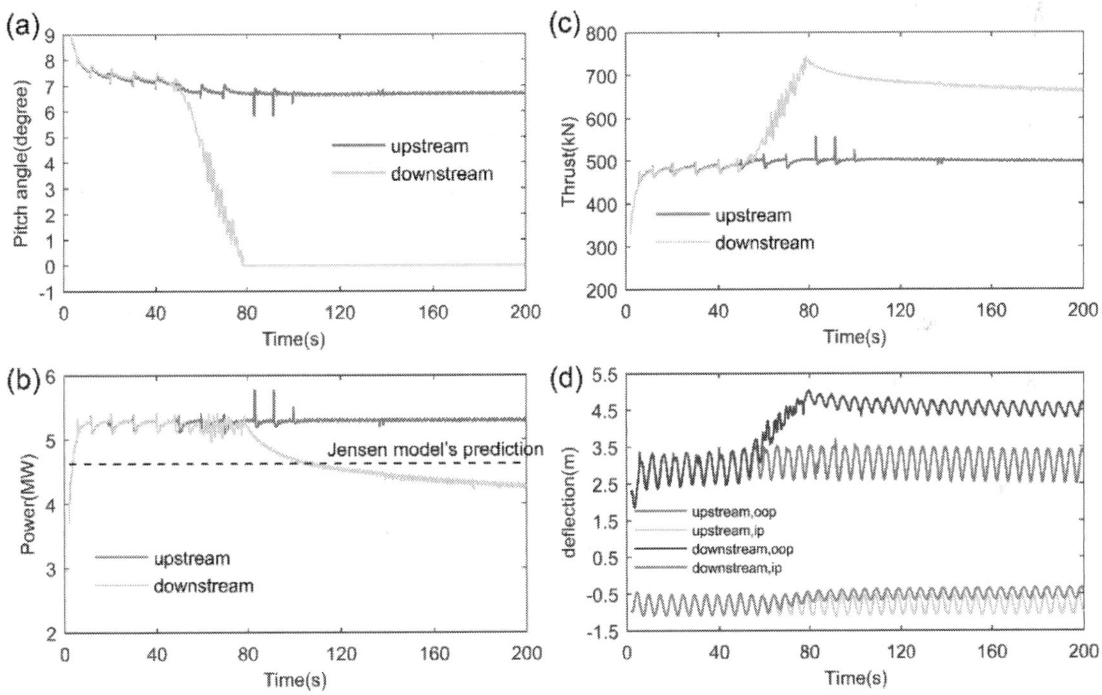

Figure 4. The simulations results for the upstream wind turbine and the downstream wind turbine. (a) the pitch angle, (b) the rotor thrust (c) the generator power (d) the tip deflection

The velocity distribution (as shown in Figure 5.) and vorticity distribution (as shown in Figure 6.) along the wind direction are extracted. As shown in Figure 5., the x axis represents the Longitude coordinate behind the upstream wind turbine and a grid in x axis also represent the inlet velocity of 14m/s. The dash line represents the velocity distribution of ABL inlet condition and the solid lines represent the velocity distribution in the wake region. Therefore, the difference between the solid lines and the dash line represents the wake effect (loss of velocity) of the wind turbine. As shown in Figure 6., the dash line represents the vorticity distribution of ABL inlet condition and the solid lines represent the vorticity distribution in the wake region.

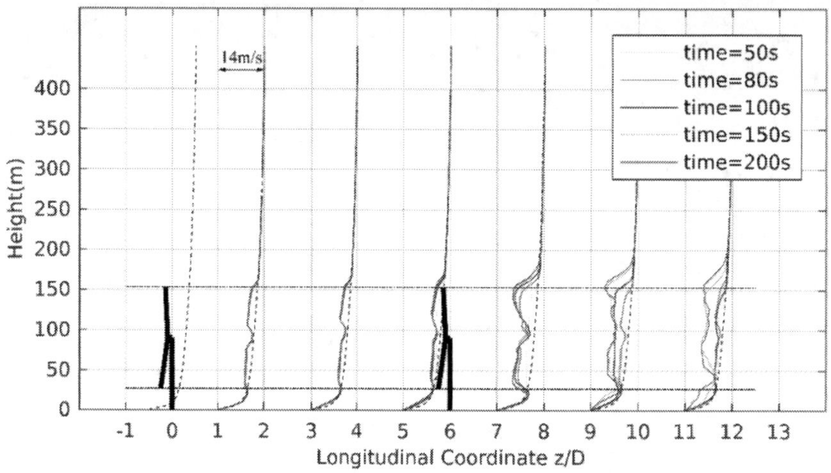

Figure 5. Velocity distribution along the wind direction

The wake results showed that:

1. The loss of velocity is mainly caused by wind turbine blades and the vorticity is mainly caused by the tip and the root of wind turbine blades. The wake effect in upper wake region is higher than the lower region.

2. The velocity of the wake region of the downstream wind turbine is much lower than the upstream one and the vorticity of the downstream wind turbine is much higher than the upstream one.

3. The wake region of the downstream wind turbine is larger than the upstream one and the velocity recovery is better.

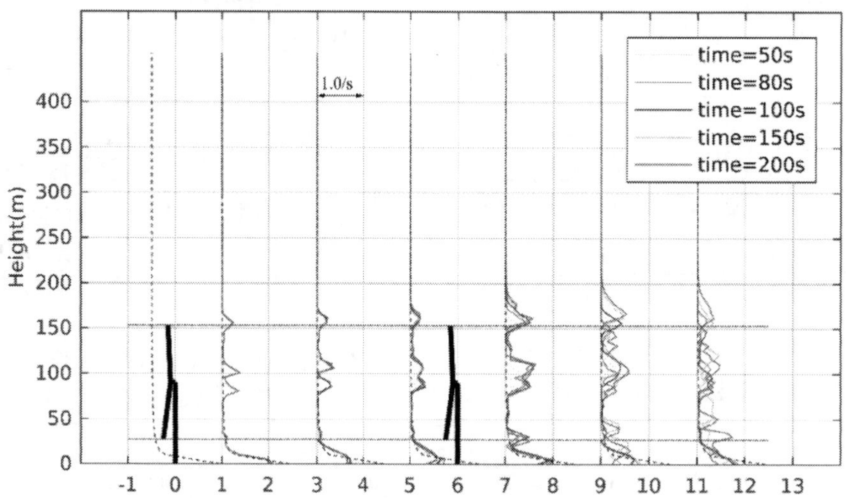

Figure 6. vorticity distribution along the wind direction

4. Conclusion

In this paper, a new validated simulation method for the aero-elastic-wake coupling behavior of wind turbine was developed and a two-wind-turbines case with power control process and over-rated inlet were studied. The conclusion is as followed:

1. The FALM method is validated by the results of the uniform inlet of 8m/s and 14m/s. The rotor power, rotor thrust, tip deflection and natural frequency result calculated by FALM agrees well with previous studies.

2. The wake of the upstream wind turbine and pitch motion will apply an oscillation to the aerodynamic and structural performances of the downstream wind turbine.

3. The wake region of the downstream wind turbine is larger than the upstream one, with lower velocity, high vorticity and better velocity recovery.

Acknowledgement
The authors gratefully acknowledge the support of the National Natural Science Foundation of China (No. 51875305).

References:

[1] J. Jonkman, S. Butterfield, W. Musial, G. Scott, Definition of a 5-mw reference wind turbine for offshore system development (2009).

[2] S. Lee, M.J. Churchfield, P.J. Moriarty, J. Jonkman, J. Michalakes, A numerical study of atmospheric and wake turbulence impacts on wind turbine fatigue loadings. Journal of Solar Energy Engineering. 135 (3) (2013) 31001.

[3] J.N. Sorensen, W.Z. Shen, Numerical modeling of wind turbine wakes. Journal of fluids engineering. 124 (2) (2002) 393-399.

[4] D. Lee, D.H. Hodges, M.J. Patil, Multi-flexible-body dynamic analysis of horizontal axis wind turbines. Wind Energy: An International Journal for Progress and Applications in Wind Power Conversion Technology. 5 (4) (2002) 281-300.

[5] A.A. Shabana, Dynamics of multibody systems., Cambridge University Press, 2013.

[6] Z. Ma, P. Zeng, L. Lei, Analysis of the coupled aeroelastic wake behavior of wind turbine. J. Fluid. Struct. 84 ((2019) 466-484.

[7] Y. Li, A.M. Castro, T. Sinokrot, W. Prescott, P.M. Carrica, Coupled multi-body dynamics and cfd for wind turbine simulation including explicit wind turbulence. Renew. Energ. 76 ((2015) 338-361.

[8] M. Jeong, M. Cha, S. Kim, I. Lee, Numerical investigation of optimal yaw misalignment and collective pitch angle for load imbalance reduction of rigid and flexible hawt blades under sheared inflow. Energy. 84 ((2015) 518-532.

[9] F.L. Ponta, A.D. Otero, L.I. Lago, A. Rajan, Effects of rotor deformation in wind-turbine performance: the dynamic rotor deformation blade element momentum model (drd–bem). Renew. Energ. 92 ((2016) 157-170.

[10] D.O. Yu, O.J. Kwon, Predicting wind turbine blade loads and aeroelastic response using a coupled cfd–csd method. Renew. Energ. 70 ((2014) 184-196.

IOP Conf. Series: Materials Science and Engineering **825** (2020) 012004 doi:10.1088/1757-899X/825/1/012004

Structure Analysis of an Innovative Vertical Axis Wind Turbine with Inclined Pitch Axes using Finite Element Method

Jia Guo, Pan Zeng and Liping Lei*

Key Laboratory for Advanced Materials Processing Technology of MOE, Department of Mechanical Engineering, Tsinghua University, Beijing 100084, China

* E-mail address: leilp@tsinghua.edu.cn

Abstract. Pitch regulation is critical to power performance of wind turbines. However, the complexity of the conventional pitch system of vertical axis wind turbines (VAWTs) retards its appliance. In the present study, an innovative pitch system of a VAWT with inclined pitch axes was proposed where a slider in the main shaft actuated upwards or downwards drives blades to be folded by linkage mechanisms. Effect of design parameters of the pitch system on blade folding movement control sensitivity was evaluated. Loads of pitch hinge, deformation of blades and stresses of components were analyzed as well as natural frequencies of the wind turbine with various azimuth and fold angles under rated operational conditions by finite element method and effect of design parameters on structural performance of the wind turbine was investigated. It was found that deformation of blades and stresses of supporting arms deteriorated but natural frequencies of the wind turbine increased when blades were folded in the negative direction. In addition, the increasing arm interval height improved structural performance with the increasing natural frequencies. In the contrary, the decreasing blade linkage length, shaft linkage length and slider height improved structural performance with the decreasing natural frequencies. This study laid the foundation for the further design optimization of the novel pitch system.

1. Introduction

Pitch regulation is important to wind turbines since it can control the power output and mitigate loads, as well as improving self-starting ability specifically for vertical axis wind turbines (VAWTs)[1]. Pitch mechanisms for horizontal axis wind turbines (HAWTs) are technically quite mature and widely applied in wind power industry[2]. Blades are connected to the hub via pitch bearings and pitch motors actuate blades to rotate around blade aerodynamic center lines. However, for VAWTs there are no widely accepted pitch mechanisms. Several pitch system designs were mentioned in the previous literature[3, 4] but these designs are complicated. Guo et al.[5, 6] proposed an innovative VAWT design with inclined pitch axes (not coincident with the aerodynamic center line) for small scale VAWTs which is structurally simple and practical. The present study mainly focuses on structural performance of the novel pitch system design.

Structural analysis is essential for VAWTs since they are typical rotary machine and may encounter structural problems during their whole service life such as resonance and fatigue failure caused by cyclic and random excitation from wind[7]. The structural performance and dynamic responses of VAWTs without pitch regulation have been investigated by numerical and experimental methods[8-

IOP Conf. Series: Materials Science and Engineering **825** (2020) 012004 doi:10.1088/1757-899X/825/1/012004

11]. In the present study, the principle design of the novel pitch system was described and design parameters were pointed out first. Then effect of design parameters on blade folding movement control sensitivity was clarified. Loads of the novel pitch system and stresses of components of the wind turbine were assessed as well as natural frequencies of the wind turbine with various azimuth and fold angles. In addition, effect of design parameters on structural performance of the novel VAWT prototype was elucidated compared to a conventional VAWT.

2. Principle design of the novel pitch system

The innovative pitch system of a VAWT with inclined pitch axes is shown below (figure 1). Blades are connected to the main shaft via supporting arms in the lower section. A slider is driven to move upwards and downwards along the main shaft and are connected to blades via shaft linkages, shaft joints (cardan joints), driving arms, blade joints (spherical joints) and blade linkages successively.

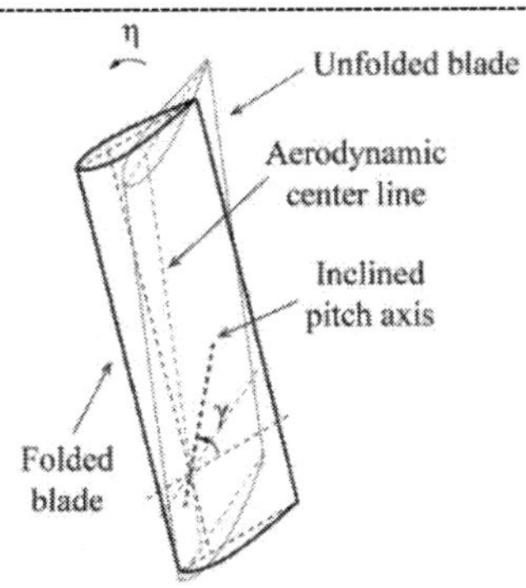

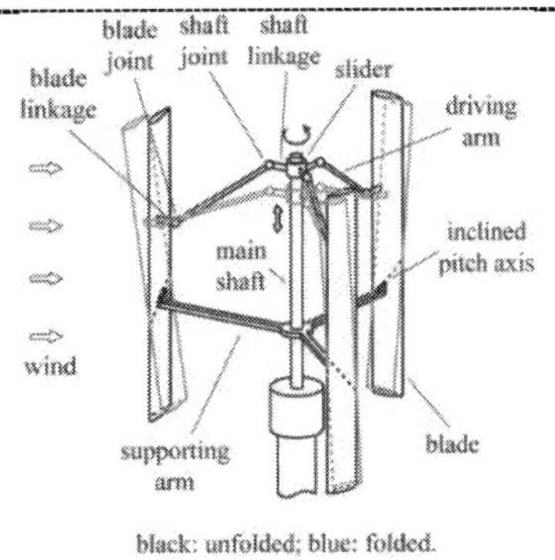

(a) (b)

Figure 1. Schematic diagrams of (a) a blade with an inclined pitch axis (b) a VAWT design using the novel pitch mechanism.

Design parameters of the novel pitch system include the inclined angle (γ), the arm interval height, the slider height, the blade linkage length and the shaft linkage length as is shown in figure 2.

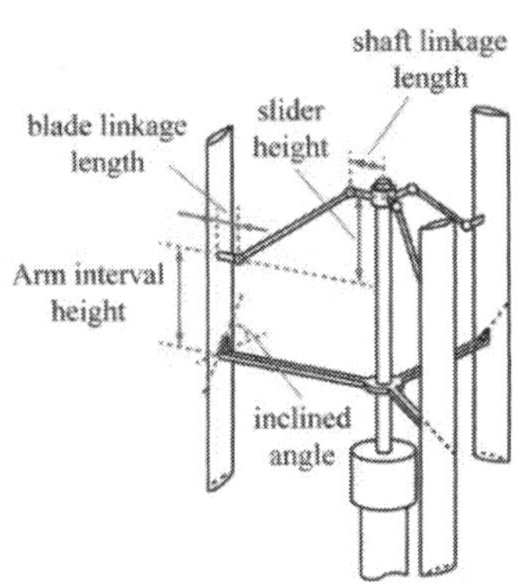

Figure 2. Schematic diagram of design parameters of the novel pitch system.

IOP Conf. Series: Materials Science and Engineering **825** (2020) 012004 doi:10.1088/1757-899X/825/1/012004

3. Kinematic analysis

Three coordinate systems are established to elaborate the kinematics of the innovative pitch system (figure 3). The first one is the inertial coordinate system (ICS), of which the origin is located at the intersection of the supporting arms and the main shaft, the x axis is in the free stream direction, the y axis is perpendicular to the free stream and the z axis is vertical upwards. The second one is the blade azimuth coordinate system (BCS), of which the origin is located at the intersection of the supporting arm and the blade, the x, y and z axis is tangential, normal and vertical, respectively. The third one is the pitched blade coordinate system (PCS), of which the origin is same as that of BCS, the x, y and z axis is in the direction of chord, thickness and span of a pitched blade, respectively.

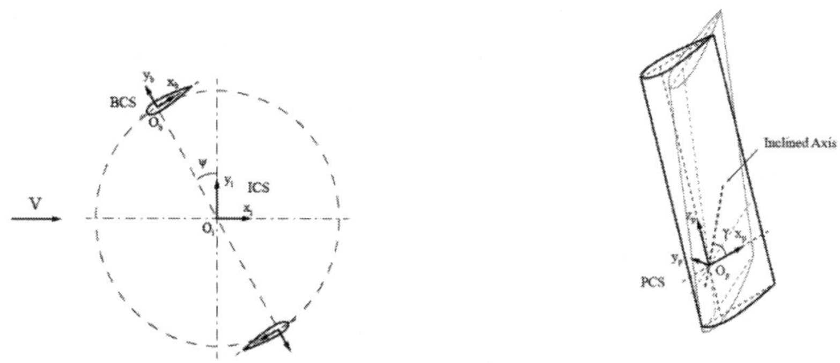

(a) (b)

Figure 3. Schematic diagrams of three coordinate systems (a) ICS and BCS (b) PCS.

Then the blade folding movement control sensitivity is calculated, namely variations of the fold angle (η) corresponding to stroke of the slider (s). The coordinate of the blade joint in PCS is

$$\boldsymbol{q_{bj,p}} = [0 \quad -l_{bj} \quad h_{bj}]^T \tag{1}$$

The transformation matrix from PCS to BCS is

$$\boldsymbol{T_{P2B}} = \begin{bmatrix} \sin^2\gamma\cos\eta + \cos^2\gamma & \sin\gamma\sin\eta & \sin\gamma\cos\gamma\,(1-\cos\eta) \\ -\sin\gamma\sin\eta & \cos\eta & \cos\gamma\sin\eta \\ \sin\gamma\cos\gamma\,(1-\cos\eta) & -\cos\gamma\sin\eta & \sin^2\gamma + \cos^2\gamma\cos\eta \end{bmatrix} \tag{2}$$

So the coordinate of the blade joint in BCS is

$$\boldsymbol{q_{bj,b}} = \boldsymbol{T_{P2B}}\boldsymbol{q_{bj,p}} \tag{3}$$

The coordinate of the shaft joint in BCS is

$$\boldsymbol{q_{sj,b}} = [0 \quad l_{sj} - R \quad h_{bj} + h_{sj} + s]^T \tag{4}$$

The length of the driving arm is constant during the blade folding movement, so

$$\left|\boldsymbol{q_{bj,b}} - \boldsymbol{q_{sj,b}}\right| = \left|\boldsymbol{q_{bj,b}} - \boldsymbol{q_{sj,b}}\right|_{s=0,\eta=0} \tag{5}$$

Variations of the fold angle to nominal stroke of the slider under conditions of different design parameters are shown in figure 4. As can be seen in figure 4(a) the blade folding movement control sensitivity varies during the whole folding process and increases when the inclined angle gets closer to 90°. As for arm interval height, the blade folding sensitivity increases when arm interval height decreases but chances are that the blade folding sensitivity becomes too high so that blade folding movement may lose control. In addition, the blade folding sensitivity increases with slider height,

IOP Conf. Series: Materials Science and Engineering **825** (2020) 012004 doi:10.1088/1757-899X/825/1/012004

blade linkage length and shaft linkage length. However, the range of the fold angle can be limited because of small slider height.

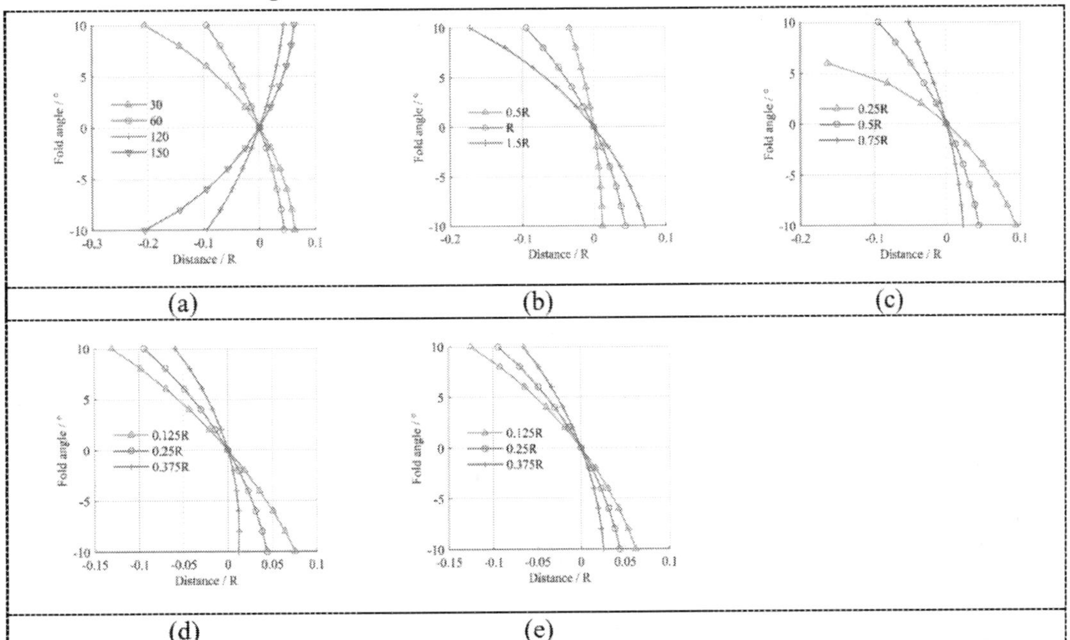

Figure 4. Effect of design parameters on blade folding sensitivity (a) inclined angle (b) arm interval height (c) slider height (d) blade linkage length (e) shaft linkage length.

4. Structural analysis

An innovative VAWT with inclined pitch axes is simulated in ANSYS as well as a conventional one as a counterpart. The model is a 10 kW prototype whose parameters and materials are shown in table 1 and table 2[12]. The MPC184 element are applied for blade joints (spherical joints), shaft joints (cardan joints) and pitch hinges of the pitch system and other components of the wind turbine are modeled by BEAM188 element. Specifically, the cross sections of blades are airfoils achieved by user-defined mesh. Appropriate mesh size is determined and mesh independence is examined. The computational models of the wind turbine are shown in figure 5.

Table 1. Summary of operational parameters of the VAWT[12]

Parameters	Value
Power / kW	10
Wind speed / (m/s)	12
Rotational speed / (r/min)	143
Diameter / m	4
Blade number	3
Airfoil	NACA 0018
Blade span / m	6
Blade chord / m	0.6

Table 2. Materials of components of VAWT[12]

Components	Material
Tower	Q235
Main shaft	40Cr
Arm	Q235
blade	GFRP

The loads of the wind turbine include inertial ones (C), gravity (G) and aerodynamic ones (F_T, F_N). Inertial loads are directly proportional to rotational speed and are around 5×10^4 N under rated condition. Gravity and aerodynamic loads are roughly one order of magnitude smaller than inertial loads. In addition, the displacements of the bottom foundation of the tower are all constrained. The loads of pitch hinges (Fsum and Msum refer to the magnitude of resultant forces and moments, respectively), displacement of blades (Ur and Uz refer to radial and vertical displacement, respectively), stresses of blades and arms and natural frequencies are assessed under conditions of various azimuth, fold angles and design parameters.

In terms of the conventional VAWT, as is shown in figure 5, loads of pitch hinges, displacement of blades and stresses of blades and arms varies in accordance with aerodynamic loads. But natural frequencies of the wind turbine remain constant.

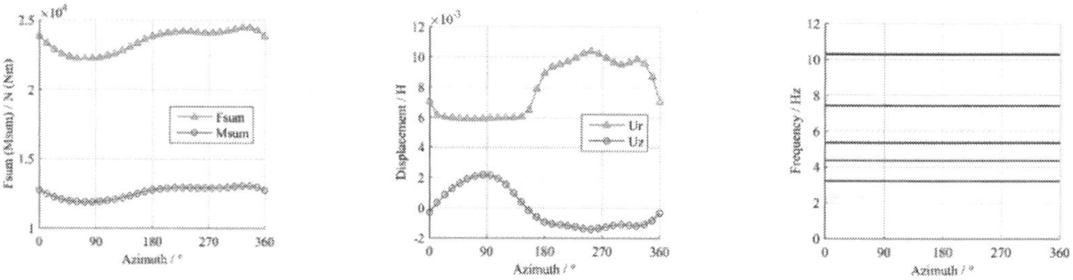

(a) (b) (c)

Figure 5. Variations of results versus azimuth (a) loads of pitch hinges (b) displacement of blades (c) natural frequencies.

In terms of the innovative VAWT, results with various fold angles are shown in figure 6. When blades are folded in the positive direction, that is to say, leading edges of blades are folded outside, the sum of forces decreases and the sum of moments increases, and vice versa. As for the displacement of blades, it stays the same when blades are folded in the positive direction, but increases rapidly when blades are folded in the negative direction. Similarly, stresses of arms increase when blades are folded in the negative direction. So does the first natural frequency of the wind turbine.

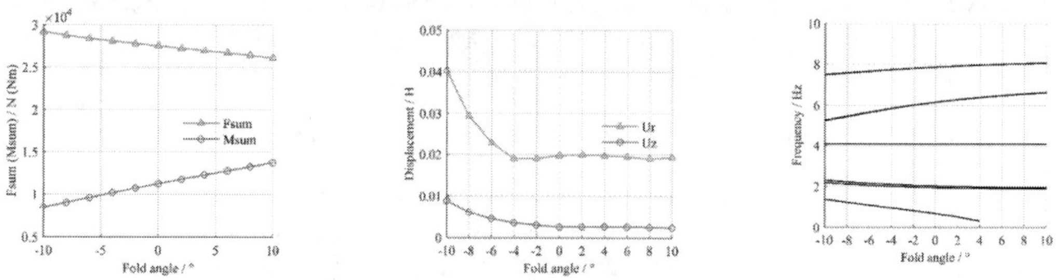

(a) (b) (c)

Figure 6. Variations of results versus fold angles (a) loads of inclined pitch hinges (b) displacement of blades (c) natural frequencies.

Then effect of design parameters is revealed and performance of the innovative VAWT is compared with that of the conventional VAWT as is shown in figure 7-10. First of all, when arm interval height increases, the sum of forces of inclined pitch hinges and the radial displacement of blades diminish. Stresses of blades and driving arms reduce while stresses of supporting arms grow. Natural frequencies increase at the same time.

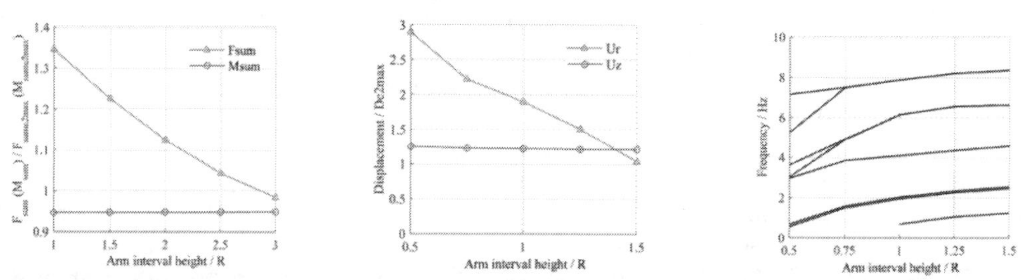

(a) (b) (c)

Figure 7. Variations of results versus arm interval height (a) loads of inclined pitch hinges (b) displacement of blades (c) natural frequencies.

IOP Conf. Series: Materials Science and Engineering **825** (2020) 012004 doi:10.1088/1757-899X/825/1/012004

As for slider height, the sum of forces of hinges, the vertical displacement of blades and stresses of arms decrease with the reducing slider height, but the first natural frequency declines as well.

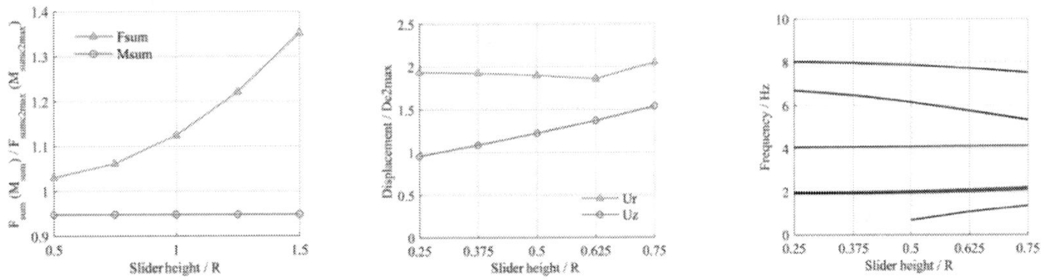

(a) (b) (c)

Figure 8. Variations of results versus slider height (a) loads of inclined pitch hinges (b) displacement of blades (c) natural frequencies.

Effect of blade linkage length is similar to that of shaft linkage length. The sum of forces of hinges, stresses of driving arms and the first natural frequency all decrease when blade linkage length and shaft linkage length decrease.

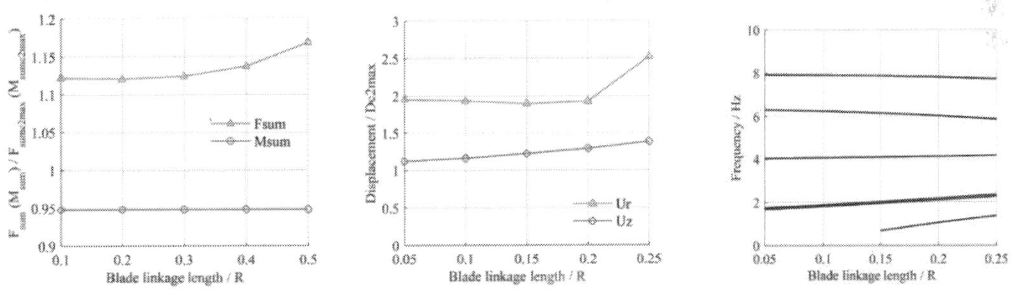

(a) (b) (c)

Figure 9. Variations of results versus blade linkage length (a) loads of inclined pitch hinges (b) displacement of blades (c) natural frequencies.

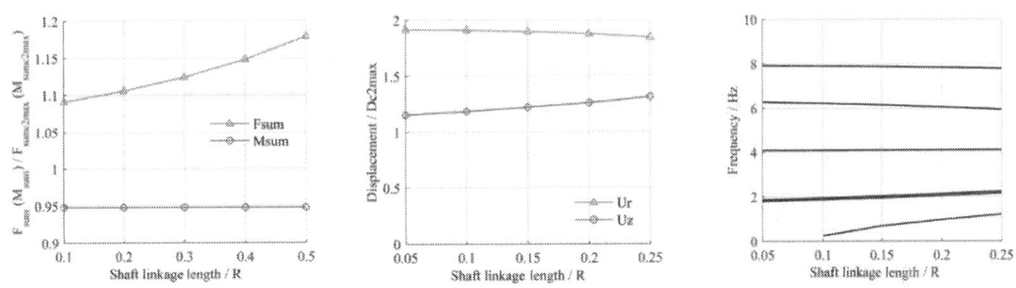

(a) (b) (c)

Figure 10. Variations of results versus shaft linkage length (a) loads of inclined pitch hinges (b) displacement of blades (c) natural frequencies.

5. Conclusion

An innovative pitch system of a straight bladed VAWT was depicted and kinematic and structural analyses were conducted. Main findings of the present study are summarized as follows. The blade folding movement control sensitivity is negatively correlated with the arm interval height while positively correlated with the slider height, the blade linkage length and the shaft linkage length. Loads of pitch hinges, displacement of blades and stresses of blades and arms varies in accordance with aerodynamic loads. Furthermore, deformation of blades and stresses of supporting arms deteriorated but natural frequencies of the wind turbine increased when blades were folded in the negative direction. In addition, the increasing arm interval height improved structural performance with the increasing natural frequencies. Nonetheless, the decrease of the other three parameters improved structural performance with the decreasing natural frequencies.

Acknowledgments
This study is supported by the National Natural Science Foundation of China (No. 51875305).

References

[1] Jain P, Abhishek A. 2016, Performance prediction and fundamental understanding of small scale vertical axis wind turbine with variable amplitude blade pitching [J]. *Renewable Energy,* **97**(97-113.
[2] Tiwari R, Babu N R. 2016, Recent developments of control strategies for wind energy conversion system [J]. *Renewable and Sustainable Energy Reviews,* **66**(268-85.
[3] Benedict M, Lakshminarayan V, Pino J, et al. 2016, Aerodynamics of a small-scale vertical-axis wind turbine with dynamic blade pitching [J]. *AIAA Journal,* **54**(3): 924-35.
[4] Elkhoury M, Kiwata T, Aoun E. 2015, Experimental and numerical investigation of a three-dimensional vertical-axis wind turbine with variable-pitch [J]. *Journal of Wind Engineering and Industrial Aerodynamics,* **139**(111-23.
[5] Guo J, Zeng P, Lei L. 2019, Performance of a straight-bladed vertical axis wind turbine with inclined pitch axes by wind tunnel experiments [J]. *Energy,* **174**(553-61.
[6] Guo J, Zeng P, Lei L. Vertical axis wind turbines with inclined pitch axes based on inclined hinged blades: China, 0669046.0 [P/OL]. 2019.10.29-.
[7] Rehman S, Rafique M M, Alam M M, et al. 2019, Vertical axis wind turbine types, efficiencies, and structural stability: A review [J]. *Wind and Structures,* **29**(1): 15-32.
[8] Lin J, Leung L K K, Xu Y-L, et al. 2018, Field measurement, model updating, and response prediction of a large-scale straight-bladed vertical axis wind turbine structure [J]. *Measurement,* **130**(57-70.
[9] Carne T G, Lobitz D W, Nerd A R, et al., SAND82-0345 [R]. Albuquerque: Sandia National Laboratories, 1982.
[10] Malcolm D J. 2019, On the structural response of two- and three-bladed vertical axis wind turbines [J]. *Wind Energy,*
[11] Bianchini A, Cangioli F, Papini S, et al. 2014, Structural analysis of a small H-Darrieus wind turbine using beam models: Development and assessment [J]. *Journal of Turbomachinery,* **137**(1):
[12] Li J. The statics analysis and dynamics analysis of 10kW H-type vertical axis wind turbine [D]. Beijing; North China Electric Power University, 2014.

IOP Conf. Series: Materials Science and Engineering **825** (2020) 012005 doi:10.1088/1757-899X/825/1/012005

Research on Automatic Manual Transmission of Pure Electric Vehicle

Qinyu Niu, Hao Men* and Rong Yang

College of Mechanical Engineering, Xi'an University of Science and Technology, Xi'an 710054, China

*Corresponding author's e-mail: menhao121@126.com

Abstract. In recent years, the electric vehicle industry has developed rapidly. Compared with traditional internal combustion engine vehicles, pure electric vehicles have the advantages of less pollution, less noise, and simple structure. At the same time, the problem of short range has become an important factor restricting the development of pure electric vehicles. This paper proposes to mount a four-speed mechanical automatic transmission in a pure electric vehicle to improve the working efficiency of the motor, thereby improving the vehicle's cruising range. CRUISE simulation software was used to build a vehicle model of a pure electric vehicle equipped with this transmission. A vehicle equipped with a fixed gear ratio reducer is used as a comparative experiment to study the impact of a mechanical automatic transmission on the economy and power of the vehicle.

1. Introduction

Cars have provided us with convenience, and at the same time brought environmental problems that cannot be underestimated. Compared with traditional fuel vehicles, electric vehicles are powered by batteries and driven by electric motors, with little or no pollution [1]. However, since the launch of pure electric vehicles, the distribution points of charging piles for electric vehicles are currently scattered, causing drivers to be unable to charge their vehicles in time at their destination. On the other hand, because the charging speed of pure electric vehicles is slower than the rate of refueling of traditional internal combustion engine vehicles, the range problem has restricted the development of pure electric vehicles to some extent [2]. Pure electric vehicles are often equipped with a fixed ratio reducer. Although the fixed ratio reducer can meet some urban road conditions, when the vehicle is driving on a relatively steep slope and more complicated road conditions, the motor cannot continue to work in the high efficiency range that conduct to loss of power [3]. The purpose of this article is to propose a transmission with multiple gears for a pure electric vehicle, so that when the motor operates under different operating conditions, it can well be more opportunities to operate in a high efficiency range, reduce power loss, and improve vehicle economics. On the other hand, the multi-stage transmission can reduce the speed and increase the torque, which improves the vehicle's power performance [4]. The introduction of automatic transmission technology for pure electric vehicles can effectively improve the comfort of the ride, improve the efficiency of shifting, and improve the overall performance of the vehicle [5]. The automatic transmission is a torque converter that replaces the clutch. It is used by most passenger vehicles because it has a small sense of frustration during gear shifting and smooth gear shifting without the need for frequent driver operations. Large, high quality, high cost, not suitable for pure electric vehicles. Combining the characteristics of high transmission

IOP Conf. Series: Materials Science and Engineering **825** (2020) 012005 doi:10.1088/1757-899X/825/1/012005

efficiency, low cost, simple structure, small size, and low failure rate of mechanical transmissions, as well as the advantages of automatic transmissions, Automatic Manual Transmissions (AMT) are widely used in pure electric vehicles [6].

2. Vehicle parameters

The parameters of the vehicle are usually defined from three aspects, the external dimensions, mass, and air resistance coefficient [7]. This paper selects the Model 3 of the American Tesla company as the research object. The parameters that need to be determined when determining the research object are shown in Table 1. The dynamic index is a key index to measure the performance of the car. The specific parameters are shown in Table 2.

Table 1. Vehicle parameters

Vehicle parameters	Parameters value
Total mass/kg	2105
Readiness quality/kg	1611
Shape size (length * width * height)/mm	4694×1850×1443
Tyre Specification and Type	235/45 R18
Wheelbase /mm	2875
Tread /mm	1580
Windward area/m^2	2.28
Wheel rolling radius/m	0.334

Table 2. Dynamic Indicators

Dynamic Indicators	Parameters value
Acceleration time (t/s)	0-50km/h < 5
Maximum Speed (km/h)	120
Maximum gradient (%)	25,20km/h

3. Transmission structure design

The structure of the AMT power transmission system is shown in Figure 1. The main components of a vehicle power transmission system are the engine, AMT, main reducer, differential, and wheels [8]. In this section, the 3D modeling of the transmission is done by SOLIDWORKS, as shown in Figure 2.

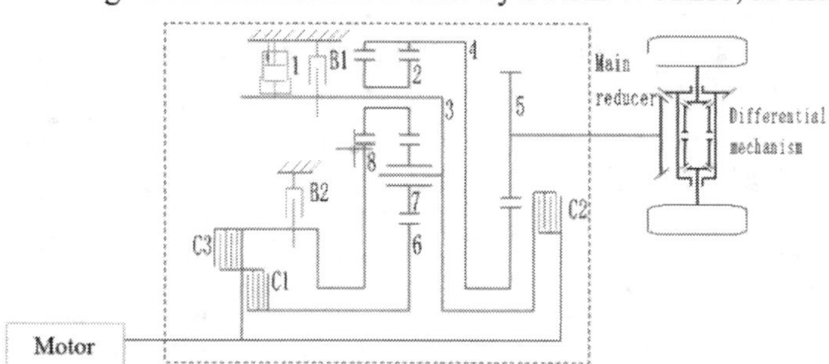

1-Unidirectional Clutch; 2-Long Planetary Gear; 3-Planet carrier; 4- Ring gear; 5-Meshing gear; 6-Small Sun Gear; 7-Short Planetary Gear; 8-large sun gear; B1, B2-Brake; C1, C2, C3-Clutch.

Figure 1. AMT power transmission structure

The structure is based on the design of double planetary gear train. It consists of two planetary wheels, small sun gear and large sun gear, long planetary gear and short planetary gear. Short

IOP Conf. Series: Materials Science and Engineering **825** (2020) 012005 doi:10.1088/1757-899X/825/1/012005

planetary gear 7 meshes with small sun gear 6, long planetary gear 2 meshes with large sun gear 8, short planetary gear 7 and ring gear 4, respectively. Ring gear as the only power output gear engages with gear 5 to ensure power output. The shift action of gearbox is completed by the cooperation of clutch C1, C2, C3 and brake B1, B2. The Planet carrier is used as input power and keeps the position of the planetary wheel. The function of unidirectional clutch is to ensure that the planetary carrier will not reverse during the operation of planetary gear train. Following is a description of the working process of four forward, reverse and neutral gears.

First gear: C1 clutch combined, small sun gear input power, B1 brake work, at same time, planetary carrier is locked, power output from the ring gear.

Second gear: C1 clutch combined, small sun gear input power, B2 brake work, large sun gear is locked, ring gear output power.

Third gears: C1 clutch and C3 clutch work at the same time, any two components of planetary gear train work in the same speed and direction, the whole mechanism is locked, and the ring gear both the input and output of power to produce direct gear.

Fourth gears: C2 clutch combined, planetary carrier input power, B2 brake working, large sun gear locked, ring gear output power.

Reverse gear: Since the electric vehicle is powered by the motor, depending on the characteristics of the motor, the motor can be reversed to achieve reverse gear.

The structural shifting feature is that the C1 clutch is always in the combined state during the shifting of the first to third gears, and the first and second gears are controlled by the B1 and B2 brakes. The C3 clutch can be combined to complete the third gear shift. Therefore, the smoothness of shifting is increased and the loss of power is reduced.

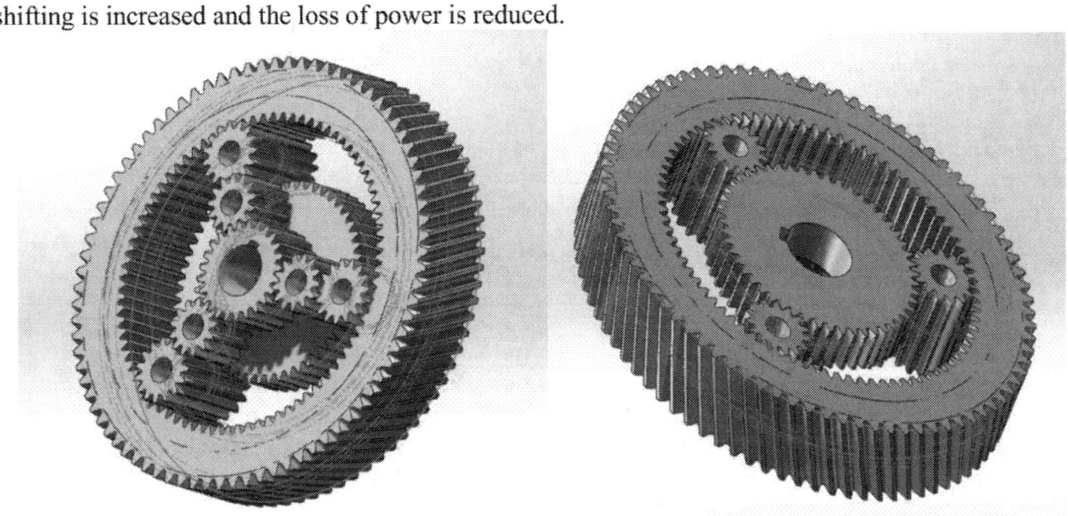

Figure 2. 3D modeling of transmission

4. Transmission ratio calculation

Reasonable transmission ratio distribution can make the drive motor have more chances to work in the high-efficiency range, so that the cooperation between the motor and the transmission can be optimized, thereby improving the performance of the entire vehicle [9]. The transmission structure is based on the design of the double planetary gears. Using the characteristic equations of the double planetary gears, the transmission ratio formula of each gear can be obtained [10] to achieve a reasonable distribution of the transmission ratios.

(1) First gear.

The characteristic equation of the rear planetary row is:

$$n_{21} - \alpha_2 n_{22} + (\alpha_2 - 1) n_{23} = 0 \tag{1}$$

n_{21}——The small sun gear speed is equal to the input speed n_i;

n_{22}——Ring gear speed, equal to the output speed n_0;

n_{23}——Planet carrier speed, this time is 0;

α_2——The ratio of the number of teeth of the ring gear to the number of teeth of the small sun gear.

Therefore, the transmission ratio can be obtained from the equation (1).

$$i_1 = \frac{n_i}{n_0} = \alpha_2$$

(2)Second gear.

$$\text{Front planetary gears meet } \alpha_1 n_0 - (1+\alpha_1) n_C = 0 \quad (2)$$

$$\text{Rear planetary gears meet } n_i - \alpha_2 n_0 + (\alpha_2 - 1) n_C = 0 \quad (3)$$

The transmission ratio obtained by the equation (2), (3) is $i_2 = \frac{n_i}{n_0} = \frac{\alpha_1 + \alpha_2}{1+\alpha_1}$.

(3) The third gear is the direct gear. So, transmission ratio is 1.

(4) Fourth gear.

According to the characteristic equation (2) of the front planetary row, the transmission ratio can be obtained as $i_4 = \frac{n_i}{n_0} = \frac{\alpha_1}{1+\alpha_1}$.

Finally, the transmission ratio calculation results are shown in Table 3.

Table 3. Transmission ratio

Transmission gear	Main reducer	First gear	Second gear	Third gear	Fourth gear
Ratio	4.3	3.18	1.84	1	0.61

5. Simulation model establishment and result analysis

The CRUISE developed by AVL includes simulations of fuel economy, vehicle dynamics, braking performance and emissions performance. It uses a comprehensive solution to increase the accuracy and speed of calculations. CRUISE software includes various components of the vehicle and vehicle model modules and is easy to operate [11]. Therefore, the entire vehicle model is built in CRUISE.

5.1. Building a vehicle simulation model

The vehicle simulation model mainly includes: vehicle module, motor module, power battery module, clutch module, transmission module, reducer module, differential module, brake module, wheel module, and cab module [12]. For the input of the vehicle parameters, refer to Table 1. The motor parameters are matched by the dynamic index, as shown in Table 2. The design range of the target vehicle is 300km, so the battery module is matched with the design range. After the vehicle model is set up, the signals are first connected to each part, and then the calculation tasks are set, that is, the maximum speed, gradient, and acceleration.The complete vehicle simulation model is shown in Figure3.

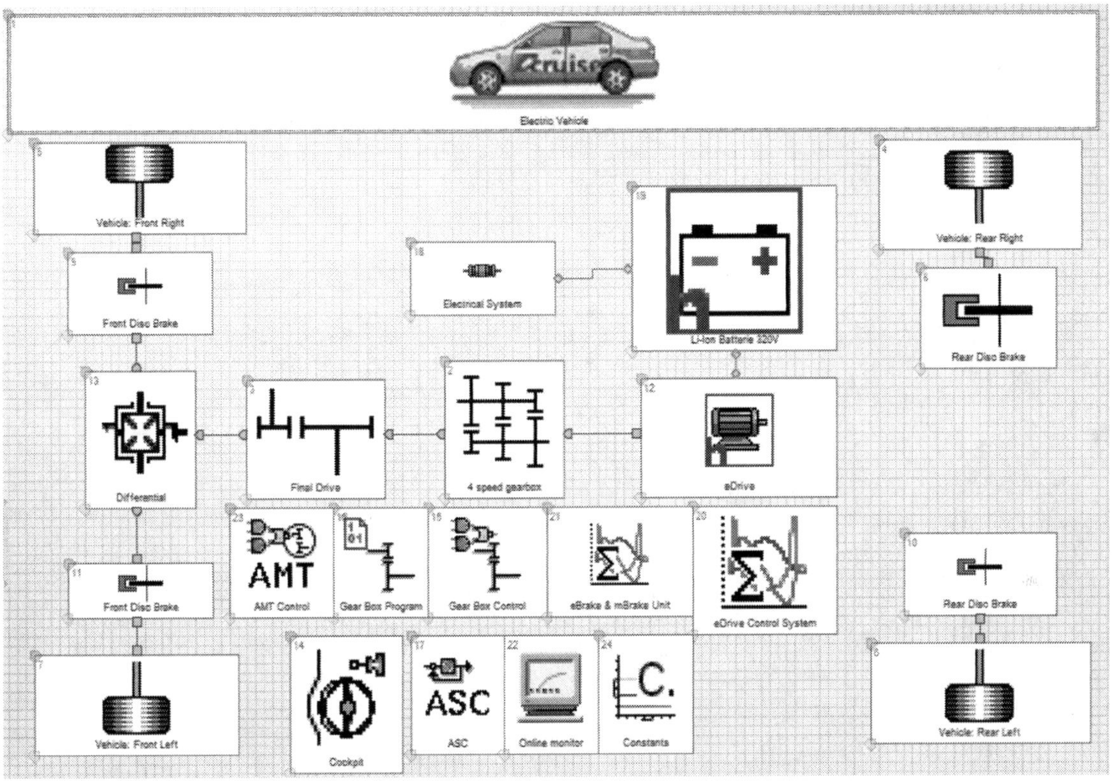

Figure 3. AMT vehicle model

5.2 Analysis of simulation results

In order to discuss the impact of AMT on the vehicle's economy and power, a pure electric vehicle model equipped with a fixed ratio reducer was selected for comparison. Figure 4 is the gradient curve of two different transmission vehicles. The blue curve is the gradient curve of the AMT vehicle, and the red curve is the gradient curve of the fixed ratio reducer vehicle. When the vehicle is traveling at 25km / h, the highest gradient of a pure electric vehicle using a fixed ratio reducer is 22.3%, and the highest gradient of an AMT vehicle can reach 26.2%, which can meet the performance requirements of the entire vehicle.

Figure 5 shows the acceleration curves of two different transmission vehicles. The blue curve is the simulation result of the AMT vehicle, and the red curve is the simulation result of the vehicle with a fixed ratio reducer. The acceleration time of the AMT vehicle from the start to 50km / h is 4.26s, which meets the acceleration performance requirements of the entire vehicle.

This paper uses the energy consumption of the two vehicles under NEDC conditions to evaluate the economic performance of the two models. Figure 6 shows the state of charge (SOC) of the vehicle under NEDC conditions. The blue curve is the SOC change curve of the AMT vehicle during driving, and the red curve is the SOC curve of the fixed ratio reducer vehicle. From the data in Figure 6 and Table 4, it can be seen that the fixed ratio reducer vehicle is in the NEDC operating state, and the battery SOC has dropped from 80% to 69.7929%. AMT vehicle battery SOC dropped from 80% to 71.1163%. In the matching scheme, the battery pack capacity is 48 kW • h, and under complete NEDC cycle conditions, the battery energy consumption of AMT vehicles is reduced from 8.883 kW • h to 7.7674 kW • h compared to fixed ratio reducer vehicles.

IOP Conf. Series: Materials Science and Engineering **825** (2020) 012005 doi:10.1088/1757-899X/825/1/012005

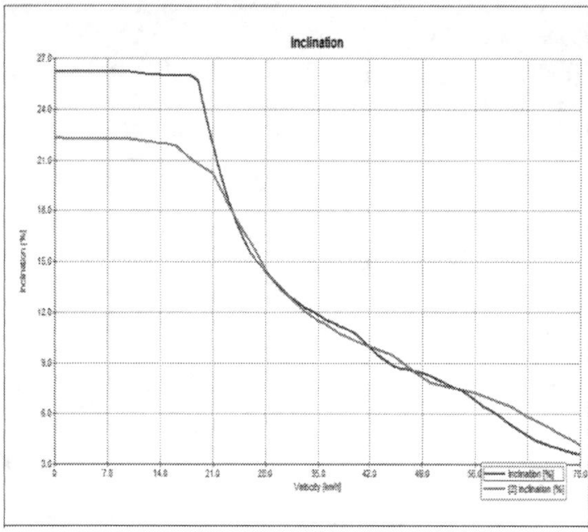

Figure 4. The climbing performance curve of the vehicle

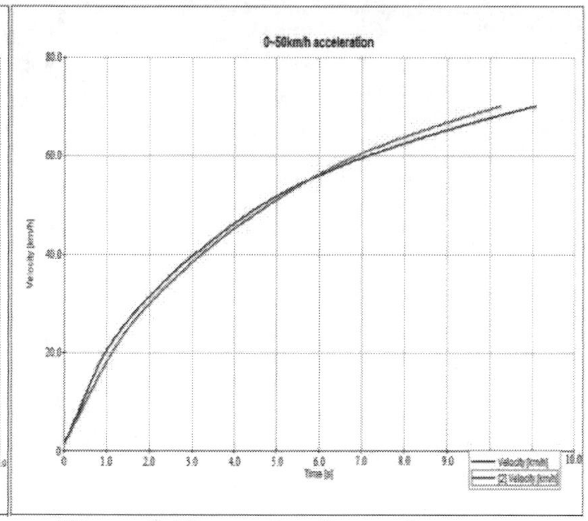

Figure 5. The acceleration performance curve of the vehicle

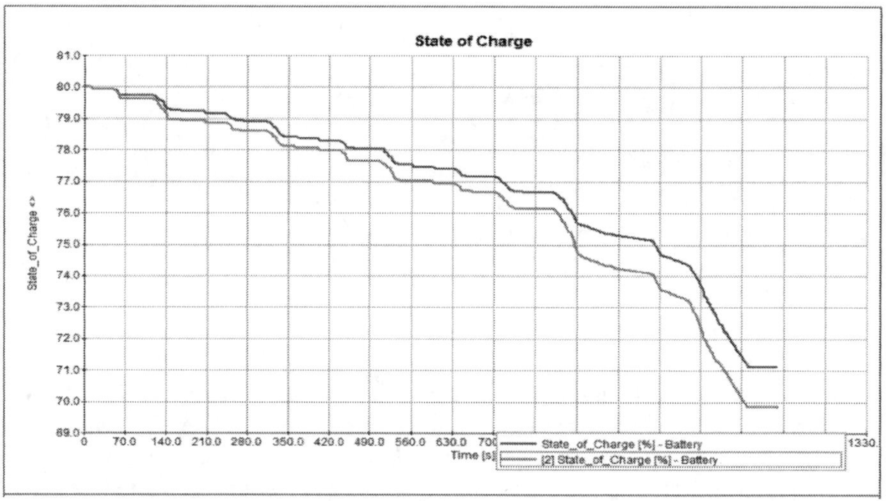

Figure 6. The contrast curve of SOC on NEDC condition

Table 4. The parameters contrast of the dynamic and economic performance index

	Performance index	AMT vehicle	fixed ratio reducer vehicle
Dynamic performance	Maximum speed (km/h)	120	114
	Maximum gradient (%)	26.2	22.3
	Acceleration time 0~50 km/h (s)	4.26	4.85
Economic performance	SOC consumption of NEDC conditions (%)	8.8837	10.2071

It is concluded from the simulation results that matching the four-speed AMT of the pure electric vehicle improves the acceleration performance and climbing performance of the whole vehicle, and reduces the power consumption of the pure electric vehicle.

6. Conclusion

In this paper, the transmission structure and transmission ratio of the pure electric vehicle transmission are designed to achieve the matching between AMT and pure electric vehicles. The CRUISE software was taked as simulation platform, and the AMT pure electric vehicle model was established. In this simulation, a pure electic vehicle model which matching a fixed ratio reducer was established as a comparison. The simulation is performed by setting related calculation tasks, and the performance of two different vehicles is compared and analyzed through the simulation results. The feasibility of matching the four-speed AMT for pure electric vehicles proposed in this paper is verified, and the economic and dynamic simulation results of two types of transmission vehicles are compared and analyzed. The results shows that the economy and power of the pure electric vehicle have improved when the vehicle matching the four-speed AMT.

References

[1] Wilberforce T Hassan Z Khatib F N Makky A A Baroutaji A Carton J G and Olabi A G 2017 Developments of electric cars and fuel cell hydrogen electric cars *International Journal of Hydrogen Energy* Vol 42 pp 25695-25734

[2] Wong Y H 2017 Electric Vehicle Technology Trends and Prospects *Automotive Technology* Vol 29 pp 278-288

[3] Yunyun Y Sen W and Zhengce C 2011 AMT Research Based on the Electric Drive System *International Conference on Electric Information & Control Engineering* United Kingdom pp 5679 - 5682

[4] Tanaka H and Wada H 1995 Fuzzy Control of Clutch Engagement for Automated Manual Transmission *Vehicle System Dynamics* Vol 24 365-376

[5] Ling L Chen Z S and Luo Y Y 2012 Study on Control Process of Automatic Mechanical Transmission System for Pure Electric Bus *Applied Mechanics and Materials* Vol 220 pp 1795-1799

[6] Lucente G Montanari M and Rossi C 2007 Modelling of an Automated Manual Transmission System *Mechatronics* Vol 17 pp 73-91

[7] Sorniotti A Pilone G L and Viotto F 2011 A Novel Seamless 2-Speed Transmission System for Electric Vehicles: Principles and Simulation Results *Sae International Journal of Engines* Vol 4 pp 2671-2685

[8] Galvagno E Velardocchia M and Vigliani A 2011 Analysis and simulation of a torque assist automated manual transmission *Mechanical System and Signal Processing* Vol 25 pp 1877-1886

[9] Ben Y Jianwu Z and Hongkui L 2015 Transmission ratio design of small pure electric vehicle transmission *Transmission Technology* Vol 29 pp 30-33

[10] Jian Z and Xuehuan Z 2018 Design of a new Lavina-type planetary gear mechanism *Mechanical Transmission* Vol 42 pp 179-181

[11] Xiaoling H and Jeffrey H 2000 Hybrid electric vehicle simulation and evaluation for UT-HEV *SAE Paper* Vol 1 pp 459-473

[12] Abdelhalim N A 2013 Use of Vehicle Power-Train Simulation with AMT for Fuel Economy and Performance *Modern Mechanical Engineering* Vol 3 pp 127-135

Method to Integrate Human Simulation into Gazebo for Human-robot Collaboration

Linxi He[1], Paul Glogowski[2], Kai Lemmerz[2], Bernd Kuhlenkötter[2] and Weimin Zhang[1,*]

[1]Advanced Manufacturing Technology Center, Tongji University, 4800 Caoan Road, Shanghai, PR China
[2]Chair of Production Systems, Ruhr-University Bochum, Bochum, Germany

[*]iamt@tongji.edn.cn

Abstract. The human-robot collaboration (HRC) takes the advantages of both the flexibility of human workers and the accuracy of robots. Simulation is an effective and economical method to verify the safety and reliability of the HRC system. This paper presents an approach to integrate the simulation of human workers into a robot simulation environment. A digital human model is built in the robot simulator Gazebo and can be controlled by data from proprietary software for human workflow planning. As a result, a holistic HRC scenario is established in Gazebo, which provides the potential for workspace monitoring and robot collision-free trajectory planning based on the Robot Operating System (ROS).

1. Introduction

Individual customization and small-batch production have become a new trend in the manufacturing industry. Shorter production cycles and complex product variants require higher flexibility of the production. Human-robot collaboration (HRC) provides an economical solution for flexible and small-batch production. The core idea of HRC is that human beings are responsible for the processes that require high flexibility and touch sensitivity, while robots take on their fast, accurate advantages to take on repetitive tasks. Such a hybrid system can meet the requirements of a higher diversity of products, fluctuating batches as well as shorter production cycles.

Different from traditional industrial robots in HRC, human operators and collaborative robots work simultaneously within a collaborative workspace. When the power to the robot's actuators is available, the physical contact between the operator and robot system can occur within a collaborative workspace [1]. Therefore, ensuring the safety of human operators in collaborative workspaces is very important for the HRC system. Due to economic and security considerations, testing and verifying the HRC system faces many difficulties in practice. Simulation can be a more efficient and economical solution for validating the HRC system before implementation. For these reasons, this paper proposes an approach to integrate human workflow planning into the robot simulation environment for a holistic HRC simulation.

The structure of the paper is as follows: Chapter 2 introduces the previous research on the HRC simulation system. Chapter 3 gives an overview of the method and describes separately the software and technology used in each part of the system. Chapter 4 introduces how to build a human model that can be simulated in Gazebo. Chapter 5 explains the implementation of the interface layer, which enables the transmission of human simulation data to the robot simulation environment. In Chapter 6 the method is validated by an HRC sample scenario.

2. Related work

In previous research, there have been some simulation systems for HRC. One of them is an application for a robot-based assistance system for welding tasks [2,3], which integrates a digital human model into a manufacturer-independent offline programming system *FAMOS*, and can make an ergonomic evaluation of welding tasks according to the OWAS analysis [4]. However, this research is only applied to HRC operation in an assistance system for welding scenarios but not for the universal HRC operation. Siemens' product lifecycle management system *Tecnomatix* also supports digital human modelling and simulation function, allowing human simulation for production process planning and ergonomic evaluation to optimize production processes and improve safety [5]. The robot simulation software *RobotStudio* from company ABB has a model library of ABB's collaborative robots, which enables to plan the path and the workflow for their collaborative robots. However, the simulation of human work is still a challenge in research and there has not yet been an integrated simulation system for the universal scenario of human-robot collaboration.

3. Architecture of the integrated HRC simulation system

As shown in figure 1, the integrated HRC simulation system [6,7] consists of three main parts: the human simulation, the interface layer, and the robot simulation environment. The main work in this paper is to build a digital human model in Gazebo and develop an interface between Gazebo and EMA. The HRC simulation will eventually be performed in Gazebo.

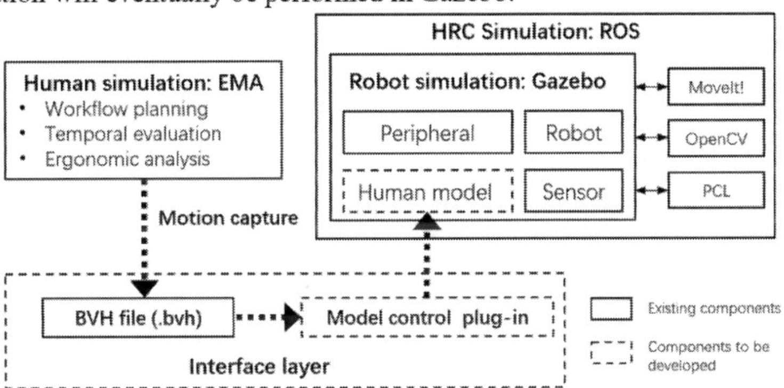

Figure 1. Structure of the integrated HRC simulation system

3.1. Human simulation

A human simulation system must be able to build the environment of most production scenarios and plan the workflows for operators in the environment [8]. In this paper, the *Editor for Manual Work Activities (EMA)* of the company imk is used to plan the human workflow. EMA is a simulation tool for human based-manual activities. It enables the holistic planning, evaluation and 3D simulation of human work in the context of a digital factory [9].

EMA also provides a digital model library of common robot systems, which can visualize the motion of the robot system. However, only limited simulation functions for robot components have been integrated [6]. In this paper, the operator motions planned in the EMA will be recorded by motion capture, and then exported in a format of *Biovision Hierarchy (BVH)* for the control of the human model in Gazebo.

3.2. Robot simulation

The robot simulation is based on the open-source software framework Robot Operating System (ROS) and the robot simulator Gazebo. As open-source software, ROS enables a cost-efficient and manufacturer-independent approach that makes the software future-proof and reusable by the use of open standards [10]. A variety of applications for the simulation of robots and sensors are integrated in ROS and a wide range of drivers for controlling sensors, actuators and entire robot systems are provided [9]. These advantages make ROS extremely suitable for the development of HRC simulation systems.

Gazebo is one of the opensource 3D robot simulators based on ROS, completed with dynamic and kinematic physics and a pluggable physics engine. As an effective and manufacturer-independent software, Gazebo offers a rich environment for rapid development and testing of complex robot system. However, Gazebo does not yet support the function of human simulation. In this paper, a human model will be built in Gazebo and an interface is developed to control the model with data from EMA.

4. Human body modelling in Gazebo

This chapter explains how to build a human body model for Gazebo. A skin file in *COLLADA* format is first built and then connected to an underlying skeleton that moves it. Through the "actor" function provided by Gazebo, the human model can be loaded into simulation environment.

4.1. Actor

By modelling a human in Gazebo, the "actor" is used. The actor is a kind of model of Gazebo, which allows adding animations to a common model. There are two main components in an actor: Skin and animation. The skin defines the visual of the actor and the underlying skeleton which moves it. The skin file is a *COLLADA* file (.dae). The animation element is an animation for the actor. The skeleton of the animation must be compatible with the skin skeleton. There can be one or more plugins in the actor, which provides access to and use of almost all aspects of the actor to control or customize the behavior of the actor.

4.2. Human body modelling by blender

In this method, the 3D computer graphics software *Blender* is used to create a skin file in COLLADA format. For the visual of the skin, a male model with a 95% Europe size is selected [11]. To ensure that the human model in Gazebo can be accurately controlled by the motion data from EMA, an underlying skeleton with the same number of joints and hierarchy as the skeleton in EMA must be established (see figure 2). Through the skinning function in Blender the skin is associated with the skeleton, so that the body of the human model can be controlled by the underlying skeleton.

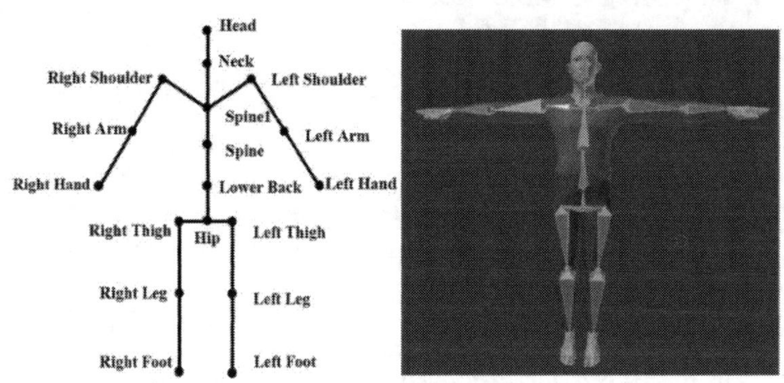

Figure 2. Left: Definition of the human underlying skeleton in EMA;
Right: Human model built in Blender

5. Implementation of interface layer

As shown in figure 1, the motion data is processed at the interface layer. An interface plug-in is developed based on the Gazebo function library, which has the following functions: Parse the motion data in BVH, calculate the world posture of each joint and add the posture data into the actor as a

IOP Conf. Series: Materials Science and Engineering **825** (2020) 012006 doi:10.1088/1757-899X/825/1/012006

skeleton animation.

5.1. Parse the BVH file

BVH is a widely used data format for motion capture. It defines the hierarchy of the skeleton and records the motion data for each joint at each moment [12]. A BVH file has two parts: a header section which describes the hierarchy and initial pose of the skeleton, and a data section which contains the motion data.

First, the hierarchical information in BVH is parsed and stored. Here, a nested if function is applied because the skeleton hierarchy in BVH is recursively defined. The hierarchical information is stored in the "skeleton" class (see figure 3). It consists of a sequence of nodes, each of which stores the original transformation information (i.e. offset) for the joint. The relationship between two nodes is described by "Parent" and "Child" pointers, thereby constructing a hierarchical relationship of the skeleton.

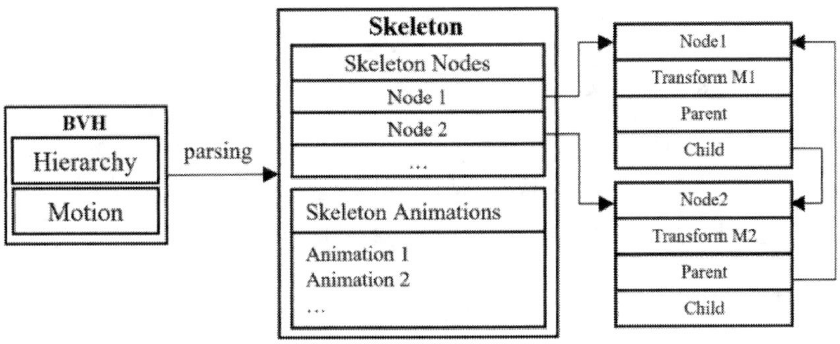

Figure 3. The standard class "Skeleton" that is used to describe the human model in Gazebo

The motion data is then read line by line and the transformation of each joint is stored in a matrix. For example, the information of a joint is recorded by three channels, which are the rotation angles around the X, Y, and Z axes:

CHANNEL 3 Zrotation Yrotation Xrotation

The rotation matrix R of the bone is calculated as in equation 1.

$$R = R_z R_y R_x \tag{1}$$

Using a homogenous matrix to represent the transformation of the bone is illustrated in equation 2. The rotational components take up the top left 3x3 cells and the translation components are the first 3 cells of the 4th column.

$$M = \begin{bmatrix} R & R & R & T_x \\ R & R & R & T_y \\ R & R & R & T_z \\ 0 & 0 & 0 & 1 \end{bmatrix} \tag{2}$$

5.2. Calculate the world posture

The transformation of the bone in the BVH file represents the rotation and translation components of the bone relative to its parent's coordinate. However, the control of joints in Gazebo requires the pose of the joints relative to the world coordinate system. To obtain a global matrix transformation for a given bone, the local transformation needs to be pre-multiplied by its parent's global transformation, which itself is multiplying its local transformation with its parent's global transformation and so on [13]. Equation 3 shows the calculation process, where n is the current bone whose parent bone is n - 1 and n = 0 is the bone at the root of the hierarchy.

$$M_{global}^{n} = \prod_{i=0}^{n} M_{local}^{i} \tag{3}$$

For example, the global transformation of bone "Left Leg" can be calculated by Equation 4. Where

M is the local transformation of each bone. After the above calculations, the global transformation of each bone can be obtained.

$$M'_{LeftFoot} = M_{Hip}M_{LeftThigh}M_{LeftLeg}M_{LeftFoot} \tag{4}$$

5.3. Add the posture data into the actor

To drive the human model in Gazebo, the calculated transformation of each joint at each moment must be sequentially added to the object "skeleton animation" of skeleton class through an *AddKeyframe* Function (see figure 4). If the "Skeleton Nodes" is compatible with the skin skeleton, the object "Skeleton Animation" will be added into the actor as the animation component, so that the human model can correctly display the animation in Gazebo.

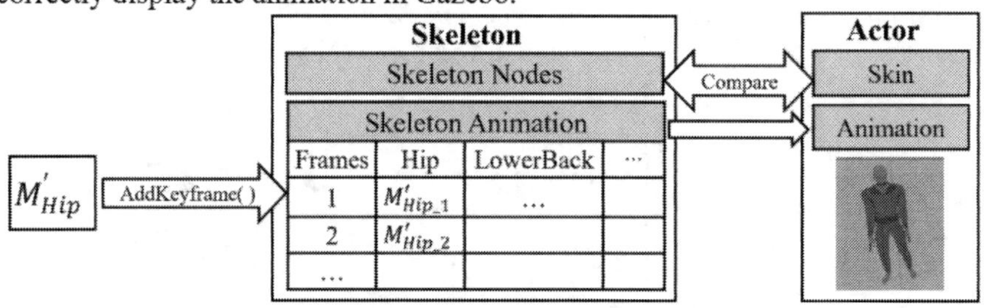

Figure 4. Process of importing posture data into actor

6. HRC simulation scenario

In this section, a sample HRC scenario is built in Gazebo. At first, a sample workflow with walking, picking up, and screwing is planned in EMA (see figure 5). The workflow planning in EMA bases on a task library. The tasks library in EMA is a collection of basic operations or human behaviours in common production. Each task can be customized through the setting of different parameters. By combining and ordering multiple tasks a complex workflow can be comprehensively defined. Then, the human work is simulated and the human motion is recorded by motion capture.

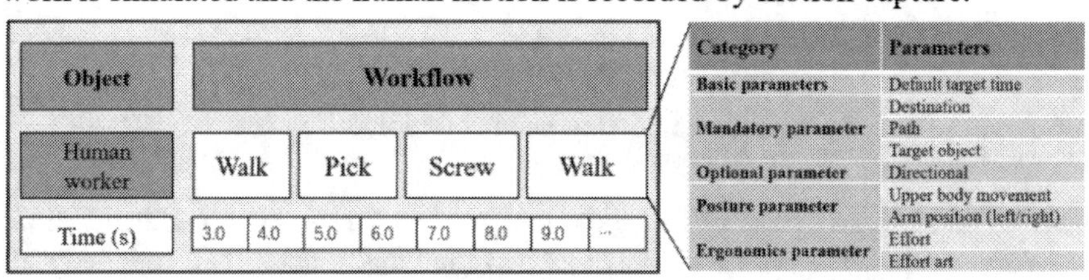

Figure 5. Human workflow planning in EMA for the HRC scenario simulation

As shown in figure 6, the previously created human body model is loaded into the Gazebo world. Through the interface *ros_gazebo_pkg* between Gazebo and ROS, a robot model can be spawned in the Gazebo world and be controlled by *MoveIt!*. The human motion data from EMA is transmitted into Gazebo through the interface layer. With the update function in the plug-in, the posture of the human model is continuously updated according to the data from EMA.

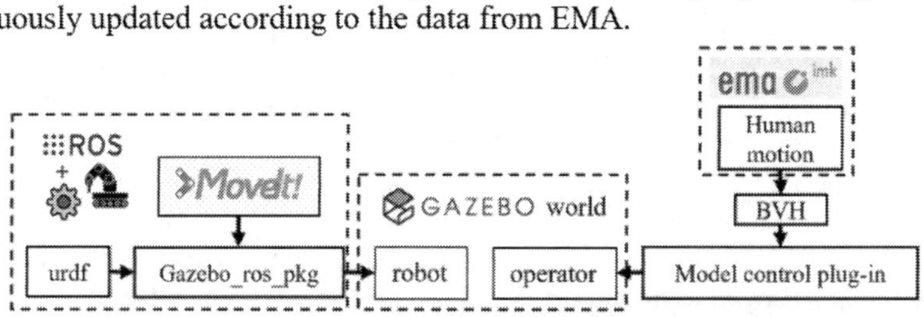

Figure 6. Method to integrate both robot and operator in a Gazebo *.world* file

As a result, the human model in Gazebo accurately repeats the workflow planned in EMA, both in terms of time and position (see figure 7). Through this method, human motion planned in EMA is simulated in the Gazebo environment, so that it can combine with the robot simulation functions in Gazebo and realize an HRC simulation.

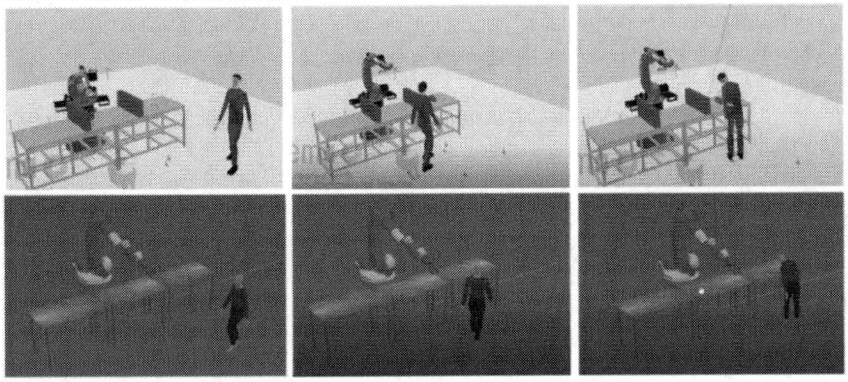

Figure 7. Comparison between the workflow simulated in EMA (above) and Gazebo (below)

7. Conclusion and future work

In this paper, a method is developed to build a digital human model in the robot simulator Gazebo. This model can be controlled by the data from the human simulation software EMA, which provides human simulation and workflow planning for HRC. This method integrates human simulation into a robot simulation environment and provides the potential for testing and verification of the HRC system. This paper introduces the way to build a human model for Gazebo, explains the concept of workflow planning in EMA, and describes the implementation of this integration method. As a result, human motion planned in EMA can be successfully simulated in Gazebo.

Concerning future work, with the rich sensor library in Gazebo, human workspace can be monitored. A collision-free trajectory planning on the collaborative robot based on workspace data can be performed through *MoveIt!* or other applications based on ROS. Besides, using real human motion data collected by cameras or other sensors to control the model in Gazebo, which will provide the potential for a more complex scenario of HRC and real-time HRC simulation, is another direction of research in the future.

Acknowledgments

The research and development project "KoMPI" is funded by the German Federal Ministry of Education and Research (BMBF) within the Framework Concept “Research for Tomorrow's Production” (fund number 02P15A060) and managed by the Project Management Agency Forschungszentrum Karlsruhe, Production and Manufacturing Technologies Division (PTKA-PFT).

References

[1] ISO/TS 15066: 2016 *Robots and robotic devices - Collaborative robots* p 2
[2] Thomas C, Matthias B and Kuhlenkötter B 2016 Human-robot-collaboration-new applications in industrial robotics *Int. Conf. on Competitive Manufacturing (Stellenbosch)* pp 293-299
[3] Thomas C, Busch F, Kuhlenkoetter B, and Deuse J 2012 Ensuring human safety with offline simulation and real-time workspace surveillance to develop a hybrid robot assistance system for welding of assemblies *Enabling Manufacturing Competitiveness and Economic Sustainability* pp 464-470
[4] Karhu O, Kansi P and Kuorinka I 1977 Correcting working postures in industry: a practical method for analysis *Applied Ergonomics* **8(4)** pp 199-201
[5] Bangsow S 2015 *Tecnomatix Plant Simulation* (Heidelberg: Springer) pp 499-541

[6] Lemmerz K, Glogowski P, Hypki A and Kuhlenkoetter B 2018 Functional integration of a robotics software framework into a human simulation system *ISR 2018; 50th International Symposium on Robotics* pp 1-8

[7] Glogowski P, Lemmerz K, Schulte L, Barthelmey A, Hypki A, Kuhlenkötter B and Deuse 2017 Task-based simulation tool for human-robot collaboration within assembly systems *Tagungsband des 2. Kongresses Montage Handhabung Industrieroboter (Springer Vieweg, Berlin, Heidelberg)* pp 155-163

[8] Schönherr D W I R 2014 *Simulationsbasierte Absicherung der Ergonomie mit Hilfe digital beschriebener menschlicher Bewegungen* pp 118-121

[9] Fritzsche L, Jendrusch R, Leidholdt W, Bauer S, Jäckel T and Pirger A 2011 Introducing ema (Editor for Manual Work Activities)–a new tool for enhancing accuracy and efficiency of human simulations in digital production planning *Int. Conf. on Digital Human Modeling (Springer, Berlin, Heidelberg)* pp 272-281

[10] Quigley M, Gerkey B and Smart W D 2015 *Programming Robots with ROS: a practical introduction to the Robot Operating System* (Sebastopol: O'Reilly Media, Inc.)

[11] Bullinger H, Angelika C and Jens M 2016 *Homo sapiens digitalis-virtuelle Ergonomie und digitale Menschmodelle (Springer Berlin Heidelberg)* pp 11-13

[12] Chung H S and Lee Y 2004 Motion capture markup language for integration of heterogeneous motion capture data *Computer Standards & Interfaces* **26(2)** pp 113-130

[13] Meredith M and Maddock S 2001 Motion capture file formats explained *Department of Computer Science, University of Sheffield* **211** pp 241-244

Lean Improvement of Product P Assembly Line based on Value Stream Map

Kai Chang[1,2], Guangchao Tan[1,2], Xiaolong Wang[1,2], Changjun Zhou[3] and Luyuan Wei[3]

[1] Shenyang Institute of Automation Chinese Academy of Sciences, Shenyang 110016, China.
[2] Institutes for Robotics and Intelligent Manufacturing, Chinese Academy of Sciences, Shenyang 110169, China.
[3] Xi'an Aerospace Engine Co., Ltd.

*E-mail: changkai@sia.cn .

Abstract. Based on the product P assembly line transformation project of a company, the main problems of the original production line are analyzed with the method of Value Stream Map. Lean improvement measures which is used in the design of new production line are put forward in the aspects of overall layout, process arrangement, production mode, material buffer, auxiliary equipment and personnel allocation to reduce some non-value added activities and wasteful behaviors. By the Value Stream Map of the improved lean production line, the evaluation indexes can be improved significantly. After the new production line is completed and put into production, the workshop stock is reduced, the production efficiency is improved, and the production cost is reduced.

1. Introduction

With the development of economy and the increasingly fierce market competition, customers have higher requirements for product personalization, diversification and delivery time. Enterprises are facing new development bottlenecks. In order to solve these problems, the production mode of enterprises must be reformed. That is to say, the past large-scale and planned production mode should be abandoned; the flexible production mode of multi variety and small batch should be developed. Enterprises should not only reduce product cost but also improve product quality. Lean production is currently recognized as the best way to meet these requirements [1] [2].

Company B is a supporting enterprise for military production. To some extent, it follows the extensive production mode of the original state-owned enterprises, such as disordered production organization, untimely planning and scheduling, massive accumulation of materials, long cycle of product manufacturing and low efficiency of personnel operation. On the one hand, facing the rapid plan change of upstream scientific research institutions, the company is overwhelmed; on the other hand, facing the short supply cycle required by downstream customers, the company can't deliver on time. So, the original mode of production organization has seriously restricted the development of enterprises.

Based on the transformation project of P product assembly line entrusted by company B, the paper analyzes the main problems in the production process of the original assembly line with the Value Stream Map. Under the theme of improving production capacity and reducing cost, in the design of the transformation scheme, the work station and procedure has been reorganized to eliminate the bottleneck in the production, production management mode has been optimized to make full use of existing people and field resources to organize production. The new assembly line will adopt the pull production mode, which makes the general assembly of products P to pull the production of parts workshops at all levels, so as to reduce the stocks of finished products and parts. The new general assembly line needs to meet the user's requirements that the production capacity can reach more than 60 sets per month (two shift system).

2. Main problems of the original assembly line of product P

Before the transformation, the general assembly line adopts the "linear" production flow layout, and the overall layout in the workshop is shown in figure 1. The original assembly line and supporting facilities occupy 3000m^2 totally, mainly including sub-assembly area, material buffer area, finished product buffer area and general assembly area. Material buffer area cached assembly parts delivered. The sub-assembly area was used to assemble and cache the parts of the overall connection station. The finished product area was used to store the finished products. The tool storage area was used to store all the tools and tooling in the assembly and testing stations. Team garden was used for rest and placement of notices Kanban. The general assembly area completed all processes from assembly to test and formed the final product P.

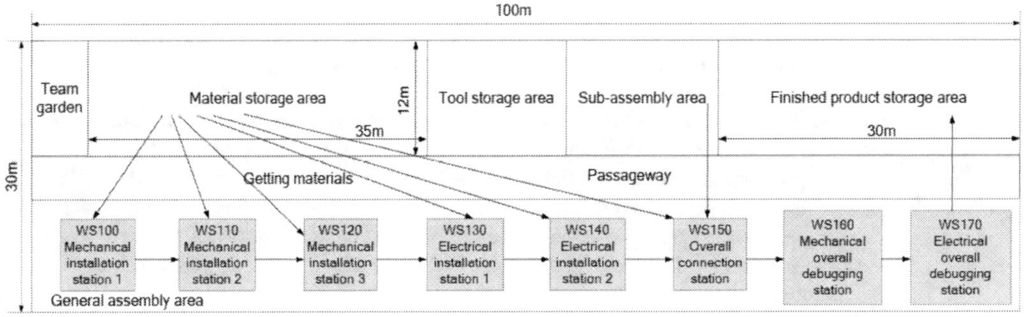

Figure 1. The layout and logistics route of original assembly line

After the investigation of the production planning management department, through the disassembly of each process in the assembly station of the original production line, operation time of each operator in each process can be recorded, and actual average operation time and waiting time of each process can be calculated. Finally, the work hours of each station can be counted by the mountain plot and the Value Stream Map can be drawn, as shown in figure 2.

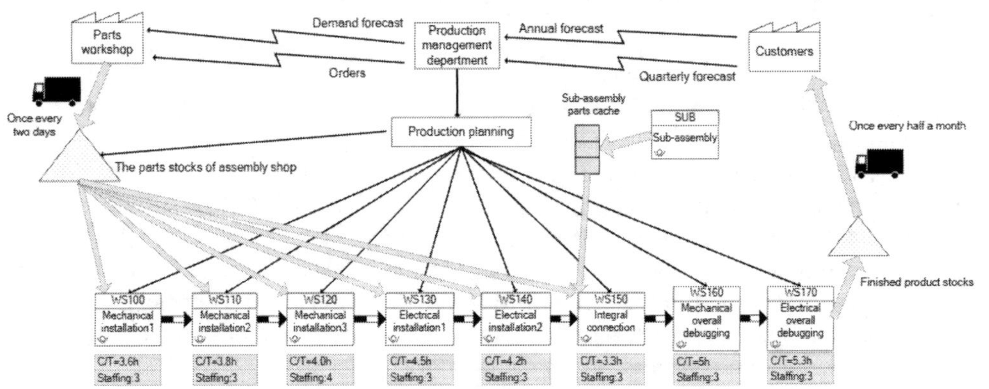

Figure 2. The Value Stream Map of original assembly line

Based on the data analysis in the Value Stream Map, value added time (AT), non-value added time (UT) and value-added rate (I) of the production process can be calculated, with the unit of hour. The specific calculation is as follows:

$$\text{AT} = \sum_{i=1}^{k} C/T_i = 3.6 + 3.8 + 4.0 + 4.5 + 4.2 + 3.3 + 5 + 5.3 = 33.7\ h \quad (1)$$

$$\text{UT} = \sum_{i=1}^{k} C/\overline{T_i} = 48 + 3 + 3 + 3 + 3 + 3 + 3 + 3 + 3 + 3 + 360 = 435h \quad (2)$$

$$\text{I} = \frac{AT}{AT+UT} \times 100\% = 7.19\% \quad (3)$$

Following, the main problems of the original assembly line are discussed from several aspects, according to the layout and the Value Stream Map.

2.1. Low balance rate of production line

Each work station of the original assembly of product P line contains multiple processes. The balance rate of the assembly line is 79.5%, and the loss rate is 20.5%. During production, some stations cannot be produced as planned, while the production capacity of other stations is relatively surplus, resulting in the idleness of some operators and devices.

2.2. Long logistics waiting time

It can be seen from figure 2 that the main logistics time between each station on the general assembly line is too long. According to the field research, there are several reasons as follows:

- Before working, that the operators of each station need to go to the material storage area and tool storage area to get corresponding materials and tools according to the production plan takes up a certain time every day.
- If type and quality problems of materials are found during the assembly process, it also takes a certain time to get and return repeatedly.
- The main parts are transported by crane between adjoining stations. The waiting time for the crane is too long, due to the following reasons. Firstly, the low balance rate of the original assembly line makes the transfer time of each station uncertain. Secondly, there is only one crown in the workshop, and some stations need auxiliary assembly by crown. Thirdly, the number of crane workers in the production department is small and the management mode of temporary deployment is adopted.

2.3. Disconnection between logistics and information flow

According to figure 2, the predictive push production mode is applied to the production process of the original assembly line, which often causes the disconnection between logistics and information flow [3]. The production management department makes the production plan according to the forecast, and then issues the periodic production order to the assembly and each parts workshop. It makes the production become a simple process of pushing forward from front to back. The forecast error will lead to the inconsistency between the actual production and the plan forecast, which will make the continuous flow impossible between the general assembly production and the production of the parts, so a large stock of materials is used to make up for above defects. It can be seen from figure 1 that material buffer area is about 420m^2 , finished product buffer area is about 360m^2, occupying nearly a quarter of the space of the production workshop. Too much stocks will cause a lot of time wasted in waiting, handling and counting.

In the field investigation, it is also found that a large number of materials and finished products cache covers the quality defects of some parts, but the loss has not attracted the attention of relevant departments of the enterprise.

3. Lean design of product P assembly line

According to the analysis results above, the modified assembly line has the following improvement objectives:

- Optimize assembly process to eliminate unnecessary non-value added activities and reduce production cost.
- Convert push production mode to pull production mode.
- Optimize production layout and reduce inventory occupation.
- Increase the balance rate of assembly line. By means of process reengineering and optimization, the cycle time of bottleneck station can be reduced, and the smoothness between processes can be improved. It can realize the continuous value flow of the whole line to improve production efficiency.

3.1. Work time balance and cycle time improvement

Among all the stations of assembly line, the operation time of mechanical and electrical overall debugging stations are 5 hours and 5.3 hours respectively, which are all bottleneck stations. After transformation, the procedures of original two bottleneck stations are disassembled into three stations arranged on the new assembly line, so that the cycle time of these three stations is about 3.5 hours. The average distance among all stations of the new assembly line which add a station is compressed from 6m to 4.8m. It makes the layout more compact and shortens the transfer distance of main parts among stations.

It is found that the process of cable detection and arrangement of the electrical installation station takes a long time, by scene research. In the design of new line, a sub-assembly 1 (The sub-assembly station of original assembly line is configured to sub-assembly station 3 of new assembly line) is added. In this new station, a special cable detection template and the actual cable binding model are used to guide the operator to finish the task. In addition, a new sub-assembly station 2 is added configured one person and special tooling to complete the drilling process of some parts which are finished in the original three mechanical installation stations. In this way, the cycle time of each station in the assembly line is reduced to about 3.5 hours, and the balance rate of the assembly line can be increased to 87.5%, as shown in the figure 3 below.

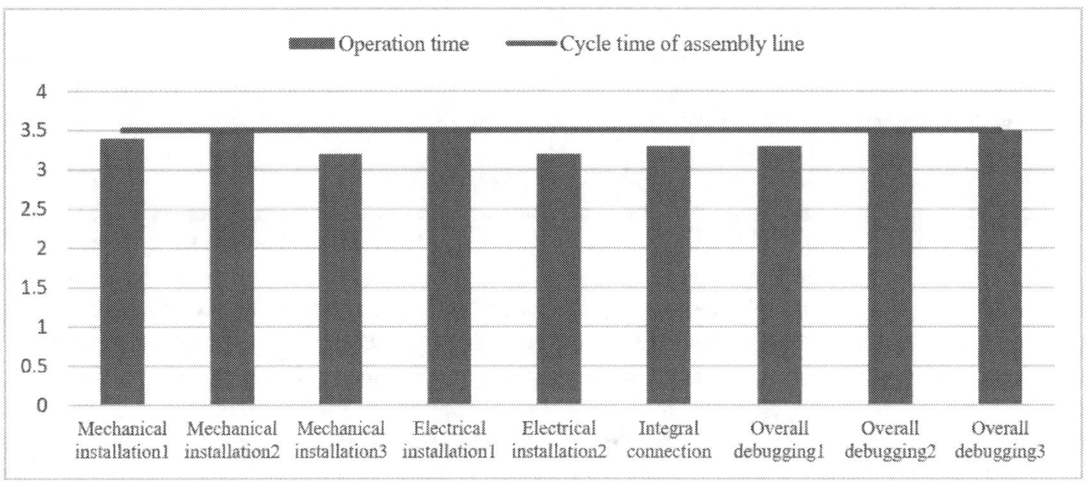

Figure 3. Comparison diagram of operation time of transformation assembly line

3.2. Adoption of pull production mode

After the cycle time of the production line is basically balanced, pull production mode is implemented to reduce the caching of materials. Taking the rationality of workshop layout and production line into account, in order to solve the problem that material buffer takes up a lot of space, the "supermarket pull system" is introduced. Supermarket system is set up between the parts production workshop and the corresponding stations of the assembly line. Picking up the goods from the "supermarket" of the

IOP Conf. Series: Materials Science and Engineering **825** (2020) 012007 doi:10.1088/1757-899X/825/1/012007

station can send the delivery instruction to the upstream part production workshop. Pulling production through the flow of "picking Kanban" and "transportation Kanban" can get rid of the predictive production of parts [4] [5]. That is to say, the production and delivery instructions of parts can be transmitted through the overall 3.5-hour cycle time of new assembly line and the picking Kanban information of each station. After receiving the delivery instruction, the parts workshop will distribute the materials to the catering area of new assembly workshop (as shown in figure 4). A fixed material rack is designed for each station according to the fixed material BOM of each station in the new assembly line. The staff in the catering area are responsible for checking, catering and delivering to the corresponding stations one cycle time ahead. It changes material picking mode of original production line and reduces waiting time in the production process. As assembly parts are distributed by Kanban and cycle time information, only one set of materials and material rack, inspection and catering space for all stations need to be reserved, instead of large amount of materials in the cache area of original line. It can reduce land occupation by nearly half, compared with the material storage area of the original assembly line.

Pulling the production of parts workshop by the production of general assembly workshop can make the enterprise transform from "predictive production" to "order production" to reduce the burden of business management [6] [7]. In addition, pulling the production of parts by the general assembly can not only reduce the dependence on stock, but also force the part workshops to improve production quality in various ways. In this way, on-time delivery is realized and the loss of the enterprise is reduced.

3.3. Optimization of overall layout of assembly line

In the layout design of the new production line, the tool storage area is cancelled and a special tool cart is equipped for each station, so that the walking distance of workers can be reduced by about 500 meters in total every day. The catering area is set up, where special persons are responsible for inspection, catering and transportation of materials, so that it can reduce the waiting time of picking up and delivering materials. A centralized control room is added in the workshop and two LED screens are installed in prominent positions, which can realize informationization and visualization management of the whole line. In the layout of new assembly line, a quality problem area is added to record and handle the quality problems found in the assembly and testing process. In this way, it can not only avoid affecting timely transportation of main parts, but also provide effective support for the improvement of product quality.

With the improvement of production efficiency, the supply cycle of finished products will be shortened. After consulting with field persons and considering the number of finished product buffer, the area of the finished product storage area was reduced to 2/3, and the sending frequency of the finished product was increased to once 10 days compared with the original one which is once half a month. The lean layout and logistics line chart is shown in figure 4. Compared with the original layout, it can save about 220m^2 space which can be used as a reserved place for scientific research.

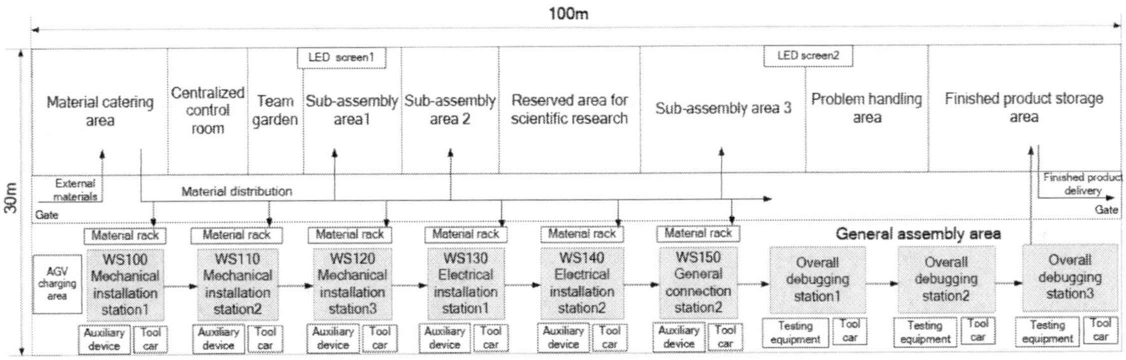

Figure 4. The Value Stream Map of new assembly line

3.4. Pulsating transfer of main logistics

The transfer of main parts was completed by crane among the stations of the original assembly line, which wastes a lot of waiting time and has security risk. The planning of new production line adopts an AGV car to realize the pulsating transfer of main logistics. The overall scheduling mode is as follows:

- In the production process, the information system counts work time for each station. When the work time of last station (comprehensive debugging 3 station) reaches 3.5 hours, AGV will arrive at the waiting area of last station from the charging area. After the workers give the instruction to finish the work, the AGV car will transfer the main parts to the finished product storage area;
- When the work time of the overall debugging 2 station reaches 3.5 hours, the AGV car will arrive at the entrance of this station. After receiving the work completion information, the AGV car will transfer the main parts to the overall debugging 3 station;
- Until the AGV car transfers the blank main parts from the material catering area to the mechanical installation station 1, the main parts of all stations are transferred to the front stations once within 3.5 hours in turn;
- The AGV car returns to charging area.

The average time of each transfer between adjacent stations is 10 minutes. As the whole line includes 9 stations, the AGV car needs to transfer 10 times in each cycle time, amounting to 100 minutes. The remaining 110 minutes in cycle time (3.5 hours) of the whole line is used for the AGV car charging. Thus, it realizes the main logistics pulsating transport in accordance with a certain time rule. It does not only makes workers improve operation skills to complete the task within cycle time, but also enables the information system grasp the actual production situation of the whole line constantly to avoid disconnection between information flow and logistics.

3.5. Reduction of station operators

In the assembly process of original mechanical installation and electrical installation stations, due to some heavy assembly parts, more people are allocated to finish the assembly task by manual handling. In the new line, these stations are equipped with some auxiliary devices to reduce labor intensity, and the number of operators working in each mechanical and electrical installation station can be reduced to 2. The original overall debugging stations composed of three stations with three persons in each station are divided into three stations with two persons in each station. The number of workers in sub-assembly station 1 remains unchanged. The newly added sub-assembly two and three stations need one person for each station, two persons added totally. Two operators are arranged in the material catering area to complete the tasks of unloading, inspection, catering and transportation. In addition, one person is allocated as administrator to realize Andon management of the whole line by information system. So, the new assembly line can reduce two persons per shift as a whole.

3.6. Value stream map of new assembly line

According to the adjusted layout above, the new product P assembly line put into production with the new equipment and production mode application. In order to make workers adapt to the new production mode, two sets of material caches are set in the "supermarket" beside each station, so as to avoid production break caused by abnormal materials. After a certain time production, the relevant data of new assembly line are counted. The Value Stream Map is drawn as figure 5 below:

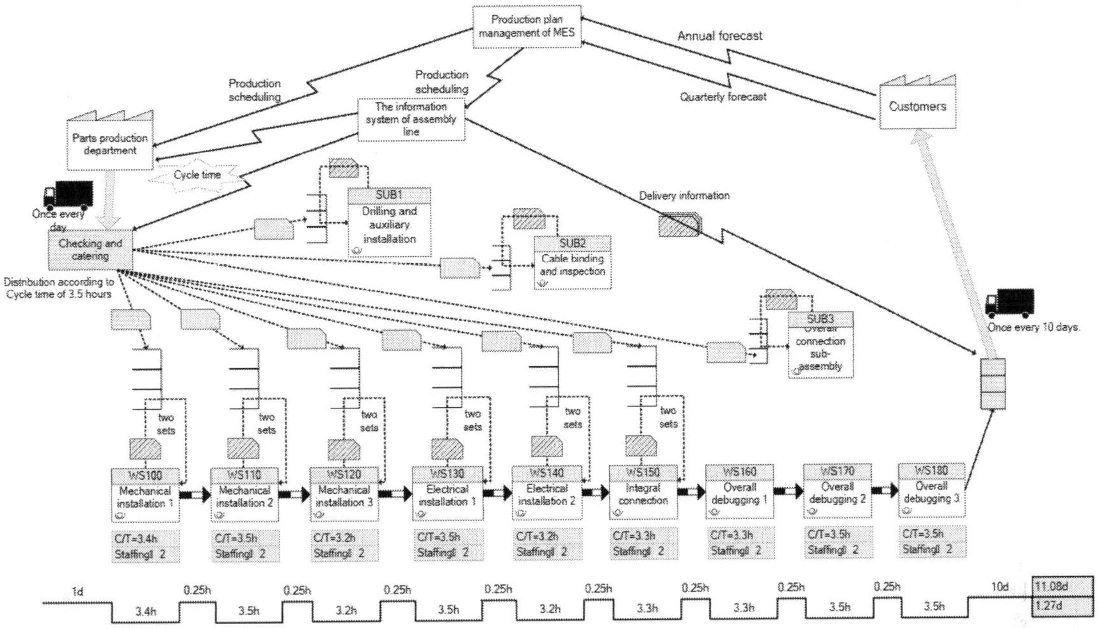

Figure 5. New assembly line layout and logistics roadmap

After the transformation, the following evaluation indexes of assembly line have been improved significantly:

- Value added time (AT), non-value added time (UT) and value add rate (I).

$$\mathrm{AT} = \sum_{i=1}^{k} C/T_i = 3.4 + 3.5 + 3.2 + 3.5 + 3.2 + 3.3 + 3.3 + 3.5 + 3.5 = 30.4\,h \quad (4)$$

$$\mathrm{UT} = \sum_{i=1}^{k} C/\overline{T_i} = 24 + 0.25 + 0.25 + 0.25 + 0.25 + 0.25 + 0.25 + 0.25 + 0.25 + 240 = 266h \quad (5)$$

$$\mathrm{I} = \frac{AT}{AT+UT} \times 100\% = 10.3\% \quad (6)$$

The value added time is shortened from 33.7h to 30.4h. The non-value added time is shortened from 435h to 266h, reduced by 58%. The value-added ratio increased from 6.72% to 10.3%, increased by 53.3%.

- The assembly production area of product P is reduced by about $220\mathrm{m}^2$.
- After the transformation, the balance rate of the production line has raised to 87.5%, increased by about 10%.
- The number of operators per shift has been reduced by 2.
- The production capacity has been increased by about 30%, which reaches 60 sets per month required by customers.

4. Conclusion

Based on the transformation project of product P assembly line, this paper analyzes some problems of the original assembly line by means of lean production principles and Value Stream Mapping. In lean research of the new production line, overall layout, process arrangement, production mode, material buffer, auxiliary equipment and staffing are improved to reduce some non-value added activities and eliminate some waste. At present, product P has been mass produced on the new assembly line and the production efficiency has been significantly improved. The application has reduces the labor intensity, saves the cost for the enterprise, and improves the staff awareness of lean production. The research provides a good demonstration for the transformation from extensive production operation to lean production operation.

Acknowledgments
This work was supported in part by the National Key R&D Program of China (No. 2018YFB1306601), the STS Program of the Chinese Academy of Sciences (KFJ-STS-QYZD-107) the Natural Science Foundation of Liaoning Province (No. 20180520016, No. 20180520008 and No. 2019-MS-343), the Project of Intelligent Manufacturing Integrated Standardization and New Model Application, and K.C.Wong Education Foundation.

References
[1] Shuhuang Liu, Jiansha Lu and Jiarao Wang 2010 *Lean Production* (Beijing: China Machine Press) pp1-22.
[2] Pengpeng Huang, Wuliang Huang and Shuangshuang Ye 2018 *Machinery Design and Manufacture* vol 3 pp 37–43.
[3] Ning Yu, Lufei Huang and Xuemei Qin 2014 *Journal of Southwest University of Science and Technology* vol 4 pp 87–91.
[4] Jian Zhou, Haofeng Xiao and Xiaohu Wang 2015 *Journal of Northeastern University* vol 10 pp 1501–1505.
[5] Qifei Pei, Nailiang Li and Xiaoming Wang 2015 *Machinery Design and Manufacture* vol 9 pp 261–264.
[6] Qinghong Xie, Xiaobin Wang and Zhixiao Pan 2013 *Industrial Engineering and Management* vol 4 pp 117–121.
[7] Lian Y H 2007 *International Journal of Production Research* chapter 4 pp 3037–3058.

IOP Conf. Series: Materials Science and Engineering **825** (2020) 012008 doi:10.1088/1757-899X/825/1/012008

Problem Solving Strategy for Product Variant Design Combined with Improved Neural Network Method

L Yang*, X Y Gu and Q Yin

School of Mechanical Engineering, Nanjing University of Science and Technology, Nanjing, Jiangsu Province, China

*ylnjust@126.com

Abstract. Aiming at the fast solution of product variant design problem, a solving strategy combined with improved neural network method was proposed. Past similar cases were extracted as training samples through case-based reasoning (CBR) technology, empirical parameters were predicted based on neural network; computational parameters were solved by the existing calculation templates. The design results visualization was realized through parametric design. To reduce the influence of artificial sample construction on neural network training effect, the sample construction strategy of the multiple input/ single output (MI/SO) structure was proposed to improve the parameter prediction quality. Finally, the feasibility and effectiveness of the proposed method was demonstrated by taking the design of a single-cylinder recuperator for example.

1. Introduction

By reusing past design resources, variant design can obtain products that satisfy users' needs in a short time on the premise of guaranteeing cost and quality, and effectively alleviate the contradictions in complexity, diversity and timeliness of current product design [1]. It is found that about 70% of the products in enterprises are variant designed on the basis of existing products according to the market demand. At present, the research on variant design mainly focuses on variant design method, variant design plan, design change propagation and so on. Xiao Xinhua et al. [2] proposed variant design technology based on modular product instances; Wu Weiwei et al. [3] studied the variant design method of mechanical product dimension based on parameterization; Du et al. [4] used game theory to optimize variant design schemes based on modules and parameters; Chen et al. [5] proposed a design change analysis method based on structural attributes and object-oriented in view of the cyclic problems of design change. In the process of variant design, some design parameters are selected combined with past design experience, because there are often no clear design criteria. Li Haiqing et al. [6] integrated fuzzy set theory and secondary-CBR technology, proposed a hybrid case-based reasoning mechanism to support feature-based product variant design and rewriting knowledge self-learning; Xu Rongzhen [7] studied the intelligent variant design method based on case-based reasoning and knowledge push; Yangliu [8] introduced neural network based parameter prediction technique based on case-based reasoning, in order to improve the design quality and efficiency for self-propelled artillery. Literature [8] gived a parameter prediction result with relatively small error, but failed to consider the influence of artificial sample selection on parameter prediction, and cannot judge whether the current prediction effect is optimal. Therefore, this paper proposes an improved neural network parameter prediction method to improve the quality of parameter prediction, which adopts the sample construction strategy of MI/SO and carries out multiple training of neural network model to obtain the best network model. On this basis, an efficient and complete solution strategy for product variant design is determined, case-based reasoning is used to extract similar past cases as training samples, empirical parameters are predicted by improved neural network method, computational parameters are solved by existing calculation models, and the output of design results is realized by parametric design method.

2. Product variant design problem solving strategy combined with neural network

According to the solve requirements of product variant design, variant design parameters can generally be divided into two categories: one is the computational parameters with clear design criteria or rules, which can be directly calculated or taken into account; the other is the empirical parameters without clear design criteria, which need to be solved consideration with design experience. In this paper, the solution strategy of product variant design problem based on neural network is defined as shown in figure 1. The solution of calculating parameters is template, combined with the template-based rapid design technology proposed in literature [9], the parameter calculation template is compiled according to the existing calculating formulas or the rules of value selection, the calculating process is simplified by calling the template to reduce the artificial calculation errors. Empirical parameters are solved by calling the neural network prediction module, similar design cases in the past are extracted as training samples of neural network. Empirical parameter prediction based on neural network is carried out to reduce the dependence of designers on relevant domain knowledge and improve the rationality of parameter selection. Finally, the result of parameter solution is transferred to the parametric design module, which drives the model regeneration and forms the three-dimensional model of the design scheme.

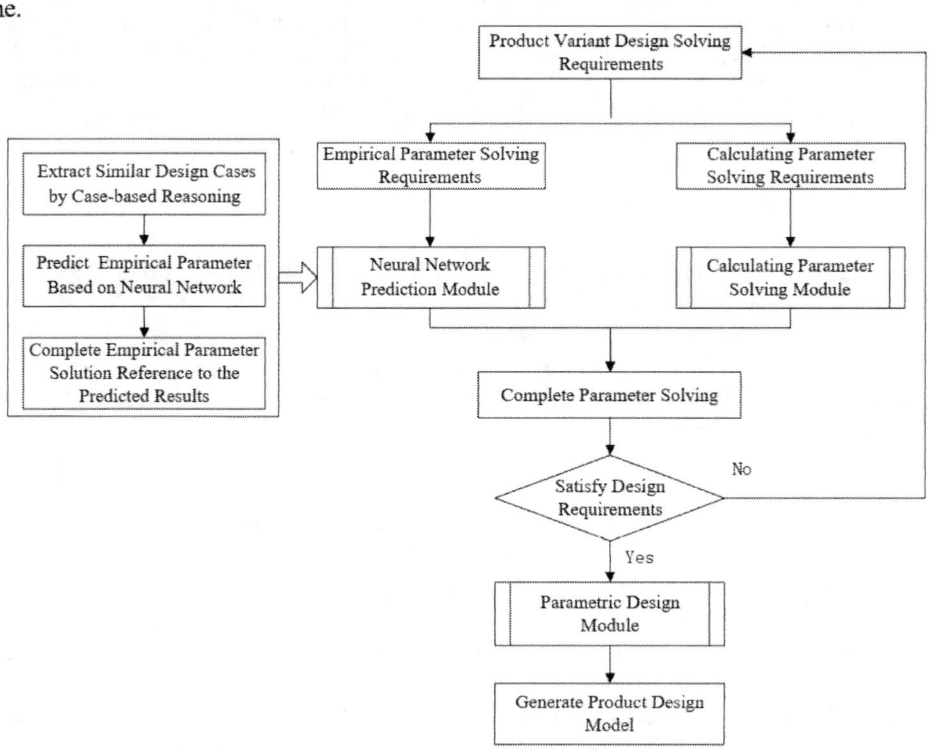

Figure 1. Product variant design problem solving strategy combined with neural network.

3. Improved neural network parameter prediction technology for product variant design

Artificial neural network is a mathematical model or computational model that imitates the structure and function of biological neural network and is used to estimate or approximate functions. As an effective method of data processing, neural network has been widely studied. The representative network models include BP, RBF, Elman network and wavelet neural network. Among them, BP network algorithm uses gradient algorithm based on error back propagation[10], makes full use of the structural advantages of multi-layer feed forward network, and the algorithm is relatively mature. Therefore, this paper chooses BP neural network to realize the empirical parameter prediction of product variant design.

3.1. BP neural network model and principle

BP neural network is a kind of multi-layer feed forward network with one-way propagation. Its main features are signal forward propagation and error back propagation. By transferring errors layer by layer, the weights and thresholds of the network are constantly adjusted to make the final output of the network as close as possible to the expected output, and the purpose of training is achieved. The design of BP network mainly includes input layer, hidden layer, output layer and transfer function between layers [11]. The number of neurons in each layer is related to the structure of training samples, the number of neurons in input layer is determined by input decision; the number of neurons in output layer is determined by output decision; and the number of neurons in hidden layer is determined by input and output layer structure. Sigmoid function or purelin function are usually used in the transmission function of BP network. Generally, sigmoid function is used as input layer transfer function, tansig or purelin function is used as output layer transfer function[12].

3.2. Training sample construction strategy of MI/SO structure

Before training, input and output items and corresponding values should be extracted from similar cases to form training sample A. The structure of training sample A is usually in the form of multiple input/multiple output (MI/MO).

The set $P=\{p_1,p_2,\ldots,p_n\}(n\geq 2,n\in N)$ is defined as the input item of training sample A, $p_i(1\leq i\leq n,i\in N)$ representing an input item of training sample A; the set $Q=\{q_1,q_2,\ldots,q_m\}(m\in N)$ is the output item of training sample A, $q_j(1\leq j\leq m,j\in N)$ representing an output item of training sample A.

The elements p_i,q_j in the set are conceptual objects, which are attributes of the input and output items, representing the set of values of the input and output items in all similar design cases. The sample **A** is $\{p_1,p_2,\ldots,p_n\}/\{q_1,q_2,\ldots,q_m\}$ (P/Q),which is the MI/MO structure of n input and m output.

The parameter prediction effect of BP network is influenced by the structure of training samples, if the input and output of the neural network are not correlated, the training effect will be greatly affected because it is difficult to form an effective functional relationship between them. In order to test the correlation between input and output in the initial sample, this paper proposes a training sample construction strategy based on MI/SO structure, studies the corresponding relationship between input and each output, reconstructs multiple training samples in multiple input/single output forms, and trains the neural network in turn for comparison. The specific strategies are as follows:

A non-empty subset $P_s=\{p_{k_1},p_{k_2},\ldots,p_{k_s}\}$ is formed by selecting s elements in the set P as input of the sample A_i, a subset $Q_i=\{q_i\}$ is formed by selecting a element q_i in the set Q as as single output of the sample A_i. The sample A_i is $\{p_{k_1},p_{k_2},\ldots,p_{k_s}\}/\{q_i\}$ (P_s/Q_i), which is the MI/MO structure of s inputs corresponding to a single output. In order to ensure that the correlation test is carried out without omission, the s values from 2 to n, the i values from 1 to m, s and i are all integers. After determining the sample scheme, the set of values represented by attribute elements is substituted into A_i, and the BP network training samples with MI/SO structure are generated after normalization.

3.3. Empirical parameter prediction method based on improved BP neural network

The empirical parameter prediction process of product variant design based on improved BP neural network is shown in figure 2. Cording to various MI/SO sample schemes, the corresponding BP network models are created and trained. According to various MI/SO sample schemes, the corresponding BP network models are created and trained. The number of layers and transfer functions of each layer are selected, and the structure of input and output layers is determined according to the sample structure. The number of hidden layer neurons is modified by step-by-step increment method, and the training model is compared to determine the number of hidden layer neurons that make the network have sufficient generalization ability and output accuracy. The BP network models of each MI/SO structure are trained and saved sequentially, and the parameter prediction results and error percentages of each model are compared according to the pre-set error upper limit, so as to eliminate the none&weakly related sample schemes. The most similar design case is used to verify the prediction effect of the model, the training effects of the remaining BP models are compared to obtain the best neural network model, so as to improve the quality of empirical parameter solution for variant design.

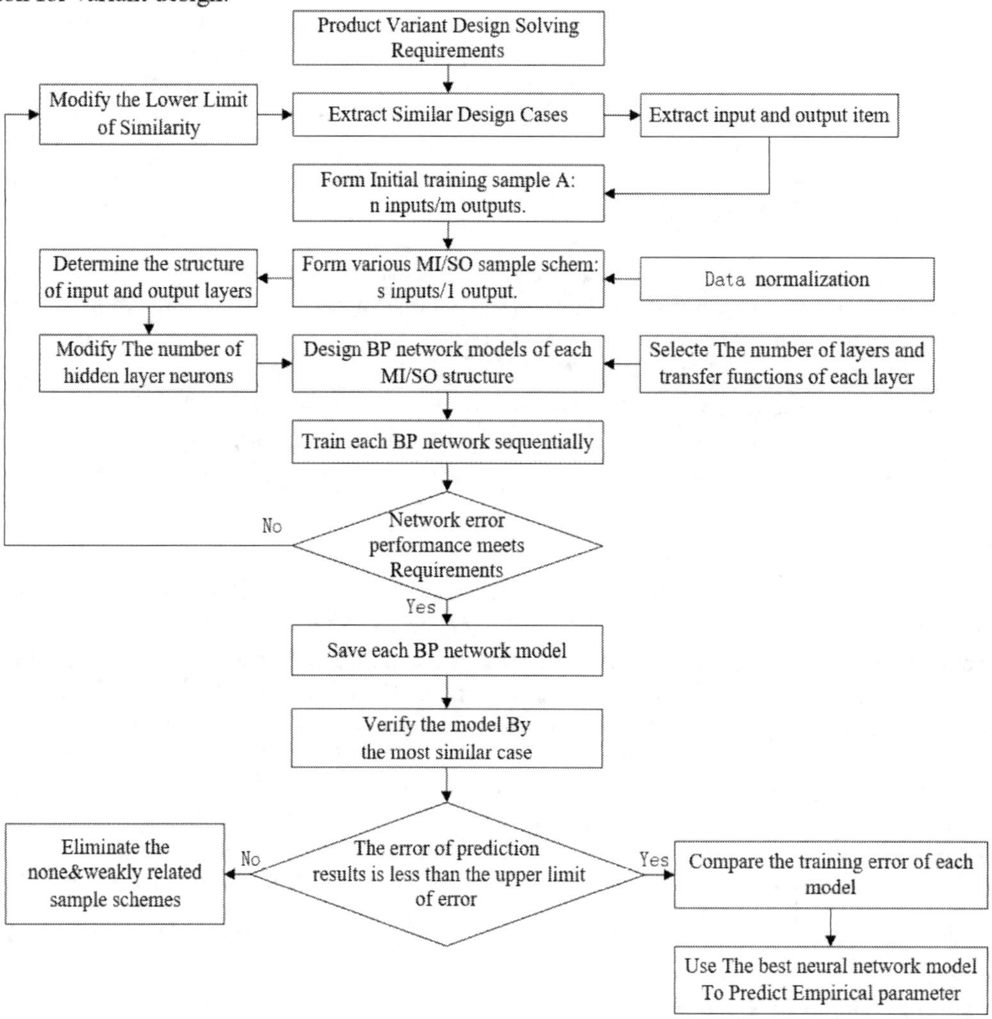

Figure 2. Empirical parameter prediction method based on improved BP neural network.

4. An example of solving variant design problem of recuperator based on improved BP neural network method

The design of recuperator in artillery recoil mechanism is a typical variant design problem. The initial pressure, compression ratio and initial volume mainly depend on the experience of the designer.initial pressure, compression ratio and initial gas volume mainly depend on the experience of the designer. In order to verify the feasibility and validity of parameter prediction technology based on improved BP neural network in solving variant design problems of products, this paper takes the design of the recuperator as an example to solve the variant design problems combined with the improved BP neural network method. Specific design requirements are as Table 1.

Table 1. Design task of recuperator.

Caliber /mm	Maximum recoil distance /mm	Maximum chamber pressure /MPa	Mass of recoiling part /kg	Efficiency of muzzle brake/%	Breech length /mm
130	950	315	2600	44.9	535

4.1. Empirical parameter prediction of recuperator based on improved BP neural network

In order to meet the variant design requirements of the recuperator, a series of past design cases which are close to the design task are obtained on the basis of case retrieval by using the neural network prediction module and selecting the lower limit of similarity. According to the requirements of current design task and the design characteristics of the recuperator, some inputs and outputs of the neural network are selected as shown in TABLE 2. According to the current design task, the most similar case obtained by case-based reasoning is shown in TABLE 3.

Table 2. Partial sample data.

Input item						Output item			
Maximum recoil distance /mm	Maximum chamber pressure /MPa	Mass of recoiling part/kg	Efficiency of muzzle brake/%	Breech length /mm	...	Initial pressure /MPa	Compression ratio	Initial gas volume /dm^3	...
820	238	392	40	360	...	3.0	1.62	3.66	...
675	255	785	58	420	...	4.8	3.62	2.07	...
1180	300	1550	45	500	...	5.2	2.30	8.23	...
1400	275	2400	0	535	...	4.5	2.38	8.30	...
950	315	2505	59.1	535	...	6.1	2.80	8.40	...
1350	315	3029	30	530	...	5.6	2.7	16.93	...
...	...	...	...	...	...	...	...	...	...

Table 3. Most similar design case.

Maximum recoil distance /mm	Maximum chamber pressure /MPa	Mass of recoiling part/kg	Efficiency of muzzle brake/%	Breech length /mm	Initial pressure /MPa	Compression ratio	Initial gas volume /dm^3
950	315	2505	59.1	535	6.10	2.80	8.40

Through the parameter prediction method based on improved BP neural network mentioned above, the correlation between input and output is checked. There are five input elements in the initial training sample. Considering that the input reorganization starts from two elements, the number of reconstructed sample schemes is too large, so the input reorganization of the recuperator starts from three elements. The initial pressure of the recuperator is choosed as the output item, and then sixteen MI/SO sample schemes are constructed. The network model was trained after normalizing the sample data. When the network error performance meets the requirements, the model training is completed and the trained BP network models are saved. Using the most similar case to verify the model, the initial pressure prediction results of each BP network repeater are obtained as shown in TABLE 4.

Preset the upper limit of error to 6%, there are there qualified sample schemes with order numbers 2, 7 and 10. By comparing the training errors of BP model of the three schemes, it is concluded that the BP network model trained by Sample Scheme 2 has the best effect. That is to say, the best sample scheme is {"maximum recoil length", "maximum bore pressure", "recoil part mass", "efficiency of muzzle brake"}/{"initial pressure"}. Therefore, the BP network is chosen as the best model to predict the initial pressure.

Taking "compression ratio" and "initial gas volume" as output terms respectively, the prediction process is the same as initial pressure prediction process. Finally, the best sample schemes are {"maximum recoil length", "maximum bore pressure", "recoil part mass", "efficiency of muzzle brake"}/{"compression ratio"} and {"maximum recoil length", "maximum bore pressure", "recoil part mass", "efficiency of muzzle brake"}/{"initial gas volume"}.

After obtaining the optimal network model, the design task of the recuperator is inputted, and the predicted results of the experiential parameters are shown in Table 5. Literature [8] also takes the design of recuperator as an example, and takes {"initial pressure", "compression ratio", "initial gas volume "} as the whole sample output to predict the parameters of the recuperator. The prediction error is 1.1% for initial pressure, 8.8% for compression ratio and 14.2% for initial gas volume. The predicted results of the parameters of the recuperator in this paper are compared with those in literature [8]. It can be found that the parameter prediction method based on improved BP neural network proposed in this paper further improves the quality of parameter prediction of the recuperator, and satisfies the requirements of variant design of the recuperator better.

Table 4. Comparison of predicted results.

Order number	Input item: Maximum recoil length	Maximum Chamber Pressure	Mass of Recoiling Part	Efficiency of Muzzle Brake	Breech Length	Model training results: Original Design Result/MPa	Parameter Prediction Results/MPa	Error
1	√	√	√	√	√	6.10	5.34	12.46%
2	√	√	√	√		6.10	6.28	2.95%
3	√	√	√		√	6.10	6.58	7.87%
4	√	√		√	√	6.10	7.12	16.72%
5	√		√	√	√	6.10	5.36	12.13%
6		√	√	√	√	6.10	5.17	15.25%
7	√	√	√			6.10	5.8	4.92%
8	√	√		√		6.10	6.47	6.07%
9	√	√			√	6.10	7.00	14.75%
10	√		√	√		6.10	6.36	4.26%
11	√		√		√	6.10	4.72	22.62%
12	√			√	√	6.10	7.43	21.80%
13		√	√	√		6.10	5.55	9.02%
14		√	√		√	6.10	7.25	18.85%
15		√		√	√	6.10	7.14	17.05%
16			√	√	√	6.10	4.63	24.10%

Table 5. The prediction results of the recuperator empirical parameters.

Empirical parameters	Initial pressure/MPa	Compression ratio	Initial gas volume/dm^3
Original design results	6.10	2.80	8.40
Parameters prediction results	6.28	2.70	8.89
Error/%	2.95%	3.57%	5.83%

4.2. Example of solving variant design problems

According to the predicted results of parameters, the empirical parameters of the recuperator are selected. The initial pressure of the recuperator is 6.28 MPa, the compression ratio is 2.70, and the initial gas volume is 8.89dm3. According to the existing formula for calculating the parameters of the recuperator, the program for calculating the parameters of the recuperator is compiled and invoked, and the calculating parameters of the repeater are solved. The results are shown in TABLE 6. The results of parameter solution are input into the parametric design program, which drives the model regeneration of the recuperator, and finally realizes the visualization of the design, as shown in figure 3.

Table 6. The values of the recuperator calculation parameters.

Diameter of recuperator rod/mm	Inner diameter of recuperator barrel/mm	Outer diameter of recuperator barrel/mm	Length of recuperator barrel/mm	Width of piston/mm	Thickness of end cap/mm	Total length of piston and rod/mm
38	85	115	2100	80	18	2200

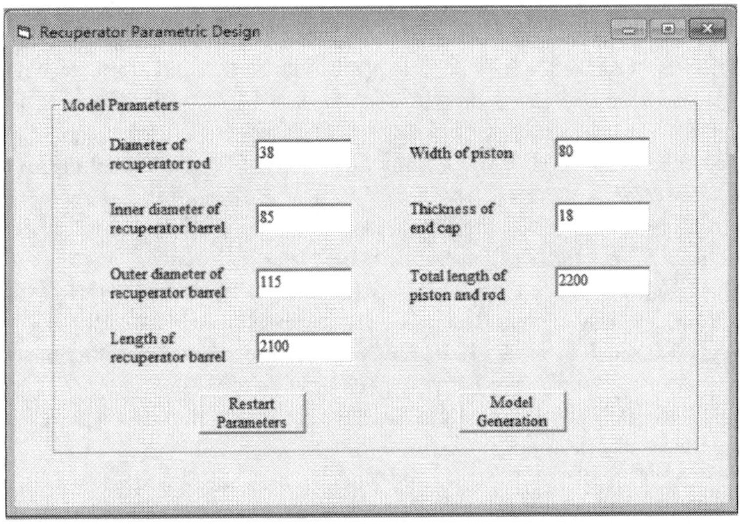

(a)

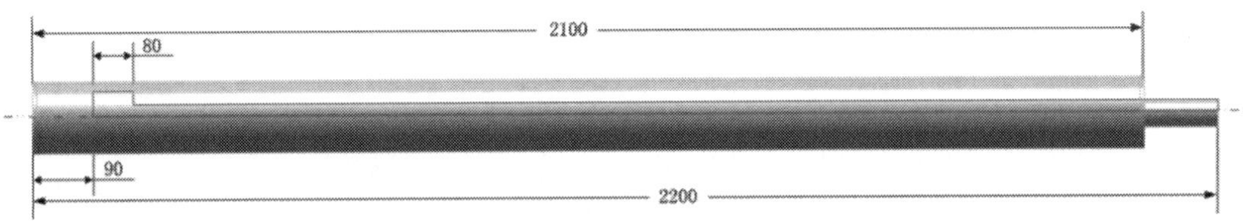

(b)

Figure 3. New variant design scheme of recuperator.

5. Conclusion

Aiming at solving empirical parameters in variant design, the technology of neural network parameter prediction is introduced on the basis of case-based reasoning and parametric design. A solution method of product variant design problem based on neural network is proposed, the solution strategy of product variant design problem combined with neural network and the empirical parameter prediction process of product variant design based on BP neural network are determined. In order to reduce the influence of irrelevant input and output caused by artificial sample construction, a training sample construction strategy based on MI/SO structure is proposed to improve the parameter prediction method of neural network in order to improve the quality of parameter prediction. Finally, taking the variant design of a single cylinder recuperator as an example, a new visual variant design scheme of the recuperator is generated. The feasibility of the proposed method is verified.

IOP Conf. Series: Materials Science and Engineering **825** (2020) 012008 doi:10.1088/1757-899X/825/1/012008

Acknowledgments
This research is partially supported by the Natural Science Foundation of Jiangsu Province, China, No.BK20180481.

References
[1] GONG J Z, QIU J, LI G X, et al. 2008 Extension approach for product module variant design. *J. Computer Integrated Manufacturing Systems*. **14** 1256-67
[2] XIAO X H, SHI M H, YANG X F, et al. 2007 Research on variant design based on modular product case. *J. China Mechanical Engineering*. **18** 803-07
[3] WU W W, TANG R Z, HOU L, et al. 2005 Research & realization on the variant design of mechanical products based on parametric techniques. *J. China Mechanical Engineering*. **16** 218-22
[4] DU G, JAO R J, CHEN M 2014 Joint optimization of product family configuration and scaling design by stackelberg game. *J. European Journal of Operational Research*. **232** 330-41
[5] CHEN C Y, LIAO G Y, LIN K S 2015 An attribute-based and object-oriented approach with system implementation for change impact analysis in variant product design. *J. Computer-Aided Design*. **62** 203-17
[6] LI H Q, YIN G F, GANG H 2007 Application of hybrid case reasoning in variant design. *J. Advanced Engineering Sciences*. **2** 164-69
[7] XU R Z 2008 Research on intelligent variant design based on CBR and knowledge recommendation. *Shandong University*
[8] YANG L,QIAN L F,DING S C,et al. 2015 Research on hybrid case-based reasoning technique for self-propelled artillery variant design. *J. Acta Armamentarii*. **36** 2190-95
[9] GU X Y, YANG L, LI J Y, et al. 2019 Study on technology of template based rapid design for recuperator. *J. Journal of Nanjing University of Science and Technology*. **43** 154-58
[10] WANG E, LI X K, P T 2014 A particle filtering algorithm based on the BP neural network. *J. CAAI Transactions on Intelligent Systems*. **9** 709-13
[11] ZHOU K L, KANG Y H 2005 Neural network model and its MATLAB simulation program design. *Beijing: Tsinghua University Press*
[12] 2010 MATLAB Chinese Forum. 30 case analysis of MATLAB neural network.*Beijing: Beihang University Press*

Design and Kinematics Analysis of a Percutaneous Needle Insertion Robotic System

Zhanfeng Zhou, Xu Zhang, Yongzhuo Gao, Dongmei Wu and Wei Dong*

State Key Laboratory of Robotics and Systems, Harbin Institute of Technology, Nangang District, Harbin, 150080, China

* dongwei@hit.edu.cn

Abstract. To overcome the limitations of conventional manual percutaneous needle insertion procedure, this paper proposes a percutaneous needle insertion robotic system with functions of biopsy and ablation. First, the robotic system, which integrates a 4-DOF positioning robotic arm and a 2-DOF RCM wrist, is designed. The 4-DOF positioning robotic arm can realize the exact position of the puncture needle and the 2-DOF RCM wrist can accurately adjust the orientation the needle. The combination of these two modules can ensure the kinematic decoupling of the position and posture of the robotic system. Second, the kinematics model of the robotic system based on classical D-H method is established, and the forward kinematics and inverse kinematics of the robotic system are analyzed. Finally, the manufacturing and assembly of the overall robotic system is performed and the robotic arm can successfully achieve percutaneous biopsy and ablation for tumor in urinary system.

1. Introduction

Prostate cancer and kidney cancer, as two kinds of urinary system tumors, have extremely high morbidity and mortality, which seriously endanger human health. Percutaneous biopsy and ablation are currently common options for these diseases. Percutaneous needle insertion is a surgical operation in which a microsurgical instrument, such as a puncture needle, penetrate through the skin into the patient's body, reach the lesion target position, and then perform biopsy, ablation, radioactive particle implantation and other surgical operations under the guidance of medical images. It has the advantages of high precision, small trauma, less complications and safe process [1,2]. Unlike endoscope-guided minimally invasive surgery, percutaneous needle insertion surgery, as a kind of minimally invasive surgery, requires external image guidance, such as ultrasound, CT and MRI.

The conventional manual percutaneous needle insertion procedure has many shortcomings: First, due to the lack of information of three-dimensional medical images, the planning of needle insertion surgery relies too much on the experience of the doctor; secondly, the needle insertion and the adjustment of the needle angle are manually operated by the doctor. This is not only dependent on the experience and skills of doctors, but also susceptible to human factors, such as the shaking of doctors' hands and the fatigue of doctors caused by too long operation time.

To overcome these problems, a variety of percutaneous needle insertion robot have been proposed and studied by researchers. Compared with manual needle insertion, robot-assisted percutaneous needle insertion surgery has higher precision and can significantly reduce the labor intensity of doctors. Stoianovici et al. [3] proposed a MINI-RCM & PAKY robot for precise needle insertion under radiological guidance, which consists of a PAKY end effector and a remote center-of-motion (RCM) actuator module. Taylor et al. [4] designed a steady-hand micromanipulation based on RCM module to extend human micromanipulation capabilities. The URobotics Brady Urological institute of Johns Hopkins University developed a compact six-degree-of-freedom robot called Acubot [5] for radiological interventions. Based on steady-hand manipulator, many robots for retinal microsurgery, such as EyeRoBot [6], EyeRoBot2 [7] and EyeRoBot2.1 [8], have been proposed by Advanced Medical Instrumentation and Robotics (AMIRo) of Johns Hopkins University. Hong [9] designed and analyzed a real-time ultrasound-guided needle insertion medical robot, which can modify the needle path in real time by using a novel ultrasonic image segmentation technique. H. Bassan et al. [10] present a new 5 degrees-of-freedom (DOF) 3D-ultrasound-guided robotic system capable of performing percutaneous needle insertion for prostate cancer.

At present, most of the percutaneous needle insertion robots adopt three-dimensional positioning sliding table to realize the position and posture decoupling function. However, after combining with the RCM mechanism, the sliding table has many shortcomings: it will reduce the flexibility and compactness of the robot; doctors cannot adjust the position of the robot during the operation, and even the percutaneous needle insertion robotic system will cause the doctor's sight to be blocked and operation inconvenience.

In this paper, a percutaneous needle insertion surgery robotic system is designed to perform the function of percutaneous needle insertion, biopsy, and ablation. The needle insertion robotic system has 6 degrees of freedom and employs a mechanical structure which can decouple the position and posture of the robot. It includes a 4-DOF positioning module and a 2-DOF RCM module. The 4-DOF positioning module can perform the position of the percutaneous needle and the 2-DOF RCM module can perform the orientation of the percutaneous needle, separately.

The rest of the paper is organized as follows: Section 2 describes the mechanical design details of the percutaneous needle insertion robotic system concerning the decoupling of the robot's position and posture. Section 3 discuss the forward and inverse kinematics modeling of the percutaneous needle insertion robotic system. Section 4 shows the manufacturing and assembly of the overall robotic system. Section 5 concludes the paper.

2. Robotic Design

2.1. Overall Design of Robotic System

Considering the principle of percutaneous needle insertion operation, the robotic system needs to be able to perform a needle insertion operation in three steps, which can improve the accuracy and convenience of the operation. First, doctors need to determine the skin needle insertion site according to the preoperative surgical plan of the robotic system, and then the puncture needle is translated to the desired skin needle insertion point for positioning. Three translational degree-of-freedom of the puncture needle are necessary to reach the desired needle position. In this step, the posture of the needle has not been adjusted yet. Second, the robot system adjusts the orientation of the needle so that the needle axis is aligned with the lesion target. Since it is not necessary to consider the rotation about the needle axis, this step requires two global rotating degree of freedom to orient the puncture needle. During the posture adjustment of the puncture needle, the position of the needle tip cannot be changed. Finally, one translational motion is needed to insert the puncture needle into the skin through the skin needle insertion point.

To achieve the step-by-step surgical procedure mentioned above, the robotic system should be able to kinematically decouples rotating and translational motions of the needle. The robotic system consists of a 4-DOF positioning robotic arm, a 2-DOF RCM wrist and a needle insertion driver. The puncture needle is fixed into the insertion driver and then mounted on the 2-DOF RCM wrist so that the tip of the needle is located at the RCM point. The 4-DOF positioning robotic arm can realize the exact position of the puncture needle. The 2-DOF RCM wrist can accurately adjust the posture of the needle. The combination of these two modules can ensure the kinematic decoupling of the position and posture of the puncture needle. Figure 1 shows the three-dimensional model of percutaneous needle insertion robotic system.

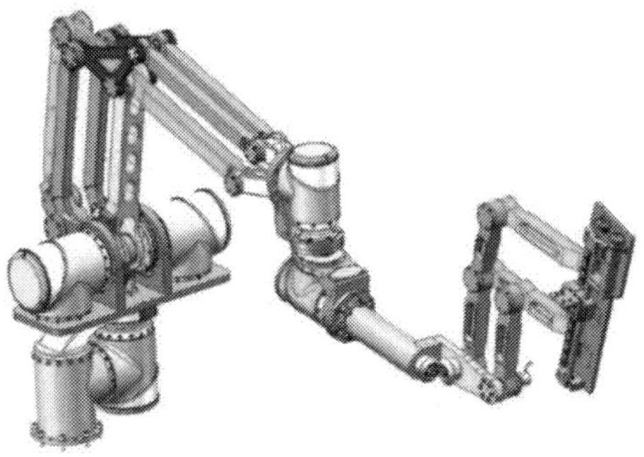

Figure 1. The three-dimensional model of percutaneous needle insertion robotic system

2.2. 4-DOF Positioning Robotic Arm

For locating the puncture needle position, it is redundant to use 4-DOF positioning robotic arm, but the redundant robotic arm can ensure that the needle can reach the desired needle insertion point at different postures. This feature enhances the flexibility of the operation. The robotic arm can bypass some obstacles to reach the target point, and can change the position of the arm according to the doctor's needs, so as to avoid the obstruction of the doctor's sight and the inconvenience of operation.

The multi-joint robotic arm must be decoupled. There are two typical decoupling schemes, one is using connecting links for decoupling and the other is using cable-driven links structure for decoupling [11]. The decoupling design using cable-driven links is simple, but it will increase the size of the structure and make it bulky. The connecting links decoupling schemes is complex, but its structure is compact and light. For medical robots, compactness and lightweight are more important, so the decoupling schemes using connecting links is adopted in this design.

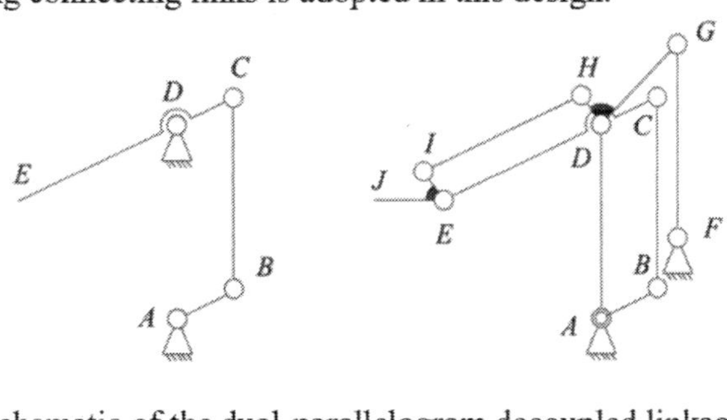

Figure 2. Schematic of the dual-parallelogram decoupled linkage mechanism

A schematic of the decoupling mechanism is shown in Figure 2. The mechanism can be regarded as a dual-parallelogram planer linkage. The degree of freedom of the mechanism is 2. The connecting rod AD and AB serve as the driving rod. From the schematic shown in Figure 2, it can be seen that the parallelogram AFGD will always ensure that the angle between the connecting rod HDG and the horizontal plane never changes, and the parallelogram HDEI will also ensure that the angle between the connecting rod IEJ and the connecting rod HDG does not change. Therefore, no matter how the driving rod AD and AB move, the linkage IEJ only produces translation without change in orientation. Consequently, after connecting with the 2-DOF RCM wrist, the decoupling of rotating and translational motions of the needle can be achieved.

The rotation angles of the active rods AD and AB are α and β, then the rotation angles of the connecting rod DE and EJ at the rotating joints D and E are

$$\begin{aligned} DE &= \beta - \alpha \\ EJ &= -\beta \end{aligned} \tag{1}$$

Based on the above dual-parallelogram decoupled linkage mechanism, the 4-DOF global positioning robotic arm is developed, as shown in Figure 3. This robotic arm not only realize global positioning for puncture needle, but also does not affect the orientation of the subsequent RCM mechanism and the puncture needle.

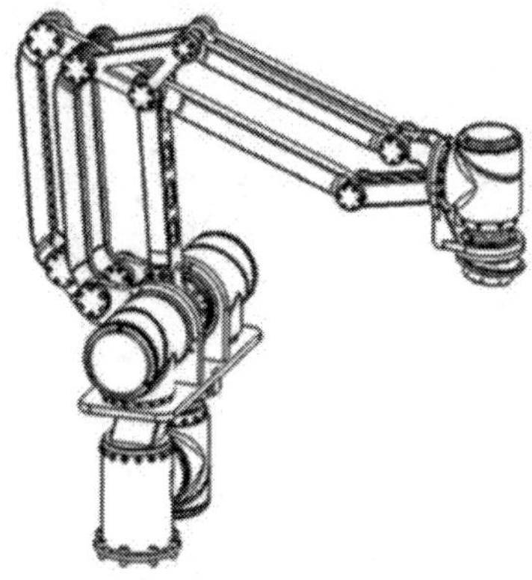

Figure 3. 4-DOF global positioning robotic arm

2.3. 2-DOF Remote Center-of-Motion Wrist

In percutaneous needle insertion procedures, the tip of the puncture needle is first translated to the skin needle insertion point by 4-DOF global positioning robotic arm, and then it needs to be oriented to point to the lesion target with the position of the needle tip unchanged. Since it is not necessary to consider the rotation of the puncture needle about its own axis, the posture adjustment of the puncture needle requires only two degrees of freedom.

The schematic of the remote center-of-motion linkage mechanism is shown in Fig. 4. This mechanism uses a dual-parallelogram linkage to achieve a fixed virtual center of rotation, which is the point O in Figure 4. If the virtual connecting rods BO and HO are connected respectively, two parallelogram linkages "ACHO" and "AEGO" can be obtained. When rod AE is the driving connecting rod, the parallelogram linkage ABCD can ensure that the joint H rotates around the virtual point O and the parallelogram linkage ABEF can ensure that the joint G rotates around the virtual point O. Thus, the connecting rod GH can rotate around the virtual point O. This realizes the function of remote motion center.

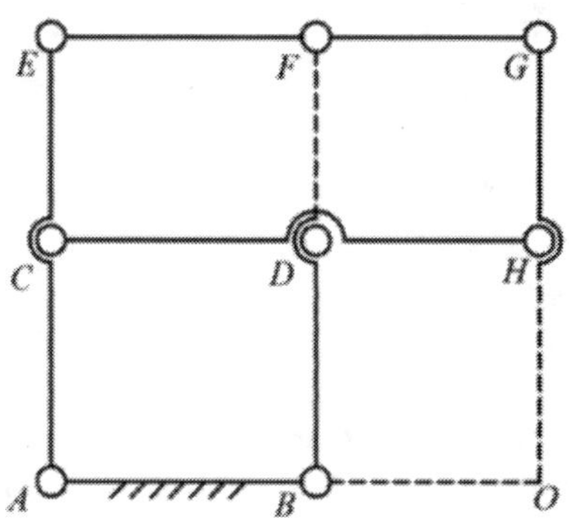

Figure 4. Schematic of the RCM Linkage Mechanism

The 2-DOF dual parallelogram RCM wrist designed in this paper is shown in Figure 5. The distal virtual point serves as the desired skin needle insertion point. It can ensure that the position of the puncture needle tip is not changed when the surgical needle is rotated about the distal virtual center point. Thus, the surgical puncture needle can be freely rotated and moved in the patient's body through a small wound.

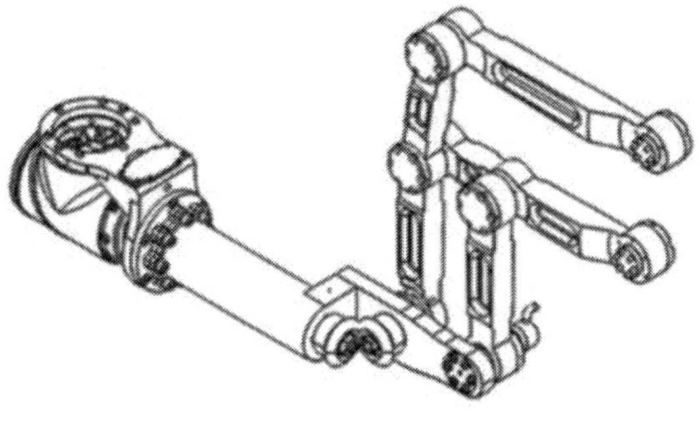

Figure 5. 2-DOF remote center-of-motion wrist

3. Kinematics Analysis of Robotic System

The robotic system consists of a mechanism with position and posture decoupling, and it also includes an RCM mechanism. So, the kinematics of such robotic system is quite different from the conventional serial robot. Now, we use the classical D-H method to derive the kinematics model of the robotic system, and then analyze the multi-solution problem of inverse kinematics according to the practical application background.

3.1. Forward Kinematics Analysis

First, we set up the D-H coordinate system of the percutaneous needle insertion robotic system, as shown in Figure 6. There is a virtual coordinate system O3' on the robot arm because of decoupled mechanism. In addition, considering the requirement of needle insertion angle for percutaneous surgery, a fixed 100-degree angle is set between joint 4 and joint 5 of the percutaneous robotic system.

The base coordinate system O_0 is placed on the mounting base of the percutaneous robotic system. The directions of the coordinate systems O_1-O_3 are arranged in parallel in space. Moreover, the Z direction of coordinate system $O_{3'}$ is consistent with the Z direction of base coordinate system O_0. The posture of the joint 4 is only rotated by a certain angle with respect to the Z axis of the base coordinate system O_0. The coordinate system O_4 is set at the intersection of the axis of joint 4 and joint 5. The coordinate system O_5 and O_6 are set at the virtual point of RCM mechanism.

Second, the parameters of the robotic system links are listed in Table 1.

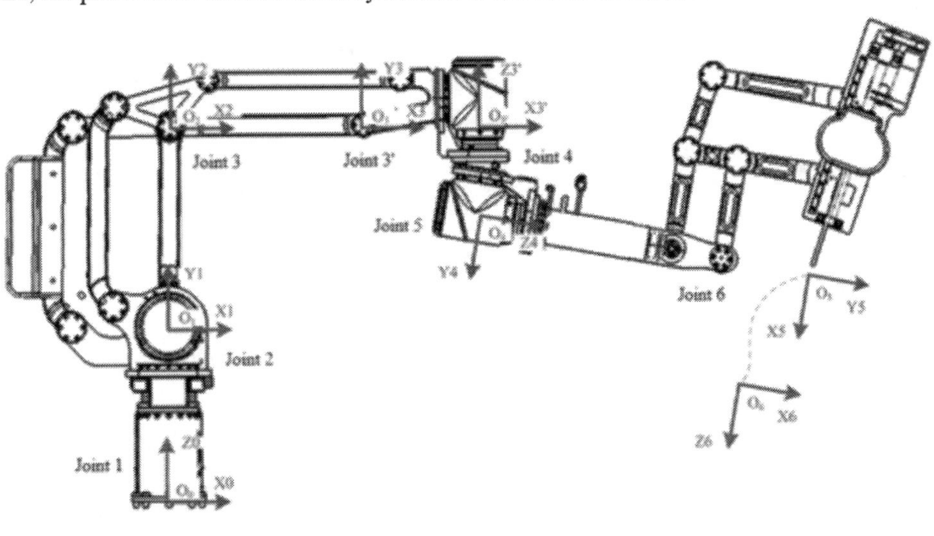

Figure 6. D-H coordinate system of the percutaneous needle insertion robot system

IOP Conf. Series: Materials Science and Engineering **825** (2020) 012009 doi:10.1088/1757-899X/825/1/012009

Table 1. Parameters of the robotic system links

link	a_i	α_i	d_i	θ_i
1	0	90°	d_1	φ_1+0°
2	a_2	0	0	φ_2+90°
3	a_3	0	0	φ_3-90°
3'	a_4	-90°	0	$\varphi_{3'}$ +0°
4	0	-100°(-α_4)	d_2	φ_4-90°
5	0	90°	d_3	φ_5+90°
6	0	90°	0	φ_6+90°

Finally, the forward kinematics analysis based on the classical D-H method is derived as follows. To make the matrix simple, we use c_θ and s_θ instead of cosθ and sinθ. The homogeneous transformation matrix between adjacent coordinate systems of the robotic system can be expressed as follows

$$A_1^0 = \begin{bmatrix} c_{\theta_1} & 0 & s_{\theta_1} & 0 \\ s_{\theta_1} & 0 & -c_{\theta_1} & 0 \\ 0 & 1 & 0 & 0 \\ 0 & 0 & 0 & 1 \end{bmatrix} \tag{2}$$

$$A_2^1 = \begin{bmatrix} c_{\theta_2} & -s_{\theta_2} & 0 & a_2 c_{\theta_2} \\ s_{\theta_2} & c_{\theta_2} & 0 & a_2 s_{\theta_2} \\ 0 & 0 & 1 & 0 \\ 0 & 0 & 0 & 1 \end{bmatrix} \tag{3}$$

$$A_3^2 = \begin{bmatrix} c_{\theta_3} & -s_{\theta_3} & 0 & a_3 c_{\theta_3} \\ s_{\theta_3} & c_{\theta_3} & 0 & a_3 s_{\theta_3} \\ 0 & 0 & 1 & 0 \\ 0 & 0 & 0 & 1 \end{bmatrix} \tag{4}$$

$$A_{3'}^3 = \begin{bmatrix} c_{\theta_{3'}} & 0 & -s_{\theta_{3'}} & a_{3'} c_{\theta_{3'}} \\ s_{\theta_{3'}} & 0 & c_{\theta_{3'}} & a_{3'} s_{\theta_{3'}} \\ 0 & -1 & 0 & 0 \\ 0 & 0 & 0 & 1 \end{bmatrix} \tag{5}$$

$$A_4^{3'} = \begin{bmatrix} c_{\theta_4} & -s_{\theta_4} c_{\alpha_4} & -s_{\theta_4} s_{\alpha_4} & 0 \\ s_{\theta_4} & c_{\theta_4} c_{\alpha_4} & c_{\theta_4} s_{\alpha_4} & 0 \\ 0 & -s_{\alpha_4} & c_{\alpha_4} & -d_4 \\ 0 & 0 & 0 & 1 \end{bmatrix} \tag{6}$$

$$A_5^4 = \begin{bmatrix} c_{\theta_5} & 0 & s_{\theta_5} & 0 \\ s_{\theta_5} & 0 & -c_{\theta_5} & 0 \\ 0 & 1 & 0 & d_5 \\ 0 & 0 & 0 & 1 \end{bmatrix} \tag{7}$$

$$A_6^5 = \begin{bmatrix} c_{\theta_5} & 0 & s_{\theta_5} & 0 \\ s_{\theta_5} & 0 & -c_{\theta_5} & 0 \\ 0 & 1 & 0 & d_5 \\ 0 & 0 & 0 & 1 \end{bmatrix} \tag{8}$$

Then the forward kinematics equation of the needle insertion arm can be obtained as follows

$$T_6^0 = A_1^0 A_2^1 A_3^2 A_{3'}^3 A_4^{3'} A_5^4 A_6^5 = \begin{bmatrix} n_x & s_x & a_x & p_x \\ n_y & s_y & a_y & p_y \\ n_z & s_z & a_z & p_z \\ 0 & 0 & 0 & 1 \end{bmatrix} = \begin{bmatrix} R & P \\ 0 & 1 \end{bmatrix} \tag{9}$$

where

$$\begin{aligned}
n_x &= c_{14}s_{\alpha_4}c_6 + s_{14}s_5s_6 - c_{14}c_{\alpha_4}c_5s_6 \\
n_y &= s_{14}s_{\alpha_4}c_6 - c_{14}s_5s_6 - s_{14}c_{\alpha_4}c_5s_6 \\
n_z &= c_{\alpha_4}c_6 + s_{\alpha_4}c_5s_6 \\
s_x &= s_{14}c_5 + c_{14}c_{\alpha_4}s_5 \\
s_y &= s_{14}c_{\alpha_4}s_5 - c_{14}c_5 \\
s_z &= -s_{\alpha_4}s_5 \\
a_x &= c_{14}s_{\alpha_4}s_6 - s_{14}c_5s_6 + c_{14}c_{\alpha_4}c_5c_6 \\
a_y &= s_{14}s_{\alpha_4}s_6 + c_{14}s_5c_6 + s_{14}c_{\alpha_4}c_5c_6 \\
a_z &= c_{\alpha_4}s_6 - s_{\alpha_4}c_5c_6 \\
p_x &= c_1\left(a_{3'} + a_3c_3 - a_2s_2\right) + c_{14}s_{\alpha_4}d_5 \\
p_y &= s_1\left(a_{3'} + a_3c_3 - a_2s_2\right) + s_{14}s_{\alpha_4}d_5 \\
p_z &= a_2c_2 + a_3s_3 + d_1 - d_4
\end{aligned} \tag{10}$$

There are seven joint variables in the forward kinematics, as in (9), but one of them is the virtual joint variable because of the mechanism decoupling. Therefore, the entire robot arm has 6 degrees of freedom, the first four degrees of freedom are redundant positioning degrees of freedom of the needle, and the last two are rotating degrees of freedom for the needle's posture. Next, to make the equations simple, six generalized variables q1-q6 are defined, which correspond to the driving of six motors of the robotic system respectively. According to the (1), the joint variables in (9) can be rewrote by generalized variables as follows

$$\begin{aligned}
\theta_1 &= \varphi_1 = q_1 \\
\theta_2 &= \varphi_2 + 90° = q_2 + 90° \\
\theta_3 &= \varphi_3 - 90° = q_3 - q_2 - 90° \\
\theta_{3'} &= \varphi_{3'} = -q_3 \\
\theta_4 &= \varphi_4 - 90° = q_4 - 90° \\
\theta_5 &= \varphi_5 + 90° = q_5 + 90° \\
\theta_6 &= \varphi_6 + 90° = q_6 + 90°
\end{aligned} \tag{11}$$

From the (9) and (11), the posture rotation matrix R contains only q_1, q_4, q_5, q_6, and the position translation matrix P contains only q_1, q_2, q_3, q_4. This clearly confirms that the percutaneous needle insertion robotic system has the characteristics of position and posture decoupling.

3.2. Forward Kinematics Analysis

Next, the closed-form solution method is used to solve the inverse kinematics of the robotic system.

Considering the element s_z in (9) and (10), we have

$$\begin{aligned}
s_5 &= -s_z / s_{\alpha_4} \\
c_5 &= \pm\sqrt{\left(1 - s_5^2\right)}
\end{aligned} \tag{12}$$

Then, the generalized variable q_5 can be obtained as

$$q_5 = \arctan 2(s_5, c_5) \tag{13}$$

Considering the s_x and s_y in (10), the simultaneous formulas can be obtained as

$$\begin{cases} s_{14}c_5 + c_{14}c_{\alpha_4}s_5 = s_x \\ s_{14}c_{\alpha_4}s_5 - c_{14}c_5 = s_y \end{cases} \tag{14}$$

Solving the (14), the sum of the generalized variable q_1 and $q4$ is given as

$$\begin{aligned} q_1 + q_4 &= \arctan 2(s_{14}, c_{14}) \\ &= \arctan 2(bc + ad, ac - bd) \end{aligned} \tag{15}$$

where

$$a = c_{\alpha_4}s_5,\quad b = c_5,\quad c = s_x,\quad d = s_y \tag{16}$$

Considering the n_z and a_z in (10), the simultaneous formulas can be obtained as

$$\begin{cases} c_{\alpha_4}c_6 + s_{\alpha_4}c_5s_6 = n_z \\ c_{\alpha_4}s_6 - s_{\alpha_4}c_5c_6 = a_z \end{cases} \tag{17}$$

Solving the (17), the generalized variable q_6 can be expressed as

$$q_6 = \arctan 2(s_6, c_6) = \arctan 2(fg + he, eg - hf) \tag{18}$$

where

$$e = c_{\alpha_4},\quad f = s_{\alpha_4}c_5,\quad g = n_z,\quad h = a_z \tag{19}$$

Considering the p_x and p_y in (10), the simultaneous formulas can be obtained as

$$\begin{cases} c_1\left(a_{3'} + a_3c_3 - a_2s_2\right) + c_{14}s_{\alpha_4}d_5 = p_x \\ s_1\left(a_{3'} + a_3c_3 - a_2s_2\right) + s_{14}s_{\alpha_4}d_5 = p_y \end{cases} \tag{20}$$

Solving the (20), the generalized variables q_1 and q_4 can be expressed as

$$\begin{aligned} q_1 &= \arctan 2(p_y - s_{14}s_{\alpha_4}d_5)/(p_x - c_{14}s_{\alpha_4}d_5) \\ q_4 &= (q_1 + q_4) - q_1 \end{aligned} \tag{21}$$

Considering the p_x, p_y and p_z in (10), the simultaneous formulas can be obtained as

$$\begin{cases} a_3c_3 - a_2s_2 = A \\ a_3s_3 + a_2c_2 = B \end{cases} \tag{22}$$

where

$$A = (p_x - c_{14}s_{\alpha_4}d_5)/c_1 - a_{3'}$$

$$or$$

$$A = (p_y - s_{14}s_{\alpha_4}d_5)/s_1 - a_{3'} \tag{23}$$

$$B = p_z - d_1 + d_4 - c_{\alpha_4}d_5 \tag{24}$$

The value of A is determined by the one where c_1 or s_1 is not close to zero. Solving the (23) and (24), the generalized variables q_2 and q_3 can be expressed as

$$\begin{aligned} q_2 &= \arctan 2(s_2, c_2) = \arctan 2(BC - AD, AC + BD) \\ q_3 &= \arctan 2(s_3, c_3) = \arctan 2(BE - AF, AE + BF) \end{aligned} \tag{25}$$

where

$$C = a_3 c_{q_3 - q_2}$$
$$D = a_3 s_{q_3 - q_2} + a_2 \tag{26}$$

All the generalized variables of the robot are obtained by the above calculation, and the multiple solutions of the generalized variables can be chosen according to the practical situation.

4. Manufacturing and Assembly of Percutaneous Robotic System

First, the manufacturing of the robotic system is performed according to the three-dimensional model in Figure 1. The first five rotating joints of the robot adopt an integrated joint module (model: RCM, from Kollmorgen Corp., Shanghai, China), which integrates a frame-less motor, a custom harmonic reducer, a high-precision absolute dual encoder, a low-voltage DC motor driver and a temperature sensor. Since the sixth rotating joint of the robot cannot adopt the integrated joint module due to structural reasons, a brushless motor module is adopted and an absolute encoder is used to measure the joint rotative angle accurately. Next, in order to achieve the precise force control of the robotic system, a specially customized single-axis SRI torque sensor is integrated at each rotating joint of robotic system. Finally, the overall machining and assembly of the robotic system is shown in Figure 7.

Figure 7. Overall machining and assembly of the robotic system

5. Conclusions

In this paper, a percutaneous needle insertion robotic system is developed with functions of biopsy and ablation. Such robotic system can overcome the shortcomings of the conventional manual percutaneous needle insertion procedure and make the percutaneous needle insertion surgery safer and more precise. The needle insertion robotic system has 6 degrees of freedom and employs a mechanical structure which can ensure the kinematic decoupling of the position and orientation of the robotic system. It integrates a 4-DOF positioning module, which can perform the position of the percutaneous needle, and a 2-DOF RCM module, which can accurately adjust the orientation the percutaneous needle. The combination of these two modules can ensure the kinematic decoupling of the position and orientation of the robotic system. The kinematics model of the robotic system based on classical D-H method is established, and the forward kinematics and inverse kinematics of the robotic system are analyzed. The multiple solutions to the inverse kinematic model can be chosen according to the actual situation. Finally, the manufacturing and assembly of the overall robotic system is performed and the robot can achieve percutaneous biopsy and ablation for tumor in urinary system.

Acknowledgments
This work has been supported by National Key R&D Program of China under Grant #2017YFB1303101.

References

[1] Y. Sun, D. Wu, Z. Du, and L. Sun, "Force-driven robotic drag control for freehand 3D

ultrasound-guided robot-assisted percutaneous surgery," 2009 IEEE International Conference on Robotics and Biomimetics, pp. 1953-1956, 2009.

[2] S. Jiang, W. Yuan, Y. Yang, D. Zhang, N. Liu, and W. Wang, "Modelling and analysis of a novel CT-guided puncture robot for lung brachytherapy," Advanced Robotics, vol. 31, no. 11, pp. 557-569, 2017. ""

[3] D. Stoianovici, L. L. Whitcomb, J. H. Anderson, R. H. Taylor, and L. R. Kavoussi, "A modular surgical robotic system for image guided percutaneous procedures," international conference on medical image computing and computer-assisted intervention, Springer, Berlin, Heidelberg, pp. 404-410, 1998.

[4] R. Taylor, et al, "A steady-hand robotic system for microsurgical augmentation," The International Journal of Robotics Research, vol. 18, no. 12, pp. 1201-1210, 1999.

[5] D. stoianovici, et al, "AcuBot: a robot for radiological interventions," IEEE Transactions on Robotics and Automation, vol. 19, no.5, pp. 927-930, 2003.

[6] B. Mitchell, et al, "Development and Application of a New Steady-Hand Manipulator for Retinal Surgery," Proceedings 2007 IEEE International Conference on Robotics and Automation, pp. 623-629, Apr 2007.

[7] A. Üneri, M. A. Balicki, J. Handa, P. Gehlbach, R. H. Taylor, and I. Iordachita, "New steady-hand eye robot with micro-force sensing for vitreoretinal surgery," 2010 3rd IEEE RAS & EMBS International Conference on Biomedical Robotics and Biomechatronics, pp. 814-819, Sep 2010.

[8] X. He, et al, "Toward clinically applicable steady-hand eye robot for vitreoretinal surgery," ASME 2012 international mechanical engineering congress and exposition, pp. 145-153, 2012.

[9] J. Hong, T. Dohi, M. Hashizume, K. Konishi, and N. Hata, "An ultrasound-driven needle-insertion robot for percutaneous cholecystostomy," Physics in Medicine & Biology, vol.49, no. 3, pp. 441, 2004.

[10] H. Bassan, T. Hayes, R. V. Patel, and M. Moallem, "A novel manipulator for 3D ultrasound guided percutaneous needle insertion," pp. 617-622, 2007.

[11] S. Wang, Q. Li, J. Ding, and Z. Zhang, "Kinematic design for robot-assisted laryngeal surgery systems," 2`006 IEEE/RSJ International Conference on Intelligent Robots and Systems, pp. 2864-2869, 2006.

Multiscale Modelling and Analysis on the Heavy-duty Hydrostatic Journal Bearing for a Precision Press Machine

Youyun Shang [1], Kai Cheng [1,2], Hui Ding [1] and Shijin Chen [1]

[1] School of Mechanical Engineering, Harbin Institute of Technology, 92 West Dazhi Street, Harbin 150001, PR China

[2] Institute of Materials and Manufacturing, Brunel University London, Middlesex UB8 3PH, UK

E-mail: shangyouyun@sina.com

Abstract. The present study aims to predict the performance of hydrostatic journal bearings during steady-state operation through multiscale modelling and analysis. An innovative multiscale modelling approach is presented for numerical modelling and analysis of the hydrostatic journal bearing as configured and designed, particularly for the precision press machine. With the challenging environment for journal bearings under hydrostatic pressures and surface features, , elastic modulus of oil and viscosity-temperature characteristics are often neglected while against heavy-loading and precision engineering requirement. This research takes into account of oil elastic modulus and viscosity-temperature characteristics to investigate the intrinsic relationship between the pressures within the hydrostatic bearing and its surface texture, feature size and locations being considered. This paper concludes with a further discussion on the potential and application of the approach in precision engineering.

Keywords: Hydrostatic Journal Bearing, Multiscale Modelling, Textured Surface, Computational Fluid Dynamics (CFD) Simulation, Precision Press Machine

1. Introduction

Hydrostatic journal bearings have been investigated through a number of simulation cases. In particular, the effects of surface roughness on Elastohydrostatic Lubrication (EHL) line contact are investigated against the smooth surface, because experimental measurement is difficult. This explains why the experiment on mixed lubrication only provides general trends of contact performance, while the detailed in-depth study on surface roughness and relevant scale is through numerical methods.

Patir and Cheng [1] presented an average flow model to solve the effects of roughness. In order to intensively study the flow factors of stochastic model of Patir and Cheng, several researchers proposed modified average Reynolds equations. Wang et al. [2,3], Epstein and Yu et al. [4], introduce flow factors in thermal elastohydrodynamic analysis of journal bearings. Based on those studies, de Kraker et al. [5, 6], presented the development of a micro–macro multiscale method. The micro flow effects for a single surface pocket was analyzed by using the Navier–Stokes equation and compared to the Reynolds solution for a similar smooth piece of surface. Gao et al. [7], presented Heterogeneous Multiscale Method (HMM) with cavitation effects to solve the EHL problem. Almqvist and Larsson [8], presented the commercial CFD software to simulate the EHL line contact issue based on the continuity equation and the energy equation. Hartinger et al. [9] used this approach to compare with the solution of Reynolds equation, the results of both approaches have good agreement. Recently Bruyere and Fillot [10] by adding the fluid thickness dimension to the EHL problem and by modelling solids with finite dimension and the thermal effects on friction have been investigated, Meng et al. [11] used Ansys Fluent software applied to isothermal cases with surface roughness effects under varied loads. Gao and Cheng et al. [12] used Ansys Fluent to investigate the influence of orifice chamber shapes of thrust bearings and optimized the shapes of the orifice.

In this paper, the hydrodynamic effect of hydrostatic bearings with surface texture and microstructure are investigated. Through the mass-conserving approach at macro-micro scales, it investigates the surface texture and the associated hydrodynamic performance of the journal bearing. ANSYS Fluent is used to simulation the bearing model and investigate the optimal location of the micro structures.

2. Development of the multiscale modelling approach

The performance of hydrostatic journal bearing depends on both local and global film height variation. Due to the oil film thickness compared with the overall bearing size is very small, the surface texture becomes important in macro-scale, and in micro-scale we should concern about the variations of oil viscosity coefficient and elasticity modulus, which in different rotating speed and oil supply pressure.

2.1. Macroscale model

The macroscale simulation describes the fluid–structure interaction in the global lubrication domain, In consideration of global surface roughness, the deterministic method still not solve 3D EHL problem. Hence, Patir and Cheng [1] presented an average rough Reynolds equation.

$$\frac{\partial}{\partial x}\left(-\phi_{px}\frac{h^3}{12\eta}\frac{\partial p}{\partial x}+\frac{(U_1+U_2)h'}{2}+\frac{\phi_s(U_1-U_2)\sigma}{2}\right)+\frac{\partial}{\partial z}\left(-\phi_{pz}\frac{h^3}{12\eta}\frac{\partial p}{\partial z}\right)=0 \tag{1}$$

With h' the mean oil film thickness, in this paper the surface roughness is Gaussian distributed, assume the surfaces to be stationary (U_2=0) in equation (1), leaving roughness deformation out of consideration, h' is equal to h, ϕ_{px} and ϕ_{pz} are pressure flow correction factors and ϕ_s is the shear flow factor and σ the surface rms roughness.

$$h(x,z)=C\left[1-\varepsilon\cos\left(\theta_{inlet}+\frac{x}{r}-\varphi\right)\right]+u(x,z)+h_\sigma(x,z) \tag{2}$$

Consider the effect of elastic deformation, the fluid film thickness h in function of the initial geometry, equation (2), by calculating the bearing on a large scale, various parameters of the bearing during the movement can be calculated accurately, especially the iterative calculation of the offset angle. This provides a reliable basis for geometric modelling for small scale calculation by CFD method.

2.2. Microscale model

Considering the influence of texture on elastohydrostatic lubrication bearing, differ with global smooth bearing, the pressure is incorrect and modified.

$$p=p_c+p_f+\bar{p} \tag{3}$$

In macroscale, with P_c the contact pressure equation (3) and P_f the fluid pressure. Two mechanisms have been pointed out contributing to the pressure generating effect of surface texture in full film lubrication. $\bar{p}$ the additional pressure between texture geometry in microscale and average pressure at macro level P_f equation (3).

$$P_c=H\bar{a}_c=\frac{H}{2}\left[1-\operatorname{erf}\left(\frac{1}{\sqrt{2}}\frac{h}{\sigma}\right)\right] \tag{4}$$

With H bearing material hardness, based on the plastic asperity model of Bowden and Tabor.

$$\overline{p}=\frac{1}{L_xL_y}\int_\Omega \max\left(p_f+p_{tex},0\right)d\Omega-p_f \tag{5}$$

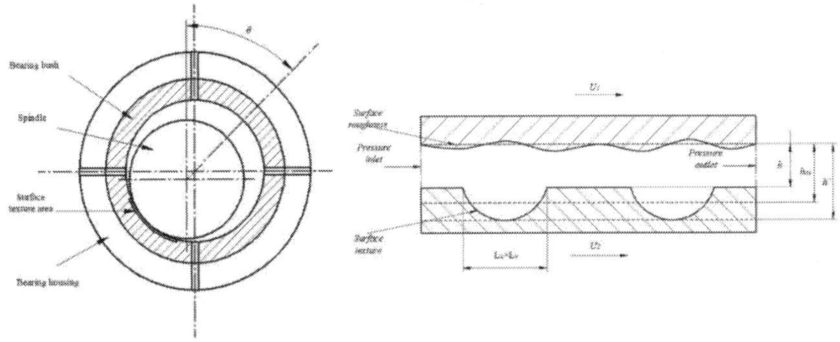

Figure 1. Schematic diagram of the journal bearing and surface microstructure

2.3. Density and Viscosity equation

The density and viscosity of oil fluid are concerned to be a constant parameter in previous studies. In reality, they are depended on the interior pressure. The linear variation of density can be written as:

$$\rho=\rho_0\left(\frac{0.6\times10^{-9}p}{1.7\times10^{-9}p+1}+1\right) \tag{6}$$

With ρ is the density of oil at any time, ρ_0 is the initial density of oil. And the relationship between viscosity and pressure can be expressed as:

$$\eta=\eta_0\exp\left\{(\ln\eta_0+9.61)\left[\left(1+5.1\times10^{-9}p\right)^k-1\right]\right\} \tag{7}$$

With η is the viscosity of oil at any time. η_0 is the initial viscosity of oil, k is the coefficient of viscosity (usually take 0.5~0.7), in this paper we take 0.6.

2.4. Cavitation and Elastic modulus of oil

To solve cavitation phenomenon, we usually assumed all negative pressures are equal zero, because the Reynolds function based on mass conservation. In this paper the macro-cavitation we use full cavitation model to interpretation in the CFD method:

$$\frac{\partial(\rho\varsigma)}{\partial t}+\nabla(\rho v\varsigma)=\nabla(\gamma\nabla\varsigma)+\beta \tag{8}$$

With ς is Volume fraction, β is Volume coefficient of vapour-oil variation. In the micro-cavitation, the cavitated area is found due to dissolved gasses that appear when the pressure drops below the vapour pressure. For more accurate calculation we introduce the elastic modulus of oil.

$$E_{oil}=700+800\log p \tag{9}$$

3. Geometry and simulation

The geometry domain is subdivided into a finite number of micro-scale, as shown in Figure 1. In order to start the iteration process, the velocity and pressure fields are approximated for the first iteration. Then these parameters are used to calculate the momentum conservation equation and the pressure of journal bearing. These values will be corrected in each iteration, until acceptable convergence of pressure and velocity is achieved as shown in Figure 2. In this research, representative hydrodynamic journal bearing that is used in the simulations.as discussed previously, the design and material parameters are shown in Table 1. The values that were used for the Young`s modulus, hardness and Poisson ratio are given.

Table 1. Main shaft bearing characteristics.

Geometry		
Bearing diameter	D	140 mm
Bearing width	B	140 mm
Radial clearance	c	30 μm
Relative eccentricity	ε	0.8-1.2
Surface rms roughness	σ	0.6 μm
Rotational speed	ω	300-1000 rpm
Viscosity at T=40.0℃	μ	0.035 $Pa \cdot s$
Operating temperature	T	40.0℃
Young`s modulus (pad)	E	0.063 MPa
Poisson`s ratio (pad)	v	0.33

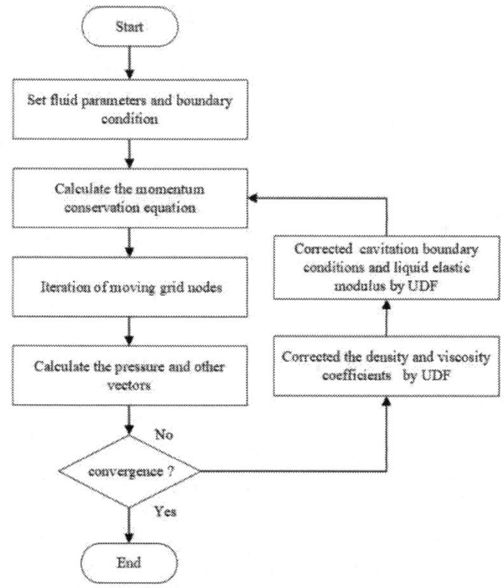

Figure 2. Schematic diagram of calculation flow

4. Results, analysis and discussion

4.1. Simulation on the oil film with surface textures

The surface microstructure of bearing is analyzed and studied by CFD method. It can be clearly seen from the pressure distribution contours of journal bearing with the rotate speed at 300 and 1,000 rpm, oil supply is 2MPa.

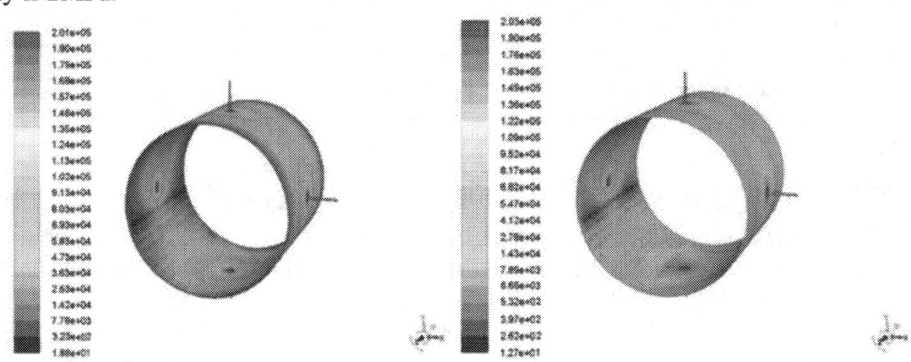

(a) (b)

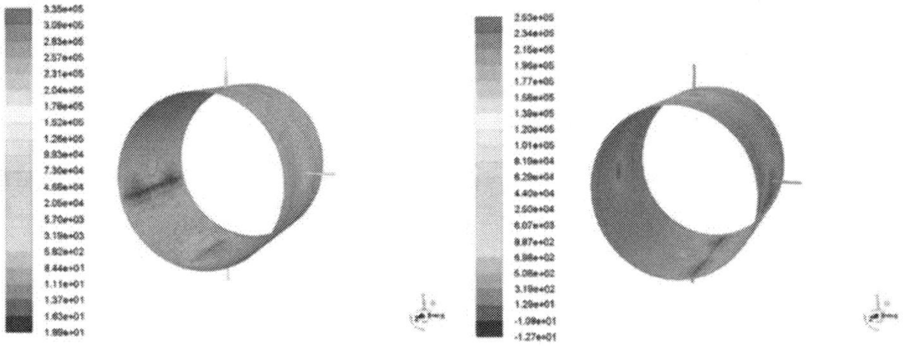

(c) (d)

Figure 3. Pressure distribution contours with surface microstructure at different rotate speed (a) Microstructure in loaded area at 300 rpm (b) Microstructure in loaded area at 1000 rpm (c) Microstructure in loaded area at 1000 rpm (d) Microstructure in unloaded area at 1000 rpm

In figure 3(a) and (b). the pressure increased at approximate contact area with multiscale method. Because the density and viscosity coefficient increased with pressure and rotate speed, and the cavation area decreased due to the elastic of oil gradually increasing. The surface deformation increase with the rotating speed, at low speed, the load capacity varies with the eccentricity. But in high rotating speed the effect is very clear, (c) and (d) at 1000 rpm.

4.2. Influence of EHL on static characteristics of the bearing

The change trend of attitude angle (left) and bearing capacity (right) are shown in Figure. 4. From top to bottom, eccentricity (e), oil supply pressure (P_0), average oil film thickness (h_0) is varied. From the first set of figures it is clear that for smaller eccentricity, a lower coefficient of friction and larger minimum film height is obtained for the same operating conditions in EHL. Reduction of the radial clearance by a factor of eccentricity the capacity of load, when the eccentricity e<0.3, the film thickness less than average surface roughness h_{av} the bearing wear and tear will be occurs. With the pressure increasing the trend of attitude angle increased obviously, especially at rotate speed 600 to 700 rpm (P_0=2 MPa).

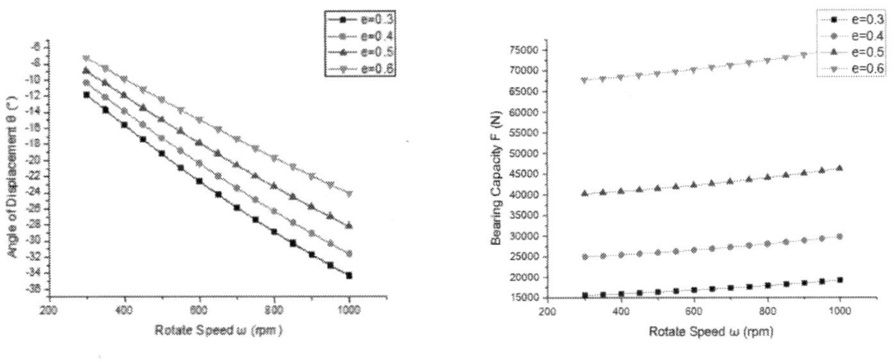

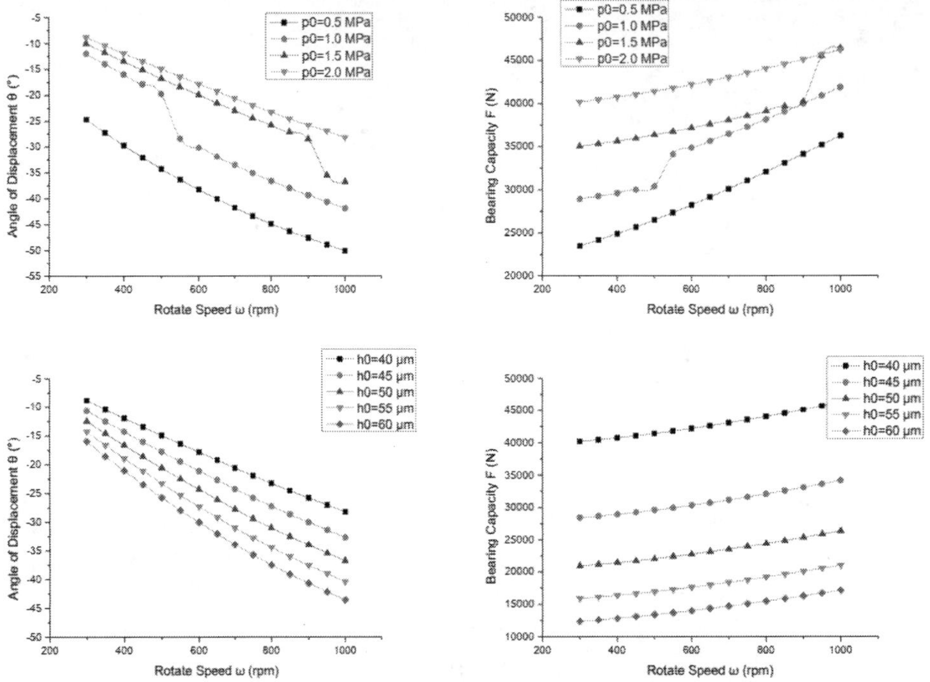

Figure 4. The change trend of attitude angle and bearing capacity

This is due to the formation of micro-cavitation inside oil at this speed, the growth of carrying capacity is also evident. In the third set of figures, the load is varied with the average oil film thickness h_0. With the deformation increasing, the contact pressure is distributed in larger area. it can be seen that the thinner of minimum film thickness the larger carrying capacity in load area.

5. Conclusions

The research presented is to study a multiscale modelling and analysis approach for hydrostatic bearings, The following conclusions are obtained through the simulation-based investigation:

- In the actual working condition of the bearing, the surface roughness of the bearing surface, elastic deformation and other factors must be taken into account.
- The bearing loading capacity is related to its speed and viscosity, so choosing the appropriate operation parameters has a great influence on the load carrying capacity of the bearing.
- The influence of the surface microstructure at the bearing surface includes the layout of the microstructure, location and the ratio of the bearing oil film thickness. The oil film thickness is smaller, the greater the influence of microstructure on the bearing capacity.

References

[1] Patir N and Cheng H S 1978 Average flow model for determining effects of 3- dimensional roughness on partial hydrodynamic lubrication *ASME J Lubri Tech* **100** 12–17

[2] Shi F and Wang J Q 1998 A Mixed-TEHD Model for Journal-Bearing Conformal Contacts—Part I: Model Formulation and Approximation of Heat Transfer Considering Asperity Contact *ASME J. Tribol.* **120** 198–205

[3] Gao S and S Srirattayawong 2012 CFD Prediction of the Effects of Surface Roughness on Elastohydrodynamic Lubrication under Rolling/Sliding Conditions *Applied Mechanics and Materials* **184-185** 86-89

[4] Epstein D, Yu T, Wang J, Keer L M, Cheng H S and Liu S 2003 An Efficient Method of Analyzing the Effect of Roughness on Fatigue Life in Mixed-EHL Contact *STLE Tribol. Trans.*, **46** 273–281
[5] A de Kraker, R A J van Ostayen and D J Rixen 2010 Development of a texture averaged Reynolds equation. *Tribol. Int* **43(11)** 2100-09
[6] A de Kraker, R A J Ostayen and D J Rixen 2007 A multiscale method modelling surface texture effects *ASME J Tribol* **129** 221–30
[7] Gao L M, de Boer GN, Hewson R W 2015 The role of micro-cavitation on EHL: A study using a multiscale mass conserving approach. *Tribol Int* **324** 331–90
[8] Almqvist T and Larsson R 2008 Thermal transient rough EHL line contact simulations by aid of computational fluid dynamics *Tribol. Int.* **41** 683-93
[9] Hartinger M, Dumont M L, Ioannides S, Gosman D and Spikes H 2008 CFD modelling of a thermal and shear-thinning elastohydrodynamic line contact *ASME J. Tribol.* **130(4)** 041503
[10] Bruyere V, et al 2012 Computational fluid dynamics and full elasticity model for sliding line thermal elastohydrodynamic contacts *Tribol Int* **46(1)** 3-13
[11] Meng F M and T Yang 2012 Preliminary study on mechanism of cavitation in lubricant of textured sliding bearing *Part J: Journal of Engineering Tribology*, **227(7)** 695-708
[12] S Y Gao, K Cheng, et al 2015 CFD based investigation on influence of orifice chamber shapes for the design of aerostatic thrust bearings at ultra-high speed spindles *Tribol Int* **92** 211–21

IOP Conf. Series: Materials Science and Engineering **825** (2020) 012011 doi:10.1088/1757-899X/825/1/012011

A Novel Comprehensive Model of Wet Clutch During the Engagement Process

Baoshan Peng[1], Heyan Li[1] and Jikai Liu[2]

[1]College of urban transportation and logistics, Shenzhen Technology University, Shenzhen, 518118, China
[2]Institute of Microelectronics, Chinese Academy of Sciences, Beijing, 100029, China

Email: loveheyanli@gmail.com

Abstract. With the development of vehicles towards higher power density and reliability, the wet clutch, which is the critical component in transmission system, always encounters the problems of ablation, wear, and breakage. To solve this problem, a novel comprehensive model is developed for the wet clutch to reveal its intrinsic working mechanism. Firstly, a dynamic clearance varying model and a torque balance model are developed for the wet clutch to explain the bearing mechanism along the axial and circumferential directions. After that, the friction coefficient, which varies depending on the interface pressure, slipping speed and temperature, is obtained. By developing the temperature prediction model, a theoretical coupling model is finally established for the engagement process of wet clutch. From the simulation results, the coupling model is sufficient to describe the changing trend of several critical parameters, including the viscous torque, the asperity torque, the contact status on the friction interface, the temperature rise and the friction coefficient, etc.

1. Introduction

As the vital important component of the transmission, the working performance of wet clutch has a directly influence on the smoothness, reliability and safety of the vehicle. During the engagement process, the clutch is always exposed to an extreme environment, such as high temperature, high pressure and high-speed difference. Many physical processes, such as the transfer of friction heat, the formation of friction torque, the wear of friction materials and the distribution of cooling oil, always exist on the friction interface at the same time and interact with each other. To reveal its intrinsic working mechanism and achieve excellent vehicle performance, a lot of researches have been done.

Wu studied the oil film characteristics of rotating annular disk during the engagement process by using the finite-difference method [1-3]. Through establishing a hydrodynamic pressure model, the changing process from the hydrodynamic fluid to the boundary lubrication is described. In 1975, Ting [4] studied the boundary lubrication stage using the porous elastic theory considering the friction surface roughness and porous surface characteristics. Furthermore, to investigate the effect of grooves, EI-Sherbiny [5] used the finite element method to study the influence of groove types on the working process of wet clutch in 1977. From 1996 to 1999, Berger [6], Yang [7], Jang [8] extends the study area form the isothermal expansion to the non-isothermal process. The results show that, because of the friction plate temperature along the radial and circumferential is different, the working condition of the friction plate changes, further impact the engagement process. With the improvement of finite element simulation methods, Jang [9] expands the clutch model into a three-dimensional model, and studies various factors such as groove, centrifugal force, and material deformation.

Owning to these considerable researches, the understanding of the working mechanism of wet clutch has been developed significantly. However, according to the actual situation, the service life the wet clutch is always less than other components in the transmission, especially for the heavy vehicles. To further improve the performance of wet clutch, a novel comprehensive model is developed.

Firstly, a dynamic model is present to describe the engagement process along the axial and circumferential directions. Then, an optimized friction coefficient model is provided. After that, a lump thermal resistance model is established to simulate the temperature variation of wet clutch. Combining the above models, a complex comprehensive model is built for the wet clutch and the simulation results are present. Finally, the obvious conclusions are drawn.

2. Dynamic model for the engagement process

The simplified structure of wet clutch is shown in Figure 1. It consists of piston, piston cylinder, steel plate and friction plate. Under the control of the oil pressure, the wet clutch engages.

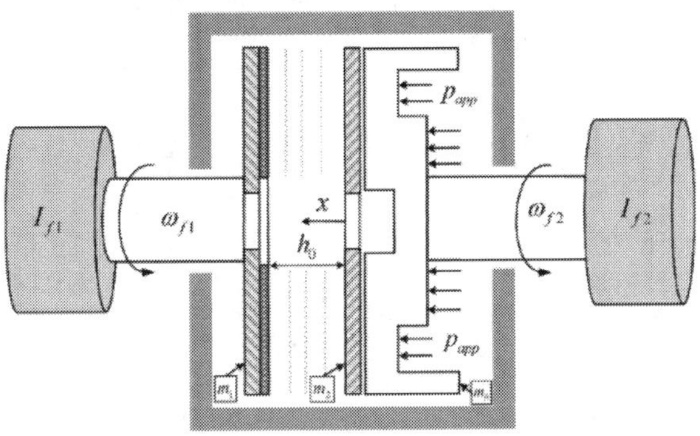

Figure 1. Simplified structure of wet clutch.

2.1 Dynamic model along the axial direction

Along the axial direction, the engagement process of wet clutch is to eliminate the clearance between the friction pair. Considering the oil film pressure and the contact pressure, the dynamic equation of the piston during the engagement process can be deduced as follows.

$$F_{app}=(m_0+m_2)\ddot{x}+(c_0+c_2)\dot{x}+F_v+F_c \tag{1}$$

$$h=h_0-x \tag{2}$$

For the oil film pressure, it is one of the fundamental parameters influencing the operation of wet clutch and can be described by the following equation.

$$F_v=\iint_{A_v}\frac{B}{4A}(r^2-R_o^2)+\frac{3\eta}{A}\frac{\partial\overline{h_t}}{\partial t}(r^2-R_o^2)+\ln\frac{r}{R_2}(\frac{B}{4A}+\frac{3\eta}{A}\frac{\partial\overline{h_t}}{\partial t})\frac{(R_o^2-R_i^2)}{\ln R_i-\ln R_o} \tag{3}$$

$$A=\phi_r h^3+12\Psi d_m \tag{4}$$

$$B=\frac{\phi_r\rho h^3}{5}(3\omega_1^2+4\omega_1\omega_2+3\omega_2^2) \tag{5}$$

$$\overline{h_t}=\frac{h}{2}[1+erf(\frac{h}{\sqrt{2}\sigma})]+\frac{\sigma}{\sqrt{2\pi}}e^{(-\frac{h^2}{2\sigma^2})} \tag{6}$$

When the oil film thickness between the friction pair reaches the same order of roughness of contact surface, the micro-convex bodies begins to contact. Then, the contact pressure generates.

$$F_c=\iint_{A_c}p_c \tag{7}$$

$$\begin{cases} p_c = K'E' \cdot 4.4086\times10^{-5}\cdot(4-H)^{6.804}, & H<4 \\ p_c = 0, & H\geq 4 \end{cases}, \quad H=\frac{h}{\sigma} \tag{8}$$

$$K' = \frac{8\sqrt{2}}{15}\pi(N\beta\sigma)^2(\frac{\sigma}{\beta})^{\frac{1}{2}} \tag{9}$$

$$\frac{1}{E'} = \frac{1}{2}(\frac{1-\upsilon_1^2}{E_1}+\frac{1-\upsilon_2^2}{E_2}) \tag{10}$$

With equations (1), (3) and (7), the changing trend of the clearance between the friction pairs can be obtained, while the rotating speed mentioned in equation (5) should be further calculated.

2.2 Dynamic model along the circumferential direction

During the entire engagement process of wet friction pair, the viscous torque generated by shearing oil film always exists because of its structure and working environment. When the oil film clearance of friction pair is less than a certain value, the micro-convex peak in the rough contact area will also generate the corresponding asperity torque. Finally, the passive part is driven to rotate.

Considering the effect of surface roughness, the viscous torque can be described as follows.

$$M_v = (1-A_{red}C)\int_0^{2\pi}\int_{R_i}^{R_o}\eta(\phi_f+\phi_{fs})\frac{r^2\Delta\omega}{h}rdrd\theta \tag{11}$$

$$C = \frac{A_c}{A_n} = \kappa\pi^2(NR_H\sigma)^2[-\frac{H}{\sqrt{2\pi}}e^{-\frac{H^2}{2}}-\frac{1}{2}(H^2+1)(erf(\frac{H}{\sqrt{2}})-1)] \tag{12}$$

ϕ_f and ϕ_{fs} are the flow factors [10].

As for the asperity torque, it is expressed as follow.

$$M_c = A_{red}C\mu\int_0^{2\pi}\int_{R_i}^{R_o}p_c r^2 drd\theta \tag{13}$$

With the obtained M_v and M_c, the rotating speed of driven part can be deduced.

$$I_{f2}\frac{d\omega_{f2}}{dt} = M_C = M_c + M_v \tag{14}$$

3. Optimized friction characteristics of wet friction pair

The wet clutch is the typical component of the application of friction phenomenon in the vehicle power transmission system. However, the value of friction coefficient μ, mentioned in equation (15), is influenced by many factors. In the previous researches, it is usually described by the slipping speed between the friction pairs, without considering the influence of the contact pressure and the temperature. In this paper, an optimized friction coefficient is presented to fill that gap in knowledge.

$$\mu = 23e^{\left(\frac{-3v}{(\ln T-3.2)\left((28.3p)^{0.44}-0.87\right)}-5.16\right)} + 0.08\left(e^{-0.005T}-1\right)\left(e^{-0.2v}-1\right) + \frac{0.008\ln(4v+1)}{e^{0.005T}} - 0.005\ln(28.3p)+0.020 \tag{15}$$

4. Lumped thermal resistance model

The Finite Element Method (FEM) is always used to analyse the thermal problems in the engineering field. However, it is difficult to predict the temperature of the component in real time. In this paper, the lumped parameter method is adopted to build a lumped thermal resistance model. As shown in Figure 2, the lubricating cooling system is divided into different parts, including the environment (E), the radiator (R), the hydraulic fluid reservoir (F), the relief safety valve (RV), the pump (P), the pressure regulating valve (V) and the clutch (C).

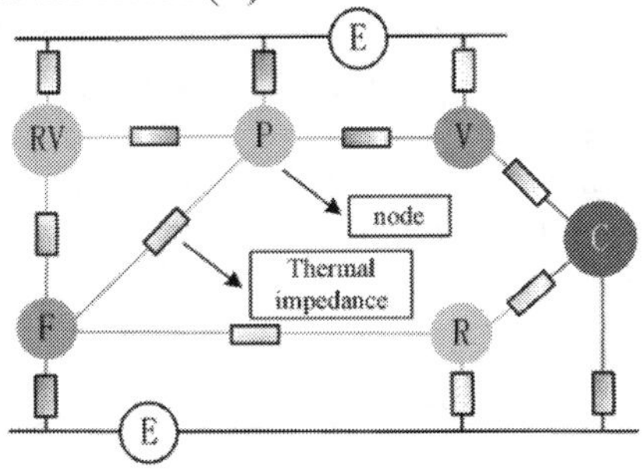

Figure 2. Heat transfer model.

For the component and the lubrication oil in each node, the temperature variations can be deduced by the following equation.

$$\begin{cases} C_i \dfrac{\mathrm{d}T_i}{\mathrm{d}t} = \Phi_i - K_{i,io}\left(T_i - T_{io}\right) - \sum\limits_j K_{i,j}\left(T_i - T_E\right) \\ C_{io} \dfrac{\mathrm{d}T_{io}}{\mathrm{d}t} = \Phi_o - K_{io,i}\left(T_{io} - T_i\right) - \sum\limits_j K_{io,jo}\left(T_{io} - T_{jo}\right) \end{cases} \tag{16}$$

in which $C_{i(o)} = \rho_{i(o)} \cdot c_{i(o)} \cdot V_{i(o)}$ and K is the equivalent heat transfer coefficient [11].

Taking the temperature variation of wet clutch as an example, the heat condition can be calculated according to the following equations.

$$\begin{cases} C_C \dfrac{dT_C}{dt} = \Phi_C - K_{C,E}(T_C - T_E) - K_{C,V}(T_C - T_V) - K_{C,oC}(T_C - T_{oC}) \\ C_{oC} \dfrac{dT_{oC}}{dt} = \Phi_{oC} - K_{oC,oV}(T_{oC} - T_{oV}) - K_{oC,oR}(T_{oC} - T_{oR}) \\ \qquad - K_{oC,E}(T_{oC} - T_E) - K_{oC,C}(T_{oC} - T_C) \end{cases} \tag{17}$$

$$\Phi_C = M_C \cdot \Delta\omega = (M_v + M_c) \cdot \Delta\omega \tag{18}$$

5. Comprehensive model and simulation results analysis

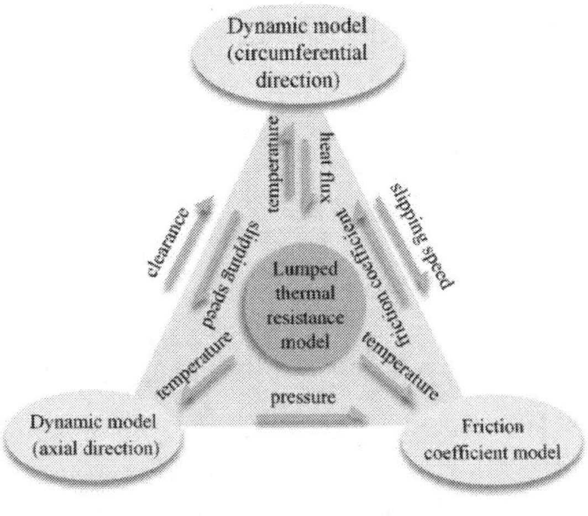

Figure 3. Coupling relationship between different models.

IOP Conf. Series: Materials Science and Engineering **825** (2020) 012011 doi:10.1088/1757-899X/825/1/012011

The above three sections build the dynamic model for the engagement process, provide an improved friction coefficient and establish a lumped thermal resistance model to predict the temperature variation of wet clutch. Different model describes different physical process and, combining these three models, a complex comprehensive model is finally obtained. Figure 3 shows the coupling relationship between different models.

The obtained simulation results are illustrated in Figure 4. Figure 4(a) presents the applied oil pressure on the piston, which is measured from the test bench. After the violent fluctuation, the value of the pressure is finally levelling out at around 1.3MPa. In Figure 4(b), when the wet clutch contacts, the generated friction torque has a sharp increase, then shows a slowly falling trend. While, at the end of the engagement process, the friction torque goes up to 109N·m and then drops to 0N·m. For the viscous torque, it reaches 17N·m at the contact moment, then decreases to 0N·m; while, for the asperity torque, it is the main composition of the friction torque, especially at the end stage of the engagement process.

Figure 4(c) shows the changing trend of the friction coefficient. It varies from 0.05 (dynamic friction coefficient) to 0.13 (static friction coefficient). As for the temperature of wet clutch, Figure 4(d) indicates that it has an increase from 65.8℃ to 87.2℃ during the entire process

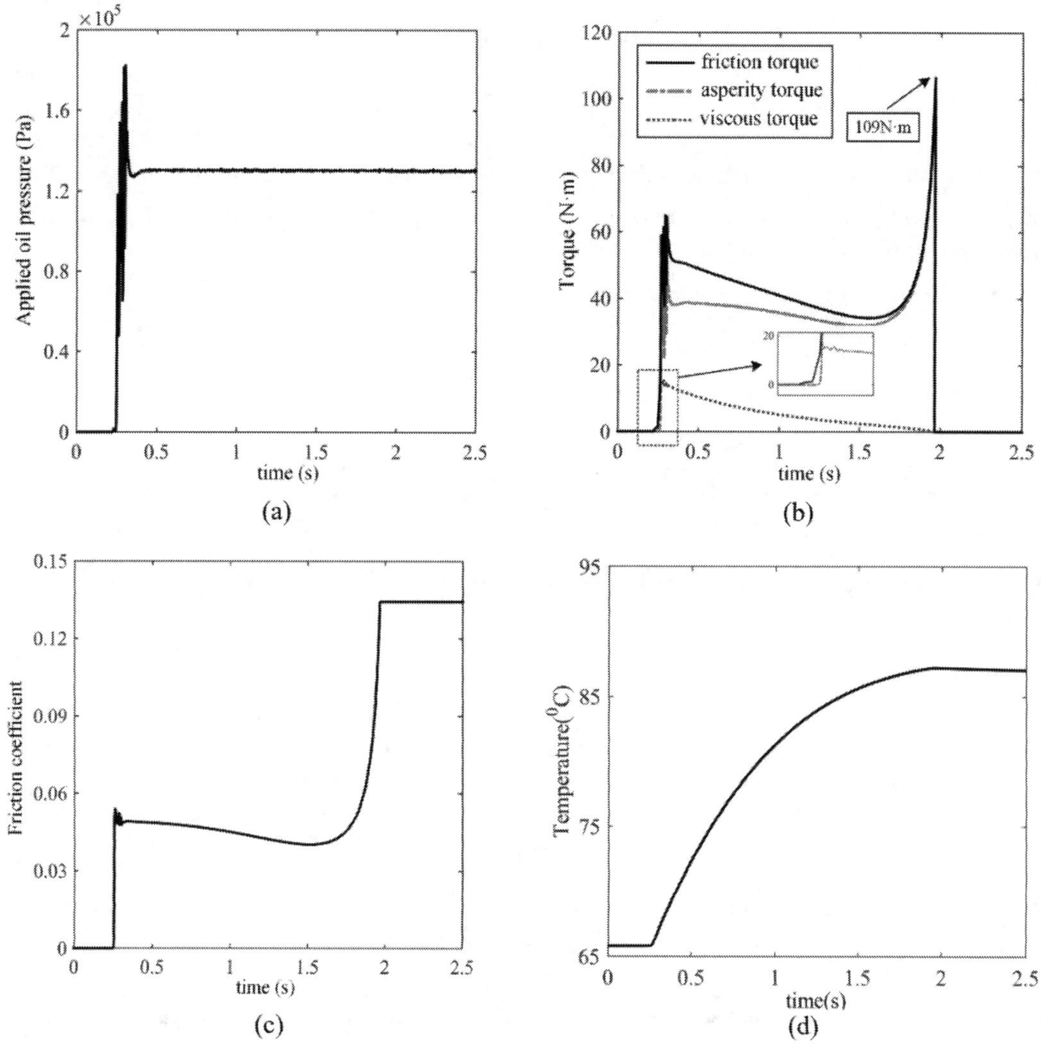

Figure 4. Simulation results.

6. Conclusions

In this paper, a novel comprehensive model is established for the engagement process of wet clutch by integrating different models, including the dynamic model for the axical and circumferential movement, the optimized fiction coefficient and the thermal resistance model for predicting the temperature variation.

Under the applied constant pressure, the friction torque goes up quickly, then shows a decreasing trend. At the end of the slipping status, the torque increases to 109N·m sharply and drops to 0N·m. The asperity torque shows a similar changing trend, while the viscous torque decreases during the entire engagement process from 17N·m to 0N·m. Under this simulation condition, the friction coefficient varies from 0.05 to 0.13 and the temperature rise of wet clutch is 21.4℃.

References

[1] Wu H 1970 Squeeze-Film Behavior for Porous Annular Disks *Journal of Lubrication Technology* **92** 593-596
[2] Wu H 1971 The Squeeze Film between Rotating Porous Annular Disks *Wear* **18** 461-469
[3] Wu H 1973 An Analysis of the Engagement of Wet Clutch Plates *Wear* **24** 23-33
[4] Ting L L 1975 Engagement Behavior of Lubricated Porous Annular Disks *Wear* **34** 159-182
[5] EI-Sherbiny M G, Newcomb T P 1977 *Numerical Simulation of The Engagement Characteristics of a Wet Clutch, Oil-immersed Brakes and Clutches* (New York)
[6] Berger E J, Sadeghi F, Krousgrill C M 1997 Analytical and Numerical Modeling of Engagement of Rough, Permeable, Grooved Wet Clutches *Journal of Tribology* **119** 143-148
[7] Yang Y B, Lam R C, Fujii T 1998 Prediction of Torque Response during the Engagement of Wet Friction Clutch *SAE transactions* **1998** 1625-1635
[8] Jang J M, Khonsari M M 1999 Thermal Characteristics of a Wet Clutch *Journal of Tribology* **121** 610-617
[9] Jang J Y, Khonsari M M, Maki R 2011 Three-Dimensional Thermohydrodynamic Analysis of a Wet Clutch With Consideration of Grooved Friction Surfaces *Journal of Tribology* **133** 011703
[10] Patir N, Cheng H S 1979 Application of average flow model to lubrication between rough sliding surface *ASME Journal of LubricationTechnology* **101** 220-229
[11] Holman J P 1986 *Heat transfer* (New York: McGraw-hill) p 521-565

Appendix

Parameters	Meanings	Parameters	Meanings
A_{red}	percentage of non grooved area	p	pressure
E	elastic modulus	r	radius
F	force	v	slipping speed
I	moment of inertia of components	$\dot{\Phi}_{i(o)}$	heat source
M	torque	Ψ	permeability of materials
N	asperity density	β	curvature radius of asperity
R	radius of friction disc	μ	friction coefficient
R_H	radius of asperity	ω	angular speed of components
T	temperature of components	η	coefficient of kinetic viscosity
V	volume	c	specific heat capacity
x	displacement	ρ	density
h	clearance	σ	the RMS surface roughness
ϕ	flow factors	υ	Poisson's ratio
d_m	thickness of surface material	κ	plastic deformation coefficient
m	weight		

IOP Conf. Series: Materials Science and Engineering **825** (2020) 012012 doi:10.1088/1757-899X/825/1/012012

Simulation Study on Flow-induced Vibration Characteristics of Multi-nozzle Ejector

Yang Yisheng[1], Liu Zongzheng[1], Yan Xiqiang[1] and Jia Hequan[1]

[1]China aerodynamics Research and Development Center, Mianyang, China

ziyaxuan.95@163.com

Abstract. The ejector is an important power source for intermittent wind tunnel. In order to accurately predict the flow-induced vibration (FIV) response, numerical simulation based on fluid-structure coupling method was carried out. Firstly, three-dimensional unsteady constant simulation on the flow field of ejector was done, using the renormalized k-ε model. The ejector inlet and outlet mass flow were basically consistent with design, and the relative error of Mach number in the third-stage nozzle was 1.79%. Secondly, in the one-way coupling simulation, the MPCCI software was performed. Thirdly, the calculated vibration spectrum was obtained. The simulation results could be found in good agreement with experiment, and the domain frequency relative error was 4.27%.

1. Introduction

The ejector is a device that transport fluid by the entrainment principle of high-speed flow, act like pump [1]. Due to the rapid response and simple structure, it's widely used in industrial field. Such as achieving rapid pressure depressurization in ejector wind tunnels, transporting media in natural gas and mining industries, and increasing thrust at low speeds in pulse detonation engines.

The internal flow of ejector is extremely complicated, which was determined by the work mechanism. For gas ejector, due to the influence of internal flow shock, vortex, airflow mixing pulsation and resonance [2], serious vibration problems may threaten the operation safety of equipment. Especially in large-injection wind tunnels and laser weapon pressure recovery systems, severe structural vibration could affect equipment performance and even cause fatigue damage [3].

With the rapid development of chemical laser weapons and pulse detonation rocket engine technology, research on ejector performance has become a hot topic in academia. A series of research results [4-6] have been published, on parameter optimization and internal flow numerical simulation of ejector. During the development of wind tunnel design and laser pressure recovery system, China Aerodynamics Research and Development Center found that the vibration of multi-nozzle ejector might show different manifestations in different applications, and the characteristics are far more complicated than ordinary wind tunnel sections. In a wind tunnel, tear occurred in nozzle support plates as the vibration concentrated. In another wind tunnel the outer shell of the exhaust ejector cracked because of vibration. For pressure recovery system, the vibration of work platform exceeded the standard, which was more difficult to accept.

For the problem of pressure recovery system, Liu Zongzheng [7] designed a damping device with a weight-coupled rubber-metal spring parallel structure, which effectively reduced the vibration amplitude. To achieve vibration reduction, Ma Yue-yin [8] optimized the transition part structure, using finite element calculation and experimental test methods. Wang [9] obtained the ejector mode of the pulse detonation rocket engine by experimental means and optimized the ejector parameters. Nie Junfeng [10,11] studied the respiratory vibration mechanism and dynamic load identification of the engine ejector, achieved good results.

Although the research in the literature [7~11] had achieved certain effects in engineering, they were just cases analysis based on specific object. All the forward research focused on the entire structural response of ejector, which were excited by multi-vibration sources. To provide technical guidance for ejector vibration reduction and structural optimization, the common problems of ejector vibration should be studied from the perspective of FIV, and the vibration mechanism should be obtained using the numerical simulation method.

2. Vibration Source Analysis

The source of ejector structure FIV could be very complex, and the vibration frequency band was wide. The vibration of ejector was mainly caused by airflow pulsation, eddy current, airflow impact caused by valve the action, resonance. When ejector work, the upstream valve opened quickly, causing severe airflow surges in ejector system tubing. Then the airflow was ejected through the nozzle, and mixed with the secondary airflow in mixing chamber. In this process, a rapid static pressure decrease occurred along the wall of ejector. The ejector principle was shown in Figure 1. On the one hand, the gas velocity in ejector could be several Mach number, the turbulent boundary layer might separate and form eddy. On the other hand, with the velocity and pressure of the mixed flow changed periodically, the pulsating airflow generated and interacted with structure continuously, which would cause system vibration. When the above three frequencies closed to the natural frequency of structure, resonance occurred.

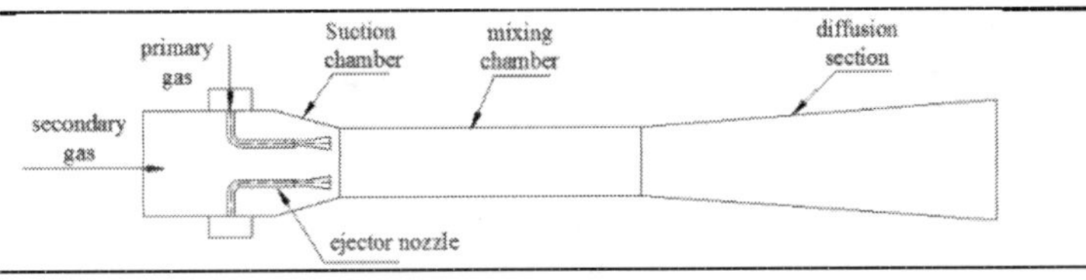

Figure 1. The schematic diagram of ejector.

A three-stage multi-nozzle ejector was chosen to be the research object, with a total length of 5 meters nearly. There were 12 nozzles in each stage, divided into two layers of circular uniformity as shown in Figure 2. The design parameters could be found in Table 1.

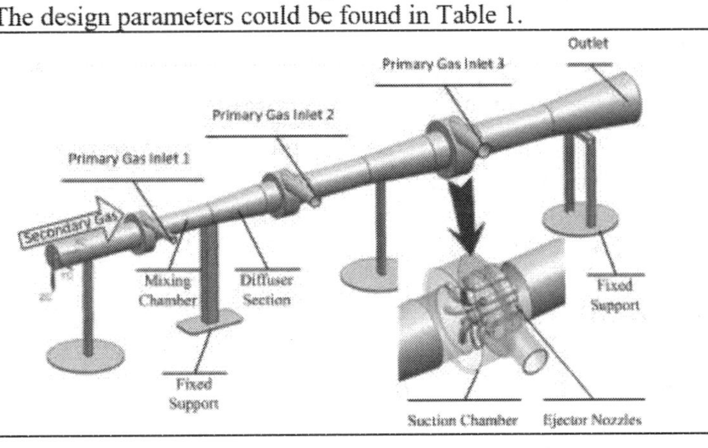

Figure 2. Three-stage ejector structural model.

Table 1. Design parameters of the ejector.

Stage \ Parameter	Inlet Pressure	Inlet Mass flow	Outlet Mach Number
First	0.2MPa	0.23kg/s	3.6
Second	0.34MPa	1.13kg/s	3.0
Third	0.97MPa	5.91kg/s	2.8

3. Internal Flow Field Simulation

The internal flow field of ejector was mainly determined by the supersonic flow emitted from nozzle. The ejector nozzle outlet was the interface of supersonic flow and subsonic flow. In the upstream of the outlet, the continuous gas with high pressure expanded and accelerated quickly, while the downstream was the intense mixing area. So, the outlet was the most complex position of the entire flow field. The nozzle part should not be simplified during model.

Since the structural size of nozzle was one or two orders of magnitude smaller than the overall structure of ejector, the computational cost would be enormous if the same grid was used. The immersion boundary method was used to generate the body mesh and Cartesian grid. The Cartesian grid was encrypted by the Octree method at the nozzle flow path and the small size of the flow feature. The total mesh size was about 1100 Million, the result of the division shown in Figure 3.

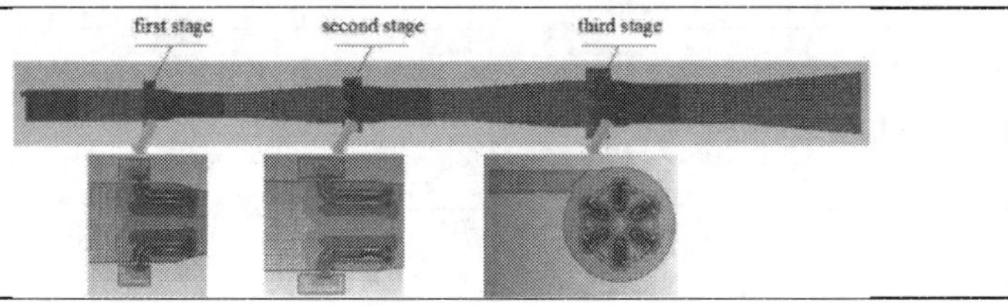

Figure 3. Fluid grid of ejector inner fluid field.

The ejector medium was dry air. In the simulation, the gas was regarded as ideal gas and all structural walls were regarded as adiabatic wall, the remaining boundary conditions were shown in Table 2.

Table 2. The table of boundary conditions.

Condition \ Point	Secondary Gas	Inlet 1	Inlet 2	Inlet 3	Outlet
Total Temperature/K	300	300	300	300	300
Total Pressure/kPa	101.5	200	340	970	101.5

The two-equation renormalization k-εmodel was used in the simulation, which was applied in fast strain, medium vortex and local transition flow filed usually. The results showed that the mass flow rate of the three-stage ejector gas collection chamber were 0.22 kg/s, 1.07 kg/s, and 5.37 kg/s, respectively, and the inlet flow rate of the venturi nozzle was 0.06 kg/s, which were basically consistent with the design value.

Figure 4 showed the pressure and Mach number distributions in the axial section of the ejector. As can be seen from Figure, the Mach number at the exit of the third stage ejector nozzle was about 2.75, and the relative error with the design was 1.79%, which could be found to be 2.8 in Table 1.

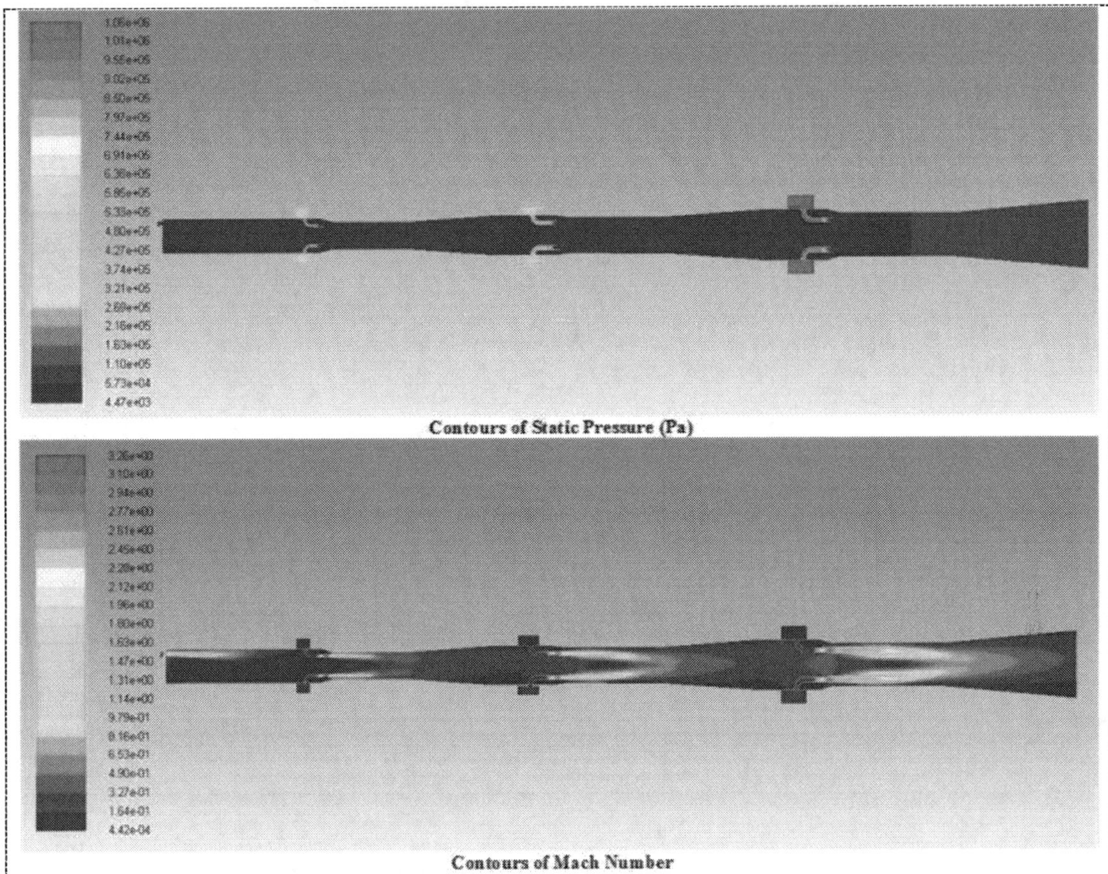

Figure 4. Pressure and Mach number contour of ejector axial section.

4. Flow-Induced Vibration Calculation

The fluid-solid coupling algorithm could be divided into one-way coupling and two-way coupling. When fluid pulsation caused solid structure deformation, if solid structure deformation affected the flow field obviously, it could be regarded as two-way coupling, if the effect could be ignored, it was one-way coupling. The ejector FIV was a one-way coupling problem. When Xi Zhide et al. studied the one-way coupling structure, the fluid-solid coupling dynamic equations of general structure were derived [12]:

$$\Gamma \boldsymbol{x}(X,t) = \boldsymbol{f}(X,t) \tag{1}$$

$$\Gamma = M\frac{\partial^2}{\partial t^2} + C\frac{\partial}{\partial t} + K + \text{r} \tag{2}$$

$$\boldsymbol{f}(\boldsymbol{X},\boldsymbol{t}) = \boldsymbol{f}_s(X,t) + \boldsymbol{f}_f(t) \tag{3}$$

$$\boldsymbol{f}_s(X,t) = -M_a\frac{\partial^2 \boldsymbol{x}(X,t)}{\partial t^2} - C_a\frac{\partial \boldsymbol{x}(X,t)}{\partial t} - K_a\boldsymbol{x}(X,t) \tag{4}$$

$$(M+M_a)\frac{\partial^2 \boldsymbol{x}(X,t)}{\partial t^2} + (C+C_a)\frac{\partial \boldsymbol{x}(X,t)}{\partial t} + (K+K_a)\boldsymbol{x}(X,t) = \boldsymbol{f}_f(t) + \boldsymbol{f}_s(X,t) \tag{5}$$

Γ was the structural operator, its concrete form depended on the shape and motion form of structure. $\boldsymbol{f}(X,t)$was the excitation vector, which could be obtained from flow equation. $\boldsymbol{f}_s(X,t)$ related to the structural motion, named as motion-related fluid force. $\boldsymbol{f}_f(t)$was fluid excitation force, which was independent of the structural motion, only determined by fluid motion. M_a, C_a, K_a was additional mass matrix, additional damping matrix and additional stiffness matrix respectively. The equations had certain applicability, but for different structures, the calculation of the relevant additional matrix still needs to be specifically modeled and analyzed.

The coupling between the fluid model and the structural model was performed using MPCCI software. The ejector structure response spectrum obtained was shown in Figure 5. As can be seen from the figure, the ejector has significant peak response with peak points of 12.2 Hz and 29.9 Hz, while the amplitude could be found as 0.065 g and 0.018 g respectively.

The test spectrum could also be found in Figure 6, the dominant frequency relative error between simulation and experiment was 4.27%. Among them, 12.2Hz close to the second-order natural frequency of the structure, 11.91Hz, and the relative error is 2.43%.

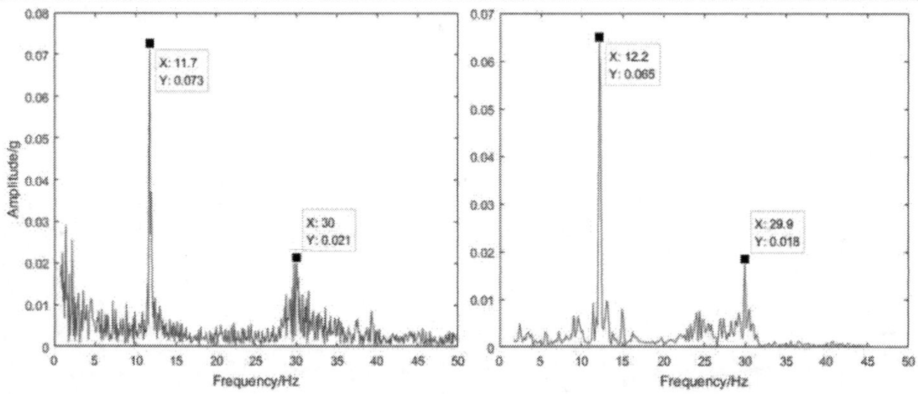

a) test spetrum b) calculated spectrum

Figure 5. FIV spetrum of simulation and experiment.

5. Conclusion

(1) In flow field three-dimensional unsteady simulation, the inlet and outlet mass flow, Mach number were basically consistent with the design. The relative error of the third stage ejector was 1.79% compared with design.

(2) The FIV frequency spectrum of ejector was obtained. The domain frequency relative error was 4.27% with experiment.

(3) The domain frequency calculated closed to the second-order structure frequency, the relative error was 2.43%.

References

[1] Zhao Jingye, Sun Houzhen, Gao Jun. Basic Principles and Applications of Ejector. *Journal of Beijing Institute of Civil Engineering and Architecture*, 2001, 17(3): 12-15.

[2] Liu Z, Shi Y, Chen Z, et al. Wind Tunnel Flow Induced Vibration Characteristic Analysis. *International Conference on Measurement, Instrumentation and Automation*. 2016.

[3] Yang Y, Liu Z, Yan X, et al. Structural vibration analysis of the multi-nozzle ejector. *American Institute of Physics Conference Series. AIP Publishing LLC,* 2017: 12-15.

[4] LIU Peiqi, WANG Haitao, WU Jintao, et al. Optimum Design and Verification of Ejector's main Structural Parameters. *Journal of Dalian University of Technology*, 2017, 57(1): 29-36.

[5] Zhang Jianqiang, Wang Zhenguo, Li Qinglian, et al. One-Dimensional Analysis for Performance of Ejector with Precooling. *Journal of National University of Defense Technology*, 2017, 39(3):1-6.

[6] Ren Zebin, Liao Daxiong, Zhang Guobiao. Aerodynamic Design of the Ejector of 2m×2m Supersonic Wind Tunnel. *Journal of Aerospace Power*, 2014, 29(10): 2288-2293.

[7] Liu Zongzheng, Guo Longde, Lu Jinlei, et al. Ejector Pressure Recovery System's Vibration Control Technology. *Journal of Tsinghua University(Science and Technology)*, 2012(1): 10-14.

[8] MA Yue-yin, WANG Yuan-xing, CHEN Wan-hua. Optimization Design of Transition Part in Pressure Recovery System. *Journal of Machine Design*, 2014(1): 51-54.

[9] Wang Yuxi, Fan Wei, Yan Chuanjun, et al. Experimental Investigation on Ejector Mode of a Pulse Detonation Rocket Engine. *Journal of Projectiles, Rockets, Missiles and Guidance*, 2008, 28(5): 139-141.

[10] NIE Junfeng, LU Qiuhai, WAN Li, et al. Experiments and Simulations for the Fracture Mechanism Induced by Respiratory Vibration of Large Ejector Tube. *Mechanics in Engineering,* 2010, 32(3): 62-67.

[11] NIE Junfeng, LU Qiuhai, ZHANG Haiquan, et al. Dynamic load identification for an ejector tube. *Journal of Vibration and Shock*, 2013, 32(18): 163-168.

[12] XI Zhide, CHEN Bingde, LI Pengzhou. Calculation Method for Flow Induced Vibration in Weak Coupled System. *Atomic Energy Science and Technology*, 2008, 42(2): 444-447.

IOP Conf. Series: Materials Science and Engineering **825** (2020) 012013 doi:10.1088/1757-899X/825/1/012013

Analysis of the Vacuum Chamber Based on the Environmental Simulation Experiment of the Lunar Surface

Yi Zhang*, **Chao He, Lei Zhang, Xiao Han, Qiang Li, Zhifei Gu and Pengsong Zhang**

Beijing Institute of Spacecraft Environment Engineering, 100094 Beijing, China

*E-mail: spcomyth@139.com

Abstract. The vacuum chamber based on the environmental simulation experiment of the lunar surface mainly provides the simulation space for the comprehensive environmental effect tests such as vacuum, low temperature, irradiation and charged dust, and provides the interface for the relevant environmental sources, in-situ testing instruments, etc. In this paper, a set of vacuum chamber for the experiment is provided, and the finite element analysis, surface temperature analysis and radiation dose analysis of the chamber are carried out. The analysis results provide a guarantee for the experiment and a reference for the related research work.

1. Foreword

With the planning and development of the future manned lunar exploration mission in China, it is necessary to carry out the research work of lunar surface environment simulation experiment technology [1]. The lunar surface environment simulation experiment is mainly used to simulate the comprehensive environment and effects of various factors on the lunar surface, such as vacuum, low temperature, electronic radiation, ultraviolet radiation, X-ray radiation, charged dust, etc, and carry out scientific research on the physical theory and effect of lunar surface dust, dust plasma, etc.

The vacuum chamber (VC) is one of the main subsystems of the experimental equipment, which provides simulation space for the comprehensive environmental effect test of the lunar surface, such as vacuum, low temperature, electronic radiation, ultraviolet radiation, X-ray radiation, charged dust, etc [2], and provides interfaces for relevant environmental sources, in-situ testing instruments. In the process of the experiment, the VC is in the environment of vacuum, low temperature, irradiation and dust, so it needs to carry out the finite element analysis, surface temperature analysis, radiation dose analysis to verify whether it meets the requirements of the experimental environment. As long as we pay attention to the protection and cleaning, the dust will not affect the VC, so this paper will not discuss.

2. Structure of VC

According to the experimental requirements, the VC is designed to be vertical, with an inner diameter of φ4m and a straight section of 5m high, supported by six supports. The test piece and personnel enter and exit through the φ3 m gate on the side. Shroud, guide rail and lifting points are installed inside the VC, and interfaces are provided for the thermal system, vacuum system, measurement and control system, relevant environmental sources, in-situ testing instruments, etc [3,4]. Its structure is shown in Figure 1.

IOP Conf. Series: Materials Science and Engineering **825** (2020) 012013 doi:10.1088/1757-899X/825/1/012013

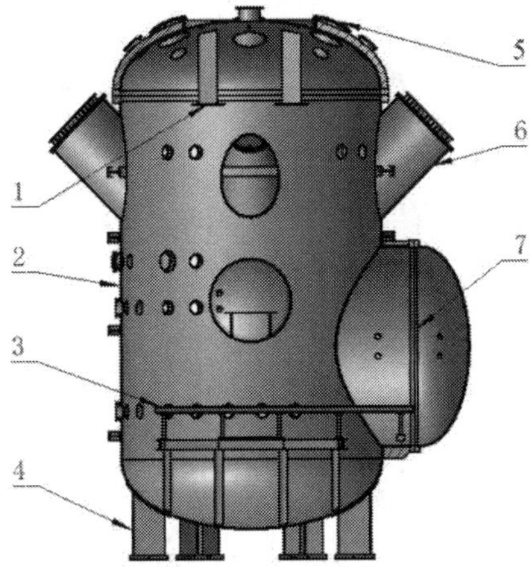

1. Lifting points;
2. Cylinder;
3. Guide rail;
4. Supports;
5. Dn500 interface flanges;
6. Dn1100 interface flanges;
7.φ3 m gate;

Figure 1. Structure of VC.

3. Finite element analysis

The structure of VC is complex, so the traditional mechanical analysis method can't be used to analyse it. However, the finite element method has its unique advantages in the analysis of stress, displacement and anti-instability ability of large VC [2].

Next, we build the finite element model of the VC, and carry out stress analysis, displacement analysis and buckling analysis.

3.1. Applied load

After modeling, set the boundary conditions of the model, constrain the six degrees of freedom of the supports, and set the load conditions of the model:

1) Apply 5 tons of test piece load evenly on the guide rail of the VC;
2) Apply 7 tons of shroud load evenly on the inner wall of the VC;
3) Apply 1 ton of low cryogenic pump load on each Dn500 interface flange;
4) Apply 1 ton load on lifting points;
5) Apply 0.8 ton load on each Dn1100 interface flange;
6) The loading directions described above are all vertical downward, and define the weight of the VC itself;
7) Since the VC is in the vacuum state during the experiment, an external pressure of 0.1MPa is applied to it.

Additional material properties for the model: the material of supports and stiffeners is Q345, and the material of other structures is S30408.

3.2. Stress analysis results

The overall stress distribution of the VC and the location of the maximum stress are shown in Figure 2. The maximum stress is 102 MPa, which appears above the joint of DN1100 interface flange and cylinder, and is lower than the yield strength of S30408 by 205 MPa [5].

3.3. Displacement analysis results

The overall displacement distribution and the position of the maximum displacement of the VC are shown in Figure 3. The maximum deformation displacement is 1.36 mm, which appears on the left side of φ3 m gate.

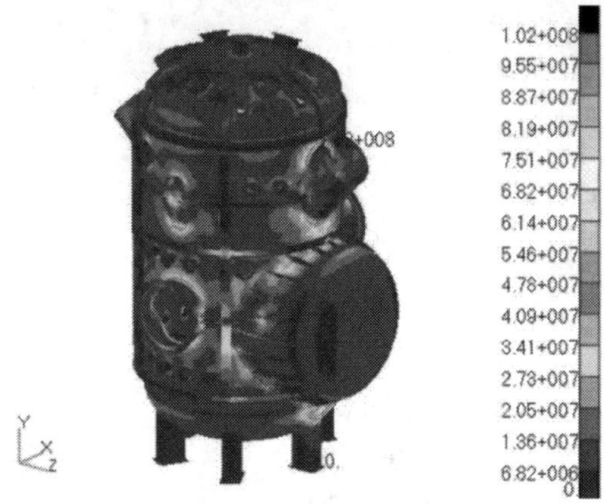

Figure 2. Stress distribution of VC.

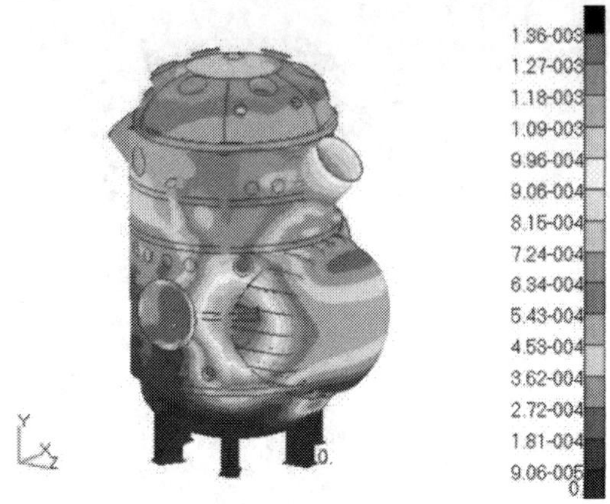

Figure 3. Displacement distribution of VC.

3.4. Buckling analysis results

The VC belongs to external pressure vessel. There are two main failure modes of external pressure vessel: one is the failure caused by insufficient strength, that is, when the compressive stress of the shell reaches the yield strength of the material, the vessel structure will be damaged, but this failure mode is rare; the other is that the shell loses its original shape due to insufficient rigidity, resulting in instability (buckling). For thin-walled vessel, the instability always occurs before the strength failure, so the stability calculation is the primary problem to be considered for the thin-walled vessel [6,7].

Under the given load of the project, extract the natural frequency and mode of the model, conduct linear buckling analysis, and obtain the corresponding critical instability factor of 26.504, as shown in Figure 4. The main load corresponding to the buckling mode is 0.1 MPa external pressure, so the critical load of the VC $P=0.1\times26.504=2.6504$ MPa. Through the linear buckling analysis above, it can be seen that the instability of the structure will not occur under the existing conditions.

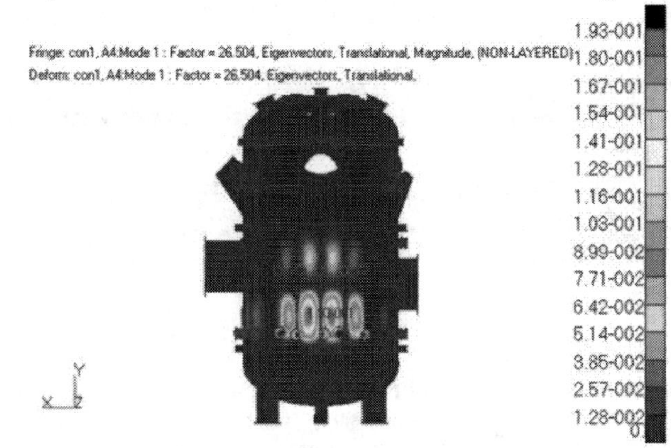

Figure 4. Linear buckling analysis of VC.

4. Surface temperature analysis

The low temperature environment is one of the essential simulation environment conditions in the lunar surface environment simulation experiment equipment [8]. The experimental equipment uses shroud to simulate the cold black environment in space, and provides 100 K low temperature background for the experiment by injecting liquid nitrogen into the shroud. There is a low temperature background formed by shroud in the lower half and bottom of the equipment: the effective space of the cylinder shroud is Φ 3500 mm*3100 mm. The structure and location of shroud are shown in Figure 5.

Next, we will analyse the temperature effect of shroud on the VC. The grid model of heat radiation is shown in Figure 6.

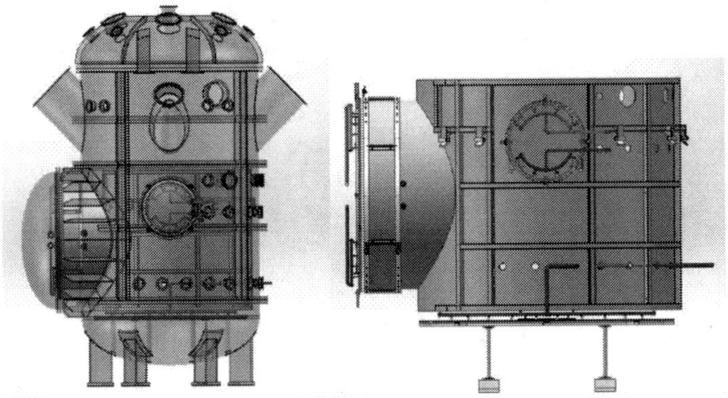

Figure 5. Structure of shroud.

Figure 6. Grid model of heat radiation.

When the natural convection heat transfer coefficient h=2 W/(M^2*K) (ventilation condition is poor) [9], the calculation results are shown in Figure 7. The surface temperature of the lower part of the VC is close to the ambient temperature of 300 K due to the radiation proof screen. The upper part faces the inner wall of the shroud directly, the surface temperature decreases obviously, and the lowest temperature is 280 K (about 7 ℃).

When the natural convection heat transfer coefficient h=5 W/(M^2*K) (the ventilation condition is general), the calculation result is shown in Figure 8. The minimum temperature of the surface of the VC is 290 K(about 17 ℃). Compared with the working condition with poor ventilation condition, the heat transfer efficiency between the VC and the environment is higher, so the temperature is improved.

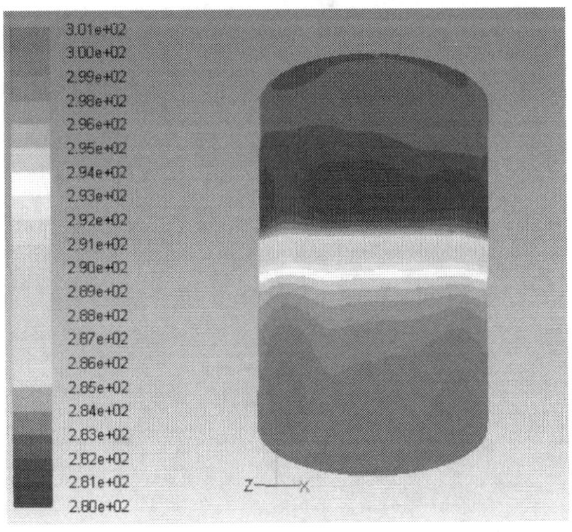

Figure 7. Cloud chart of thermal radiation effect, h=2 W/(M^2*K).

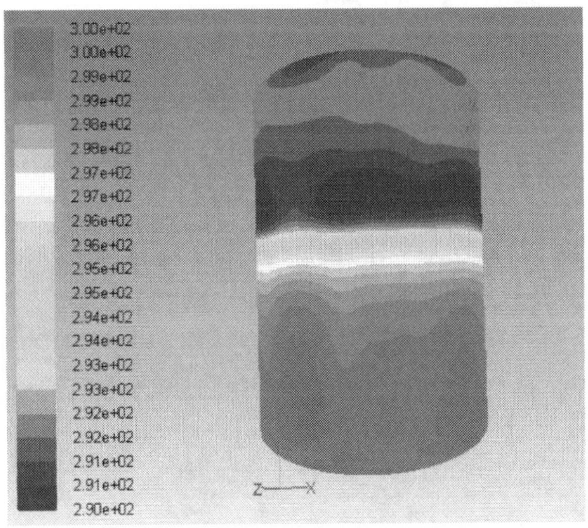

Figure 8. Cloud chart of thermal radiation effect, h=5 W/(M^2*K).

According to the above analysis, when the ventilation condition is poor, the minimum temperature of the VC is 280 K (about 7 ℃); when the ventilation condition is general, the minimum temperature of the VC is 290 K (about 17 ℃).

IOP Conf. Series: Materials Science and Engineering **825** (2020) 012013 doi:10.1088/1757-899X/825/1/012013

5. Radiation dose analysis

5.1. Set conditions for the model

In the simulation experiment of lunar surface environment, the radiation source is 200 keV electron source and 80 kev X-ray. Because the dose rate of X-ray is three orders of magnitude lower than that of the electron source, this paper only analyses the electron source with high dose rate.

The electronic source is installed on the side of the VC, and the angle between its axis and the axis of the VC is 45 °. The simulation of the irradiation area is shown in Figure 9.

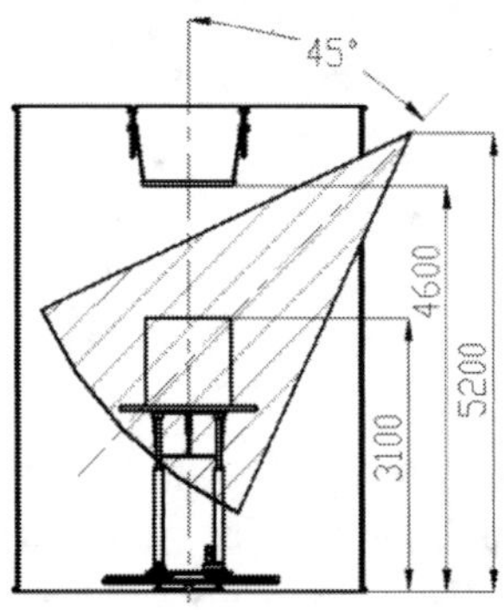

Figure 9. Irradiation area of electron source.

The model conditions of the electronic source are:

1) The range of electron energy is 30-200 kev, and the error of output energy stability is less than ± 0.5%;

2) The maximum current intensity is 15 mA, and the minimum regulation current precision of filament is ≤ 0.1 mA;

3) The size of irradiation surface is 1000 mm * 1000 mm, which is about 3500 mm away from the exit of electron beam.

5.2. Dose rate distribution

The ionizing radiation produced by the electron source is calculated by FLUKA software, and the secondary radiation produced by the electron source is also calculated. Three different ways are used to represent the radiation dose rate distribution inside and outside the VC.

The first is the longitudinal section, and the dose rate distribution of the longitudinal section where the incident angle is located is shown in Figure 10.

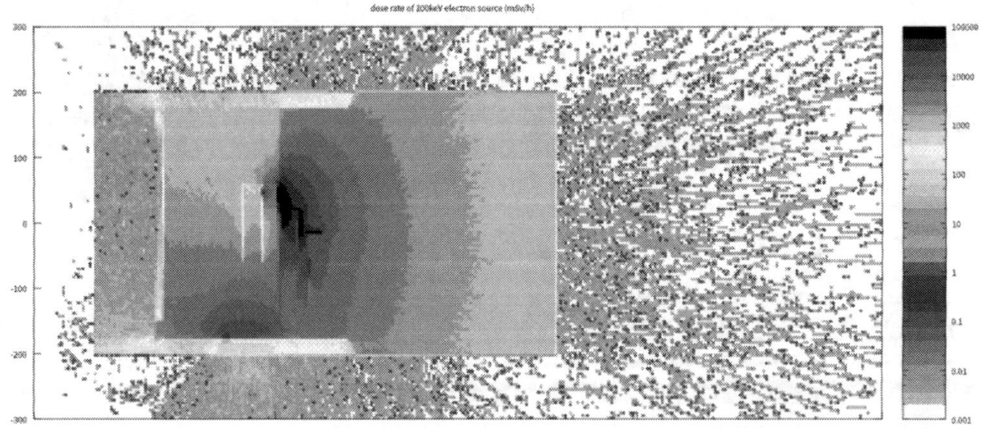

Figure 10. Dose rate distribution of longitudinal section, mSv/h.

The second is the transverse section. Take the transverse section of the key position of the VC for analysis. See Figure 11 for the detailed position and table 1 for the dose rate distribution of each height.

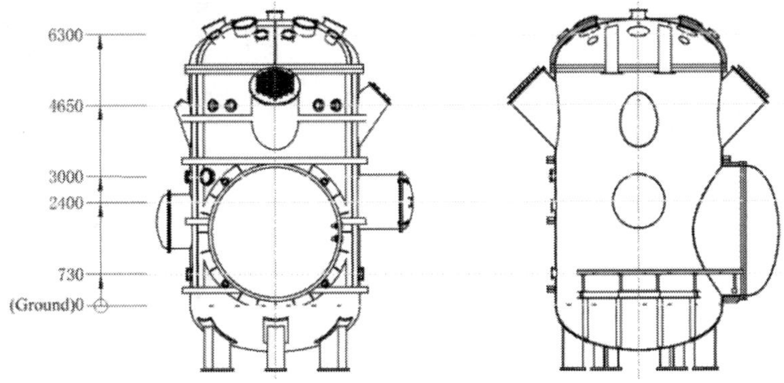

Figure 11. Location of transverse section.

Table 1. Dose rate distribution of transverse section.

Height (mm)	Dose rate outside VC (mSv/h)	Dose rate in VC (Sv/h)	Dose rate distribution
730	0-10	1-10	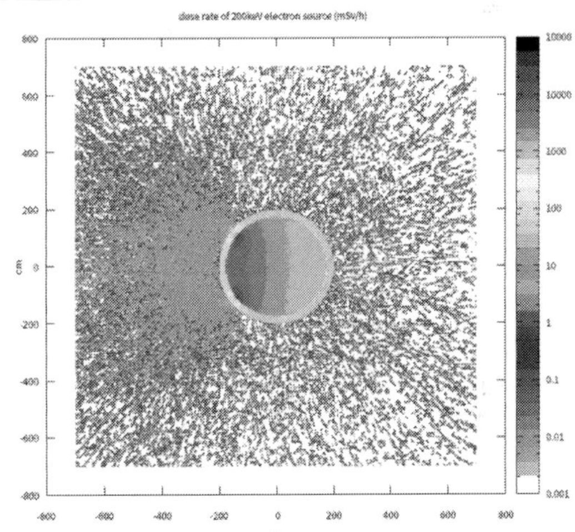
2400	0-10	1-100	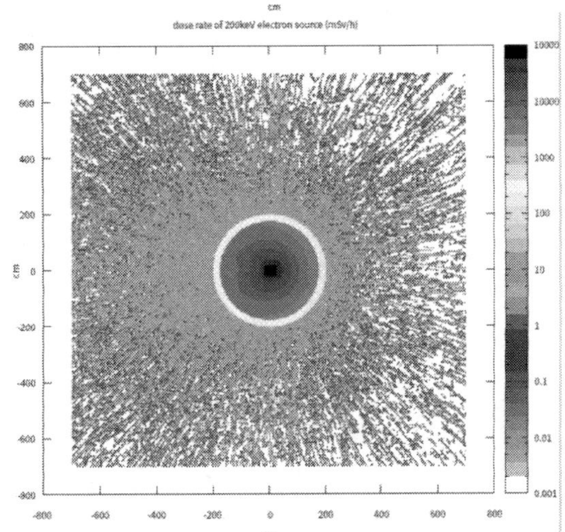

3000	0-10	1-100	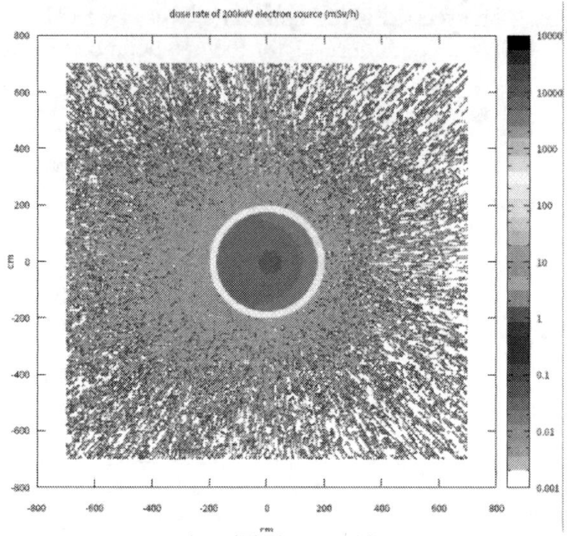
4650	0-1	1	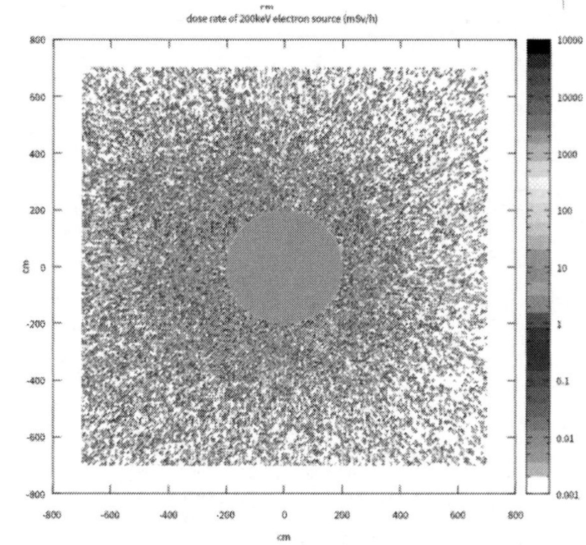
6300	0-1	1	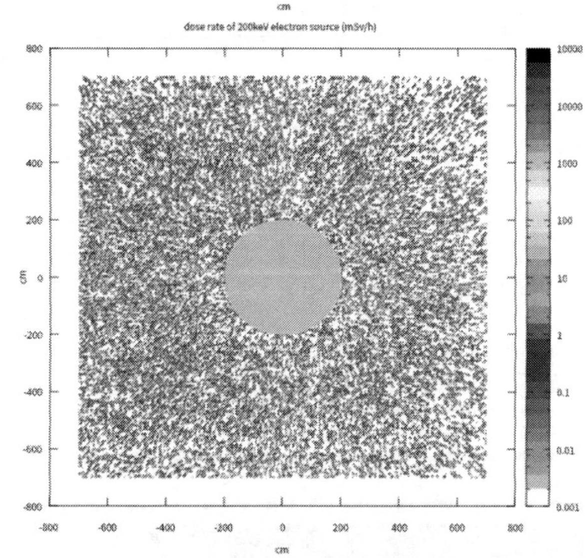

The third is the circumferential section. The dose rate distribution in different radius sections is shown in Figure 12-Figure 14. In the figure, the x-axis is the height and the y-axis is the angle.

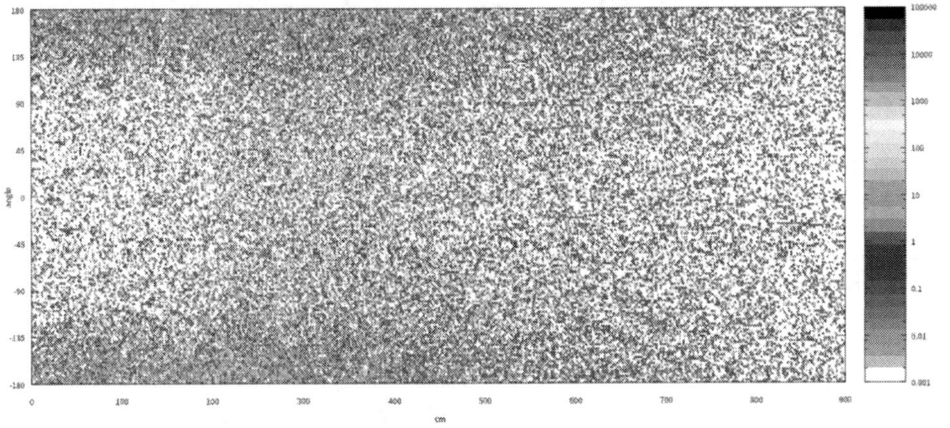

Figure 12. Dose rate distribution of R=7000 mm, mSv/h.

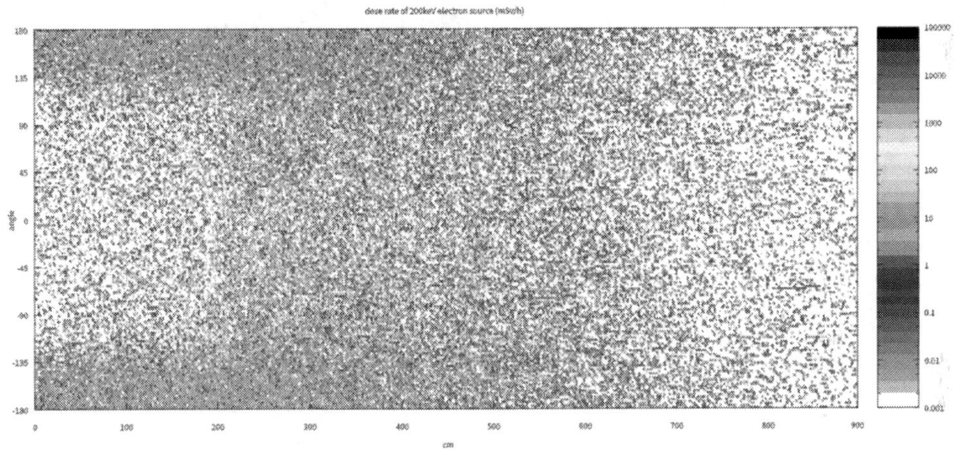

Figure 13. Dose rate distribution of R=5000 mm, mSv/h.

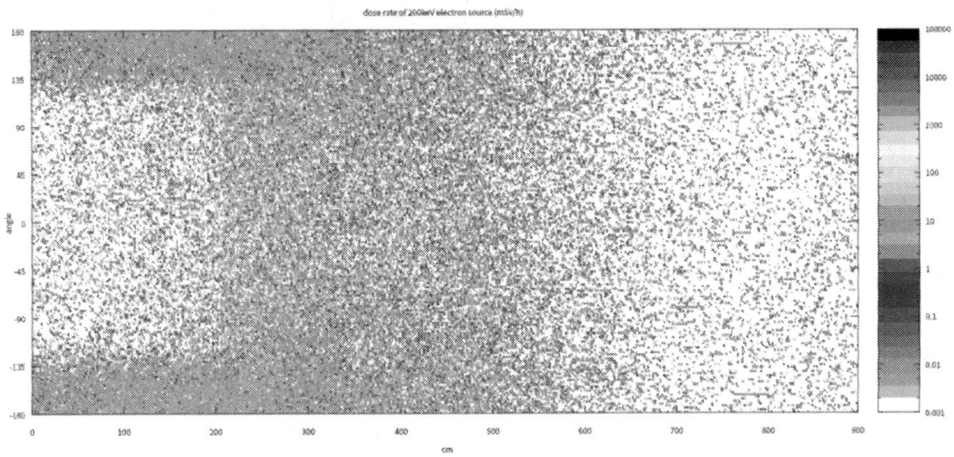

Figure 14. Dose rate distribution of R=3500 mm, mSv/h.

IOP Conf. Series: Materials Science and Engineering **825** (2020) 012013 doi:10.1088/1757-899X/825/1/012013

5.3. Analysis of calculation results

The radiation of VC mainly comes from 200 keV electron source, and the radiation dose rate level in VC is 1-100 Sv/h. Because the stainless steel wall of the VC attenuates the radiation of 200 keV electron greatly, the radiation dose rate level outside the VC is 1-10 mSv/h.

6. Conclusion

This paper introduces a set of VC used in the environmental simulation experiment of the lunar surface, and analyses it. The analysis contents include: stress analysis, displacement analysis, buckling analysis, surface temperature analysis, radiation dose analysis. The analysis results are as follows:

1) The maximum stress of the VC is 102 MPa, which appears above the joint of DN1100 interface flange and cylinder, which is lower than the yield strength of S30408 by 205MPa;

2) The maximum deformation displacement is 1.36 mm, which appears on the left side of φ3 m gate;

3) The critical load of the VC is P=2.6504 MPa, which is greater than the main load of 0.1 MPa external pressure. Therefore, under the existing working conditions, the structural instability will not occur;

4) When the ventilation condition is poor, the minimum temperature of the VC is 280 K (about 7 ℃);

5) The radiation dose rate level in the VC is 1-100 Sv/h, and the radiation dose rate level outside the VC is 1-10 mSv/h.

In conclusion, the VC meets the requirements of the environmental simulation experiment of the lunar surface, which provides a guarantee for the experiment, and the analysis results provide a reference for the related research work [10].

References

[1] Yaoqi Feng, Zhong Yi, Xiyuan Li, and et al 2018 *Key technologies in lunar surface environmental simulation test for manned lunar exploration* (*Spacecraft Environment Engineering* vol 1) p 1-6

[2] Bencheng Huang and Jinming Chen 2005 *Space Vacuum Environment & Vacuum Technology* (Beijing: National Defense Industry Press) p 17-68

[3] Bomin Ding and Renliang Cai 1992 *Pressure Vessel Design* (Beijing: China Petrochemical Press) p 80

[4] Xiao Han, Yan Qi and Botao Liu 2010 *Structural design and optimization for door flange of a large vacuum vessel* (*Spacecraft Environment Engineering* vol 4) p 493-495

[5] GB 150.2-2011 2012 *Pressure vessels-Part 2: Materials* (Beijing: Standards Press of China)

[6] Shizeng Lv, Xiao Han and Yan Qi 2016 *Reliability optimization design of vacuum chamber for extra-large space environmental simulation equipment* (*Chinese Journal of Vacuum Science and Technology* vol 1) p 38-43

[7] Xinming Wang 1986 *Design and calculation of engineering pressure vessel* (Beijing: National Defense Industry Press) p 186

[8] Shouquan Ke, Bencheng Huang, Chuanda He and et al 1993 *Spacecraft environmental engineering and simulation test* (Beijing: Astronautic Press) p 138

[9] Shiming Yang and Wenshuan Tao 1998 *Heat Transfer* (Beijing: Higher Education Press) p 5

[10] Jingyu Tong, Man Li, Yu Bai and et al 2013 *Research for lunar dust effects and its ground simulation methods* (*Chinese Space Science and Technology* vol 2) p 78-83

IOP Conf. Series: Materials Science and Engineering **825** (2020) 012014 doi:10.1088/1757-899X/825/1/012014

Influence of the Clearance of the Transmission Mechanism of the Transmission Characteristics

Jian XU and Siyi FAN

China Academy of Launch Vehicle Technology, Beijing, 100076

776773695@qq.com

Abstract. The transmission mechanism is composed of a bearing cover, a rocker arm cover, a rocker arm, a shaft sleeve, a sleeve, a bushing, and a bearing. For a small space and weak rigidity servo drive mechanism, its clearance has a greater impact on static servo characteristics such as swing angle response and loop width. In this paper, the clearance of the transmission mechanism and the play clearance of the bearing are analysed, and the influence on the servo characteristics is obtained.

1. Introduction

The transmission mechanism is composed of a bearing cover, a rocker arm cover, a rocker arm, a shaft sleeve, a sleeve, a bushing, and a bearing. For a small space and weak rigidity servo drive mechanism, its clearance has a greater impact on static servo characteristics such as swing angle response and loop width. In this paper, the clearance of the transmission mechanism and the running clearance of bearing are analysed, and the theoretical calculation is verified by simulation to obtain the influence of the transmission system clearance on the servo characteristics of the transmission mechanism with small space and weak stiffness.

2. Introduction to Design of Transmission Mechanism

The rudder surface transmission mechanism's schematic diagram is shown in Figure 1. The length of the two connecting axes of the rocker arm is 77.4mm[1]. When the rudder deflection angle is 0 °, the angle between the rocker arm and OYVL in the XVLOYVL plane is 10.3 °.

2.1. Key and keyway size tolerance

Rudder shaft flat key width tolerance is $8^{0}_{-0.01}$mm, key height tolerance is $7^{0}_{-0.01}$mm. The key groove width tolerance on the rudder shaft is $8^{+0.01}_{-0.005}$mm, and the upper hub width tolerance of the flat tail rocker arm is $8^{+0.015}_{0}$mm. The hub and key are matched with a clearance of less than 0.01mm.

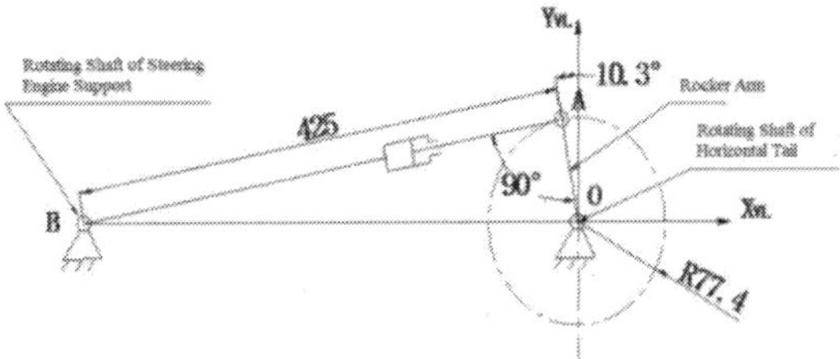

Figure 1. Schematic diagram of transmission mechanism

2.2. Dimension tolerance of electromechanical actuator interface

The electromechanical actuator and the rocker arm are connected to the lugs, the electromechanical actuator and the actuator support are connected to the lugs are all single-ear piece lugs, which are connected by a pin assembly, the pin assembly includes a pin, a nut, and an opening Pin with lug width $10^{0}_{-0.1}$ mm. Uniform specification of connecting pin, pin diameter $\phi 12^{-0.01}_{-0.016}$ mm. The diameter of the hole connecting the steering gear to the rocker arm is $\phi 12^{0}_{-0.008}$ mm, and the diameter of the hole connecting the steering gear to the steering gear base is $\phi 12^{0}_{-0.005}$ mm.

The tolerance diameter of the double ear holes of the pin of the steering gear and the double ear holes of the rocker arm is $\phi 12^{0}_{-0.005}$ mm. The width of the double ears of the rocker arm and the electromechanical actuator base is $12^{+0.2}_{0}$ mm. The centre of the double ear holes is more than 18mm from the inside. The specific size chart is shown in Figure 2.

3. Clearance analysis

Drive steering system clearance composition is shown in Figure 3,it includes[2]:

a) Assembly clearance between rudder shaft, key and rocker arm;
b) Assembly clearance between rocker arm, pin shaft and steering gear;
c) Assembly clearance between steering gear, pin and steering gear base.

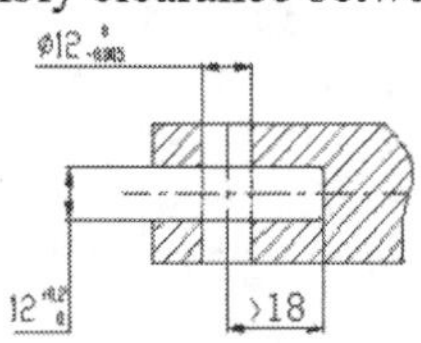

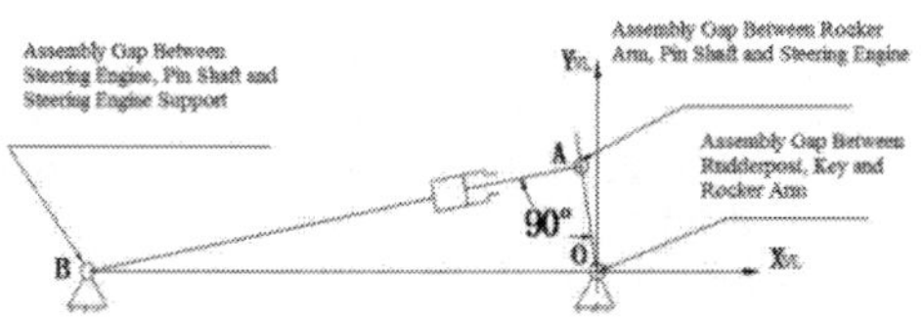

Figure 2. Ear holes dimensional diagram **Figure 3.** Schematic diagram of gap composition

3.1. Assembly clearance between rudder shaft, key and rocker arm

The maximum clearance between key and keyway on the rudder shaft is 0.02mm, the theoretical maximum clearance between key and upper hub of the rocker arm is 0.025mm, the theoretical overall maximum clearance is 0.045mm. During actual assembly process, the key and the hub are matched, and the matching requirement is less than 0.01mm, the maximum gap between shaft key and hub is 0.01mm, the actual overall maximum gap is 0.03mm.

3.2. Assembly clearance between rocker, pin shaft and steering gear

The maximum clearance between the pin shaft and the rocker arm is 0.016 mm, and the maximum radial clearance between the pin shaft and the end bearing of the rudder joint bearing is 0.016 mm.

3.3. Assembly clearance between steering gear, pin and steering gear base

The maximum clearance between the pin and the rudder base is 0.016mm, and the maximum radial clearance between the pin and the single lug end of the rudder is 0.016mm.

3.4. Clearance calculation of servo drive system

3.4.1. The angular deviation caused by the gap of the servo drive system. According to the triangular relationship of the transmission system, the angle caused by the clearance between the pin and the steering gear base and steering gear is: $\theta_1 = arcsin\theta_1 = arcsin(\frac{0.016+0.016+0.016+0.016}{77.4}) = 0.0474°$;The angle caused by the clearance between the key and the rudder shaft and rocker arm is: $\theta_2 = \frac{0.03}{14} = 0.123°$. The angular deviation due to the gap is $\theta_p = 0.0474 + 0.123 = 0.1704°$.According to the clearance analysis of the servo drive system, the limit angular deviation due to the clearance is 0.1704 °.

*3.4.2. Angle deviation caused by load:*The rudder shaft system is shown in Figure 4, where A and B represent the intersection of the cross section and the axis of the rudder shaft support bearing, A is the inner bearing, B is the outer bearing, C is the intersection of the cross section of the rocker arm and the axis, and D is Control the origin of the local coordinate system of the rudder surface[3].

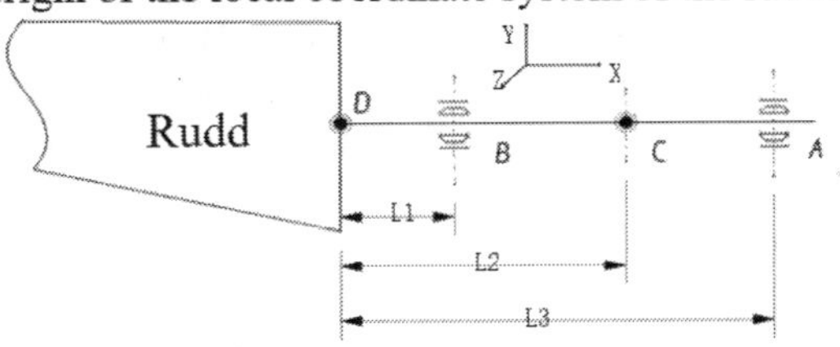

Figure 4. Rudder axis system and rudder axis coordinate system

When a torque is applied through a steel plate and an angular displacement sensor is used to measure the rotation angle, the state of the rudder shaft and bearing is shown in Figure 5.

Table 1. Six-component load.

Items	Fx (N)	Fy (N)	Fz (N)	Mx (Nm)	My (Nm)	Mz (Nm)
1	0	0	0	0	0	380

As shown in Figure 5, point D is the origin of the local coordinate system of a certain control rudder surface. When the load shown in Table 1 is taken, the forces on the internal bearing point A, the external bearing point B, and the point C of the rocker arm are shown in Table 2.

Table 2. Force of rudder shafting

Items	Point A		Point B		Point C	
	Fy (N)	Fz (N)	Fy (N)	Fz(N)	Fy (N)	Fz (N)
1	3133	645	1876	451	-5036	-951

Due to the existence of the gap, the structure is deformed under the load of point D under the load of points A, B, and C. Due to the large stiffness of the rudder shaft, the relative deformation of A, B, C, and D is finally formed as shown in Figure 6.

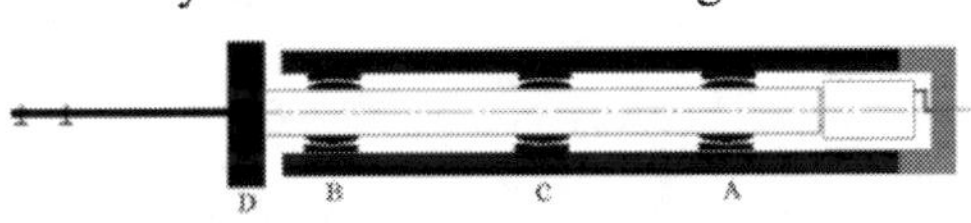

Figure 5 Unloaded rudder shaft and bearing **Figure 6** Deformation rudder shaft and bearing

When the AB distance is 70mm, if the A point does not move, the B point is deformed by 1mm along the Y or Z direction due to the gap, and the rudder axis deflection angle is 0.82 °. Therefore, the gap of the transmission mechanism becomes larger due to loading[4], which will cause the swing angle of the rudder shaft relative to the theoretical position, resulting in excessive deviations in indicators such as the loop width.

3.5. summary

According to the clearance analysis of the servo drive system, the limit angular deviation due to the clearance is 0.1704°. Due to the large gap of the transmission mechanism due to loading, the rudder axis deflection angle was 0.82 °.The total deflection angle caused by the clearance of the transmission mechanism is 0.9904 °.

IOP Conf. Series: Materials Science and Engineering **825** (2020) 012014 doi:10.1088/1757-899X/825/1/012014

4. Analysis of influence of bearing play clearance on measurement angle

4.1. Theoretical analysis

In the case of shaft deflection, the direction of the axial travel gap is uncertain, and the internal and external bearings of the radial clearance are reversed[5]. Based on the local coordinate system identified in Figure 4, the clearance and tolerances affect the shaft deflection. The calculation results of the first 4 items in Table 3 are theoretical conditions. The main factors include the axial and radial clearance of the inner and outer bearings. The axial clearance of the outer bearing is 0.25 mm, the axial clearance of the inner bearing is 0.2 mm, and the radial bearing of the outer bearing The gap is 0.06mm, and the inner bearing is 0.05mm. The theoretical maximum gap between the inner bearing and the mount is 0.036mm. The axial clearance can move positively or negatively along the X axis of the local coordinate system, so there are four cases after permutation and combination.

If the outer bearing has excessive axial running clearance, take 2mm as an example, according to the normal maximum radial running clearance of 0.06mm, double the normal maximum radial running clearance of 0.12mm, and the normal maximum radial running clearance. Calculate 3 times 0.18mm and 4 times 0.24mm of normal maximum radial play clearance. At the same time, the maximum theoretical clearance between the inner bearing and the mounting seat is 0.036mm, and the calculation is shown in Table 3.

Table 3. Rudder shaft deflection calculation

Items	External Bearing		Internal Bearing		Off-axis angle	Radial clearance of outer bearing
	X	Y	X	Y		
1	-2	0.06	70.3	-0.086	0.12°	0.06mm
2	-2	0.06	70.7	-0.086	0.12°	0.06mm
3	2	0.06	70.3	-0.086	0.12°	0.06mm
4	2	0.06	70.7	-0.086	0.12°	0.06mm
5	-2	0.12	70.3	-0.086	0.16°	0.12mm
6	-2	0.12	70.7	-0.086	0.16°	0.12mm
7	2	0.12	70.3	-0.086	0.17°	0.12mm
8	2	0.12	70.7	-0.086	0.17°	0.12mm
9	-2	0.18	70.3	-0.086	0.21°	0.18mm
10	-2	0.18	70.7	-0.086	0.21°	0.18mm
11	2	0.18	70.3	-0.086	0.22°	0.18mm
12	2	0.18	70.7	-0.086	0.22°	0.18mm
13	-2	0.24	70.3	-0.086	0.26°	0.24mm
14	-2	0.24	70.7	-0.086	0.26°	0.24mm
15	2	0.24	70.3	-0.086	0.27°	0.24mm
16	2	0.24	70.7	-0.086	0.27°	0.24mm

As shown in Table 3, when considering the play clearance, the deflection angle of the rudder shaft should be about 0.12 °-0.27 °, which will cause the rudder shaft to sway from the theoretical position, resulting in excessive deviations in indicators such as the loop width.

4.2. Simulation analysis
The effect of bearing play clearance on the measurement angle was calculated by simulation analysis[4].The overall finite element model is used as the analysis object, and the influence of the body structure on the stiffness and force transmission characteristics of the inner and outer bearing supports is fully considered. The local structure of the bearing support is also simulated by shell elements.

There are three cases to discuss the effect of bearing play clearance. In each case, bearing deformation is not considered[3]. By default, the rigidity of the bearing itself is much greater than the overall rigidity of the support and the body structure.

4.2.1. Without considering the influence of the rudder shaft.. Without considering the influence of the rudder shaft, the maximum deformation occurs at the inner bearing support, and the maximum deformation is 0.189mm, as shown in Figure 7.

4.2.2. With the effect of rigid rudder axis. Temper the rudder shaft and calculate the deformation. The analysis result is shown in Figure 8. The maximum deformation of the structure appears at the outer edge web of the inner bearing, and the maximum deformation is 0.048mm.

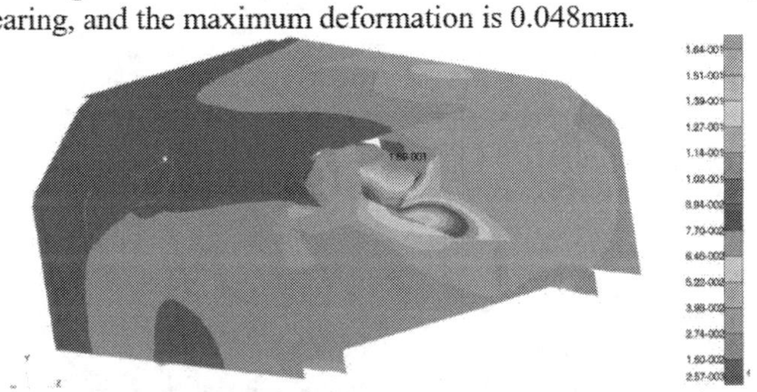

Figure 7. Deformation without rudder shaft influence

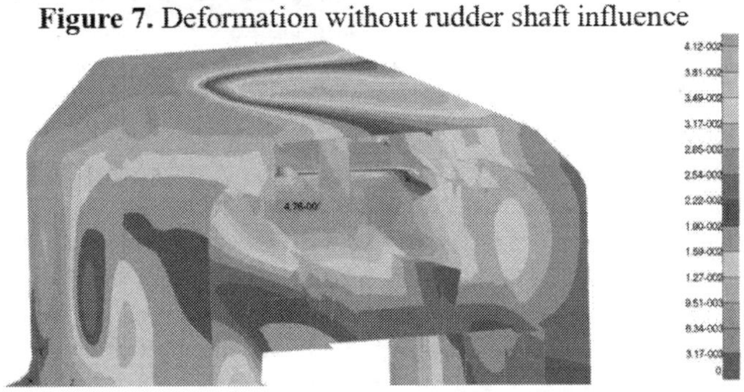

Figure 8. Deformation of rudder shaft

4.2.3. With the influence of elastic rudder axis. Considering the influence of the rudder shaft stiffness, the analysis is performed. The rudder shaft cross section is simulated with a variable cross section. The outer diameter of the inner bearing is 40 mm, the inner diameter is 26 mm, and the length is 35 mm. 52mm, at the connection with the outer bearing, the outer diameter is 50mm, the inner diameter is 26mm, and the length is 47.5mm. The calculation result is shown in Figure 9. The structure's maximum deformation value 0.05mm occurs at the outer edge of the inner bearing.

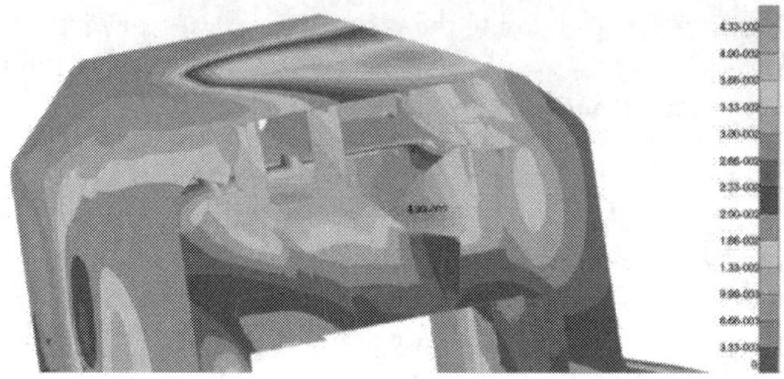

Figure 9. Structural deformation affected by elastic rudder shaft

4.3. Summary

Combining the deformation of the inner and outer shaft positions under the three analysis conditions, it is concluded that the corresponding rudder shaft deflection angle should be about 0.15 ° -0.32 °, which is basically consistent with the numerical analysis. This deflection angle caused by the running clearance of the bearing will cause a sway angle of the rudder shaft relative to the theoretical position, which will lead to excessive deviations in indicators such as the width of the loop.

5. Conclusion

In this paper, a clearance analysis is performed according to the design requirements, bearing characteristics, selection, and design of the transmission mechanism. The transmission clearance of the entire system is calculated. The measurement limit angular deviation caused by the clearance is analyzed. Come to the following conclusions:

1) For a transmission mechanism with stiffness in a small space, the clearance of the transmission system is small, which translates into a change of angle of 0.9904 °

2) For a transmission mechanism with a rigidity in a small space, the play clearance on one side of the bearing has less deformation on the rudder shaft, which translates into a change in the angle of 0.15 ° -0.32 ° and an overall change of 0.3 ° -0.64 °

Based on the above two angle changes, the angle change caused by the clearance of the transmission system and the running clearance of the selected bearing is 1.2904 ° -1.6304 °, which causes the rudder shaft to produce visible vibrations, and has impact on static servo characteristics such as swing angle and loop width.

References

[1] Zhao Shanshan, Comparative Analysis on the Aircraft Stage Separation Based on Rigid Guided-Contact Model, Missiles and Space Vehicles, 2013

[2] Yang Li,Modeling and Optimization of Composite and Servo-driven Actuator for Blank-holder Force Control,Journal of Mechanical Engineering,2015

[3] Zhang Peng, Design and analysis on dynamic characteristic of servo mechanism's load simulation system, Journal of Beijing University of Aeronautics and Astronautics, 2014

[4] Zhou Yanhua, Type Synthesis and Optimization of Main Driving Mechanism for Servo-punch Press, Journal of Mechanical Engineering, 2015

[5] XU Zhenhua, Analysis of Elbow Servo Crank Mechanism, Equipment Manufacturing Technology, 2017

Immersive Virtual Reality for Fire Safety Behavioural Skills Training via Gesture-based Technology

Yaqin Fu[1] and Qi Li[1]

[1]School of Art and Design, Shanghai University of Engineering Science, China

qili@sues.edu.cn

Abstract. The fire safety can be one of the biggest problems in metropolitan area. To reduce injuries in a fire, the virtual reality based behavioural skills training (BST) has been used to train people to survive in emergency circumstance. VR-based BST training has been demonstrated to effectively help people of different ages to acquire safe escape skills in a virtual environment. However, most of people have encountered difficulty to train in a right gesture or poor body performance, such as running at unsuitable body position. Gesture-based technology can effectively help people to learn the correct gestures in order to enhance training efficiency. Thus, this paper proposes a theoretical framework of "VR-Gesture BST", which is expected to improve behavioural skills training more effectively for fire safety training. It adopts motion capture device obtaining user's body data to evaluate the body performance during the BST training. The outcomes of the study provide guideline and suggestions for VR based design in practice of the behavioural skills training adopted gesture-based technology.

1. Introduction

Fire hazard in high buildings is determined by the structure, size of building, and the complexity of domestic electric power etc. One of the effective ways to reduce damages from a fire is to adopt behavioural skills training to improve the people's fire emergency management and escape ability [1]. Research has demonstrated that behavioural skills training (BST) can be effective for developing a range of safety skills, by employing instruction, modelling, rehearsal, praise and corrective feedback [2]. BST program is also applied "for teaching personal safety skills to individuals from various age groups and skill levels for abduction, firearm safety, sexual harassment, wild animal injuries, and fire safety" [3-5]. However, some studies argue that the skills developed through behavioural skills training cannot be used effectively in hazardous situations [6]. Other studies suggest more innovative approaches are required to teach behavioural skills for safety [6].

BST training requires as close to the real environment as possible during the process of training and VR can create a "real" environment for behavioural skills training. The "real" environment enables situational learning. Situated learning theory emphasizes the interaction between knowledge and situation, and it suggests that learners are actively construct knowledge within a real environment [7]. For example, VR technology has been used as an innovative way to create an effective learning environment for behavioural skills training for teaching children the fire safety skills [8-9].

The learning environment created by VR technology can be used as a "real" learning environment, which "contains rich and colourful learning resources and presents more interactive activities" [7]. Compared to the traditional training methods, the virtual reality learning environment is more likely to highly attract the attention of the trainees and provides them positive feedback in real time [10]. It effectively enhances the motivation of the trainees as they continue to learn, in particular in family situation [11].

However, fire safety training faces some problems to obtain effective and satisfying training, such as incorrect body gesture or poor body performance during the behavioural skills training. The training may take more time to complete and it also lacks an effective method to evaluate the body performance. Therefore, it is critically important to improve body correct gesture during fire safety skills training.

2. Virtual reality Based Fire Safety Training

Traditional fire safety training can be dangerous and a high cost. The advantage of VR is to simulate the fire scene, so that the trainees can immerse themselves into virtual environment. The trainees can actively interact with the learning environment through exploring the virtual environment, which makes easier for the trainees to construct knowledge [12]. The "reality" of VR allows for immersion, interaction and imagination, which in turn facilitate trainees' active construction of knowledge [13]. The knowledge construction supports the obtaining of knowledge and its transformation to solving practical problems.

The learning environment provided by virtual reality has been used to train professionals to work in complex environments, such as fire safety in mines [14-15]. Smith and Ericson [16] suggest that VR technology was not only used to simulate "real" fire scenes, but also to help the trainees learn about knowledge of fire safety. Interactive design has been used to encourage trainees to explore the virtual world. However, there are few studies on fire safety skills training for evaluate body performance in a virtual environment. Therefore, this study applies gesture based technology along with virtual reality technology to fire safety training and also applies behavioural skills training to improve the trainee fire safety ability and correct performance.

2.1. Behavioral Skills Training

Behavioural skills training are considered as an effective way to correct behavioural response. It consists of four steps: instruction; modelling; rehearsal; and feedback (Figure 1). The instruction involves trainers demonstrating the correct behaviour to the learner. Modelling provides targeted description of specific activities. Rehearsal has learners practice the correct behaviour after the instruction and modelling have been completed. This should be carried out as far as possible in a real situation or in a real simulated environment, to promote the behaviour being generalised. Feedback is provided to the learner on the outcome of rehearsal, which should include praise and corrective enhancements, including corrections to the errors if it is necessary. Further instruction on how to improve is provided.

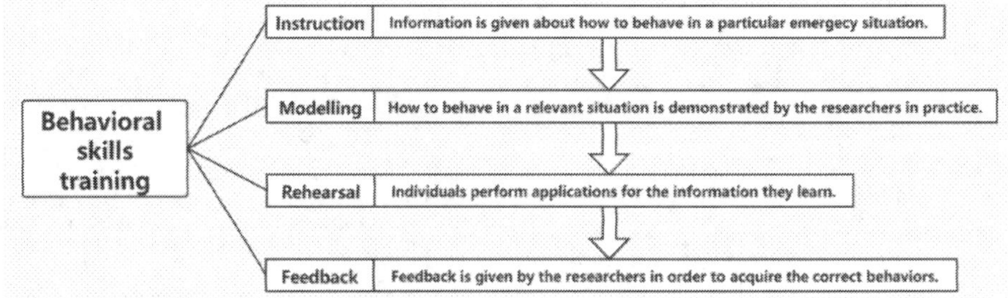

Figure 1. The behavioural skill training model

Behavioural skills training emphasizes the repeatability of training and requires continuous instruction, modelling, rehearsal and feedback processes until the correct behaviour is able to be demonstrated consistently.

2.2 VR-BST Fire Safety Training

Çakiroğlu and Gökoğlu [17] propose a behaviour skills training (VR-BST) model based on virtual reality technology to teach children basic behavioural skills for fire safety. The four steps formed the basis of design; VR is a technical means to provide children with the perception of real experience and opportunities for self-learning, using In situ training (IST) and In situ assessment (ISA). The study adopted design-based research methods including analysis, design and evaluation [17]. The results showed that the fire safety training of virtual reality technology developed within the framework of VR-BST method, could demonstrably improve the behaviour skills of children, and that this positive effect was enhanced when VR-BST was combined with IST. Most children were able transfer the skills they acquired in VR-BST training to the real world [18].

Although the VR-BST model may effectively improve fire safety behaviour skills for children, it lacks a method to evaluate trainee body performance, such as correct gesture, speed, reaction and ability to handle fire equipment. Therefore, this paper proposes to add gesture based technology to VR based behavioural skills training to train fire safety skills.

3. Virtual reality and gesture-based technology

Gesture-based technology with VR is associated with entertainment, education, immersion, interaction and imagination (Figure 2). Gesture-based technology has the advantage of facilitating and enhancing interactivity and immersion, in particular in manipulating with body gesture. It can enhance the learning experience through highly immersive human-computer interaction with a simulated "real" environment. Additionally, gesture-based technology enables trainee manipulating the avatar with their body movement, which is considered as natural movement. It also can be correct the trainee's body gesture and position, which are required from the professional fire safety training. The correct body performance in behavioural skills training improves the effective of the training. Therefore, this training method, which combines VR technology with gesture-based technology, has advantage over traditional training methods. This VR-Gesture framework needs to design the behavioural skills training structure, and then apply it to an immersive virtual environment afforded by VR devices, to provide scenarios for the practice of fire safety self-rescue procedures in high-rise apartments.

In this paper, VR technology is used to design and develop a virtual environment. Under the framework of behavioural skills training, it provides the fire safety knowledge and training required for high-rise apartments. It can help stimulating learners' motivation by providing an enhanced learning experience through situational learning and promoting knowledge transfer to the handling of a fire safety emergency.

The development of virtual reality technologies has made it possible to develop highly realistic fire escape environment for high-rise apartment in metropolitan areas. It suggests that some new models could be adopted to construct a learning environment based on virtual reality in order to provide fire safety training. With this in mind, this study will apply behavioural skills training methods to teach correct behaviour in special circumstances. As the framework for design and implementation, a model of "VR-Gesture BST" is proposed for fire safety skills training. This model can be used to most of trainees, including both professional and amateur fire safety training.

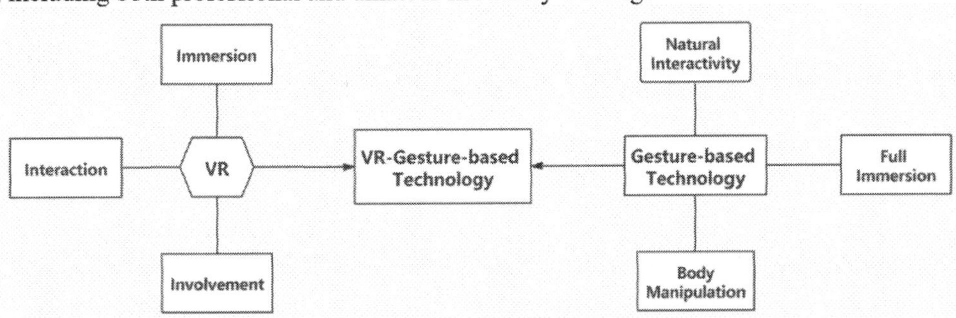

Figure 2. The framework of virtual reality-Gesture based technology.

In this study, Perception Neuron motion capture device from Noitom was adopted and tested [19]. The research methods and outcomes of evaluation will be discussed in the following.

4. Research Methods

The fire safety training on the VR-BST model through gesture-based technology was designed to transform educational theory into teaching practice, using design research to apply theory to the development of a real environment for education and training. McKenney, Nieveen and Akker [20] argue for a process of analysis, design and evaluation during the development of the fire safety training. The three steps are a cycle that initially requires analysis in order to apply theory to a problem, followed by a design stage, involving modelling, testing, evaluating and revising the approach for an improved product and stronger learning outcomes. The final step involves summative evaluation of the whole process.

The behavioural skills training proposed through the VR-Gesture BST framework was tested with ten adults who were all living in high apartments in Shanghai, China. The study adopted qualitative research methodology with interview and questionnaires methods to collect data. A Pilot study was implemented to test the users' immersion and perception in the VR and motion capture device.

4.1 Participants

Ten adults aged between 18 and 50 years old were recruited for the study. Computer games, virtual reality and fire experience were considered as main elements when designed and implemented the proposed VR-Gesture BST framework. Participants were selected based on their experience of fire escape, prior knowledge of fire safety training, body movement ability and health conditions. Participant information is listed in Table 1.

Table 1. Information about the participants

Participants	Gender	Age	Fire safety training experience
P1	F	20	Yes
P2	F	21	No
P3	M	20	No
P4	M	19	No
P5	M	30	Yes
P6	F	28	Yes
P7	M	21	No
P8	M	25	No
P9	F	34	Yes
P10	M	41	Yes

4.2 Interview

Interview has been used to collect data on trainee's experience with the VR-BST approach with motion capture device. The experience of participants with their perceptions of the VR-BST was digitally voice recorded and each interview was about 20 minutes. The interview questions were designed for perception of interactive with the virtual environment for immersion.

4.3 Pilot study

The pilot study was conducted with 20 participants. Participants received training for fire safety and then they were asked to complete some tasks. Participants were advised to explore the virtual environment and find the emergency alarm, moving to the stairs as soon as possible. Participants' perceptions of the virtual learning environment was evaluated a questionnaire. After the completed the tasks, a questionnaire was used to evaluate the quality of immersion within the environment. Witmer, Jerome and Singer [21] developed a "Presence Questionnaire" (PQ) as a method for testing perceived levels of immersion in virtual reality environments. The term "Presence" refers to, "the effect felt when controlling real-world objects remotely as well as the effect people feel when they interact with and immerse themselves in VEs" [22]. It is an ideal way to test the participants' perception of the VR environment. The outcome of the questionnaire suggests that all participants had a high degree of presence in the virtual environment. They stated that they had perceived a sense of realism.

4.4 Data collection

Data from the Perception Neuron motion capture system were collected for analysis of body performance. The results of the study show that most of the trainees have a correct performance after the training. However, their fire safety skills can be improved based on these body data. Therefore, it suggests that gesture-based technology can effectively collect data of the trainee performance. It could be used for further training, in particular in professional fire fighter training.

5. Conclusion

This study proposed a framework combined VR-BST approaches adopted motion capture device to teach the behavioural skills. The results from the study demonstrated that behavioural skills training can be enhanced through VR based design within the framework of a VR-Gesture BST approach. The study showed that motion capture system has positive outcomes in improving training performance. The training could improve the correct responses and action to identify fire hazards in a real life situation. The immersion generated by the virtual environment and the sense of presence can be critical factors in enhancing behaviour skills. The results of the study demonstrated that the proposed framework of the fire safety virtual reality technology could be an alternative approach to increasing the learning ability through behavioural skills training. The motion capture device used in VR based BST training not only demonstrates more accuracy in natural body manipulation but also help to evaluate the action and body gesture for the improvement of the performance. It suggests that the findings from this study can be applied to further research on virtual learning environment using motion capture device under the proposed theoretical framework.

Acknowledgement

This project is sponsored by Shanghai Pujiang Program (18PJ068)

References

[1] Wang, B. (2009) The current situation of fire safety and fire protection of the elderly apartment. Fire Science and Technology,28 (07):537-540.

[2] Himle, M. B., & Miltenberger, R. G. (2004). Preventing unintentional firearm injury in children: The need for behavioural skills training. Education & Treatment of Children, 27(2), 161–177.

[3] Rosales, R., Stone, K., & Rehfeldt, R. A. (2009). The effects of behavioural skills training on implementation of the picture exchange communication system. Journal of Applied Behavior Analysis, 42(3), 541–549.

[4] Mazo, A. D. (2014). Using behavioural skills training to teach medication safety skills to children (Unpublished master's thesis).USA: Saint Louis University.

[5] Houvouras, A. J., & Harvey, M. T. (2014). Establishing fire safety skills using behavioural skills training. Journal of Applied Behavior Analysis, 47(2), 420–424. https://doi.org/10.1002/jaba.113.

[6] Beck, K. V., & Miltenberger, R. G. (2009). Evaluation of a commercially available program and in situ training by parents to teach abduction-prevention skills to children. Journal of Applied Behaviour Analysis, 42, 761–772.

[7] Lian Technology. (2018), Statistics of National Fire Accidents from January to August 2018. https://www.sohu.com/a/254299018_582914 Accessed on July 10, 2019.

[8] Stewart, K. K., Carr, J. E., & LeBlanc, L. A. (2007). Evaluation of family-implemented behavioural skills training for teaching social skills to a child with Asperger's disorder. Clinical Case Studies, 6(3), 252–262.
[9] Kelso, P. D., Miltenberger, R. G., Waters, A., Egemo-Helm, K., & Bagne, A. G. (2007). Teaching skills to second and third grade children to prevent gun play: A comparison of procedures. Education & Treatment of Children, 30(3), 29–48.
[10] Park, K.-M., Ku, J., Choi, S.-H., Jang, H.-J., Park, J.-Y., et al. (2011). A virtual reality application in role-plays of social skills training for schizophrenia: A randomized, controlled trial. Psychiatry Research, 189(2), 166–172.
[11] Liu, H., & Liang, G. (2011) Discussion on the Technical Performance of Family Fire Emergency Escape Equipment. Fire Science and Technology. 30(05):425428.
[12] Li, L., Wang, P. & Zeng, R. (2018) Comparative study on VR devices for virtual environment. Journal of Hefei University of Technology, 41 (2).
[13] Zhou, K. (2013) Media Construction of Virtual reality: a Perspective of the History of Media Technology. Modern Communication: Journal of Communication University Of China, (8), 29/33.
[14] Qiu, S., Ren, H., & Yin, J. (2018) Ship life-saving training system based on virtual reality technology. China Navigation, 41 (2).
[15] Zhu, H., Qin, X., Yang, C., and Nan, C. (2014) The development of mine fire rescue VR training system based on Virtools. Mining research and development, 34 (04):95-98.
[16] Smith, S., & Ericson, E. (2009) Using immersive game –based virtual fire reality to teach fire-safety skills to children. Virtual Reality, *13*(2) , 87-99.
[17] Çakiroğlu,Ü., Gökoğlu, S. (2019). Development of fire safety behavioural skills via virtual reality. COMPUTERS & EDUCATION,133,56-68.
[18] Zhu, D. (2018) A preliminary study on the development prospect of the combination of virtual reality technology and educational games. The Contemporary Film, (2):164-167.
[19] Noitom, (n.d.) https://www.noitom.com.cn/PN/195.html
[20] McKenney, S. Nieveen, N. and Van den Akker, J. (2006). Design research from a curriculum perspective. In J. Van den Akker, K. Gravemeijeijer, S. McKenney, & N. Nieveen (Eds.). Educational design research (pp. 110-143). London: Routledge.
[21] Witmer, B. G., Jerome, C., & Singer, M. J. (2005). The factor structure of the presence questionnaire. Presence, 14(3), 298–312. https://doi.org/10.1162/105474605323384654.
[22] Sheridan, T,B (1992) Musing on telepresence and virtual presence. Presence: Teleop Virt 1:120-125

IOP Conf. Series: Materials Science and Engineering **825** (2020) 012016 doi:10.1088/1757-899X/825/1/012016

Modeling and Analysis of Micro-cantilever Plate Piezoelectric Energy Harvester with a Tip Mass

Li Hua Chen[1], Chao Ran Hou[1], Jiang Tao Xue[1], Lu Tong Shan[1] and Feng Hong Yang[2,*]

[1] College of Mechanical Engineering, Beijing University of Technology, Beijing, 100124, PR China
[2] School of Statistics and Mathematics, Central University of Finance Economics, Beijing, 100081, PR China

[*] Corresponding author: yangfh_cufe@163.com

Abstract. With the development of MEMS (Micro-electro-mechanical Systems) techniques, research interest has deepened in micro piezoelectric harvesting device, which can supply electrical power for wireless sensors from ambient vibration. This paper proposes a nonlinear analysis of a micro-scale PEH (piezoelectric energy harvester), which is modeled by a micro-cantilever plate with two layers and a tip mass attached to the free end. Considering size effect and nonlinear curvature, based on strain gradient and inextensible plate theory, nonlinear dynamic equations of the micro cantilever piezoelectric energy harvester with tip mass are established by virtue of the Hamilton's principle. The maximum output voltage of piezoelectric energy harvester with different tip mass is calculated respectively using MATLAB software. Comparative analysis confirms that an increase of K (the mass ratio of the tip mass to piezoelectric plate) reduces the natural frequency of energy harvester and significantly enhances the nonlinear phenomenon. In addition, when the mass ratio K exceeds 1, obvious superharmonic resonance phenomena generates, which considerably improves the output voltage of the proposed PEH in a lower frequency band. This study proposes an approach to enhance the output performance of micro-cantilever plate energy harvesters and broaden the bandwidth of PEH to adapt low-frequency ambient vibration.
Keywords: Micro-cantilever plate, Piezoelectric energy harvester, Strain gradient, Nonlinear

1. Introduction

In recent years, MEMS technique has developed rapidly. Many researchers have focused on micro-size energy harvester, which has the advantages of long service life and low cost [1]. The additional tip mass is an effective way to improve the performance of the energy harvester, which can significantly increase the amplitude and generate more electricity through the electro-mechanical coupling effect of piezoelectric materials. In addition, the tip mass has the function of adjusting the natural frequency of the structure to adapt to a wide range of external excitation environments [2].

Although many researchers have studied Piezoelectric energy harvesters with cantilever plate structure [3,4], there is still little work to analyze the effects of size and nonlinear dynamic behaviors for micro-plate piezoelectric energy harvester based on strain gradient theory. Therefore, this paper proposes the model of the micro-cantilever plate PEH with tip mass. Then the natural frequencies, phase figures, bifurcation diagrams and frequency-output voltage curves are given to discuss the nonlinear behaviors of the structure, such as bifurcation and superharmonic resonance, which improves the performance of micro-size device by increasing output voltage and broadening frequency bandwidth.

2. Dynamic model of the harvester

x, y, z	position on the coordinate axis
u, v, w	displacement in the x, y, and z directions
κ_{ij}	nonlinear curvature of mid-plane
$\varepsilon_{xx}, \varepsilon_{yy}, \gamma_{xy}, \gamma_{yx}$	strain in plate
$\sigma_{xx}^s, \sigma_{yy}^s, \tau_{xy}^s, \tau_{yx}^s$	stress in the base layer
$\sigma_{xx}^p, \sigma_{yy}^p, \tau_{xy}^p, \tau_{yx}^p$	stress in the piezoelectric layer
h^s, h^p	thicknesses of the base layer and piezoelectric layer
a^s, a^p	lengths of the base layer and the piezoelectric layer
b	width of the plate
m_t	mass of the tip mass
R	total resistance in the circuit
w_0	amplitude of base excitation
Ω	frequency of base excitation
l	material scale parameter according to Ramezani [5]
ρ^s	density of base layer
ρ^p	density of piezoelectric material

Structure diagram of the piezoelectric energy harvester is shown in figure.1, which consists of two layers, the lower is the base cantilever plate and the upper is the PVDF piezoelectric material.

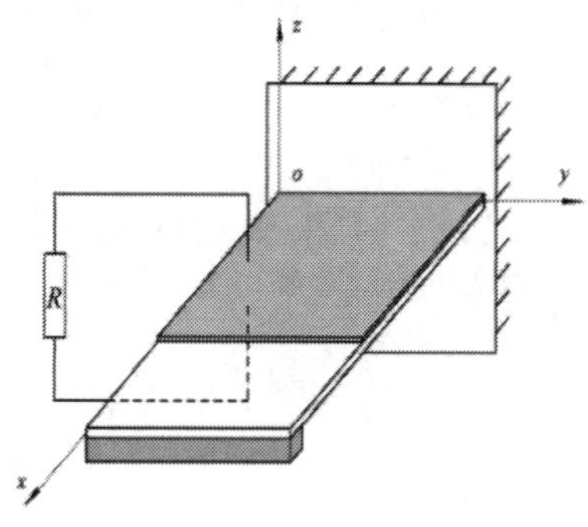

Figure 1. schematic diagram of PEH.

2.1. Potential energy

According to the nonlinear curvature is given by Dowell [6], The strain field is represented as:

$$\begin{aligned}
\varepsilon_{xx} &= -z\frac{\partial^2 w}{\partial x^2}[1+\frac{1}{2}(\frac{\partial w}{\partial x})^2+\frac{1}{2}(\frac{\partial w}{\partial y})^2] \\
\varepsilon_{yy} &= -z\frac{\partial^2 w}{\partial y^2}[1+\frac{1}{2}(\frac{\partial w}{\partial x})^2+\frac{1}{2}(\frac{\partial w}{\partial y})^2] \\
\gamma_{xy} = \gamma_{yx} &= -2z\frac{\partial^2 w}{\partial x \partial y}[1+\frac{1}{2}(\frac{\partial w}{\partial x})^2+\frac{1}{2}(\frac{\partial w}{\partial y})^2]
\end{aligned} \tag{1}$$

And the stress field of the basic layer and piezoelectric layer can be obtained on the basis of the constitutive equation respectively:

$$\sigma_{xx}^{s} = \frac{E_s}{1-v_s^2}\left(\varepsilon_{xx} + v_s\varepsilon_{yy}\right)$$
$$\sigma_{yy}^{s} = \frac{E_s}{1-v_s^2}\left(\varepsilon_{yy} + v_s\varepsilon_{xx}\right) \tag{2}$$
$$\tau_{xy}^{s} = \tau_{yx}^{s} = \frac{E_s}{2(1+v_s)}\cdot\gamma_{xy} = \frac{E_s}{2(1+v_s)}\cdot\gamma_{yx}$$
$$\sigma_{xx}^{p} = \frac{E_p}{1-v_p^2}\left(\varepsilon_{xx} + v_p\varepsilon_{yy}\right) + d_{31}\cdot D_3$$
$$\sigma_{yy}^{p} = \frac{E_p}{1-v_p^2}\left(\varepsilon_{yy} + v_p\varepsilon_{xx}\right) + d_{32}\cdot D_3 \tag{3}$$
$$\tau_{xy}^{p} = \tau_{yx}^{p} = \frac{E_p}{2(1+v_p)}\cdot\gamma_{xy} = \frac{E_p}{2(1+v_p)}\cdot\gamma_{yx}$$

where D represents the tensor of electric displacement and d is the electrical polarization coupling coefficient.

The strain energy of the plate can be expressed as according to continuum mechanics theory:

$$\bar{U}^{\varsigma} = \iiint_{V^{\varsigma}} \frac{1}{2}(\sigma_{xx}^{\varsigma}\varepsilon_{xx} + \sigma_{yy}^{\varsigma}\varepsilon_{yy} + \sigma_{xy}^{\varsigma}\varepsilon_{xy})dxdydz \quad (\varsigma = s, p) \tag{4}$$

Considering the size effect of micro-size plate, the strain gradient η can be expressed as:

$$\eta_{ijk} = \eta_{ikj} = \varepsilon_{jk,i} = \varepsilon_{kj,i}\left(i = x, y, z \quad j,k = x, y\right) \tag{5}$$

The expressions of strain energy density and corresponding strain energy based on strain gradient are given by Mindlin [6]:

$$\hat{W}^{\varsigma} = L_1^{\varsigma}\eta_{iik}\eta_{kjj} + L_2^{\varsigma}\eta_{ijj}\eta_{ikk} + L_3^{\varsigma}\eta_{iik}\eta_{jjk} + L_4^{\varsigma}\eta_{ijk}\eta_{ijk} + L_5^{\varsigma}\eta_{ijk}\eta_{kji}$$
$$\hat{U}^{\varsigma} = \iiint_V \hat{W}^{\varsigma}dxdydz \qquad (\varsigma = s, p) \tag{6}$$

where $L_i^{\varsigma}\left(i = 1,2,3,4,5\right)$ are material parameters given by Ramezani [5]:

$$L_2^{\varsigma} = \frac{1}{2}l^2\frac{E_{\varsigma}v_{\varsigma}}{(1-2v_{\varsigma})(1+v_{\varsigma})}$$
$$L_4^{\varsigma} = l^2\frac{E_{\varsigma}}{2(1+v_{\varsigma})} \qquad (\varsigma = s, p) \tag{7}$$
$$L_1^{\varsigma} = L_3^{\varsigma} = L_5^{\varsigma} = 0$$

According to strain gradient theory, one obtains the expression of Potential energy of micro cantilever plate PEH:

$$U = \bar{U}^{s} + \bar{U}^{p} + \hat{U}^{s} + \hat{U}^{p} + \frac{1}{2}E_3\cdot D_3 \tag{8}$$

where:

$$E_3 = a_{33}\cdot D_3 + d_{31}\cdot\varepsilon_{xx} + d_{32}\cdot\varepsilon_{yy} = -\frac{V_1}{h^p} \tag{9}$$

a_{33} refer to the dielectric constant; V_1 is the out-put voltage.

2.2. Kinetic energy

Kinetic energy of the system consists of three parts:

$$T = T^{s} + T^{p} + T_0 \tag{10}$$

where T^s and T^p represent the kinetic energy of base layer and piezoelectric layer respectively, and T_2 represents kinetic energy of the additional mass.

$$T^s = \iiint_{V^s} \frac{\rho^s}{2}\left(\frac{\partial w}{\partial t} - w_0\Omega\cos(\Omega t)\right)^2 dV^s$$
$$T^p = \iiint_{V^p} \frac{\rho^p}{2}\left(\frac{\partial w}{\partial t} - w_0\Omega\cos(\Omega t)\right)^2 dV^p \tag{11}$$

$$T_0 = \frac{m}{2}\left(\frac{\partial}{\partial t} w\big|_{x=a} - w_0\Omega\cos(\Omega t)\right)^2 \tag{12}$$

2.3. Hamilton principle

Virtual work done by damping can be written as:

$$\delta W_r = -\int_{V^s} r^s \cdot \frac{\partial w}{\partial t} \cdot \delta w dV^s - \int_{V^p} r^p \cdot \frac{\partial w}{\partial t} \cdot \delta w dV^p \tag{13}$$

Substitute the equation (8), and equation (10)- equation (13) into the equation of Hamilton principle and rearrange:

$$\begin{aligned}
&\int_{V^s} (\sigma_{ij}^s \delta\varepsilon_{ij} + \tau_{ijk}^s \delta\eta_{ijk}) dV^s - \frac{1}{4} m \cdot \left(\frac{\partial^2}{\partial t^2} w\big|_{x=a} - w_0\Omega^2 \sin(\Omega t)\right) \\
&-\int_{V^s} \rho^s \cdot \left(\frac{\partial^2 w}{\partial t^2} - w_0\Omega^2 \sin(\Omega t)\right)\delta w dV^s + \int_{V^p} (\sigma_{ij}^p \delta\varepsilon_{ij} + \tau_{ijk}^p \delta\eta_{ijk}) dV^p \\
&-\int_{V^s} r^s \cdot \frac{\partial w}{\partial t} \cdot \delta w dV^s - \int_{V^p} \rho^p \cdot \left(\frac{\partial^2 w}{\partial t^2} - w_0\Omega^2 \sin(\Omega t)\right)\delta w dV^p \\
&-\int_{V^p} r^p \cdot \frac{\partial w}{\partial t} \cdot \delta w dV^p = 0
\end{aligned} \tag{14}$$

2.4. Electrical equations

Electric displacement can be expressed by equation (9):

$$D_3 = \frac{E_3 - (d_{31} \cdot \varepsilon_{xx} + d_{32} \cdot \varepsilon_{yy})}{a_{33}} \tag{15}$$

According to Kirchhoff's law, the electrical governing equation can be obtained as:

$$\frac{S^p}{a_{33} h^p} \dot{V}_1 + \frac{V_1}{R} + \frac{1}{a_{33} h^p} \int_{V^p} \left(d_{31} \cdot \dot{\varepsilon}_{11} + d_{32} \cdot \dot{\varepsilon}_{22}\right) dV^p = 0 \tag{16}$$

2.5. Establishment of model

Modal function of plate is regarded as a combination of fixed-free beam function $X_i(x)$ in x direction and free-free beam function $Y_j(y)$ in y direction, and the third order dispersion of deflection is described as:

$$w(x,y,t) = w_1(t) \cdot X_1(x) Y_1(y) + w_2(t) \cdot X_1(x) Y_2(y) + w_3(t) \cdot X_2(x) Y_1(y) \tag{17}$$

where the first two order of $X_i(x)$ and $Y_j(y)$ is given as follows:

$$X_1 = \sin\frac{1.9x}{a^s} - \sinh\frac{1.9x}{a^s} - 0.7\left(\cosh\frac{1.9x}{a^s} - \cos\frac{1.9x}{a^s}\right)$$
$$X_2 = \sin\frac{4.7x}{a^s} - \sinh\frac{4.7x}{a^s} - \left(\cosh\frac{4.7x}{a^s} - \cos\frac{4.7x}{a^s}\right) \tag{18}$$

$$Y_1 = 1$$
$$Y_2 = (3)^{\frac{1}{2}}\left(\frac{1-2y}{b}\right) \tag{19}$$

The expansion expression of the deflection w (equation (17)) is applied to the equations of the theoretical dynamics model. Dynamic ordinary differential equations can be derived by equation (14) and electrical ordinary differential equation can be obtained by equation (16). Then the generalized coordinate w_1, w_2, w_3 and V_1 can been solved using equation (20).

$$\begin{aligned}
\ddot{w}_1 &= a_1w_1 + a_2w_1^3 + a_3w_2^3 + a_4w_3^3 + a_5w_1w_2^2 + a_6w_1w_3^2 + a_7w_2w_1^2 \\
&+ a_8w_2w_3^2 + a_9w_3w_1^2 + a_{10}w_3w_2^2 + a_{11}w_1w_2w_3 + a_{12}\dot{w}_1 + a_{13}V_1 \\
&+ a_{14}w_1^2V_1 + a_{15}w_2^2V_1 + a_{16}w_3^2V_1 + a_{17}w_0\sin(\Omega t) \\
\ddot{w}_2 &= b_1w_2 + b_2w_1^3 + b_3w_2^3 + b_4w_3^3 + b_5w_1w_2^2 + b_6w_1w_3^2 + b_7w_2w_1^2 \\
&+ b_8w_2w_3^2 + b_9w_3w_1^2 + b_{10}w_3w_2^2 + b_{11}w_1w_2w_3 + b_{12}\dot{w}_2 + b_{13}V_1 \\
&+ b_{14}w_1^2V_1 + b_{15}w_2^2V_1 + b_{16}w_3^2V_1 + b_{17}w_0\sin(\Omega t) \\
\ddot{w}_3 &= c_1w_3 + c_2w_1^3 + c_3w_2^3 + c_4w_3^3 + c_5w_1w_2^2 + c_6w_1w_3^2 + c_7w_2w_1^2 \\
&+ c_8w_2w_3^2 + c_9w_3w_1^2 + c_{10}w_3w_2^2 + c_{11}w_1w_2w_3 + c_{12}\dot{w}_3 + c_{13}V_1 \\
&+ c_{14}w_1^2V_1 + c_{15}w_2^2V_1 + c_{16}w_3^2V_1 + c_{17}w_0\sin(\Omega t) \\
\dot{V}_1 &= d_1w_1^2\dot{w}_1 + d_2w_1^2\dot{w}_2 + d_3w_1^2\dot{w}_3 + d_4w_2^2\dot{w}_1 + d_5w_2^2\dot{w}_2 \\
&+ d_6w_2^2\dot{w}_3 + d_7w_3^2\dot{w}_1 + d_8w_3^2\dot{w}_2 + d_9w_3^2\dot{w}_3 \\
&+ d_{10}\dot{w}_1 + d_{11}\dot{w}_2 + d_{12}\dot{w}_3 + d_{13}V_1
\end{aligned} \tag{20}$$

where parameters $a_i, b_i, c_i, d_i (i = 0,1...17)$ are the coefficients about w_1, w_2, w_3 and V_1 after integrating equation (14) and equation (16), which relate to geometric dimensions and material parameters. Because the expanded formula is complex and space is limited, letters are used instead of specific expansion expressions to make the form of the final expression more intuitive.

3. Results and discussion

Parameters of the device are given in Table 1.

Table 1. Parameters of the structure.

Physica	Value	Physical	Value
a^s	450 μm	E^s	185 Gpa
a^p	225 μm	E^p	2 Gpa
b	2225 μm	v^s	0.29
h^s	1.2 μm	v^p	0.36
h^p	0.6 μm	R	$10^6\Omega$

Based on Rayleigh-Ritz method, the natural frequencies of structures with different mass ratios are obtained as follows and K refer to mass ratio of the additional tip mass to piezoelectric plate.

Table 2. Natural frequencies of structure.

Natural frequency (KHz)	$K=0$	$K=1$	$K=2$	$K=3$
First order	9.928	4.501	3.344	2.762
Second order	41.72	21.82	16.52	13.84
Third order	57.54	43.31	42.53	42.27

From Table 2, it can be seen that the natural frequency of the structure will decrease accordingly with the tip mass of the cantilever plate increasing, and the reduction of the first order natural frequency is more obvious than higher order ones. When K =3, the first order natural frequency decreases by 72%, and the third order natural frequency decreases by 27%.

Figure.2 shows the comparison of the maximum output voltage of PEH between different mass ratios, and the abscissa represents the range of external excitation frequency. Figure.3 illustrates the relationship between nonlinear behavior and linear behavior. Figures.4 presents the nonlinear phenomenon of the proposed structure through phase diagram and bifurcation.

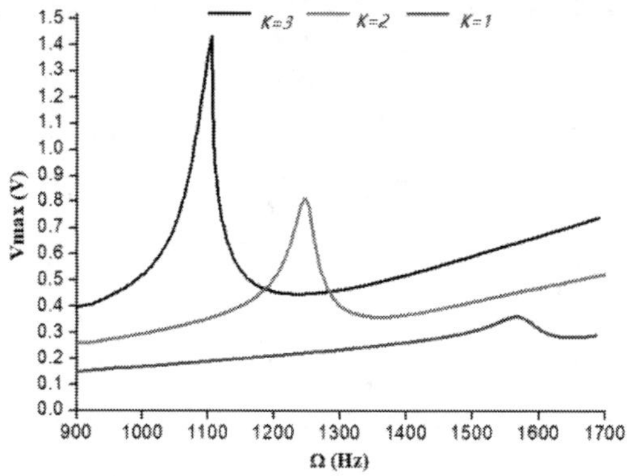

Figure 2. Maximum output voltage for different mass ratio.

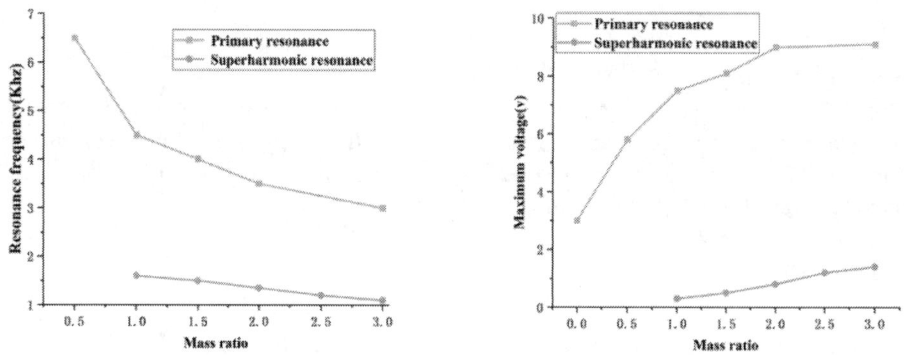

Figure 3. natural frequency and maximum voltage for superharmonic and primary resonance.

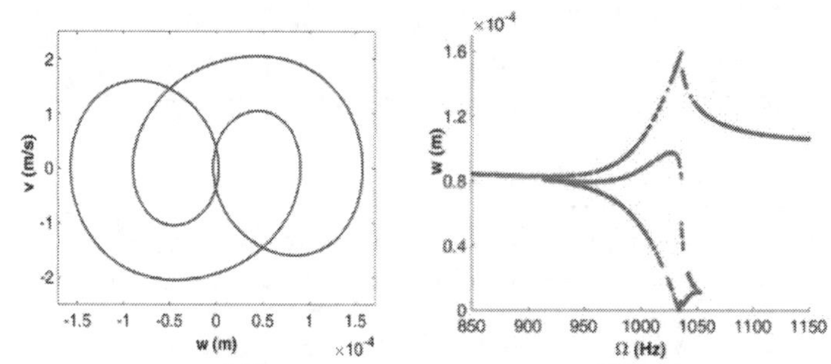

(A) (B)

Figure 4. for $K = 3$: (A) phase diagram at 1040 Hz; (B) bifurcation diagram.

When the ambient vibration frequency Ω is equal to the first order natural frequency of the PEH p_1, primary resonance occurs. When $3\Omega= p_1$, three times superharmonic resonance with nonlinear characteristics occurs. As is presented in figure.3, when the mass ratio K<1, only the primary resonance occurs and there is no superharmonic resonance. The result can be interpreted as the stiffness of the structure is relatively large, which accordingly makes the deflection is small and the nonlinear phenomenon is not obvious. When the mass ratio K >= 2, the phenomenon of three times superharmonic resonance becomes more obvious with the increasing of additional mass, which

IOP Conf. Series: Materials Science and Engineering **825** (2020) 012016 doi:10.1088/1757-899X/825/1/012016

enables the proposed PEH to output high voltage in a lower frequency ($p_1/3$) and expands the frequency bandwidth of the energy harvester.

Through the analysis of simulation results above, we can find that the improvement of the performance of the micro-size PEH is significant both in the increasing of output voltage and broadening the bandwidth of low frequency, which is due to the superharmonic resonance caused by the nonlinearity of the structure. Therefore, the bifurcation diagrams of amplitude with frequency at K =3 and the phase diagrams at $p_1/3$ are given in figures.4 to further analyze this nonlinear phenomenon.

Figures.4 show that, with the increase of K, the system appears bifurcation and superharmonic resonance, which means that vibration amplitude and output voltage are also significantly enhanced in superharmonic resonance region except for the primary resonance. At this time, the system can capture energy from the ambient vibration in the low frequency band through the super harmonic resonance with nonlinear characteristics, which explains the improvement of two key performances of the proposed micro-size PEH in detail, namely, the output voltage and the working bandwidth.

4. Conclusion

This paper proposes a micro-size piezoelectric energy harvester consisting of a cantilever plate with a tip mass on the free end. The dynamic model of the PEH is established considering electromechanical coupling effect and size effect. Based on the nonlinear dynamic equations, the performance of PEH is discussed with different additional mass ratios. Simulation results indicate that bifurcation and superharmonic resonance occurs with the increase of mass ratio K. This nonlinear phenomenon enables the device to work in a lower bandwidth and capture more power from ambient vibration, which observably improves harvesting efficiency. The research direction of this paper is to proposes an approach to enhance micro-size PEH performance in a lower bandwidth, and how to obtain an optimal mass ratio to meet the requirement of the structural characteristics in practical applications remains to be further studied.

References

[1] ROUNDY S, WRIGHT P K, RABAEY J M. Energy scavenging for wireless sensor networks[M]. Springer US, 2004.

[2] Marzencki M, Basrour S, Charlot B, et al. Design and fabrication of piezoelectric micro power generators for autonomous microsystems[C]//Proc. of DTIP'05. 2005: 299-302.

[3] C. W. S. To VIBRATION OF A CANTILEVER BEAM WITH A BASE EXCITATION AND TIP MASS. Journal of Sound and Vibration (1982) 83(4), 445-460

[4] X.Q. Zhao, Z.G. Shang, G.X. Luo, L.C. Deng, A vibration energy harvester using AlN piezoelectric cantilever array, Microelectron. Eng. 142 (2015) 47–51.

[5] RAJABI F, RAMEZANI S. A nonlinear microbeam model based on strain gradient elasticity theory[J]. Acta Mechanica Solida Sinica, 2013, 26(1): 21-34.

[6] Mindlin, R.D., 1964.Micro-structure in linear elasticity.Arch.Rational Mech.Anal.16, 51–78.

Study on Vibration Characteristics of Cantilever Plate with Annular Mass Frame

Li Hua Chen[1], Chao Ran Hou[1], Shou Jie Cui[1], Lu Tong Shan[1] and Feng Hong Yang[2,a]

[1] College of Mechanical Engineering, Beijing University of Technology, Beijing, 100124, PR China
[2] School of Statistics and Mathematics, Central University of Finance Economics, Beijing, 100081, PR China

[a] Corresponding author: yangfh_cufe@163.com

Abstract. The principle of piezoelectric energy harvester is to convert vibration energy into electric energy, which can supply power for other electronic devices. The actual environment vibration is broadband, random and multi-directional, and the working frequency band of piezoelectric energy harvester needs to be increased, so as to raise energy capture efficiency. In order to increase the numbers of resonance frequencies and widen working band, a new type of cantilever plate structure with ring mass frame is designed. This structure is composed of two anti-symmetrically placed cantilever plates with variable cross-section and additional annular mass. Firstly, the natural frequencies and mode shapes of the system are obtained by modal analysis using the finite element method. Based on modal diagrams, the bending and torsional of cantilever plate and the motion of rigid ring mass frame are considered comprehensively, then the theoretical modal functions of the structure are given. Furthermore, the natural frequencies of the structure are obtained by Rayleigh method. Natural frequencies and theoretical modal functions are verified by comparing with the finite ones, which provides theoretical basis for the analysis of the piezoelectric energy harvester in the future.
Keywords: Rayleigh method, modal functions, cantilever plate, mass frame.

1. Introduction

The problems of high resonance frequency and narrow working frequency band are universal in the vibration piezoelectric energy harvester based on cantilever structure, which leads that it can not be widely applied in practical engineering. How to reduce the resonance frequency, widen frequency band and improve the effective energy capture efficiency is the main research direction of many scholars at present [1-6].

For the sake of reduce the natural frequency and widen the frequency band, our group proposes a cantilever plate structure with annular mass frame. The structure is composed of two variable cross-section cantilever plates and additional rigid annular mass frames. The rigid annular mass frames not only rotate around the center of mass, but also translational motion with the center of mass. So, the bending and torsional of the cantilever plate is combined with the motion of the rigid body, which makes that the vibration mode of the system is the combination of the bending and torsional of cantilever plate and the motion of rigid. Eventually, three lower natural frequencies were obtained to achieve the purpose of widening the bandwidth.

The natural frequencies and mode shapes of the combination structures are obtained by finite element method. Although the finite element software developed and successfully applied to various fields of engineering, it is inconvenient in the extreme to analyze complex non-linear behaviors such as bifurcation, chaos and non-linear vibration controlling, so theoretical analysis method is needed. Based on the modal diagram, the theoretical modal function expression is gotten, and then Rayleigh method is used to calculate the natural frequencies of the system. The theoretical modal is verified by the finite element method, which provides theoretical basis for the subsequent analysis of the piezoelectric energy harvester.

2. Thostheoretical vibration modes

Figure 1 shows the schematic diagram of the cantilever plate structure with annular mass frame. The structure consists of two anti-symmetrically placed variable section cantilever plates. Their ends are bonded to the annular mass frame. In Figure.1, the upper and bottom plates are named plate 1 and plate 2, respectively.

Firstly, the finite element method is applied to simulate the structure our group designed. The first three natural frequencies obtained by finite element method are 32Hz, 54Hz and 76Hz, respectively. And the principal modes are shown in Figure. 2. It should be noted that the red area represents the fixed end of the cantilever plate. And the green color describe that the deformation of structure is small, and the blue and red colors represent the larger deformation at different directions. The results of finite element simulation show that the structure proposed by our group can achieve the function of widening bandwidth.

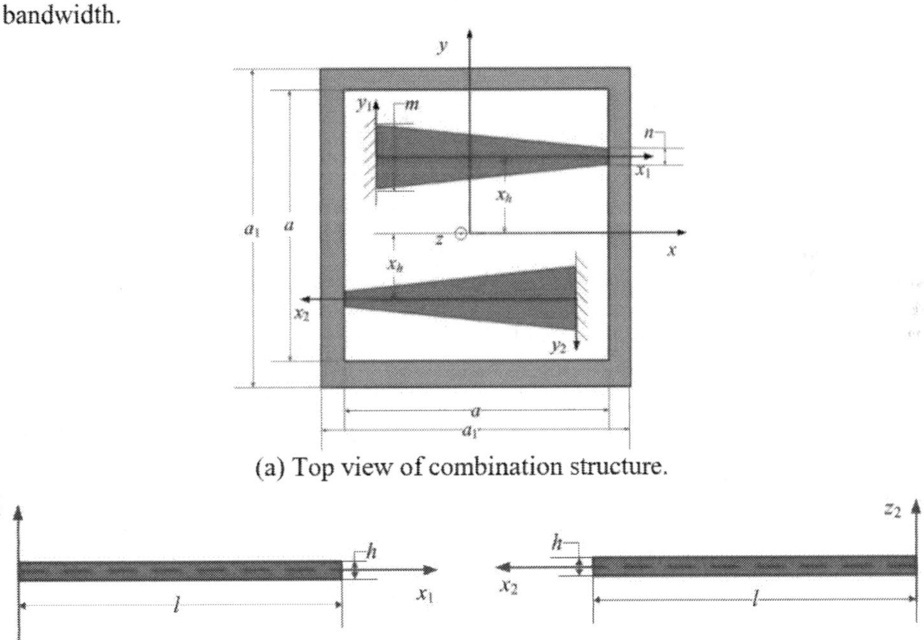

(a) Top view of combination structure.

(b) Side view of plate 1. (c) Side view of plate 2.

Figure 1. Structural schematic diagram.

Secondly, based on the mode diagram, the modal functions of structural vibration are expressed in the form of trigonometric functions. The middle surface of plate 1 and plate 2 are set as x_1oy_1 and x_2oy_2, respectively, and the z axis is perpendicular to the middle plane of the plate. The vibration modes of plates can be assumed as a combination of beam functions. The expressions are as follows:

$$\begin{aligned} w_1(x,y,t) &= x_1(t)\cdot X_1^1\cdot Y_1^1 + x_2(t)\cdot X_2^1\cdot Y_1^1 - x_3(t)\cdot X_1^1\cdot Y_2^1 \\ w_2(x,y,t) &= -x_1(t)\cdot X_1^2\cdot Y_1^2 + x_2(t)\cdot X_2^2\cdot Y_1^2 + x_3(t)\cdot X_1^2\cdot Y_2^2 \end{aligned} \tag{1}$$

where the trial functions $X_i^1, Y_i^1 (i=1,2)$ represent the i th order mode of the cantilever plate 1 in the x

IOP Conf. Series: Materials Science and Engineering **825** (2020) 012017 doi:10.1088/1757-899X/825/1/012017

and y directions, respectively. $X_i^2, Y_i^2 (i=1,2)$ represent the i th order mode of the cantilever plate 2 in the x and y directions, respectively. w_i (i = 1, 2) represents the mode function of plate 1and 2, respectively. x_i (i=1,2,3) is the undetermined coefficients about time.

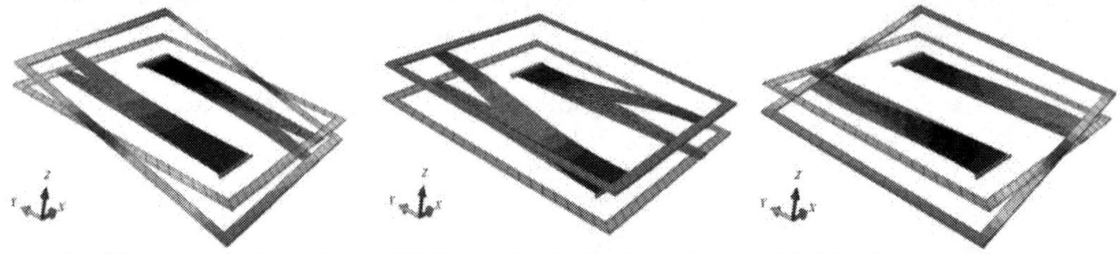

(*a*) First principal mode. (*b*) Second principal mode. (*c*) Third principal mode.

Figure 2. Structural principal mode by finite element method.

Thirdly, how to solve the concrete expressions of the trial functions ($X_i^1, Y_i^1, X_i^2, Y_i^2$ $(i=1,2)$) of cantilever plates 1 and 2 in *xyz* coordinate system? Observing Figure.2, we find that the trial functions(Y_i^1, Y_i^2) of the cantilever plate in the y direction are relatively simple. For the first and second principle modes, there is no obvious deformation on the plate 1 and 2 in the y direction, and they twist around the axis of y = 0 in the third principle mode diagram. Therefore, we can assume that the expressions of the trial functions of plate 1 and 2 in the y direction are written as follows:

$$Y_1^1 = 1$$

$$Y_2^1 = (3)^{\frac{1}{2}} \left(x_h + \frac{m}{2} \right)^{-1} y \tag{2}$$

$$Y_1^2 = 1$$

$$Y_2^2 = -(3)^{\frac{1}{2}} \left(x_h + \frac{m}{2} \right)^{-1} y \tag{3}$$

For the trial functions(X_i^1, X_i^2) of plate 1 and 2 in x direction can be obtained using the idea that the expression of the trial functions (X_i^{p1}, X_i^{p2}) in the $x_1y_1z_1$ and $x_2y_2z_2$ coordinate system can be obtained first, and then the mode in the *xyz* coordinate system can be obtained by coordinate transformation. In the $x_1y_1z_1$, $x_2y_2z_2$ coordinate systems, the expression of the trial functions in x direction can be set as follows:

$$X_i^{p1}(x_1) = A_i \sin k_i x_1 + B_i \cos k_i x_1 + C_i \sinh k_i x_1 + D_i \cosh k_i x_1 \tag{4}$$

$$X_i^{p2}(x_1) = A_i' \sin k_i x_1 + B_i' \cos k_i x_1 + C_i' \sinh k_i x_1 + D_i' \cosh k_i x_1 \tag{5}$$

The undetermined coefficients (A_i, B_i, C_i, D_i) and k_i can be gotten using different boundary conditions, and then expressions (X_i^{p1}, X_i^{p2}) can be obtained.

Next, the solve method of the expression (X_i^{p1}) of plate 1 is describe as follows. Observing the first and second principle modes, we find that the boundary condition of the fixed end (the red area) are the same, but the boundary conditions at the free end of the cantilever plate are different. The main reason why the boundary conditions at the free end of a cantilever plate are different is that the motion of rigid annular mass frames is different in the first two principle mode. In the first principle mode, the rigid annular mass frames rotates around the axis of x = 0, which means that the free end of the cantilever plate is free. However, in the second principle, the frame moves up and down along the z axis, which makes that the rotation angle of the free end of the plate is equal to zero. So, the boundary conditions of the fixed end in the first two principle mode are different and they are expressed as follows:

$$X_1^{p1}(0)=0 \quad \frac{\partial X_1^{p1}(0)}{\partial x}=0$$
$$\frac{\partial^2 X_1^{p1''}(l)}{\partial x^2}=0 \quad \frac{\partial^3 X_1^{p1}(l)}{\partial x^3}=0 \tag{6}$$

$$X_2^{p1}(0)=0 \quad \frac{\partial X_2^{p1}(0)}{\partial x}=0$$
$$\frac{\partial X_2^{p1}(l)}{\partial x}=0 \quad \frac{\partial^3 X_2^{p1}(l)}{\partial x^3}=0 \tag{7}$$

we can get the equations about A_i, B_i, C_i, D_i by taking equations(6) and (7) into equation (4) and (5), respectively. The condition for the homogeneous equations that exist non-zero solutions is that the determinant of the coefficient matrix is equal to zero. The frequency equation of the first and second principle mode can be obtained as follows:

$$\cos(k_1 l)\cosh(k_1 l)=-1 \tag{8}$$

$$-\sinh(k_2 l)\cos(k_2 l)-\sin(k_2 l)\cosh(k_2 l)=0 \tag{9}$$

Using the drawing method, the equations (8, 9) can be solved and $k_1 l=1.875$, $k_2 l=2.635$. Putting the $k_1 l$ and $k_2 l$ into the equations about A_i, B_i, C_i, D_i, the expressions of $X_i^{p1}(x_1)$ can be obtained as follows:

$$X_1^{p1}(x_1)=\cosh k_1 x_1-\cos k_1 x_1+C_1(\sinh k_1 x_1-\sin k_1 x_1) \tag{10}$$

$$X_2^{p1}(x_1)=\cosh k_2 x_1-\cos k_2 x_1-C_2(\sin k_2 x_1-\sinh k_2 x_1) \tag{11}$$

Using the same method, the expressions of $X_i^{p2}(x_2)$ can be obtained as follows:

$$X_1^{p2}(x_2)=\cosh k_1 x_2-\cos k_1 x_2+C_1'(-\sin k_1 x_2+\sinh k_1 x_2) \tag{12}$$

$$X_2^{p2}(x_2)=\cosh k_2 x_2-\cos k_2 x_2-C_2'(\sin k_2 x_2-\sinh k_2 x_2) \tag{13}$$

Finally, the expression of X_i^1, X_i^2 can be gotten by the coordinate transformation that is the expressions of $X_i^{p1}(x_1)$ and $X_i^{p2}(x_2)$ in the coordinates of $x_1y_1z_1$ and $x_2y_2z_2$ are transformed into X_i^1 and X_i^2 in the coordinates of xyz. After transformation, the expressions of X_i^1, X_i^2 are written as follows:

$$X_1^1=X_1^{p1}\left(x_1+l-\frac{a}{2}\right),$$
$$X_2^1=X_2^{p1}\left(x_1+l-\frac{a}{2}\right) \tag{14}$$

$$X_1^2=X_1^{p2}\left(-x_2+l-\frac{a}{2}\right),$$
$$X_2^2=X_2^{p2}\left(-x_2+l-\frac{a}{2}\right) \tag{15}$$

Bring the equations (14, 15) into the equation(1), the expression of $w_1(x,y,t), w_2(x,y,t)$ can be gotten.

3. Verification of vibration modes

Using the expression of mode function $w_1(x,y,t), w_2(x,y,t)$, and applied the MAPLE software to draw the principle modes of the structure. Figure 3 exhibitions the theoretical modal function diagram, and the red area describes the fixed end of the cantilever plate. From the Figure.3, it can be clearly see that the principle modes is formed by the bending and torsional deformation of the cantilever plate and the rotation of the frame. The mode diagrams drawn by theoretical mode function coincide with that drawn by finite element software, which qualitatively verifies the correctness of theoretical mode function.

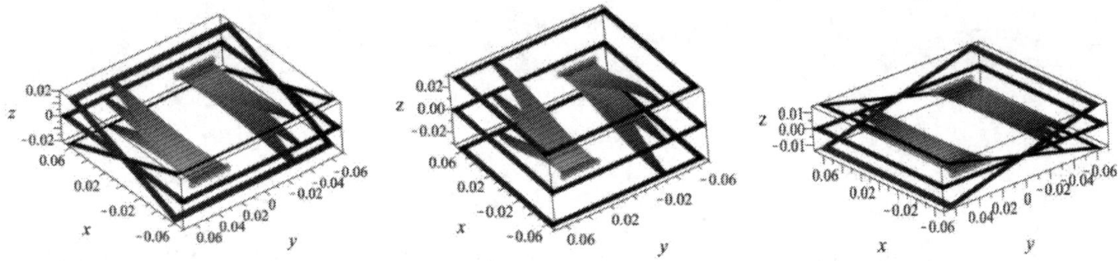

(a) First principal mode. (b) Second principal mode. (c) Third principal mode.

Figure 3: Structural principal mode by theoretical method.

4. Verification of nature frequencies

In this part, the natural frequencies of the system are solved by Rayleigh method. Using the mode functions, we can set the displacements functions of the cantilever plates 1 and 2 at any time can be set as follows:

$$w_1 = \left(x_1 \cdot X_1^1 \cdot Y_1^1 + x_2 \cdot X_2^1 \cdot Y_1^1 - x_3 \cdot X_1^1 \cdot Y_2^1\right)\sin\left(pt + \varphi\right)$$
$$w_2 = \left(-x_1 \cdot X_1^2 \cdot Y_1^2 + x_2 \cdot X_2^2 \cdot Y_1^2 + x_3 \cdot X_1^2 \cdot Y_2^2\right)\sin\left(pt + \varphi\right) \tag{16}$$

The strains at any point in the plate 1 and 2 can be expressed as:

$$\varepsilon_{xx}^{\xi} = \hat{\varepsilon}_{xx}^{\xi} + z\kappa_{xx}^{\xi}$$
$$\varepsilon_{yy}^{\xi} = \hat{\varepsilon}_{yy}^{\xi} + z\kappa_{yy}^{\xi} \tag{17}$$
$$\varepsilon_{xy}^{\xi} = \hat{\varepsilon}_{xy}^{\xi} + z\kappa_{xy}^{\xi}$$

where $\hat{\varepsilon}_{xx}^{\xi}, \hat{\varepsilon}_{yy}^{\xi}, \hat{\varepsilon}_{xy}^{\xi}$ and $\kappa_{xx}^{\xi}, \kappa_{yy}^{\xi}, \kappa_{xy}^{\xi}$ describe the stains and curvatures of the middle surface, respectively. As well as $\xi = 1,2$. Assuming that the stains of the middle surface are equal to zero, and the inextensible theory is applied to get the expression of curvatures as follows:

$$\kappa_{xx}^{\xi} = -\frac{\partial^2 w_{\xi}}{\partial x^2}[1 + \frac{1}{2}(\frac{\partial w_{\xi}}{\partial x})^2 + \frac{1}{2}(\frac{\partial w_{\xi}}{\partial y})^2]$$
$$\kappa_{yy}^{\xi} = -\frac{\partial^2 w_{\xi}}{\partial y^2}[1 + \frac{1}{2}(\frac{\partial w_{\xi}}{\partial x})^2 + \frac{1}{2}(\frac{\partial w_{\xi}}{\partial y})^2] \tag{18}$$
$$\kappa_{xy}^{\xi} = -2\frac{\partial^2 w}{\partial x \partial y}_{\xi}[1 + \frac{1}{2}(\frac{\partial w_{\xi}}{\partial x})^2 + \frac{1}{2}(\frac{\partial w_{\xi}}{\partial y})^2]$$

The constitutive equations of cantilever plates are described as follows:

$$\sigma_{xx}^{\xi} = \frac{E^{\xi}}{1 - v^{\xi 2}}(\varepsilon_{xx}^{\xi} + v^{\xi}\varepsilon_{yy}^{\xi})$$
$$\sigma_{yy}^{\xi} = \frac{E_S^{\xi}}{1 - v^{2}}(\varepsilon_{yy}^{\xi} + v^{\xi}\varepsilon_{xx}^{\xi}) \tag{19}$$
$$\tau_{xy}^{\xi} = \frac{E^{\xi}}{2(1 + v)}\gamma_{xy}^{\xi}$$

where E, v represents the Young modulus and Poisson's ratio, respectively.

Based on the theory of continuum mechanics, the total potential energy of two cantilever plates can be written as follows:

$$U = \iiint_V \frac{1}{2}\left(\sigma_{xx}^{\xi}\varepsilon_{xx}^{\xi} + \sigma_{yy}^{\xi}\varepsilon_{yy}^{\xi} + \tau_{xy}^{\xi}\gamma_{xy}^{\xi}\right)dxdydz \tag{20}$$

The expression of kinetic energy of cantilever plates is written as follows:

$$T_0 = \iiint_V \frac{\rho}{2}(\dot{w}_1^2 + \dot{w}_2^2)dV^{\xi} \tag{21}$$

According to the mode diagram, the kinetic energy of the additional mass frame consists of two parts, i.e. translational kinetic energy and rotational kinetic energy, as follows:

$$T_m = \frac{1}{2}J\left(\frac{\dot{w}_1|_{x=l} - \dot{w}_2|_{x=l}}{a_1}\right)^2 + \frac{1}{2}m_0\left((\dot{w}_1|_{x=l})^2 + (\dot{w}_2|_{x=l})^2\right) \tag{22}$$

where J is the moment of inertia of the frame and m_0 is the mass of the frame, and

$$J = \frac{1}{12}m_0\left(a_1^2 - a^2\right) \tag{23}$$

So, the total kinetic energy of the system is expressed as:

$$T = T_0 + T_m \tag{24}$$

Maximum kinetic and potential energy can be calculated as

$$T_{\max} = p^2\left(g_1x_1^2 + g_2x_2^2 + g_3x_3^2 + g_4x_3x_2 + g_5x_1x_2\right) \tag{25}$$

$$U_{\max} = s_1x_1^2 + s_2x_2^2 + s_3x_3^2 + s_4x_3x_1 + s_5x_3x_2 + s_6x_1x_2 \tag{26}$$

where g_1-g_5 and s_1-s_6 represents the material parameters of the cantilever plates.

The maximum kinetic energy is equal to the maximum potential energy, and the expression of natural frequency is obtained as follows:

$$p = \sqrt{\frac{\left(s_1x_1^2 + s_2x_2^2 + s_3x_3^2 + s_4x_3x_1 + s_5x_3x_2 + s_6x_1x_2\right)}{\left(g_1x_1^2 + g_2x_2^2 + g_3x_3^2 + g_4x_3x_2 + g_5x_1x_2\right)}} \tag{27}$$

When the first natural frequency is the dominant vibration of the system, i.e. $\{x_1, x_2, x_3\} = \{1, 0, 0\}$, the first nature frequency is 36Hz. The second nature frequency was calculated as 67Hz when it was the dominant vibration, i.e. $\{x_1, x_2, x_3\} = \{0, 1, 0\}$ .when $\{x_1, x_2, x_3\} = \{0, 0, 1\}$, the third one is 94Hz.

Table 1. Comparison of natural frequencies.

Order	Finite element method	Theoretical results	Deviation	Relative error
The first	32	35	3	9.4%
The second	54	60	6	11.1%
The third	76	88	12	15.8%

Table 1 clearly describes the natural frequencies obtained by theoretical and finite element methods. From Table 1, the error of the first natural frequency is 9.4% within the allowable range (10%), but the second and third natural frequencies are quite different. The reason is that the accuracy of the first order natural frequency calculated by Rayleigh method can be guaranteed. When the order is higher, the calculation error is too large to compare. Therefore, the theoretical natural frequencies are accordant with those calculated by finite element method, which proves the suitability of the theoretical model.

5. Conclusion

In this paper, a novel cantilever plates structure with an additional ring mass frame is proposed, which can obtain three lower and close natural frequencies. Firstly, the natural frequencies and the first three principle modes are obtained by the finite element method, and the idea is verified. Next, the theoretical vibration modal expressions are obtained by referring to the mode gotten by the finite

IOP Conf. Series: Materials Science and Engineering **825** (2020) 012017 doi:10.1088/1757-899X/825/1/012017

element method. The theoretical mode are consistent with simulation results of finite element method, which verified the correctness of the solution of modal expression. The theoretical natural frequencies of structures are obtained by Rayleigh method. The comparison results of natural frequencies shows that our theoretical model is suitable. It provides theoretical basis for the follow-up study of piezoelectric energy harvester.

Acknowledgements
The authors express gratitude for the supports provided by the National Natural Science Foundation of China under Grants No. 11472019 for this research work.

References
[1] J.W. Xu, Study on micro inextensible plate piezoelectric energy harvester with mass. Master's thesis. Beijing university of technology, China, 2018.
[2] Q.C. Guan, B. Ju, J. W. Xu, et al. Improved strain distribution of cantilever piezoelectric energy harvesting devices using H-shaped proof masses. American Journal of Veterinary Research. 2013, 24(9): 1059-1066.
[3] W. G. Li, S. He, S. Yu. Improving Power Density of a Cantilever Piezoelectric Power Harvester Through a Curved L-Shaped Proof Mass. IEEE Transactions on Industrial Electronics. 2010, 57(3): 868-876.
[4] K.M. Tsuruta, D.A. Rade, R.M.F Neto, et al. Experimental evaluation of a cruciform piezoelectric energy harvester. Mechanical Systems & Signal Processing. 2016, 79: 141-148.
[5] A. Erturk, J. M. Renno, D. J. Inman. Modeling of Piezoelectric Energy Harvesting from an L-shaped Beam-mass Structure with an Application to UAVs. Journal of Intelligent Material Systems & Structures. 2009, 20(5): 529-544.
[6] H. Hu, C. Zhao, S. Feng, et al. Adjusting the resonant frequency of a PVDF bimorph power harvester through a corrugation-shaped harvesting structure[J]. Ultrasonics Ferroelectrics & Frequency Control IEEE Transactions on. 2008, 55(3): 668-674.

IOP Conf. Series: Materials Science and Engineering **825** (2020) 012018 doi:10.1088/1757-899X/825/1/012018

High Definition Map Based Motion Plan and Control of Autonomous Vehicle on Structured Road

K Zhang[1,*], S J Wang[2], L Ji[3] and C Wang[1]

[1]E/E department, Brilliance Automobile Engineering Research Institute, shen yang 110141, China

[2]School of Mechanical Engineering, Shenyang University of Technology, shen yang, 110023, China

[3]School of Mechatronics Engineering, Shenyang Aerospace University, shen yang, 110136, China

[*] hit_zhangkun@126.com

Abstract. A high definition (HD) map based motion plan and control method of autonomous vehicle on structured road is proposed in this paper. The system is designed in a multi-layer structure: a motion planner and a motion controller, furthermore, the motion planner is consisted of a global path planner and a local trajectory planner. The inputs of the global path planner are HD map, start and goal state, Dynamic Program (DP) is used to plan a global path that connect these two states based on the lane-level structures and traffic rules extracted from HD map. The inputs of local motion planner are the planned global path, current host vehicle state and the surrounding dynamic object information, several candidate trajectories are generated in Frenet frame, and then the optimal one is chosen by a multi-term cost function calculation. Taking the vehicle dynamics under high speed into consideration, Model Predictive Control (MPC) is used in motion controller to track the planned target trajectory. Results of the conducted simulation tests show that the designed motion plan and control method works well in velocity keeping, vehicle following, stopping and going, and vehicle passing scenarios.

1. Introduction

Great progress has been made in autonomous driving in the past 3 decades, especially since the DARPA urban challenge which was held in 2007. It has been widely accepted that the autonomous driving software designed in a abstracted layered structure, including sensor interface layer, perception and localization layer, decision and planning layer, and vehicle interface layer. Decision and planning layer is further divided into route planner, behaviour executive layer, motion planner and motion controller. To limit the scope of this paper, we focus on the motion planner and motion controller.

Regarding the motion planner, a mass of methods have been proposed. These methods can be divided into 3 fields: variational methods, graph-search methods and incremental search methods. Variational methods represent the path as a finite-dimensional vector function, and the optimal path is found through optimizing over the parameter vector using non-linear continuous optimization techniques, this method can rapidly convergence to local minimum[1]. Graph search methods represent the configuration space of the vehicle as a graph or a multi-dimensional lattice grid, and the target path is found through a multi-term cost function, many of the popular used method like A*, anytime A*, D* falls into this criteria[2, 3]. Incremental search methods build a reachability tree by sampling the configuration space, once the graph grows large enough that the goal is reached, the target path is found, RRT, RRT*, RRTX belongs to this criteria[4, 5].

The motion control of autonomous vehicles, depending on the vehicle models used, can be divided into 3 classes: geometric model[6, 7], kinematic model[8, 9] and dynamic model[10]. Instinctively, the lateral and longitudinal of autonomous vehicle should be coupled controlled to achieve more accurate path tracking. On the other hand, the relative high speed on structured road and rapid changing characteristic of the road users need the controller to be able to response quickly.

Thus, in this paper the whole time used for motion plan and motion control is limited to less than 100ms. Furthermore, the comfort of passenger are taken into consideration during the controller design. Based on these design requirements, an abstract layered motion plan and control method of autonomous vehicle on structured road is proposed in this paper, in which the global planner is implemented based on HD map information, local optimal lateral and longitudinal trajectory is generated and then tracked through MPC by the motion controller.

2. System structure

The designed system structure diagram is shown in figure 1, where the proposed motion plan and motion control system is designed in a multi-layered form, with a motion planner and a motion controller and the motion planner is further consisted of a global path planner and a local trajectory planner.

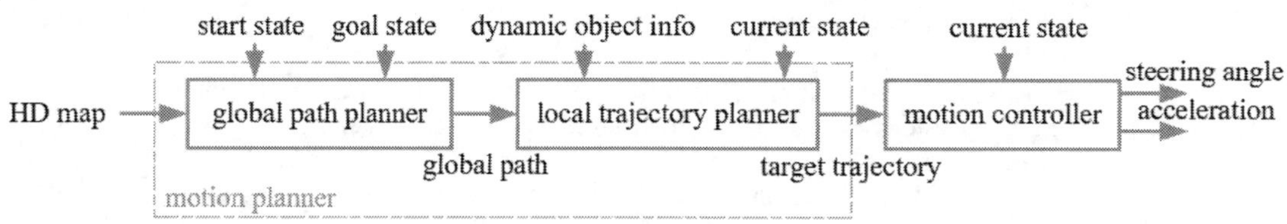

Figure 1. System structure diagram

The inputs of the global path planner are HD map, start and goal state, and the output is a global path that connect these two states based on the HD map information. The HD map information includes the lane structure and traffic rules, such as the speed limit, traffic sign, et al., and is usually provided by the map supplier. The start and goal state are consisted of position, orientation and movement info, which is supplied by the behaviour executive layer or directly assigned by the customer through human machine interface (HMI).

The inputs of local trajectory planner are the planned global path, current host vehicle state and the surrounding dynamic objects info, and the output is a target trajectory. The current host vehicle state is provided by the GNSS/INS coupled localization system or a 3D/2D SLAM system. Several candidate trajectories are generated in Frenet frame, and then the optimal one is chosen through a predefined multi-term cost function, which should take safety, comfort and smoothness into consideration.

The target trajectory is then tracked by the MPC motion controller, whose inputs are the target trajectory and current vehicle state, and the outputs are the corresponding steering and acceleration control signal.

3. Motion planner

3.1. Global path planner

Depending on the driving scenarios, different global planning methods are used, such as A* algorithm and its variants are often used in the off-road or the parking lot conditions. These methods are able to get optimal or sub-optimal result but need much computation, and thus are not suitable for high speed structured road application.

The global path plan method proposed in this paper focus on real-time dynamic application, so DP is used to generate a feasible global path from directly from HD map, which is light weight and do not need much computation. Meanwhile, because the lane structure extracted from HD map is already smooth enough and the embedding traffic rule has been taken into consideration, this planned global path is considered to be feasible and conformed to the traffic rule.

The detailed procedure of DP based global path planner is: first, the lane centre is extracted and potential smooth work are made; second, a search tree of all the feasible path along the road is built from the start position abiding the traffic rule, meanwhile a predefined cost is calculated for every path, this step stops when the goal is reached; third, the path with the minimum cost is chosen as the optimal global path, as is shown in figure 2.

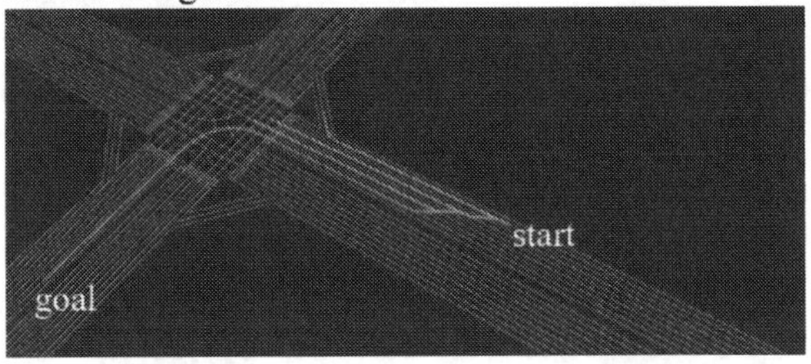

Figure 2. DP based Global path plan

3.2. Local trajectory planner

The target of local trajectory planner is to generate a safe, feasible and comfortable local trajectory along the provided global path. Based on this definition, the following requirements need to be satisfied: the planned trajectory should not collide with other dynamic and static obstacles; the planned trajectory should be smooth enough that is trackable by a non-holonomic vehicle with angular and linear velocity/acceleration constraints; the planned trajectory should be comfort enough for the passengers, thus sharp lateral and longitudinal acceleration should be avoided; planning computation needed should be small enough so as to realize real-time control. Thus, a two-step method is proposed, including a candidate trajectories generation step and an optimal calculation step.

3.2.1. Candidate trajectories generation. A jerk term that was proposed by Moritz Werling [11] is used in this paper to ensure the smooth and comfort of the generated lateral and longitudinal trajectories, which is defined as $\dddot{d}$ and $\dddot{s}$ separately. The jerk optimal solution of a start state $\begin{bmatrix} d_0 & \dot{d}_0 & \ddot{d}_0 \end{bmatrix}$ and end state $\begin{bmatrix} d_1 & \dot{d}_1 & \ddot{d}_1 \end{bmatrix}$ can be represented by a quintic-polynomials.

For lateral control, its meaningful to set the end state as $\begin{bmatrix} d_1 & 0 & 0 \end{bmatrix}$, and then d_1 and the prediction time is sampled with a predefined step. Given the start and sampled different end states, a set of candidate trajectories are generated, then the validation of these trajectories is checked by curvature, lateral velocity and lateral acceleration, and the invalid ones are filtered out, as is shown in figure 3.

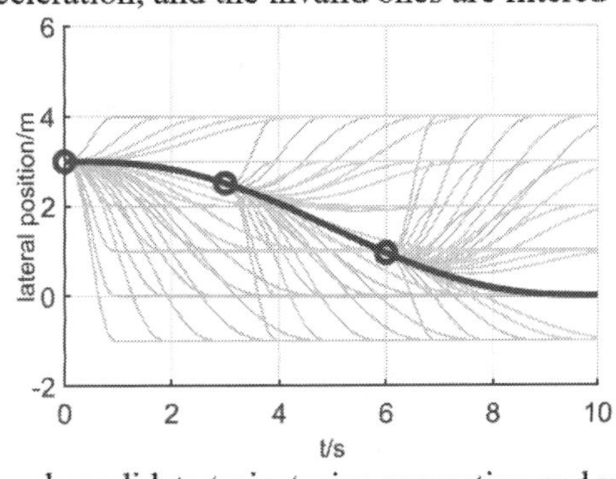

Figure 3. Lateral candidate trajectories generation and validation check

IOP Conf. Series: Materials Science and Engineering **825** (2020) 012018 doi:10.1088/1757-899X/825/1/012018

As is shown in figure 3, the invalid trajectories are drawn in grey and the valid ones are drawn in green, and the optimal is drawn in blue.

In contrast to lateral control, longitudinal control is divided into 2 conditions, one accounts for the velocity keeping scenario and the other accounts for stop and going, vehicle following and merging scenarios. The end state of the first condition is designed as $[\dot{s}_1 \quad 0]$, and $[s_1 \quad \dot{s}_1 \quad 0]$ for the second one. Then, the same sampling and validation checking step as lateral control is conducted, as is shown in figure 4.

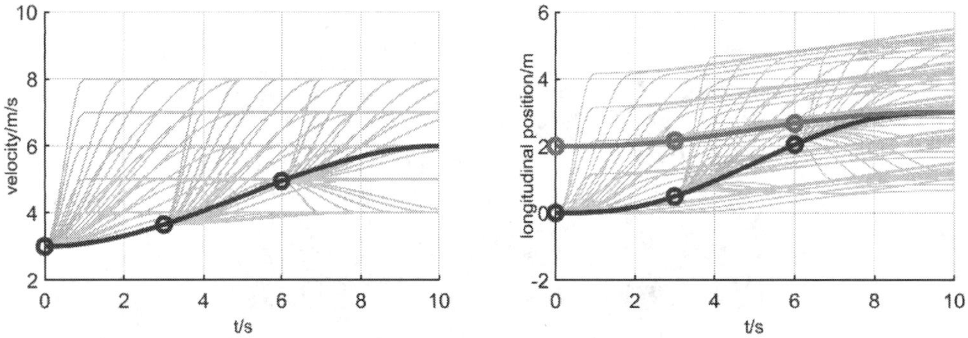

Figure 4. Longitudinal candidate trajectories generation and validation check

The longitudinal trajectories generated for velocity keeping are shown in left part of figure 4, where the invalid trajectories for velocity keeping are drawn in grey and the valid ones are drawn in green, and the optimal is drawn in blue. On the other hand, the longitudinal trajectories generated for stop and going, vehicle following and merging are shown in left part of figure 4.

*3.2.2. Optimal trajectory calculation.*Then for every left validated trajectory, the collision check is first made based on the Separating Axis Theorem (SAT). After this procedure, those trajectories that are in collision with other obstacles are all masked out, left the safety ones. Then, in order to decide the optimal trajectory, several cost functions are defined considering trajectory smooth, comfort and execution time.

For lateral control, the cost function C_{lat} is given by

$$C_{\mathrm{lat}} = K_{\mathrm{latj}} \int_{t_0}^{t_1} \dddot{d}^2 d\tau + K_{\mathrm{latt}} \left(t_1 - t_0 \right) + K_{\mathrm{latd}} d_1^2 \tag{1}$$

Where K_{latj}, K_{latt} and K_{latd} is weights for lateral jerk, execution time and end state lateral position.

For the first condition of longitudinal control, which is speed keeping, a cost function is given by

$$C_{\mathrm{long}} = K_{\mathrm{longj}} \int_{t_0}^{t_1} \dddot{s}^2 d\tau + K_{\mathrm{longt}} \left(t_1 - t_0 \right) + K_{\mathrm{longvel}} \left(\dot{s}_1 - \dot{s}_{\mathrm{target}} \right)^2 \tag{2}$$

Where K_{longj}, K_{longt} and K_{longvel} is weights for longitudinal jerk, planning time and end state longitudinal velocity.

For the second condition of longitudinal control, which includes stop and going, vehicle following and merging, a cost function is given by

$$C_{\mathrm{long}} = K_{\mathrm{longj}} \int_{t_0}^{t_1} \dddot{s}^2 d\tau + K_{\mathrm{longt}} \left(t_1 - t_0 \right) + K_{\mathrm{longpos}} \left(s_1 - s_{\mathrm{target}} \right)^2 \tag{3}$$

Where K_{longpos} is weights for the end state longitudinal position.

Then the total cost of lateral and longitudinal trajectories are combined by

$$C_{\text{total}} = K_{\text{lat}} C_{\text{lat}} + K_{\text{long}} C_{\text{long}} \tag{4}$$

Where C_{total} is the total cost, K_{lat} and K_{long} are weights for lateral and longitudinal cost.

4. Motion controller

The most commonly used dynamic vehicle steering model is called the bicycle model, as is shown in figure 5, is consisted of two wheels that were rigidly linked, and is made under the following assumptions: (1) Ignore the suspension system, the vehicle is only allowed travel on a plane without pitch and roll movement. (2) Ignore the steer system, the front wheel is directly controlled to rotate about an axis normal to the driving plane within some range constraints. (3) The vehicle speed at a specific time is thought to be constant. (4) The lateral acceleration of the vehicle is below 0.4g.

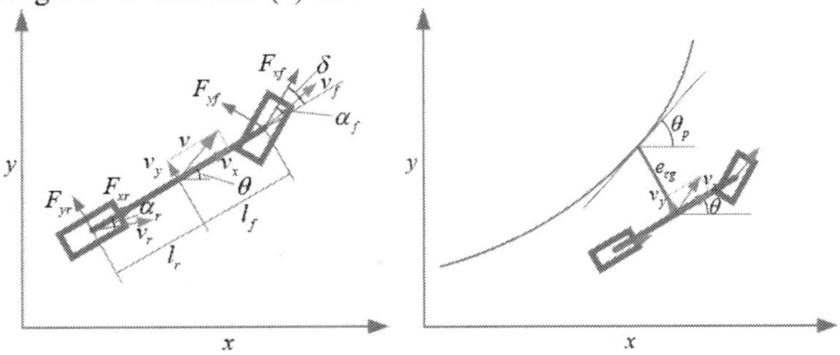

Figure 5. dynamic vehicle steering model

The left part of figure 5 illustrated the relationship between the force and lateral movement, where F_{yf} and F_{xf} denote the lateral and longitudinal force of the front wheel, F_{yr} and F_{xr} denote the lateral and longitudinal force of the rear wheel, v_f and v_r is the velocity of the front and rear wheel, v is the velocity at the vehicle central gravity. α_f and α_r denote the angle between velocity and tire heading, which are referred as lateral slip angle. The heading θ is an angle describing the direction of the vehicle with respect to the x axis. The right part of figure 5 illustrated the dynamic vehicle steering model with respect to the reference path, which is expressed as a function of its length s, and define θ_p as the angle between the path tangent and the global x axis.

Lateral and yaw movement of the vehicle are a primary concern in steering control, given by

$$\begin{cases} F_{yf}\cos\delta + F_{xf}\sin\delta + F_{yr} = m(\dot{v}_y + v_x\omega) \\ l_f(F_{yf}\cos\delta + F_{xf}\sin\delta) - l_r F_{yr} = I_z\dot{\omega} \end{cases} \tag{5}$$

Where m is the mass of the vehicle, ω is the yaw rate, I_z is the inertial moment, l_f and l_r are the length of front and rear half axis. When the slip angle is small enough, the lateral force of the front and rear wheel is linearly proportional to the slip angle, which is given by

$$\begin{cases} F_{yf} = \text{-}c_f\alpha_f = \text{-}c_f\left(\arctan(\dfrac{v_y + l_f\omega}{v_x}) - \delta\right) \\ F_{yr} = \text{-}c_r\alpha_r = \text{-}c_r\arctan(\dfrac{v_y - l_r\omega}{v_x}) \end{cases} \tag{6}$$

Where c_f and c_r is the corner stiffness of the front and rear wheels.

Define $c(s)$ as the curvature along the path, the derived yaw rate $\omega(s)$ and lateral acceleration $\dot{v}_y(s)$ under constant longitudinal velocity assumption are given by $\omega(s)=c(s)v_x$ and $\dot{v}_y(s)=c(s)v_x^2$. The lateral distance error of the vehicle central gravity and heading angle error should satisfy

$$\begin{cases} \dot{e}_{cg} = v_y + v_x \sin\theta_e \\ \dot{\theta}_e = \omega - \omega(s) \end{cases} \tag{7}$$

Substitute Eqs.5 and Eqs.6 into Eqs.7, and represent it in state space form, yields

$$\dot{x} = Ax + Bu + C\dot{\omega}(s) \tag{8}$$

Where $x=\begin{bmatrix} e_{cg} & \dot{e}_{cg} & \theta_e & \dot{\theta}_e & s & \dot{s} \end{bmatrix}^T$, $u=\begin{bmatrix} \delta & a \end{bmatrix}^T$,

$$A=\begin{bmatrix} 0 & 1 & 0 & 0 & 0 & 0 \\ 0 & -\frac{c_f+c_r}{mv_x} & \frac{c_f+c_r}{m} & -\frac{l_fc_f-l_rc_r}{mv_x} & 0 & 0 \\ 0 & 0 & 0 & 1 & 0 & 0 \\ 0 & -\frac{l_fc_f-l_rc_r}{I_zv_x} & \frac{l_fc_f-l_rc_r}{I_z} & -\frac{l_f^2c_f+l_r^2c_r}{I_zv_x} & 0 & 0 \\ 0 & 0 & 0 & 0 & 0 & 1 \\ 0 & 0 & 0 & 0 & 0 & 0 \end{bmatrix}, B=\begin{bmatrix} 0 & 0 \\ \frac{c_f}{m} & 0 \\ 0 & 0 \\ \frac{l_fc_f}{I_z} & 0 \\ 0 & 0 \\ 0 & -1 \end{bmatrix}, C=\begin{bmatrix} 0 \\ -\frac{l_fc_f-l_rc_r}{mv_x}-v_x \\ 0 \\ -\frac{l_f^2c_f+l_r^2c_r}{I_zv_x} \\ 0 \\ 1 \end{bmatrix}$$

This designed dynamic model is then transferred into discrete form, meanwhile the reference local trajectory generated by the motion planner can be represented by as sequence of reference states x^*. MPC is then used to utilize path tracking, within a prediction horizon H, define the cost function as

$$J=\sum_{t=1}^{H}\left(\left(x_t-x_t^*\right)^T Q\left(x_t-x_t^*\right)+u_t^T R u_t\right) \tag{9}$$

Where Q and R is the weight matrix, furthermore, Q is designed as a semi-definite diagnose matrix, R is designed as a definite diagnose matrix. Then given a set of constraints on u, some of which are derived from the vehicle dynamic constraints, others directly from the HD map, the optimal control during the prediction horizon can be solved by

$$u_t^* = \text{argmin}(J) \tag{10}$$

5. Simulation and results

In order to test the effect of this work, a set of simulation tests are conducted. The simulation results show that the designed motion plan and control method works well on structured road, which can handle common driving scenarios like velocity keeping, vehicle following, stop and going, passing, etc., some of these scenarios are shown in figure 6.

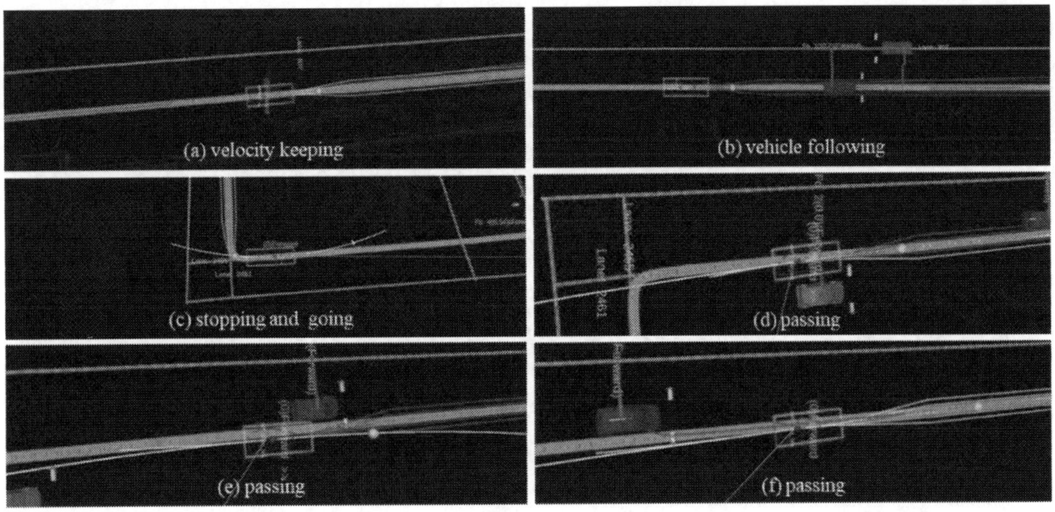

Figure 6. dynamic vehicle steering model

As is shown in figure 6: (a) when there is no other central in-path vehicles (CIPV), the host vehicle will keep a predefined target speed; (b) when there exist a lower speed CIPV, the host vehicle will follow it with a self-adaptive space gap; (c) when approaching a turning corner or a stopping line, the host vehicle will stop at the defined position and waiting for the signal and then start moving again; (d-f) when there exist a slow moving CIPV and the speed gap is less than a predefined threshold, the host vehicle will passing it and then merging back to the original lane. Limited by the space of this paper, only a few representable common driving tasks are illustrated here.

6. Conclusion

A HD map based autonomous driving motion plan and control method for structured road condition is proposed in this paper, which is consisted of a motion planner and a motion controller, and the motion planner is further divided into a global planner and a local planner. DP is used to generate a global path based on HD map information. Several candidate trajectories are generated in Frenet frame, and the optimal is chosen through a predefined cost function. Then MPC is used in motion controller to track the reference trajectory. Simulation results show that the designed motion plan and control method is able to handle common scenarios occurred on structured road.

Acknowledgments

This work is supported by the China National key research and development plan key special project of intergovernmental international science and technology innovation cooperation development project "automotive passive and active collaborative protection technology for pedestrian safety" (No. 2018YFE0192900).

References

[1] Darby C L, Hager W W and Rao A V 2011 An hp-adaptive pseudo spectral method for solving optimal control problems *Optimal Control Applications & Methods* vol 32(4) pp 476-502.

[2] Likhachev M, Ferguson D, Gordon G, Stentz A and Thrun S 2005 Anytime Dynamic A*: An Anytime, Replanning Algorithm *Proceedings of the International Conference on Automated Planning and Scheduling (ICAPS)* pp 262-71.

[3] Ferguson D and Stentz A 2006 Using interpolation to improve path planning: The Field D* algorithm *Journal of Field Robotics* vol 23(2) pp 79-101.

[4] Karaman S and Frazzoli E 2010 Optimal Kinodynamic Motion Planning using Incremental Sampling-based Methods *Proceedings of the 49th IEEE Conference on Decision and Control, CDC 2010* pp 7681-87.
[5] Webb D J and Berg J V D 2013 Kinodynamic RRT*: Asymptotically optimal motion planning for robots with linear dynamics *IEEE International Conference on Robotics & Automation. IEEE* pp 5054-61.
[6] Thrun S, Montemerlo M and Palatucci M 2009 Stanley: The Robot that Won the DARPA Grand Challenge *Journal of Field Robotics* vol 23(9) pp 661-92.
[7] Urmson C, Anhalt J, Bagnell D, et al 2008 Autonomous Driving in Urban Environments: Boss and the Urban Challenge *Journal of Field Robotics* vol 25(8) pp 425–66.
[8] Luca A D, Oriolo G and Samson C 1998 Feedback control of a nonholonomic car-like robot *In Robot Motion Planning and Control* pp 171–249.
[9] Campbell S F 2007 Steering control of an autonomous ground vehicle with application to the DARPA Urban Challenge *Massachusetts Institute of Technology* pp 25-42.
[10] Guo J H, Hu P, Li L H, et al 2012 Automatic steering controller design for vision-based unmanned vehicle *J Dalian Univ of Tech* vol 52(3) pp 437-42.
[11] Moritz W, Julius Z, Sören K and Sebastian T 2010 Optimal Trajectory Generation for Dynamic Street Scenarios in a Frenet Frame *Proceedings - IEEE International Conference on Robotics and Automation* pp 987-93.

IOP Conf. Series: Materials Science and Engineering **825** (2020) 012019 doi:10.1088/1757-899X/825/1/012019

Transient Quality Performance Evaluation of Multi-stage Flexible Manufacturing Systems

Shihong Liu[1,2], Shichang Du[1,2,*] and Lifeng Xi[1,2]

[1]Department of Industrial Engineering and Management, School of Mechanical Engineering, Shanghai Jiaotong University, Shanghai, 200240, China
[2]State Key Lab of Mechanical System and Vibration, Shanghai, 200240, China

[*]Corresponding author e-mail: lovbin@sjtu.edu.cn

Abstract. This paper aims at evaluation of the quality performance of multi-stage manufacturing system during transients. A Markovian method is constructed in order to evaluate transient propagation of quality at multi-stage manufacturing systems which have remote quality information feedback. With the derived Markovian model, analytical expressions of estimating quality metrics in transients are provided, such as settling time and dynamical quality. Case study at a factory is given to demonstrate the derived approach. The conclusions drawn illustrate the applicability of transient quality evaluation of multi-stage systems.

1. Introduction

1.1. Motivation

Steady state analysis of manufacturing systems has been numerously researched in the past few decades. It is effective in mass volume production since the transients can be ignored compared to the entire production. However, steady state analysis may become ineffective in flexible manufacturing systems, in which production is conducted with small volume and thus transients may become dominant in the manufacturing process. In flexible manufacturing systems multi-types of products are manufactured with batches on the same line. During the production, product quality is dominant by locating accuracy of flexible fixtures. Product changes may result in defect items caused by frequent fixture relocation errors. After product changeover, system quality suffer from transients due to condition such as the fixture relocating errors. Quality transients display the quality before arriving at steady state and are important. System quality mean is unstable during transients and different from steady state. However, compared to the huge results of steady state analysis, transient quality analysis of manufacturing systems maintains generally unexplored. Few analytical approaches regarding the relation between manufacturing systems and quality propagation during transients are researched in current works. This work aims at contributing to this part.

1.2. Literature Review

Numerous research has been put on flexible manufacturing systems which are widely applied in modern industry. It has been shown in recent research flexibility has effects on quality [1]. Among various modeling methods of manufacturing systems, the application of Markov models to manufacturing and service systems has been largely published recently (see [2]-[4]). However, these

IOP Conf. Series: Materials Science and Engineering **825** (2020) 012019 doi:10.1088/1757-899X/825/1/012019

results are largely focused with steady state of manufacturing systems and transient analysis is neglected. Recently an effort is paid to transient analysis in terms of throughput of manufacturing systems. The production throughput of lines with Bernoulli reliability machines in transients were studied using Markov model. Later on more reports are published in multi-stage Bernoulli systems, batch-based lines, assembly systems, closed production lines in [5]-[10]. Despite of these efforts, transient quality analysis of flexible systems is not fully understood. [11] deals with the special two-stage case of quality evaluation during transients. This work addresses the transient system quality in multi-stage systems.

2. Models and problem formulation

2.1. Descriptive Model

The assumptions of system quality transition in multi-stage manufacturing systems are as follows.

1. The manufacturing system has n stages with an inspection station at last stage.
2. Time slot equals to machine cycle time and only consider working period of system machines.
3. The product quality by stage $M_i(i \geq 2)$ not only relies on states of M_i but coming part quality from upstream. There are both quality correction and quality degradation in the manufacturing system and quality may get better or worse after a certain stage.
4. Denote the stage of $M_i(i = 1,2, \ldots, n)$ which is in good state g_i or defective state d_i if producing a good or defective product.
5. Quality of coming parts in stage $M_i(i \geq 2)$ at time t relies on upstream product quality state at time $(t-1)$. The good state g_{i-1} or defective d_{i-1} of stage M_{i-1} means a good or defective coming part for M_i at time t respectively.
6. When M_1 is in good state g_1, the probability to transit to good state g_1 is $(1-\alpha_1)$ and to defective d_1 is α_1. If M_1 is in defective d_1, the probability to transit to defective d_1 is $(1-\beta_1)$ and to good state g_1 is β_1.

In terms of good coming part, if $M_i(i \geq 2)$ is in good state g_i, the probability to transit to good state g_i is $(1-\gamma_i)$ and to defective d_i is γ_i. If M_i is in defective state d_i, the probability to defective d_i is $(1-\mu_i)$ and probability to transit to good state g_i is μ_i.

In terms of defective coming part, if $M_i(i \geq 2)$ is in good g_i, it has probability $(1-\eta_i)$ to transit to good state g_i and probability η_i to defective state d_i. If M_i is in defective d_i, it has probability θ_i to transit to g_i and probability to d_i is $(1-\theta_i)$.

2.2. Mathematical Model

Under the previous assumptions, transient system quality are developed in two-stage condition and then expanded for multi-stage condition.

At certain time, the two-stage manufacturing system contains four quality states. (1) State g_1g_2, both M_1 and M_2 produce good products; (2) state g_1d_2, M_1 produces good product and M_2 is producing defective part; (3) state d_1g_2, M_1 produces defective product and M_2 is producing good; (4) state d_1d_2, both M_1 and M_2 produces defective products.

A Markov chain with above four states describe the two-stage system. At time t the Markov states of the system in matrix are

$$Z_2(t)=\left[P(g_1g_2,t) \quad P(g_1d_2,t) \quad P(d_1g_2,t) \quad P(d_1d_2,t)\right]^T \tag{1}$$

The system state transits between the four quality states according to the transition probability. Set the state transition probabilities together so as to build the transition matrix.

$$B_2=\begin{bmatrix} (1-\alpha_1)(1-\gamma_2) & (1-\alpha_1)\mu_2 & \beta_1(1-\eta_2) & \beta_1\theta_2 \\ (1-\alpha_1)\gamma_2 & (1-\alpha_1)(1-\mu_2) & \beta_1\eta_2 & \beta_1(1-\theta_2) \\ \alpha_1(1-\gamma_2) & \alpha_1\mu_2 & (1-\beta_1)(1-\eta_2) & (1-\beta_1)\theta_2 \\ \alpha_1\gamma_2 & \alpha_1(1-\mu_2) & (1-\beta_1)\eta_2 & (1-\beta_1)(1-\theta_2) \end{bmatrix} \tag{2}$$

Define $P(g_2,t)$ as the system probability to produce good products.

$$P(g_2,t)=P(d_1g_2,t)+P(g_1g_2,t) \tag{3}$$

$$Z_2(t+1)=B_2Z_2(t) \tag{4}$$

The evolution of $P(d_2,t)$ and $P(g_2,t)$ are calculated as

$$y_2(t)=\begin{bmatrix}P(g_2,t)\\P(d_2,t)\end{bmatrix}=\begin{bmatrix}1&0&1&0\\0&1&0&1\end{bmatrix}Z_2(t) \tag{5}$$

3. Transient quality performance evaluation

3.1. Two-stage Case

For two-stage systems under definition of assumptions above, it is shown by the mathematical model that B_2 is transition probability matrix of the Markovian chain. The matrix B_2 is transformed to a diagonal matrix with a non-singular T matrix.

$$TB_2T^{-1}=diag\begin{bmatrix}1&\lambda_2&\lambda_3&\lambda_4\end{bmatrix}$$

Introduce the following substitution (6) and substitute equation (6) into equations (4)-(5),

$$\tilde{Z}_2(t)=TZ_2(t) \tag{6}$$

$$\tilde{Z}_2(t+1)=\tilde{B}_2\tilde{Z}_2(t) \tag{7}$$

$$y_2(t)=\tilde{C}\tilde{Z}_2(t) \tag{8}$$

The evolution of system states, $P(g_2,t)$ can be denoted as

$$\tilde{Z}_2(t)=\tilde{B}_2{}^{t}\tilde{Z}_2(0)=diag\begin{bmatrix}1&\lambda_2^t&\lambda_3^t&\lambda_4^t\end{bmatrix}\tilde{Z}_2(0) \tag{9}$$

$$P(g_2,t)=P(g_2)_{SS}\left(1+\frac{\tilde{C}_{12}}{\tilde{C}_{11}}\tilde{Z}_{2,2}(0)\lambda_2^t+\frac{\tilde{C}_{13}}{\tilde{C}_{11}}\tilde{Z}_{2,3}(0)\lambda_3^t+\frac{\tilde{C}_{14}}{\tilde{C}_{11}}\tilde{Z}_{2,4}(0)\lambda_4^t\right) \tag{10}$$

Equation (9) shows that Markov chain $\tilde{Z}_2(t)$ arrives at the steady state denoted as exponential functions of λ_i. Equation (10) shows that system quality $P(g_2,t)$ are described by the exponential factors and eigenvalues λ_i.

3.2. Multi-stage Case

The proposed method from section 3.1 can be expanded to multi-stage case. The iteration process is described by Figure. 1. System quality of two-stage M_1-M_2 is developed using the Markov method proposed from section 3.1. Then stage M_1 and M_2 are merged to a combined stage M_2'. And then system quality of fresh two-stage $M_2'-M_3$ is developed, after which stage M_2' and M_3 are merged to a merged stage M_3'. Conduct the iteration method and the first ever $(n-1)$ system stages are aggregated into a merged stage M_{n-1}'. Finally we build the model of system quality in final two-stage $M_{n-1}'-M_n$. The iteration method is as follows.

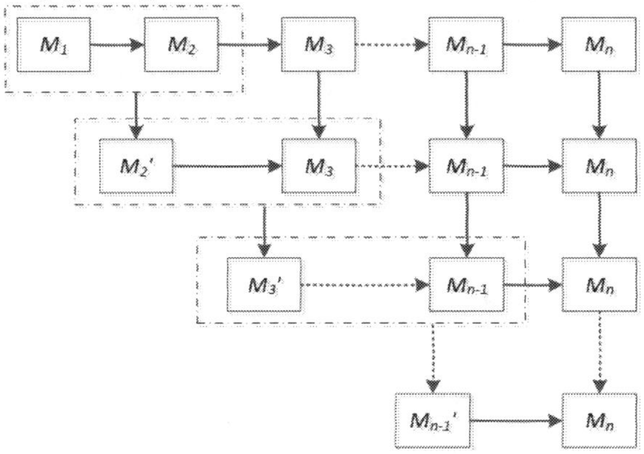

Figure 1. Iteration method of multi-stage systems.

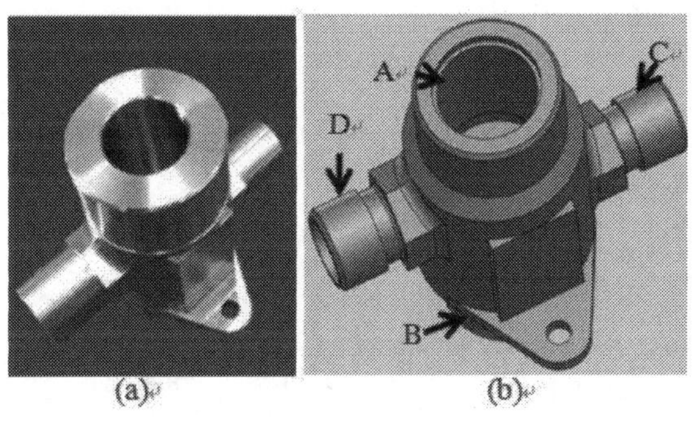

Figure 2. A type of valve shell in case study.

First, any two-stage systems $M_i' - M_{i+1}$ has a total of six system transition parameters. These consist of parameters$\alpha_i'(t), \beta_i'(t)$ representing the characteristics of the new merged stage M_i' and $\gamma_{i+1}, \eta_{i+1}, \mu_{i+1}, \theta_{i+1}$ representing the characteristics of the stage M_{i+1}. The parameters $\alpha_i'(t), \beta_i'(t)$ are calculated as follows.

$$\alpha_i'(t) = \frac{P(g_{i-1}g_i,t)\gamma_i + P(d_{i-1}g_i,t)\eta_i}{P(g_{i-1}g_i,t) + P(d_{i-1}g_i,t)} \tag{11}$$

$$\beta_i'(t) = \frac{P(g_{i-1}d_i,t)\mu_i + P(d_{i-1}d_i,t)\theta_i}{P(d_{i-1}d_i,t) + P(g_{i-1}d_i,t)} \tag{12}$$

Second, put transition probabilities in matrix form

$$B_{i+1}(t) = \begin{bmatrix} (1-\alpha_i'(t))(1-\gamma_{i+1}) & (1-\alpha_i'(t))\mu_{i+1} & \beta_i'(t)(1-\eta_{i+1}) & \beta_i'(t)\theta_{i+1} \\ (1-\alpha_i'(t))\gamma_{i+1} & (1-\alpha_i'(t))(1-\mu_{i+1}) & \beta_i'(t)\eta_{i+1} & \beta_i'(t)(1-\theta_{i+1}) \\ \alpha_i'(t)(1-\gamma_{i+1}) & \alpha_i'(t)\mu_{i+1} & (1-\beta_i'(t))(1-\eta_{i+1}) & (1-\beta_i'(t))\theta_{i+1} \\ \alpha_i'(t)\gamma_{i+1} & \alpha_i'(t)(1-\mu_{i+1}) & (1-\beta_i'(t))\eta_{i+1} & (1-\beta_i'(t))(1-\theta_{i+1}) \end{bmatrix} \tag{13}$$

The matrix of system state probabilities at certain time is described as follows,

$$Z_{i+1}(t) = \left[P(g_ig_{i+1},t) \quad P(g_id_{i+1},t) \quad P(d_ig_{i+1},t) \quad P(d_id_{i+1},t)\right]^T \tag{14}$$

The evolution of $Z_{i+1}(t)$ can be described by the linear equation,

$$\begin{aligned} &Z_{i+1}(t+1) = B_{i+1}(t)Z_{i+1}(t) \\ &P(g_ig_{i+1},t) + P(d_ig_{i+1},t) + P(g_id_{i+1},t) + P(d_id_{i+1},t) = 1 \end{aligned} \tag{15}$$

In multi-stage systems, final system quality equals to that M_{i+1} produces a good quality part. Thus we define $P(g_{i+1}, t)$ as system probability to produce good quality.

$$P(g_{i+1},t) = P(d_ig_{i+1},t) + P(g_ig_{i+1},t) \tag{16}$$

3.3. Settling time

Settling time denotes system lasting time of transients for quality performance evaluation in transients. The settling time is defined when quality performance arrives at and keeps in $\pm 3\%$ of steady state metric. Calculate the settling time as below.

$$t_S = \left| \frac{P(g_n,t) - P(g_n)_{SS}}{P(g_n)_{SS}} <= 3\% \right| \tag{17}$$

4. Case study

Here demonstrates a case study in manufacturing system for valve shell to verify the effectiveness of the proposed model. The valve shell is shown in Figure. 2. The production process has five stages, i.e., OP10 to OP50 in Figure. 3. These five stages are correlated. For example, the flatness variation in OP10 may put effects on clamping precision of downstream OP30. The hole in the upstream OP10 can be degraded or corrected in downstream OP50. Quality propagation of products is explored in the five-stage system.

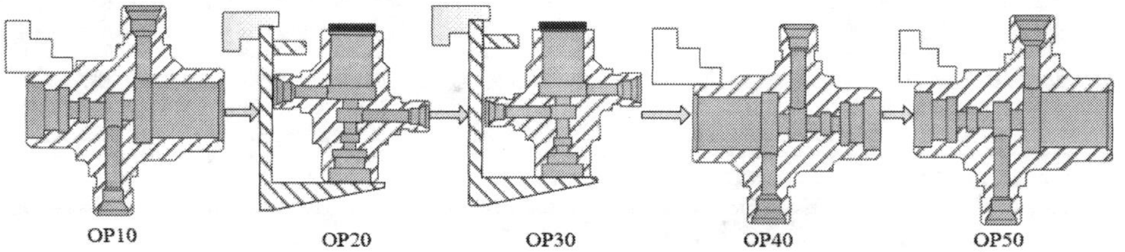

Figure 3. Valve shell manufacturing system.

After statistical analysis of processing data, transiting data of this five-stage system are given by quality failure and quality repair probabilities. The transition probability parameters are calculated $\alpha_1 = 0.05$, $\beta_1 = 0.9$, $\gamma_i = [0.05, 0.1, 0.05, 0.05]$, $\mu_i = [0.8, 0.8, 0.9, 0.9]$, $\eta_i = [0.5, 0.5, 0.4, 0.5]$, $\theta_i = [0.4, 0.3, 0.2, 0.4]$.

Using transient analysis of quality in this paper, we can calculate the transient quality performance, settling time and quality of steady state in this multi-stage system. The quality of steady state of each two-stage merged system are 91.36%, 84.85%, 88.25%, 89.07%, respectively. Settling time lasts 8 slots. And during transients the evolution of system quality $P(g_i, t)$ of each two-stage merged system are plotted in Figure. 4. The results are consistent to real data at the plant floor and prove the effectiveness of the constructed model.

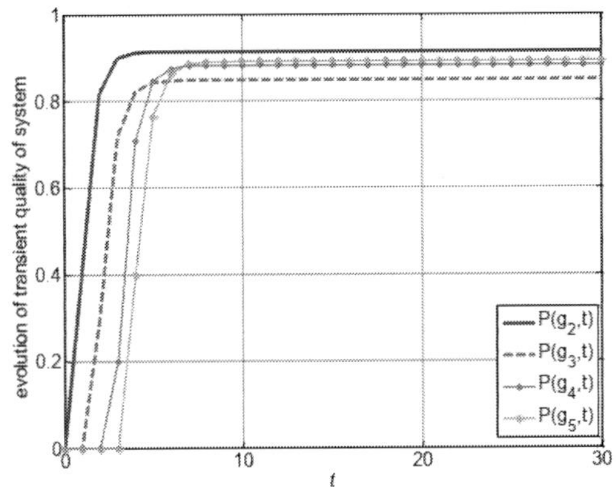

Figure 4. Dynamics of quality of every two-stage system in transients.

The evolution of system states of the final two-stage merged system $M_4' - M_5$ is illustrated in Figure. 5 (a). The dynamics of transition probability parameters corresponding to the merged stage M_4' are shown in Figure. 5 (b). It indicates that transition probabilities of merged stage also converge as time evolves. In the future analysis of monotonicity and sensitivity will be carried out and investigate how system parameters changes will pay influence on quality performance in multi-stage systems during transients, which will provide the guidance for quality improvement.

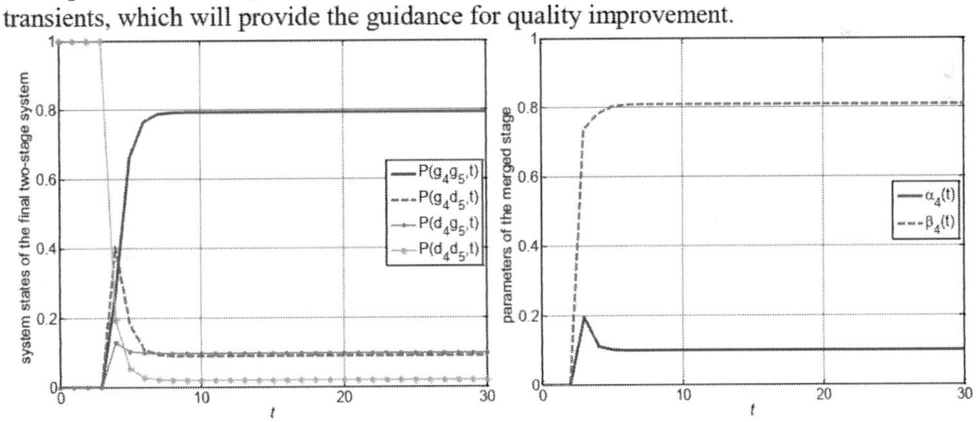

(a) system states of final two-stage system (b) transition parameters in merged stage

Figure 5. Transients of system states and transition parameters in merged stage.

5. Conclusion

This paper aims at evaluation of the quality performance of multi-stage manufacturing system during transients. A Markovian method is constructed in order to evaluate transient propagation of quality at multi-stage manufacturing systems which have remote quality information feedback. With the derived Markovian model, analytical expressions of estimating quality metrics in transients are provided, such as the settling time and dynamical quality. Case study at a factory is given to demonstrate the derived approach. The conclusions drawn illustrate the applicability of transient quality evaluation of multi-stage systems. Future works can be pointed to continuous improvement through analysis of structural properties, and extension of the approaches to modeling and evaluation of multi-types of products during transients.

Acknowledgments
This work was supported by the National Natural Science Foundation of China (Grant No. 51535007, 51775343).

Reference

[1] Zhao C, Li J and Huang N 2016 *IEEE T. Autom. Sci. Eng.* **13(1)** 105-121.
[2] Park K and Li J 2019 *International Journal of Production Research* **57(2)** 470-487.
[3] Papadopoulos C T, Li J and O'Kelly M J 2019 *Computers & Industrial Engineering* **128** 219-44.
[4] Kang Y and Ju F 2019 *IISE transactions* **51(5)** 501-516.
[5] Wang F, Ju F and Kang N 2019 *IISE Transactions* **51(7)** 709-728.
[6] Kang Y, Yan H and Ju F 2019 *IEEE Transactions on Automation Science and Engineering* 1-11.
[7] Chen G, Wang C and Zhang L 2016 *IEEE Transactions on Automatic Control* **61(4)** 877-891.
[8] Jia Z and Zhang L 2017 *IEEE Robotics and Automation Letters* **2(4)** 1893-1900.
[9] Ju F, Li J and Horst J A 2017 *IEEE Transactions on automatic control* **62(2)** 694-707.
[10] Ge Y , Li L and Wang Y 2017 *Journal of Manufacturing Systems* **44** 22-41.
[11] Liu S, Du S and Xi L 2018 *Computers & Industrial Engineering* **117** 262-281.

A Health Status Assessment Approach of Intelligent Manufacturing System based on Fuzzy Analytic Hierarchy Process

Yu-Yun Kang[1,2,*], Gui-Fang Feng[2] and Jie Sun[1]

[1]School of Mechanical Engineering, Shandong University, China
[2] Linyi University, China

[*]kyygood@163.com

Abstract. In order to timely identify, analyze the health status of an intelligent manufacturing system, an approach based on fuzzy analytic hierarchy process (FAHP) is proposed. In the method, the random perturbation is divided into several classes and each is subdivided to several events. Then an evaluation model is constructed and it has two-layer criteria system, which can evaluate every perturbation class and the whole system at the same time. The monitoring data is expressed with 10 scales and the evaluation accuracy is more improved. FAHP is used to calculate the criteria weights. Finally, the proposed method is detailed illustrated by an example, and the method is verified.

1. Introduction

It is very important to evaluate and identify the health status of intelligent manufacturing system, which can deal with some unexpected disturbance as early as possible and keep the whole system healthy operation [1] [2]. But in current evaluation methods, monitoring data is described in binary, or you can only evaluate the health status of the entire manufacturing system, not a single class of perturbation. The evaluation system is not flexible enough to deal with and detect random potential damages timely [3]. Therefore, in this paper, the monitoring data is subdivide to 10 scales and a two-layer evaluation model is constructed, and the classification and accurate evaluation of the health status of the intelligent manufacturing system is realized.

2. Disturbance identification

The random perturbation of the manufacturing system are divided into several classes, and each is used as evaluation criteria and described by $At_1, At_2, ..., At_m$. Each type of perturbation contains n events, so each is described by the vector as follows,

$$At_i = [At_{i1}, At_{i2}, ..., At_{ij}, ..., At_{in}], i = 1, 2, ..., m. \quad (1)$$

The normal value of detected items of the evaluation criteria is in a database and it is described as follows:

$$v_i^{'} = \left[v_{i1}^{'}, v_{i2}^{'}, ..., v_{ij}^{'}, ..., v_{in}^{'}\right] = \left[\alpha, \beta, ...\alpha, ...\alpha\right], i = 1, 2, ...m. \quad (2)$$

Where α and β are the values of the criteria variables and they are between 0 and 1.

The monitoring data detected and collected by the monitoring system in real time is processed into values with 10 scales between 0 and1. It is described as:

$$v_i = [v_{i1}, v_{i2}, ..., v_{ij}, ..., v_{in}] = [\mu, \mu, ...v, ...\mu], i = 1, 2, ..., m. \quad (3)$$

Where μ and v are the monitoring values, which between and 0 and 1.

Detected values will be matched with the data in the database, if there is a difference and the disturbance At_{ij} is generated, as shown in Figure.1.

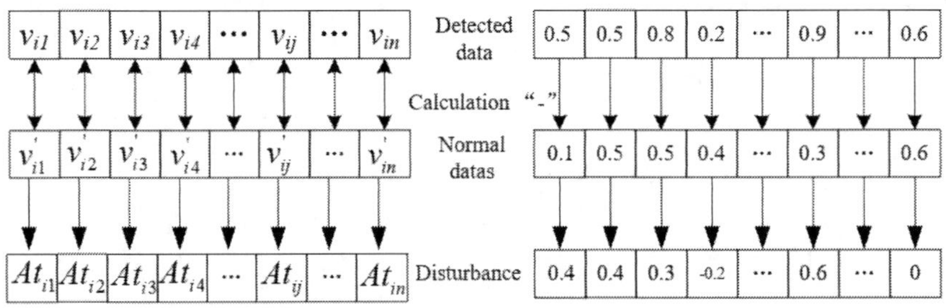

Figure 1. Generation of the disturbance.

3. Health status evaluation approach

Firstly, an evaluation model is constructed, and secondly the FAHP is used to calculate the weight of each criteria and its event. Thirdly, the evaluation value of each criteria is calculated by multiplying and adding the events' value and the events' weight. Finally, the evaluation index of the manufacturing system is calculated by multiplying and adding the criteria's values and its weight.

3.1 Evaluation model

An example of evaluation model is shown in Figure. 2. There are five criteria, such as planning task, material resource, manufacturing process, manufacturing resource, personnel changes.

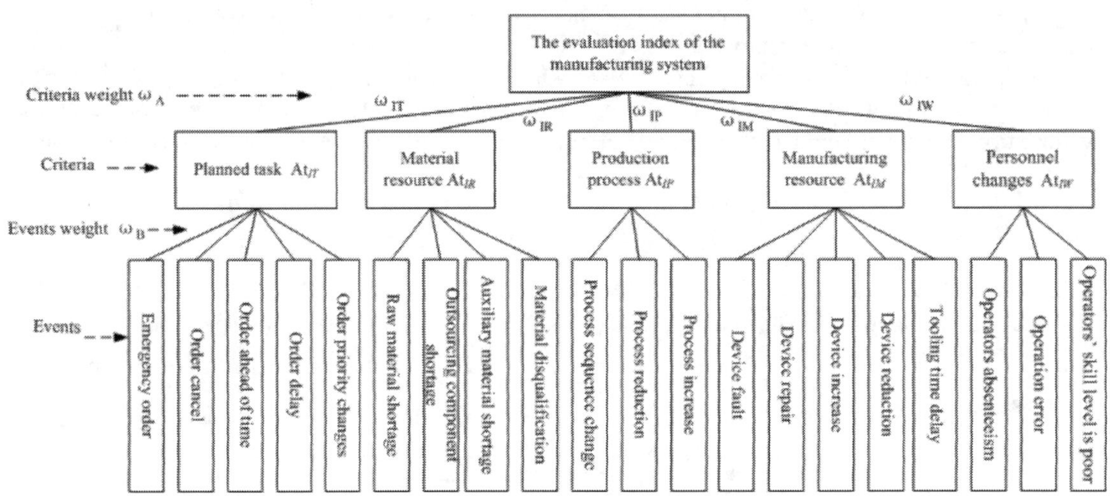

Figure 2. Evaluation model.

In the Figure. 2, the criteria is described by a vector shown as follows,

$$At = [At_{IT}, At_{IR}, At_{IP}, At_{IM}, At_{IW}]. \quad (4)$$

Each criteria is subdivided into several events, which are shown as follows,

$$At_{IT} = [At_{IT1}, At_{IT2}, At_{IT3}, At_{IT4}, At_{IT5}], \tag{5}$$

$$At_{IR} = [At_{IR1}, At_{IR2}, At_{IR3}, At_{IR4}], \tag{6}$$

$$At_{IP} = [At_{IP1}, At_{IP2}, At_{IP3}], \tag{7}$$

$$At_{IM} = [At_{IM1}, At_{IM2}, At_{IM3}, At_{IM4}, At_{IM5}], \tag{8}$$

$$At_{IW} = [At_{IW1}, At_{IW2}, At_{IW3}]. \tag{9}$$

3.2 Calculation of the weight

The theory of fuzzy mathematics is applied to the analytic hierarchy process, so it can provide a basis for evaluation and decision-making through quantitative analysis [4] [5]. For the fuzzy complementary judgment matrix $A=(a_{ij})_{n\times n}$, the matrix $\overline{A}=(\overline{a}_{ij})_{n\times n}$ is called the synthesis matrix of the matrix A. It is calculated by $\overline{A}=\lambda_1 A_1 \oplus \lambda_2 A_2 \oplus \cdots \oplus \lambda_s A_s$. $r_i=\sum_{k=1}^{n} a_{ik}$, $(i=1,2,...,n)$ is used to calculate the sum of each row in matrix $\overline{A}=(\overline{a}_{ij})_{n\times n}$, and according to the formula $r_{ij}=\frac{r_i-r_j}{2(n-1)}+0.5$, then get fuzzy consistent matrix $R=(r_{ij})_{n\times n}$. Finally the unit eigenvectors of this fuzzy consistent matrix are obtained by the normalizing rank aggregation (NRA), and it is calculated by $\omega_i=\frac{\sum_{j=1}^{n} a_{ij}+\frac{n}{2}-1}{n(n-1)}$, $i=1,2,...,n$. Then the sort vector is weight vector.

3.2.1 Calculation of evaluation value of each criteria. The evaluation value of each criteria is described by f_i calculated by the follow formula,

$$f_i = |At_i|.\omega_{Bi} = \sum_{j=1}^{n} |At_{ij}| \times \omega_{Bij}, i \in [1,2,...,m], \tag{10}$$

where f_i indicates the index of the i criteria. ω_{Bij} is the weight of each event of the i criteria. m is the quantity of evaluation criteria.

3.2.2 Calculation of evaluation index of the entire manufacturing system. The weight vector of criteria is expressed as:

$$\omega_A = [\omega_1, ..., \omega_i, ..., \omega_m]^T, \tag{11}$$

where $\sum_{i=1}^{m} \omega_i = 1$, ω_i is the weight of i criteria.

$f=(f_1, f_2, ..., f_m)$ is defined to describe the evaluation index vector. The index E_T is the evaluation value of the entire system, which is calculated as follows:

$$E_T = f \cdot \omega_A = \sum_{i=1}^{m} f_i \times \omega_i, \tag{12}$$

The smaller the value of E_T, the smaller the disturbance. The greater its value, the greater the damage of the system.

4. Application example

4.1 Calculation of the weight

Two manufacturing experts are hired to construct the fuzzy complementary matrix. They have equal weight and each is 0.5. Therefore, the two fuzzy complementary matrices are obtained as shown,

$$A_1 = \begin{matrix} & IT & IR & IP & IM & IW \\ IT \\ IR \\ IP \\ IM \\ IW \end{matrix} \begin{bmatrix} 0.5 & 0.3 & 0.2 & 0.4 & 0.5 \\ 0.7 & 0.5 & 0.7 & 0.4 & 0.5 \\ 0.8 & 0.3 & 0.5 & 0.7 & 0.6 \\ 0.6 & 0.6 & 0.3 & 0.5 & 0.4 \\ 0.5 & 0.5 & 0.4 & 0.6 & 0.5 \end{bmatrix}, \tag{13}$$

$$A_2 = \begin{matrix} & IT & IR & IP & IM & IW \\ IT \\ IR \\ IP \\ IM \\ IW \end{matrix} \begin{bmatrix} 0.5 & 0.6 & 0.7 & 0.5 & 0.7 \\ 0.4 & 0.5 & 0.4 & 0.5 & 0.4 \\ 0.3 & 0.6 & 0.5 & 0.3 & 0.5 \\ 0.5 & 0.5 & 0.7 & 0.5 & 0.4 \\ 0.3 & 0.6 & 0.5 & 0.6 & 0.5 \end{bmatrix}. \tag{14}$$

A_1 and A_2 are respectively transformed into the fuzzy consistent matrix R_1 and R_2. Then the fuzzy uniform judgment matrix $\overline{R} = \lambda_1 R_1 \oplus \lambda_2 R_2$ is acquired and shown as follows:

$$\overline{R} = 0.5R_1 \oplus 0.5R_2 = \begin{matrix} & IT & IR & IP & IM & IW \\ IT \\ IR \\ IP \\ IM \\ IW \end{matrix} \begin{bmatrix} 0.5000 & 0.49375 & 0.4875 & 0.49375 & 0.49375 \\ 0.50625 & 0.5000 & 0.49375 & 0.5000 & 0.5000 \\ 0.5125 & 0.50625 & 0.5000 & 0.50625 & 0.50625 \\ 0.50625 & 0.5000 & 0.49375 & 0.5000 & 0.5000 \\ 0.50625 & 0.5000 & 0.49375 & 0.5000 & 0.5000 \end{bmatrix}. \tag{15}$$

The weight of each criteria is obtained by using the NRA in the $\overline{R}$, and it is

$$\omega_A = [\omega_{IT}, \omega_{IR}, \omega_{IP}, \omega_{IM}, \omega_{IW}] = [0.19843, 0.2, 0.20157, 0.2, 0.2]. \tag{16}$$

In the same way, the weight of each event for every criteria is acquired and shown as follows,

$$\omega_{BIT} = [0.2375 \;\; 0.2275 \;\; 0.2215 \;\; 0.1685 \;\; 0.145]^T, \tag{17}$$

$$\omega_{BIR} = [0.1325 \;\; 0.3225 \;\; 0.1208 \;\; 0.4242]^T, \tag{18}$$

$$\omega_{BIP} = [0.2432 \;\; 0.525 \;\; 0.2318]^T, \tag{19}$$

$$\omega_{BIM} = [0.12\ \ 0.2195\ \ 0.3125\ \ 0.221\ \ 0.127]^T, \tag{20}$$

$$\omega_{BIW} = [0.237\ \ 0.337\ \ 0.426]^T, \tag{21}$$

where ω_{BIT}, ω_{BIR}, ω_{BIP}, ω_{BIM}, ω_{BIW} represent respectively the weight vector of each event.

4.2 Calculation of evaluation index

The disturbance detected at a moment is:

$$At_{IT} = [At_{IT1}, At_{IT2}, At_{IT3}, At_{IT4}, At_{IT5}] = [0.6, 0.8, -0.5, 0.7, 0.7], \tag{22}$$

$$At_{IR} = [At_{IR1}, At_{IR2}, At_{IR3}, At_{IR4}] = [0, 0.3, 0, 0], \tag{23}$$

$$At_{IP} = [At_{IP1}, At_{IP2}, At_{IP3}] = [0.2, 0, 0.1], \tag{24}$$

$$At_{IM} = [At_{IM1}, At_{IM2}, At_{IM3}, At_{IM4}, At_{IM5}] = [0.2, 0, -0.1, 0.1, 0.1], \tag{25}$$

$$At_{IW} = [At_{IW1}, At_{IW2}, At_{IW3}] = [0.2, 0, 0.1]. \tag{26}$$

The disturbance evaluation value of the plan task criteria can be calculated by

$$f_{IT} = |At_{IT}| \times \omega_{BIT} = [0.6, 0.8, |-0.5|, 0.7, 0.7] \times \begin{bmatrix} 0.2375 \\ 0.2275 \\ 0.2215 \\ 0.1685 \\ 0.145 \end{bmatrix} = 0.6547. \tag{27}$$

Similarly, we can calculate the evaluation value of other criteria. Therefore the evaluation index vector is

$$f = (f_{IT}, f_{IR}, f_{IP}, f_{IM}, f_{IW}) = [0.6547, 0.09675, 0.07182, 0.09005, 0.09]. \tag{28}$$

The evaluation index vector f is multiplied by its weight ω_A and then added together to get the system index E_T, and the formula is

$$E_T = f.\omega_A = [0.6547, 0.09675, 0.07182, 0.09005, 0.09] \times \begin{bmatrix} 0.19843 \\ 0.2 \\ 0.20157 \\ 0.2 \\ 0.2 \end{bmatrix} = 0.199747. \tag{29}$$

E_T is the disturbance index, so I_T is defined to described the healthy index, and its value is calculated by the formula

$$I_T = 1 - E_T = 0.800253. \tag{30}$$

Therefore, the system as a whole is healthy, but AT_{IT} is more damage to the system

IOP Conf. Series: Materials Science and Engineering **825** (2020) 012020 doi:10.1088/1757-899X/825/1/012020

5. Conclusions

The detected data is detailed to improve the detection accuracy. The evaluation model is constructed to realize the double-layer monitor. It can not only monitor the health status of single class indicators, but also monitor the health status of the whole system. The disturbance of the whole system can be intervened in advance, so as to ensure the healthy operation of the intelligent manufacturing system.

In the next work, the hardware should be designed and the real-time data detected by the hardware should be processed. The software system will be completed to realized the method. The method of automatically responding to the disturbance should be researched.

Acknowledgments

This research was supported by Shandong Province Natural Science Foundation (No. ZR2016EEM47), Shandong Province Higher Educational Science and Technology Program (No. JB18B018), Linyi University Project of Doctor Research Foundation (No. LYDX2015BS014).

References

[1] Abdelhamied M, Badr I, Sami M and Drwaish A 2015 An Agent-Based Method for Automating Disturbance Handling for FMS *International Conference on Computer Engineering & Systems (Heliopoilis Cairo)* vol 1 pp. 191-196

[2] Bokrantz J, A., Skoogh T, Ylipää and Stahre J 2016 Handling of Production Disturbances in the Manufacturing Industry *Journal of Manufacturing Technology Management* **27** pp 1054-1075

[3] Li Zh, Wang H F, Liu T T, and Fei Y Zh 2013 Design of Real-Time Monitoring System of Workshop for Manufacturing Process *Mechanical Design and Manufacturing* **3** pp 256-259

[4] Felea I, Dzitac S, Vesselenyi T and Dzitac I 2014 Decision Support Model for Production Disturbance Estimation *International Journal of Information Technology & Decision Making* **13** pp 623–647

[5] McLean R, Walker A J and Bright G 2017 An Artificial Neural Network Driven Decision-Making System for Manufacturing Disturbance Mitigation in Reconfigurable Systems *IEEE International Conference on Control & Automation (Ohrid)*, vol 1 pp 695-700

Restoration and Reuse Design of Industrial Heritage based on Virtual Reality Technology

Chen Jue[1] and Wang Chen[2,*]

[1]School of Arts and Design, Guangdong University of Finance & Economics, No. 21, Luntou Road, Haizhu District, Guangzhou, Guangdong Province, China
[2]School of Design, South China University of Technology, University Town, Panyu District, Guangzhou, Guangdong Province, China

*E-mail: 20181032@gdufe.edu.cn

Abstract. Based on the practical work of revitalizing the old industrial heritage of Zhengzhou No.2 Abrasive Wheel Factory, the method and technical feasibility of the restoration and reuse design of industrial heritage based on virtual reality technology are discussed in the paper. The research is based on MARS virtual reality technology platform, which is compatible with traditional modeling software, such as SKETCHUP and 3DMAX, and can directly use related software for modeling and import, reduce repetitive work and greatly shorten working time. In addition, the platform has good post-scene simulation function and is suitable for restoration and creative conceptual design in complex environment. The paper is divided into three parts. The first part is a brief introduction of the project background and task. The second part discusses the virtual reality technology solution proposed according to the project needs, and further analyzes the implementation process and technical characteristics of the solution. The final part evaluates the role of virtual reality technology in the restoration and reuse design of industrial heritage, so as to provide some inspiration for relevant designs.

1. Introduction

The project base of Zhengzhou No.2 Abrasive Wheel Factory is located in the former No.2 Abrasive Wheel Factory in the west of Huashan Road and South of Zhongyuan Road, Zhongyuan District, Zhengzhou City, with a total area of about 0.56km2. Built in 1953 with the aid of the Democratic Republic of Germany, it was the largest Abrasive Wheel Factory in China. The project is close to Zhongyuan Road, the main axis of the east-west development of Zhengzhou, and close to the west third ring road, enjoying good surrounding business atmosphere, perfect supporting facilities, and good accessibility, key development location toward west, as well as good development potential. The base is rich in industrial heritage and historical resources, and has a high degree of scarcity and historical and cultural significance. There are seven cultural relics buildings in the factory, which are listed as city level, namely, Ceramic Abrasive Wheel Manufacturing Workshop, Emery Sand Warehouse and Binder Treatment Workshop, Refractory Materials and Binder Processing Workshop, Spare Parts Tools and Lubricating Oil Warehouse, Abrasive Wheel Finished Products Warehouse and Delivery Room, Rubber Abrasive Wheel Manufacturing Workshop and Factory Office Building. Among them, the Rubber Abrasive Wheel Manufacturing Workshop is a single-storey workshop with a construction area of about 3,000m2 (100m long and 30m wide) (Figure 1). It is made of reinforced concrete roof frame, and has large clearance at the bottom, high artistic value, reuse value, and excellent potential to transform and reuse industrial heritage buildings.

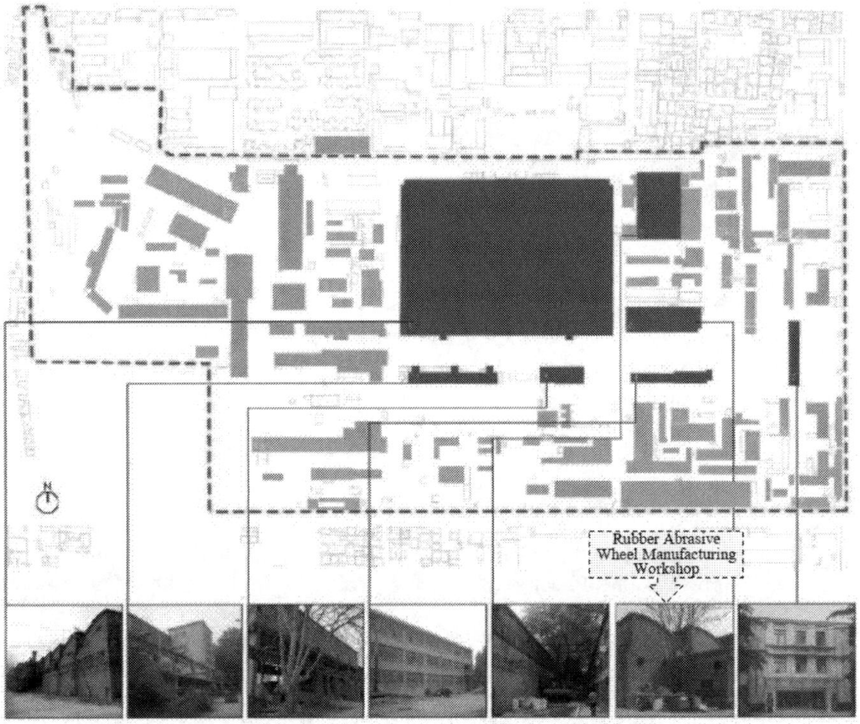

Figure 1. Plane of Seven First -level cultural relic

2. Project analysis

2.1. Tasks

The project aims to restore the original appearance of the former industrial heritage buildings, focus on the city and industry of Zhengzhou, find the core driving force of the project development, create a multi-functional compound place with diverse needs of people, and form a cultural and creative industrial base with the continuous vitality of the whole life cycle combining functions and prominent driving force.

2.2. Design Scope and Technical Indicators

The design scope of this case is the remains of original equipment in and around the Rubber Abrasive Wheel Manufacturing Workshop. The building structure is well preserved. As the second largest building in the construction park, it is 9.5m in height, 35.4m in length from north to south, and 105m in length from east to west. Due to its huge volume, we hope to take this case as an experimental project and copy its experience after its success, so as to prepare for the overall design.

2.3. Guiding Thought and Design Principles

The case aims to retain the primitive historical sense of the original factory building, restore the original appearance of its cultural relics building, integrate the internal and external resources of the project, and make full use of the transportation convenience, make good use of the external environment, and improve the cultural quality. The design style is inclined to modern fashion, but peculiar exaggerated shape is not advocated so as to moderately control cost.

IOP Conf. Series: Materials Science and Engineering **825** (2020) 012021 doi:10.1088/1757-899X/825/1/012021

2.4. Difficulties in the Design
The main problems encountered in the design mainly include the technical restrictions on the restoration of the original building appearance and the protection of industrial heritage buildings and cultural relics, as well as the feasibility study on the implantation of new space functions.

2.4.1 Technical Requirements for the Restoration of Industrial Heritage. Since the project has been updated and changed in the course of its use, it is now necessary to restore its original shape and structure based on the information. The comparison of surface shape and original shape of each part is shown in the following table1:

Table1. Contrast between the status quo and the original

Part	The Status Quo	The Original Shape and Structure
Skylight	Upper skylight is loose, ventilated, polluted and broken	The same to the status quo
Facade	Rot, deformation and broken glass is common 	Wooden western-style window, and glass is installed inside the window 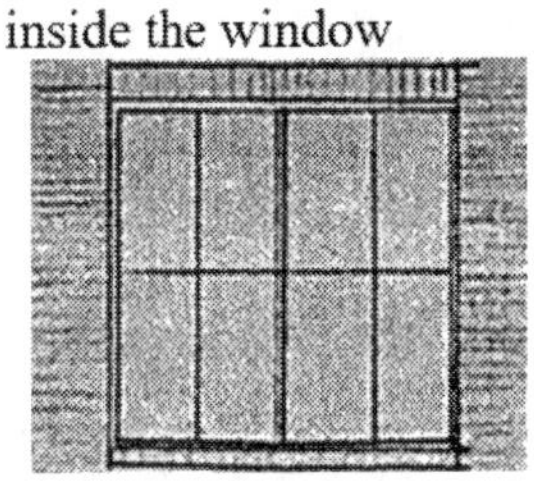
Roof	Aging, leakage and other phenomena	The same to the status quo
Doorway	North facade: specification of 1#, 2#, 3# and 4# is 1500x2000; specification of 5# is 2500x3200; South facade: specification of 6# is 1000x2200; specification of 7# and 8# is 2500x3000; West facade: specification of 9# is 2000x2600. The current situation is all iron doors; 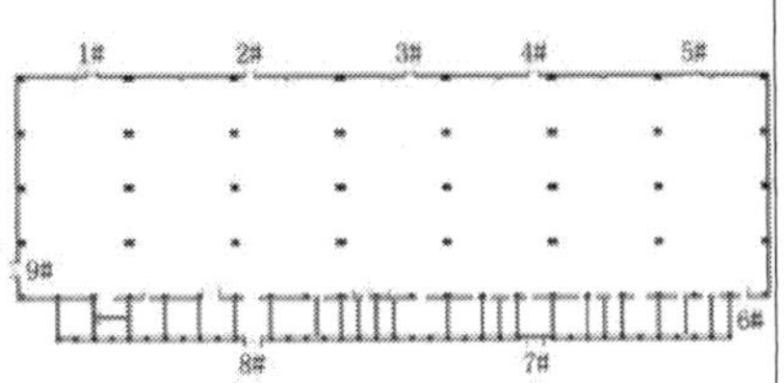 	North facade: specification of 1#, 2#, 3# 4#, and 5#is 1500x2000; South facade: no doorway for 6#; specification of 7# and 8# is 2500x3000; West facade: specification of 9# is 1000x2000. The reasons for the discrepancy between the present situation and the original shape and system are all transformation by later generations. The original doorway is made of iron. 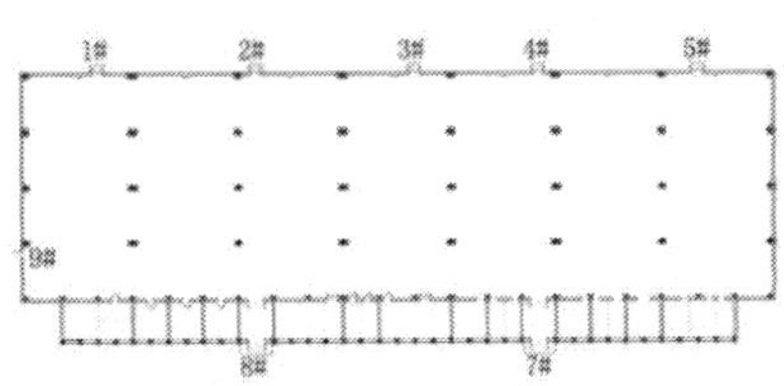

2.4.2 Implantation of New Space Functions. According to the general design policy of the creative base, it is necessary to put the show function (show, assembly, exhibition and other functions of large flow of people) into the building. The following problems should be considered: a. The number and width of door openings in the original shape system do not meet the evacuation requirements of large flow of people, and the demand for access channels of large equipment. b. Therefore, according to the evacuation needs of the crowd and the actual use needs, combined with the existing outer eaves door size, it is advisable to completely retain the size of the door and combine with the overall effect, try to make use of the current situation of the door, and appropriately add door holes. c. Before meeting the premise of safety and use, it is necessary to respect history as much as possible, refer to the facade style of original door leaf, and make and install according to the size of hole, but ensure that the facade effect of door leaf is in harmony with the original shape. d. The added equipment and pipes shall not be destructive to the original building structure and shall meet the recoverable principle, ensuing future demolition shall not affect the structure.

3. Virtual Reality Technology Solutions

One goal of virtual reality prototyping is the substitution of physical prototypes by using virtual models [1] Two aspects need to be addressed:

- augmentation of physical prototypes with (interactive) virtual content
- virtual simulations of physical prototypes [2]

To solve the technical problems that restoration of industrial heritage complex construction environment and building reuse may involve, evaluation and validation are conducted. The virtual reality technology is introduced into the design. It is designed based on the MARS virtual reality researched and developed by "shining city". The recovery and reuse design platform of industrial heritage based on virtual reality technology is compatible with commonly-used architectural design software such as SKETCHUP and 3Dmax, and can directly use the related software for modeling, reduce duplication of effort and shorten working hours. The design process is divided into three stages: (Table2)

- The first stage is to generate a 3D model by measuring the existing environment.
- The second stage is to restore the shape and structure of the original buildings and restore the original appearance of the heritage buildings based on the current 3D model.
- The third stage is to bring the status model into the scene design, and verify the possibility of reuse of the building by inserting roles and formulating functional ways. In the stage, the virtual reality technology plays an important role, which enables people not only to experience industrial heritage buildings after reduction through the roaming space, but also more vividly feel environment after transformation and utilization in the existing cultural relics under the premise of building construction and not influence. Besides, it can provide explanatory design scheme for decision makers to develop project development plans.

Table 2. Three stages and different work contents

Stages	Contents
I Generate a 3D model by measuring the existing environment	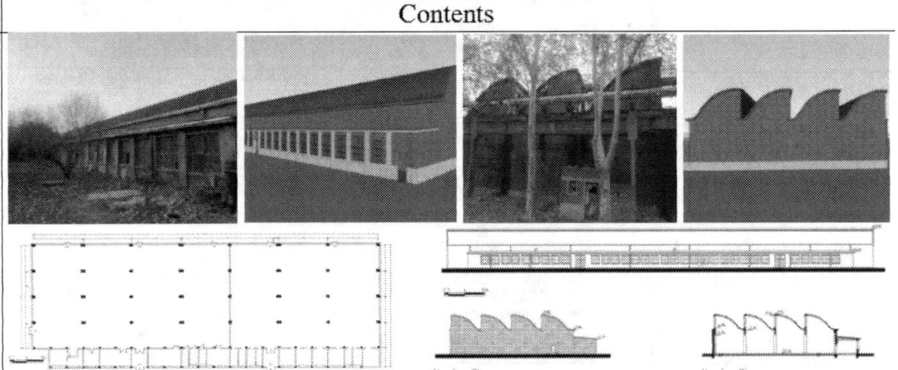
II Restore the original appearance of the heritage buildings	

IOP Conf. Series: Materials Science and Engineering **825** (2020) 012021 doi:10.1088/1757-899X/825/1/012021

III Bring the status model into the scene design, and verify the possibility of reuse of the building by inserting roles and formulating functional ways	

4. Technical Evaluation

VR also defined as three Dimensional (3D) graphical simulation models, where a user can control the viewpoints, motion and interact intuitively in real time.[3] Compared with 3Dmax, SketchUp and other commonly-used 3D simulation software, the most prominent feature of MARS virtual reality technology is the addition of 3D glasses and digital handlebars to realize 3D real-time graphics display and other functions. It has a good performance in immersion. Users can feel the space distance, speed, including sound distance and field characteristics in the 3D virtual environment created by computer.[4] In terms of interaction, MARS can real-time control and grasp objects through functional handles. However, due to constraints of hardware and network, it has weak interactive experience, cannot simulate the free grasping experience of virtual objects by human hands, nor accurately perceive the weight of objects. MARS provides certain possibilities on imagination. In the multi-dimensional information space of the virtual world, users can obtain more knowledge and feelings than traditional design methods by relying on their own perception ability, which plays a positive role in solving problems (Figure 2,Figure 3) .

Figure 2. User Experience

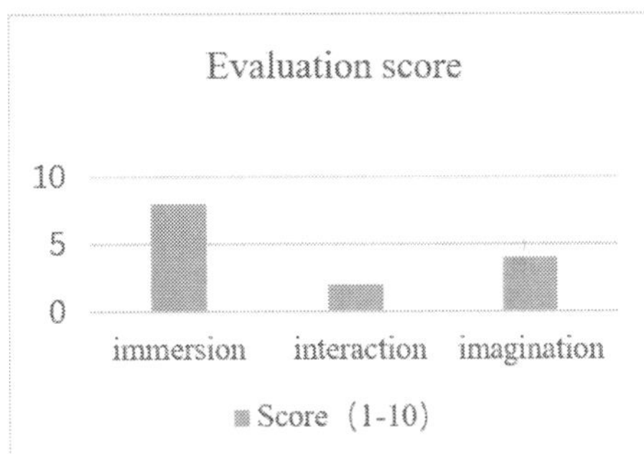

Figure 3. post occupancy evaluation

5. Conclusion

By adopting MARS virtual reality design system, the technical problems of old building restoration and new space function implantation and research are solved in the design of industrial heritage restoration and reuse of Zhengzhou No. 2 Abrasive Wheel Factory. VR technologies are a promising tool,[5] compared with the traditional 3D design software, MARS virtual reality technology has higher work efficiency, provides users with better immersion and imagination experience in the environmental rehabilitation experience. However, in terms of interaction, it is not essentially superior to the traditional design methods. Therefore, there is still a great space for technological breakthroughs and development.

Acknowledgement

Funds support: General Project on Humanities and Social Science Research of Chinese Ministry of Education,2017. (Project Approval NO. 17YJC760087). "13th Five-Year Plan" for Philosophy and Social Sciences of Guangdong Province,2017 (Project Approval NO.GD17CYS09),

References

IOP Conf. Series: Materials Science and Engineering **825** (2020) 012021 doi:10.1088/1757-899X/825/1/012021

[1] Bordegoni, M.; Ferrise, F.; Lizaranzu, J.: Use of interactive virtual prototypes to define product design specifications: a pilot study on consumer products. In: Proceedings of the ISVRI 2011, pp. 11–18. IEEE, New York (2011).

[2] Günther, Tobias; Kammer, Dietrich; Wojdziak, Jan; Lorenz, Sebastian; Groh, J.: Smart prototyping Improving the evaluation of design concepts using virtual reality. In: Virtual, Augmented and Mixed Reality - 8th International Conference, p48. (2016).

[3] J. M. Moshell and C. E. Hughes, "Virtual Reality as a Tool for Academic Learning," in Handbook of Virtual Environments: Design, Implementation and Applications, K. M. Stanney, Ed.:Lawrence Erlbaum Associates, 2002, pp. 893-910.

[4] H. Neale and S. Nichols, "Theme-based content analysis: a flexible method for virtual environment evaluation," Int. J. Human-Computer Studies, vol. 55, pp. 167-189, 2001.

[5] W. Winn, "A Conceptual Basis for Educational Applications of Virtual Reality [online]," Seattle, University of Washington, Human Interface Technology Laboratory 1993.

Research and Design for Hotel Security Experience for Women Traveling Alone

Hong-Jiao Wang[1] and Chi-Hua Wu

[1]School of Art and Design Guangdong University of Technology 729 DongFeng Street East, YueXiu District, GuangZhou City, GuangDong Province, China

Email: 1047958379@qq.com

Abstract. Hotel security has been a main issue for tourism. It is urgent for solo female travelers since more than 50% of women worldwide had travelled alone, and Chinese women take account for 42% of the business travels. Focusing on this issue, the main goal of this study is to study individual female traveler's experience in terms of sense of security when staying in hotels in China.The study consist of two parts. Firstly, preliminary interviews and questionnaires are used to explore solo female traveler's perception on hotel security and potential factors influencing such perception. In-depth interviews were adopted to achieve better understanding on the factors. Secondly, future-oriented techniques are adopted to provide directions for design. Main factors and corresponding scenarios are gathered and illustrated using customer journey map. Mirroring approach is then used to flip the problematic customer journey into wishful one from which design opportunities could be extracted. Main outputs include a facial recognition system for elevator, statistical identification system for the number of people entering the room, rescue word alarm system and intelligent door defense system. Subsequent design outcome has won UXPA award, indicating certain utility and value for future development.

1. Introduction

In recent years, the voice of women's safety protection on the Internet is growing. Hotel is one of the places where women are prone to danger. According to China's emergency response law, safety accidents in the hotel industry are divided into four types: accident disaster, public health event, social security event and natural disaster. According to the data of China tourism safety report, the number of hotel social safety accidents in 2011-2015 accounted for the highest proportion among the four categories of safety accidents. Among them, social security incidents mainly include: robbery, rape, theft, violence between the guest and the host, homicide, violence in the guest room and so on[1].Such incidents are particularly dangerous for women. On April 3, 2016, a woman living alone in Beijing Wangjing 798 HEVI hotel was confronted with violence such as kidnapping and forced dragging by strange men, until a passer-by came to the rescue. As soon as the case happened, there was a heated discussion on the Internet[2]. How to ensure the personal safety of the guests in the hotel attracts people's attention. Women are in a weak position in the confrontation with men, and do not have an advantage in strength. They bear more safety risks in the process of traveling alone. Fifty-four percent of Chinese women have traveled alone. During the tour, their safety concerns mainly focus on hotel accommodation and transportation. [3].

2. Literature review

2.1. Increase in the proportion of women traveling alone

Female tourists are mainly divided into four categories: women who are on leisure travel alone, women who are on business trips, women who are travelling with their spouses or relatives and friends, and women who participate in tour groups. Among them, women on business trips are a new market with obvious growth trend in recent years[4].This is closely related to women's frequent participation in social work. At present, the business trip population in China is mainly concentrated in the middle management level, mainly the people born in the 1980s and 1990s, of which women account for 58%[5].Leisure travel alone has always been one of the important ways for women to explore themselves. British Airways released the Global Solo Travel Study and more than 50% of women in the world have chosen to travel alone, and 75% of them have plans to travel alone in the future. 59% of Chinese female tourists have experienced 2-5 independent trips[6]. Especially when traveling alone,women on their own have to be alert to all kinds of possible dangerous scenes. They are afraid of the threat of male violence. When going out alone at night, women's fear of the outside world will also rise. It can be seen that time limit is also one of the important factors affecting the activities of women[7].

Although the number of women has increased significantly in the tourism market, there is a lack of research on the safety experience of women traveling alone in China. Women's psychological sensitivity to unfamiliar environment is more sensitive, and the experience of travel security is more demanding. Therefore, the research on the experience of women's sense of security is more critical.

2.2. Existing hotel security service system

At present, in the hotel environment, monitoring and access control are widely used in the hotel security. The traditional Chinese hotel security system is still rely on personnel, equipment and system, and low management efficiency. The relevant security products are not able to give early warning to the dangerous situation in the hotel, so they can only provide the event information at that time after the accident[8].But with the improvement of modern people's safety awareness, people's demand for security is also growing. Therefore, the hotel security needs to change to the intelligent security direction, to ensure the safety of hotel guests and establish a safe and reliable brand image.

3. Research purpose and significance

The purpose of this study is: to design the hotel security service system for independent women. Improve the safety problems in the independent women's Hotel, provide complete hotel security services, make design innovation based on the existing technology, make it a set of practical service system, explore its realizability, patent and diversified business value.

Based on the investigation of relevant literature and case analysis, this study aims at young women aged 18-26. Young women are active in social or school life, and they are the main group of business travel and leisure travel. And easy to become the target of criminals. The independent women in this paper generally refer to this kind of young women.

4. Research process and methods

From the perspective of user experience. Sort out the potential safety hazards in the hotel, collect the concerns of independent women about the safety of the hotel and the experience of staying in the hotel, and design the scheme. The research flow is as follows (as shown in the figure 1):

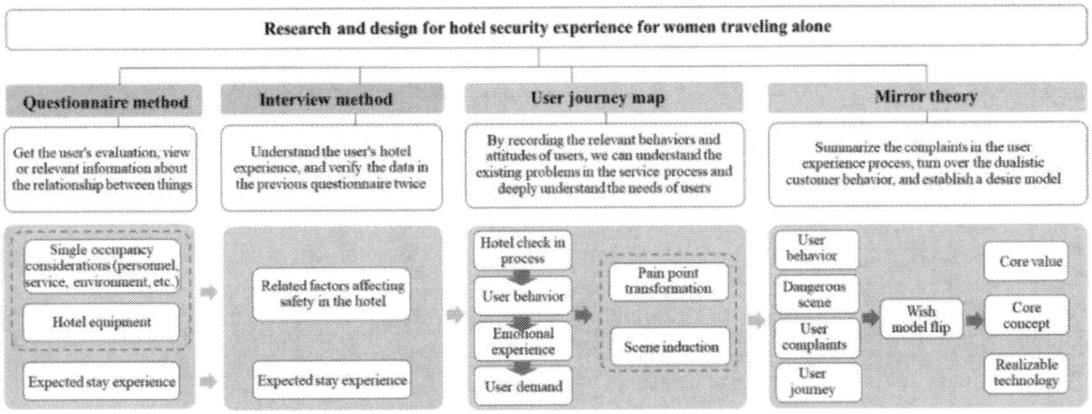

Figure 1. Research framework

The methods used are summarized as follows:

- Questionnaire (Introduction) interview: The questionnaire part of this study mainly focuses on the factors to be considered when staying alone, the equipment and facilities that have the greatest impact on the safety of the hotel, and the expectation of staying experience of lone women. Get the user's evaluation of the hotel security service.
- Interview method: In this study, face-to-face semi-structured interviews were used to collect information in the form of chatting, and the user data in the questionnaire results were verified twice through in-depth interviews[9].
- User journey map: The user journey map is used to describe the user's stage status in the process of consuming or experiencing a service. In this study, users' activities after entering the hotel are divided into different stages for analysis, mainly including: user behaviour, emotional experience, pain points, user needs, opportunity points, etc[10].
- Mirror theory: Mirror theory is a thinking tool (as shown in the figure 2). Including systematic operating procedures, which can effectively guide the design direction. This study uses the mirror theory to flip the pain points in the user experience process and find out the expectation of users for hotel safety in different stages. Combined with the corresponding technology, build an ideal hotel security service system[11]

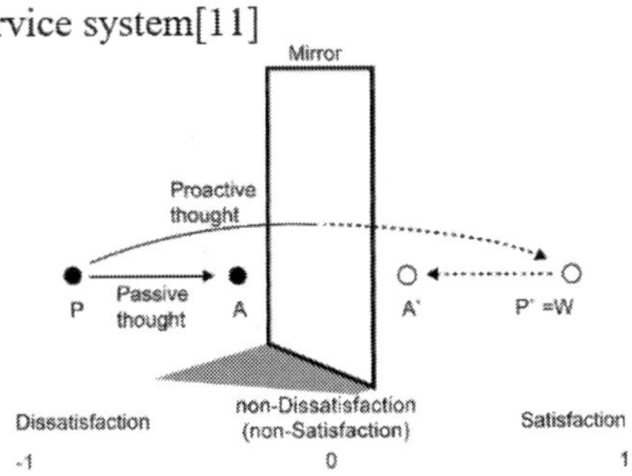

Figure 2. Mirror theory

5. Research result

5.1. Questionnaire

In this study, 216 questionnaires were sent out on the Internet, and 181 valid questionnaires were recovered. The main issues focus on age, whether there is a single occupancy experience, considerations of single occupancy, facilities and equipment that have a great impact on the safety of the hotel, and expected occupancy experience.

The statistical results of the questionnaire are as follows:

- The respondents are young women aged 18-26, whose income is directly proportional to the hotel price they choose.
- More than 60% of women have experience of living in the hotel alone.
- The main factors to be considered when staying in the hotel are sanitary conditions, geographical location, hotel price, security system, equipment and facilities.
- The highest proportion of the facilities that affect the safety of the hotel is room lock, emergency call facilities, monitoring system and hotel elevator.
- For single occupancy, the most desirable experience for women is safety and convenience, accounting for 73.04%.

Combined with the analysis of questionnaire results and related literature survey, three factors of women's travel alone are summed up: business travel, travel, learning. The business travel is a passive travel, women's independent choice is small, can not choose hotels according to personal needs or preferences. The hotel can be selected according to personal itinerary and preference, mainly focusing on learning and travel.

5.2. In-depth interview
In order to confirm the authenticity of the user needs in the questionnaire and describe the user experience more specifically, 16 young women aged 18-26 were randomly selected for in-depth open-ended interviews. The interview focuses on the links of staying in the hotel, the relevant factors affecting the safety of the hotel and the expected experience of staying in the hotel.

5.3. User journey
Based on the results of questionnaire and interview, the user's journey is drawn (as shown in the figure 3), and the customer's concerns and needs are integrated into the process and scene sequence of staying in the hotel, which is sorted out. It is divided into five stages: check-in, go to the guest room, enter the guest room, receive visitors and go out for a short time, and draw the user's emotional fluctuation experience curve to provide direction for subsequent design.

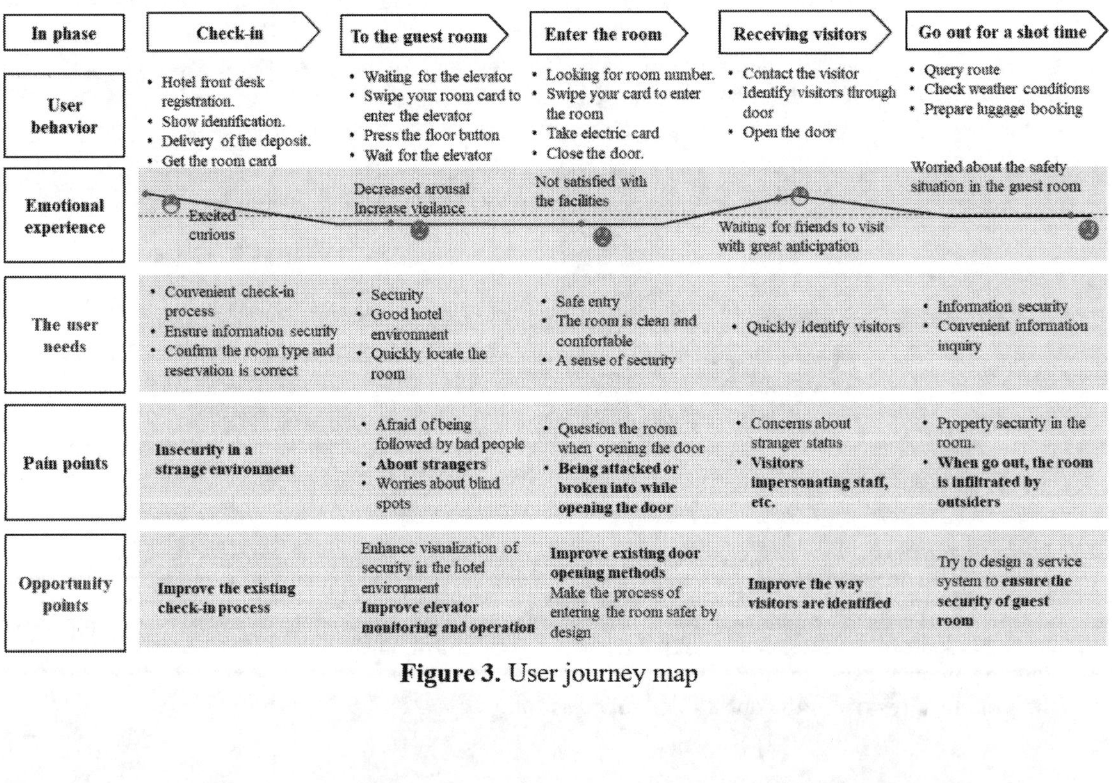

Figure 3. User journey map

This study found that in all aspects of the existing hotel service, users have varying degrees of uneasiness. The main expectation of users is that the hotel can provide timely rescue in case of danger rather than obtain evidence after the accident. According to the deficiencies in the existing security and the pain points in the user's journey, we can summarize the problem scenarios into five categories.

- Check in. When women check in at the hotel alone, they are easy to be the targets of criminals.
- Encounter a dangerous situation in the elevator. It is easy to be violated in elevator confined space.
- Encounter a dangerous situation in the corridor. Strangers impulsive crime, forced drag women into the guest room.
- Stalking. Women who return to their rooms alone tend to be targets of stalking and violence.
- Attacked indoors. When encountering a premeditated ambush and violence, they can't ask for help from the outside world in time. For single occupancy, the most desirable experience for women is safety and convenience, accounting for 73.04%.

5.4. Mirror theory

The dangerous scene is sorted out and the user's complaints about the dangerous scene are collected. The ideal wish is flipped by using the "mirror theory"[11] (as shown in the figure 4.5).

Activities	Problems	Complaints	Journey	Wish Model
Check-in	Check-in	Full of insecurity in unfamiliar environment	Unable to quickly register and view room information in real time	Fast and convenient registration mode, real-time confirmation of room information
Take the elevator	Encounter a dangerous situation in the elevator	Fear of being attacked in the elevator	Can't confirm if the person taking the elevator is a hotel guest	Ensure that all passengers taking the elevator are registered
Going to guest rooms	Encounter a dangerous situation in the corridor	Afraid of being dragged by bad people	Don't know how to escape when being pulled by criminals	In case of danger, inform the security personnel immediately for rescue
Enter guest room	Stalking	Afraid of being followed into the room	Assaulted in the room unable to ask for help from the outside world	Security personnel can find out the dangerous situation immediately
Receiving Visitors	Attacked indoors	Worried about strangers posing as workers	Unable to confirm the real identity of the visitor	Be able to confirm the real identity information of visitors in real time
Go out for a shot time	Attacked indoors	When go out, the room is infiltrated by outsiders	Unable to call for help in time in the room	In the room can also timely and effective to the outside world for help

Figure 4. Mirror theory

Wish Model	Core value	Core concept	Available Techniques
Fast and convenient registration mode, real-time confirmation of room information	Safe and convenient check-in experience	More convenient check-in process, real-time confirmation of room information	Face recognition equipment
Ensure that all passengers taking the elevator are registered	Establish safety area and isolate external personnel	Set up the right to take the elevator. Non registered personnel are not allowed to take the elevator	Electronic display screen, elevator face recognition equipment
In case of danger, inform the security personnel immediately for rescue Security personnel can find out the dangerous situation immediately	Ensure the personal safety of residents	Enhance the security intensity of public areas and the timeliness and flexibility of dealing with dangerous situations	Corridor portrait monitoring, corridor voice monitoring
Be able to confirm the real identity information of visitors in real time	Prevent identity impersonation	Guests can confirm visitor information in real time	Mobile app remote confirmation
In the room can also timely and effective to the outside world for help	Accidents in non-public space can also be effectively called for help	Set up emergency call equipment to ensure the safety of guests in the guest room	Door lock pull alarm

Figure 5. Mirror theory--- Core value

From the flipped conclusion of the mirror theory, the user's most expected wish model is obtained, and the real core value behind it is explored through the wish model. Core values mainly include:

- Safe and convenient check-in experience: women are easy to become targets of criminals when they register alone. The hotel can use face recognition equipment for registration, making the check-in process more convenient for users. When criminals commit crimes, they will deliberately avoid the input of face information, so that they cannot enter the hotel, which has the effect of preventing criminals from entering the hotel.
- Establish a visual security area to isolate the danger: women can't confirm the identity of fellow travelers in the process of taking the elevator alone, which is easy to cause tension. The hotel shall conduct registration verification at the front desk and isolate the unregistered personnel.
- When the danger occurs, it can be stopped in time. The existing hotel security is mainly based on monitoring, and can't find the dangerous situation in time. In case of danger in public areas, users usually pull with criminals or cry for help. Therefore, the first response of the user in an emergency can be taken as the starting point of the design.
- Accidents in non-public space can also be effectively called for help. Guests in the room may also face risks, such as intentional ambush of criminals. At this time, the women are usually coerced by criminals, so they can't effectively inform the security personnel when the danger occurs. According to the questionnaire survey, the hotel door lock is the key point of the safety facilities that users think. In case of indoor danger, users need to escape from the door, and the door lock is also the high-frequency contact point in the process of escape.

6. Security system design

By combing the scene extraction and core value, combined with the corresponding technology and product support, a visual and all-round hotel security service system is designed-RUNNING ANT hotel security system. It mainly includes four security systems, five types of security facilities and six security lines. The specific design is shown in the table below (as shown in the table 1):

Table 1. Security system design.

<table>
<tr><th>Potentially dangerous scenarios</th><th>6 security lines</th><th>Category 5 security facilities</th><th colspan="2">Available Techniques</th></tr>
<tr><td>Check-in</td><td>Face recognition registration</td><td rowspan="2">Face recognition equipment</td><td rowspan="2" colspan="2">Safety protection system for passengers in elevator</td></tr>
<tr><td>Encounter a dangerous situation in the elevator</td><td>Permission authentication of being allowed to take the elevator</td></tr>
<tr><td>Encounter a dangerous situation in the corridor</td><td>Corridor voice monitoring</td><td>Corridor voice monitoring equipment</td><td rowspan="2" colspan="2">Security monitoring system in corridor</td></tr>
<tr><td>Stalking</td><td>Identification of the number of people on the corridor</td><td>Corridor portrait recognition equipment</td></tr>
<tr><td rowspan="2">Attacked indoors</td><td>Door lock pull alarm</td><td>Intelligent protective door lock</td><td rowspan="2">Door lock safety alarm system</td><td rowspan="2">Visitor management system</td></tr>
<tr><td>Visitor identification</td><td>Hotel Service app software</td></tr>
</table>

- Face recognition registration: after the guest enters the hotel, the personal information of the front desk will be registered through the face recognition equipment to obtain the access rights such as the elevator door lock in the hotel. Then, the front desk transmits personal data to the mobile app, where the guest can immediately confirm the room information, personal identity, visitors and other information.
- Authorization authentication of the allowed to take the elevator: when the resident enters the elevator, the face recognition equipment automatically recognizes the portrait information. If

there are unregistered personnel entering, the elevator will remain in the non operating state. At the same time, the display panel inside the elevator will display the portrait of unregistered personnel to prompt other residents.

- Corridor voice monitoring: when a lone woman is dragged by criminals, she can shout for help words to help herself. The corridor voice monitoring equipment will recognize the keywords that can trigger the alarm and send the distress signal to the security department.
- Identification of number of people in the corridor: When the guest is followed by others, the corridor portrait identification equipment recognizes that the number of people entering the door does not conform to the number of people registered. The monitoring center transmits the room information to the security center, and the security center's telephone system automatically dials the corresponding room phone number. If the guest is hijacked and fails to answer, the alarm message will be played automatically, and the security personnel will be sent to visit.
- Door lock pull alarm: when a lone woman is in danger in the guest room, she can pull the smart door lock. When the pull sensor reaches the alarm threshold, it will automatically send a distress signal to the security system. (as shown in the figure 6.)
- Visitor identity authentication: visitors register information at the front desk. The front desk transmits the visitor data to the hotel service app. After confirming the information is correct, the visitor obtains the authority of the visitor in the hotel. When visitors arrive at the door of the guest room, they touch the doorbell of the smart door lock. The resident can view the visitor image in real time through the app to verify the identity of the visitor.

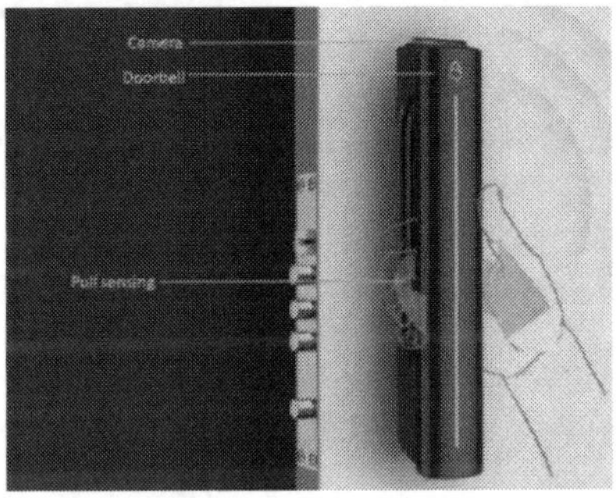

Figure 6. Door lock pull alarm

7. Conclusion

Enhancing the sense of psychological security and reducing the crime rate have an important impact on improving the travel experience. This study summarizes the important scenes of potential danger of lone women in the process of staying in the hotel, and puts forward practical solutions for each link. Based on the current market and technical situation, a forward-looking and preventive security design is extended to reduce the possibility of potential hazards and ensure the personal safety factor and psychological safety of users. This study also summarizes the important influencing factors of Chinese independent women's security experience, and then constructs an innovative hotel security system through the guidance of mirror theory. It points out that the development of the hotel security management system is possible, and the achievement has won the 11th user experience design award, hosted by UXPA China (user experience professionals' Association). The third prize in China, among which the relevant hardware has applied for one utility model patent. This system is not only used in the hotel scene, but also can be used in residential, hospital, office building and other places that need security, which has certain market potential and commercial value.

References

[1] Limei, L.et (2018, 4 3). Distribution Pattern of Hotel Accidents in China and their Precautions. *Journal of Xiamen University of Technology*, pp. 40-45.

[2] *A girl in Yihe hotel in Beijing is dragged by a strange man*. (2016, 4 6). Retrieved from OBSERVER: https://www.guancha.cn/FaZhi/2016_04_06_356117.shtml

[3] *Tourism consumption analysis report of women tourists in TUNIU 2018-2019*. (2019, 3 11). Retrieved from Cntgol.com: http://news.cntgol.com/dyzd/20190311/218052.html

[4] Jocelyn Siah Chee Haoa, C. O. (2014). A study of preferences of business female travelers on the selection. *Procedia - Social and Behavioral Sciences* (pp. 176-186). Elsevier Ltd.

[5] *Travel questionnaire for contemporary business people*. (2018, 6 21). Retrieved from THE PAPER news network: https://www.thepaper.cn/newsDetail_forward_2207686

[6] *British Airways - Global Solo Travel Study*. (2018, 11 1). Retrieved from Sohu News: http://www.sohu.com/a/272738315_652544

[7] Erica Wilson, D. E. (2008). The Solo Female Travel Experience: Exploring the. *Current Issues in Tourism*, 167-186.

[8] Yang, L. B. (2015, 10 15). Current situation and future of hotel security. *China Public Security*, pp. 70-72.

[9] User interview Questionnaire survey. (2014). In D. U. School of Industrial Design and Engineering, *Delft Design Guide: Design strategies and methods* (pp. 50-53). Wuchang: Huazhong University of Science and Technology Press.56.

[10] Li, F. (2019, 5 25). Omni-channel Customer Journey Experience Map: Integration Research Based on Personas, Customer Experi[...]. *Technology Economics*, pp. 46-56.

[11] Luh, D. B. (2014). Classification and solution of positive solution in method paper. In D. b. Luh, *Positive Creativity - The Mirror Theory* (pp. 23-33 89-100). Taiwan: De Wing Internation Co., LTD.

Motion Analysis-based Design of Wall Painting Tool Storage

Tang Ruoyue[1, a], Chen Luwei[2, b] and Wang Jingya[3, c]

[1]Faculty of Industrial Design, Zhejiang Gongshang University Hangzhou College of Commerce
[2]Faculty of Industrial Design, China Jiliang University College of Modern Science and Technology
[3]Industrial Designer, Hangzhou Jiuyuan Network Technology Co., Ltd.

[a]tangruoyue_hit@163.com; [b]Frie1118@hotmail.com; [c]931204848@qq.com

Abstract. While the wall painting is applied frequently in interior design and decoration, many problems exist in the wall painting tools usage and transportation. With the interview method conducted in the investigation of the wall painters, based on the motion analysis and the economic principle of the related action, the design prototype of the wall painting tool storage was proposed. Conducting the multiple actual working scenes motion test, this design was continuously improved aiming to offer an ergonomic optimization with the appearance and function balanced design.
Keywords: Wall painting tool storage; Motion analysis method; Product design

1. Introduction

In the decoration of clubs, exhibition halls, restaurants, bars, departments, wall painting has become a preferred choice its economic efficiency. This trend has become increasingly popular in recent years. While there are many problems existed in the process of usage and carrying of the wall painting tools. For instance, the toolbox is always bulky with less function, most of which can only be carried by hand inconveniently. During the painting operation, wall painting staff need to repeatedly search the targeted tools. Obviously, this process is not integrated designed to offer an efficient usage experience. The targeted problems of this design solution is how to use and carry these various painting tools. Furthermore, this design process is aimed to explore the optimization motion path during wall painting process.

Gilbreth proposed the Therblig action classification (Table 1), which decomposes human work into three categories of eighteen elements. The first category is the necessary action to complete the work, and the second one is the auxiliary action. The third category is redundant action. Gilbreth also proposes the principle of obtaining the most efficient action with minimal fatigue, seeking the most reasonable action [1,2]. Through the improvement of Ralph, Baenes and others, this principle has evolved to an action economic principle system. Some of these principles, such as: effective use of limbs, making movements as economical as possible, making movements appropriate to human motion characteristics, rational allocation of movements, avoiding static and continuous exertion, etc., can be used not only for production management, but also for improving product design[3,4]. The application of the action classification and the economic principles in product design is instructive for solving the problems of wall painting process faced by wall painters.

Table 1. The Therlig Action Classification of Gilbreth

category	Motion element
1	Transport empty [unloaded] (TE) Group (G) Transport loaded (TL) Position (P) Assemble (A) Disassemble (DA) Use (U) Release load (RL)
2	Search (Sh) Find (F) Select (St) Inspect (I) Plan (Pn) Preposition (PP)
3	Hold (H) Unavoidable delay (UD) Avoidable delay (AD) Rest in peace (R)

2. Design analysis and prototyping

2.1. Design research

More than 30 wall painters were interviewed with the following questions.

i. What is the most troublesome in the usual wall painting?
ii. What is the most afraid of encountering in the process of wall painting?
iii. What tools are needed? How to take the tools to the workplace?
iv. Is your current toolbox easy to carry and what problems exist?
v. What tools are used and replaced most frequently during the wall painting process?

Through the analysis of interview results, the existing toolbox is very inconvenient to use and carry. The mainly reasons include its bulky size, less function, fixed internal structure, and the limited place for tool placement. Furthermore, as the acrylic paints are difficult to clean, wall painters are most afraid to soil their clothes. The use and replacement of brushes is also considerably frequent during the operation, which brings a lot of trouble to the wall painter. There are two situations about the tools' transportation to the workplace. Firstly, when time permits, the wall painter will express the bulky paint directly to the workplace, and carry the small tools such as brushes by himself. Secondly, in the case of tight time, painters have to carry cumbersome tools and heavy paints to the workplace by themselves[5].

Through the on-site observation of the wall painter's work process, the actions in the wall painting process are divided into three categories according to the Gilbres' Therblig action classification (Table 2)

Table 2. Motion classification in the wall painting process

category	Motion element
1	Rise hand Hold the pen Move

	Position
	Bow
	Rise head
	Wave hand
	Place
	Find
	Choose
	Change
2	Think
	Pre-place
	Wash the brush
	Color mixing
	Grip
3	Upper and lower scaffolding

According to the economic principle of motion, the third type of motion should be avoided, and second type of motion should be reduced. And with optimization of the first type of motion, a relatively higher efficiency should be achieved by minimum fatigue.

2.2. Proposal of design prototype
According to the design survey and analysis above, the design prototypes are shown in Figure 1 and Figure 2.

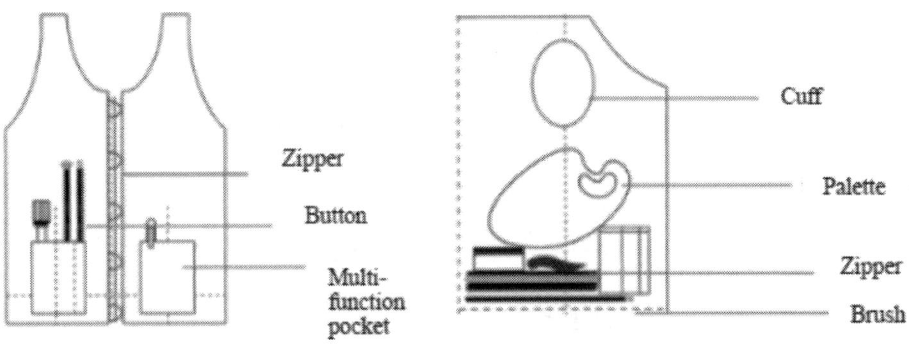

Figure 1. "vest-Handbag" section.

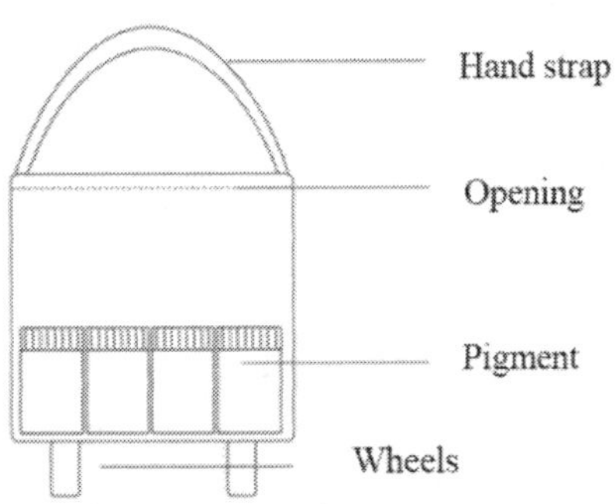

Figure 2. Large capacity paint bag with wheels

The entire design solution consists of two parts. The first part (Fig. 1) can be used as both a vest (Fig. 1 left) and a handbag (Fig. 1 right). This section is used to store brushes, artboards, and other common

IOP Conf. Series: Materials Science and Engineering **825** (2020) 012023 doi:10.1088/1757-899X/825/1/012023

wall painting widgets. The multi-function pocket on the vest can be inserted into various tools to facilitate replacement and access during wall painting. The second part (Figure 2) is a larger capacity pigment bag that is designed to hold a large amount of pigments. There are foldable wheels at the bottom to make the entire product possible to move on the ground.

3. Design verification and improvement

3.1. Design proofing and verifications

Based on the relevant ergonomic roles[6] to determine the specific dimensions (Figure 3), the samples were made for the real user scene test (Figure 4).

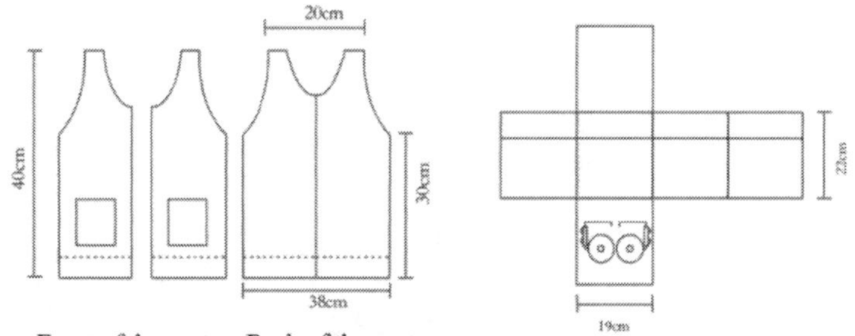

Front of the vest Back of the vest

Figure 3. the dimensions of the sample

Figure 4. Product sample

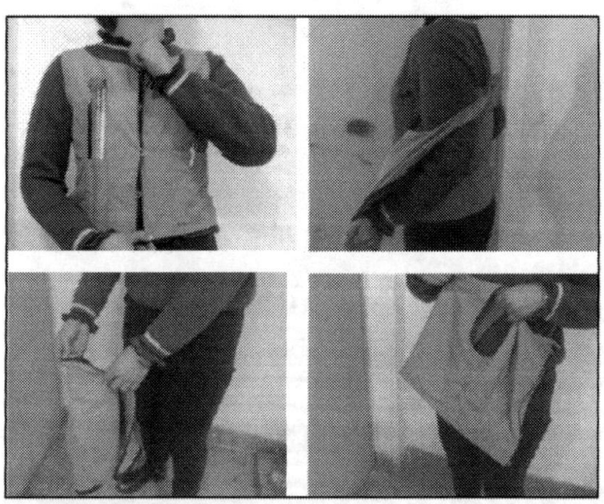

Figure 5. Transformation between the vest mode and handbag mode

The grey and waterproof nylon was chose as the main material of the product sample, which is tested in the real usage scene. Particularly, the carry-on handbag can transform to a wearable vest when opening the zipper on the side (Figure 5).

Selecting two wall paintings with roughly the same area and similar difficulty, one wall painting team conduct the same painting project. The first time, this team works with previous wall painting tools. In order eliminate the unrelated influence factor, such as fatigue and painting adapting, the second experiment was conducted several weeks later. In the next time, same team works with this product sample instead of previous ones. The number of times of each motion occurred were recorded by a 5-minute period over every 25 minutes. Record the average value of each motion in 8 recording periods within 4 hours, as shown in Table 3.

Table 3. Comparison of the times of occurrences of each motion in the average 5 minutes wall painting process

category	Motion element	With sample	Without sample
1	Rise hand	70	62
	Hold the pen	always	always
	Move	5	3
	Position	9	3
	Bow	64	62
	Rise head	always	always
	Wave hand	2	2
	Place	64	62
2	Find	3	0
	discover	3	0
	Choose	3	3
	Change	3	3
	Think	3	0
	Pre-place	3	0
	Wash the brush	2	0
	Color mixing	2	2
3	Grip	3	0
	Upper and lower scaffolding	3	0

According to the data above, the third and second types of motions are significantly reduced by using this product sample. This product design reduces the occurrence of unnecessary motions by a certain extent, improves the work efficiency, which conforms the human movement characteristics[7].

At the same time, some problems were found during the previous experiment. Firstly, the fabrics is too light and thin, which is easily deformed during the shipping, causing abrasion and even damage to the internal pigment boxes. Secondly, the size of the clothes is too small while it feels tight when wearing. Thirdly, the amount of brushes that can been insert into the pocket of the vest is not enough[8].

3.2. Design evaluation and development

According to the problems above, this product design is further adjusted. The waist circumference of the vest is expended to 76cm, and the chest circumference is expended to 95cm while whose length is expended to 90cm. The brush pocket is increased from the original two to three. Considering the material used is too soft, with re-examination of a variety of materials for comparison, final material selection is nylon 420D (D - Daniel = g / L * 9000, expressed the weight of the wire within the unit length[9]) . This fabric is moderately soft and has good water resistance and a certain breathability. The soft crease-resistant nylon 75D was selected as the inner lining, and used in the second proofing experiment (Fig. 6, Fig. 7).

Figure 6. Second experiment, vest-handbag secton

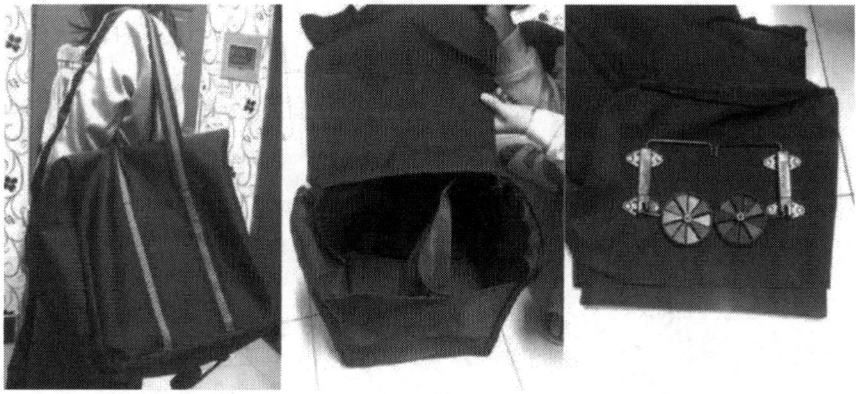

Figure 7. Second experiment, Large capacity paint bag with wheels

The result of second sample experiment was considerably improved. Finally, the appearance of the product was redesigned. Choosing the cloth with splashing ink on the white background, this solution is aims to solve the problem of inadvertently staining during the wall painting process, while balance the aesthetics aspect. The final product is shown in Figure 8.

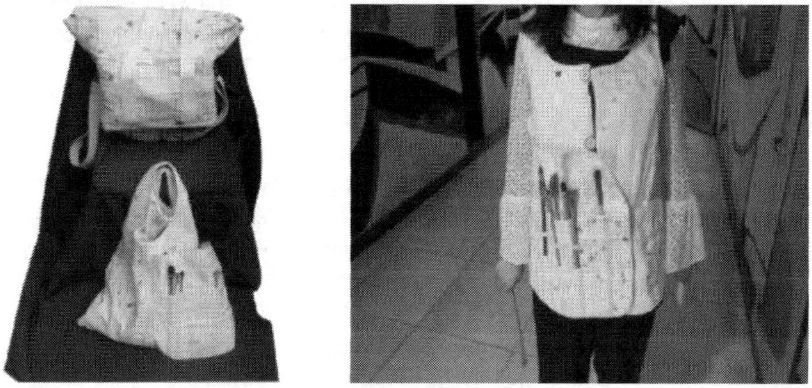

Figure 8. Wall painting storage design

4. Conclusion

Based on the motion analysis method, and detailed investigation and analysis conducted into the working process of the wall painters, this wall painting tool was designed. Through the continuously proofing and trial to evolve the solution, a shape and function balanced product design which satisfies the economic principle and ergonomic optimization was developed. This design has got the appearance and utility patent, and won the second prize of the 9th "Flash Cast Cup" industrial design competition in Zhejiang Province. It has also won the 2019 China Packaging Creative Design Competition as the shortlisted product design.

Acknowledgments
Related Fund: VR Visual Industry Double-Master Dual-Energy Faculty Team Construction, 2018, Industry-University Collaborative Education Project, Ministry of Education, China (No. 201802172011) Research of Application of Individualized Teaching Mode in the Product Design Course under the Background of Pan-entertainment, the first batch of teaching reform research projects of the 13th Five-Year Plan of Higher Education of Zhejiang Province.

References
[1] Wang Luo, Gilbres 2013 The father of action research (Contemporary Power Culture) 80-81
[2] Henry S. Dreyfuss 2003 *Designing for People* (NewYork: Facts) 42-46
[3] Hu Wendan, 2019 Analysis of the principle of action economy (Science and Technology) 12-12
[4] Pieter M. A. Desmet. 2001 *Measuring Emotions* (London: The Design Council) 33-34
[5] Alan Cooper 2015 About Face 4: The Essentials of Interaction Design (Beijing: Electronic Industry Press) 124-128
[6] Hu Haiquan 2013 Industrial design and application of ergonomics (Shenyang: Liaoning Science and Technology Press) 24-26
[7] Hu Xiaotao, Zhao Jie, 2011 Research on job analysis design method based on motion analysis. (Development and innovation of mechanical and electrical products) 59-60, 55
[8] Su Yanli, Hu Xiaotao 2008 *Research on Theory and Method of Product Function Analysis in Industrial Design. Development and Innovation of Electromechanical Products* 6.
[9] Yin Xiaochen, Zhang Liang, Wei Yanli. 2009 *Product Design Materials and Processes*(Hefei: Hefei University of Technology Press) 72-79

IOP Conf. Series: Materials Science and Engineering **825** (2020) 012024 doi:10.1088/1757-899X/825/1/012024

Innovative Design Analysis of Storage Cabinet Based on Traditional Palindrome Symbol

Kun Li[1], Ying Yu[2] and Tianrui Zeng[1]

[1]School of Mechanical and Power Engineering, Guangdong Ocean University, Zhanjiang, China
[2]College of Fashion and Design, Donghua University, Shanghai, China

Email: 55025869@qq.com

Abstract. Based on fundamental research and case analysis of traditional semiotics, the observation to life scenes discovers the humanization and personalization's demand to storage products. Fret patterns are integrated and reorganized, and then delivered in structure and application of storage lockers. The consumers would experience the fret patterns' aesthetic connotation of uniform and flouring. The traditional semiotics are applicated into moulding design culture with Chinese characteristics. The semiotics are transformed to design elements. The attractive products according to human qualities will be designed.

1. Introduction

Nowadays, the application of Chinese traditional culture and creativity has been valued increasingly, and moreover, it has become a consensus to trace back to the source of culture. The organic integration of design art and traditional culture is not only a historical trend but also one of the main directions for the development of design art in the future. The combination of traditional culture and design helps to enhance the added value of the product. While meeting the functional needs of the product, it also helps to arouse the consumer's identity and form the brand personality of the product. Some products, however, still have the following problems in the current product design expression:
(I) The application of Chinese traditional culture is very superficial in product design, valuing the form over the connotation.[1]

(II) The decoration of creative elements is greater than practicability, and the low point of agreement between creativity and products and the farfetched additions influence the integrity of the product.

(III) The creative elements are mostly "primary elements", lacking thinking and deep processing. Many traditional Chinese cultural elements have been transplanted into tourist souvenirs by means of copying, and there is a lack of systematic and multi-faceted step-by-step process in the extraction process. Fail to "understand" first, then "extend the meaning", "pass on", and finally "transcend".[2]

(IV) Severe imitation of design originality and single angle. With the cultural system diversity nowadays, the understanding and application of traditional culture shall also adopt the developing, dialectical, and pluralistic criteria.

(V) The perceptual intuition and communicability expressed by the designers through the products are insufficient, thus the users can't immediately resonate with the products.

As the social economy develops and the people's living standard is improved largely, people's needs are more individualized, more and more attention are paid to the humanized design of products.[3] Humanized design refers to adjust based on the basic functions of the original design through people's behaviour, mental condition as well as way of thinking, so that it is more suitable for people to use. The quickening pace of life makes people have to face the pressures brought by the work, study and life and run around for life, while the most relaxing place at this time is a clean and comfortable room.[4] Therefore, the collection of room items is very important, especially for the groups who rent houses in the long term or only have small houses. Since the living space is very limited, they are increasingly inclined to the flexible, convenient, multi-functional user-friendly way of collection. However, traditional storage cabinets are difficult to meet people's needs of individualization and humanization, and therefore, the flexible and portable design of storage cabinets with traditional cultural connotation is very important in today's society.[5]

2. Interpretation of the traditional fretwork symbol

2.1. The connotation and meaning of the fretwork

The fretwork first appeared on bronzes, used only as a decorative pattern, with the passage of time, it had been slowly given many auspicious meanings by the ancients. The fretwork pattern is mainly composed of circular curls or square folds, and is composed of continuous "return" lines. It has a variety of forms, such as monomers, connection of positive and negative, or in pairs, and a continuous strip-shaped form, which can be connected from the head to the tail and painted continuously. In the folk it has always been an auspicious symbol, the characteristics of its pattern also endow good meanings of "life and growth in nature", "nine divide by nine is one", "it's perfection itself", as well as "neat and uniform". The meaning of fretwork is also different in different areas. In the Mongol nationality, the fretwork symbolizes firmness, with the implied meaning of the sun's rotation and the bed of roses in four seasons.

2.2. Aesthetic overview of the fretwork

The composing form of the fretwork is a kind of repetition and continuity of the word "return" (HUI in Chinese character). It is composed of simple repetition with simple lines, neat and uniform, showing the solemn and orderly beauty in form. The pattern of the fretwork is simple and plain, but it contains rich aesthetic contents. It not only meets the aesthetic requirements of brevity and movement, but also contains the philosophical meaning of unity of opposites.

Modern furniture products are also designed to be neat and uniform and to be orderly aesthetic. As the collection furniture, the cabinet is to keep the room's items neatly and orderly, and therefore, the aesthetic meaning of the fretwork perfectly agrees with the principle of the collection design.[6]

2.3. Analysis of application of the fretwork in products

As a typical representative of Chinese traditional culture symbol, the fretwork's form is simple and generous, which conforms to the law of design aesthetics. Moreover, the fretwork is often used as decorative pattern in product design and used in decorative pattern of ceramics, cloth as well as furniture. In modern design, the fretwork is still widely used in daily clothing, advertising, furniture and other fields to show the products' texture, structure, decorative patterns and so on. In the ceramic design of oriental decoration, as shown in Figure 1, the decorative function of the fretwork is changed, it is used as the texture of the ceramic surface to provide visual and tactile enjoyment. In addition to be used as decorative patterns in the household products, the fretwork can also be used as a structure to make it three-dimensional, as shown in Figure 2. From the above analysis, we can see that the fretwork can not only be used as a decoration, but also be applied to the product through its graphic features, so that people have a better understanding of its meaning and connotation.

Figure 1. The Application of the Fretwork in Ceramics

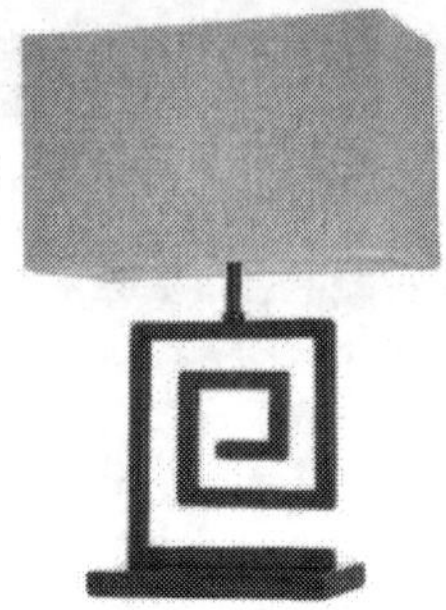

Figure 2. The Application of the Fretwork in the Product Structure

3. Thoughts on the application of the fretwork symbol in the design of storage cabinet

3.1. Redesign basis of the fretwork symbol

The fretwork symbol is a complex collection of symbols, which not only carries the symbolic information of multi-cultural symbols, but also has the structure of symbols. Therefore, it is necessary not only to use the innovative mechanism of the symbol to create new symbols by applying the corresponding design concept in the redesign of the fretwork symbols, but also to adjust the symbol structure of the fretwork itself.

The designer Wang Yang who once studied in Germany said that "Chinese things are very implicit, rather than superficial, so I hope to be able to reveal the connotation of Chinese culture under a fashionable appearance, as well as the traditional charm."

Traditional cultural symbols are embodied not only in tangible material culture, but also in intangible spiritual culture. The intangible spiritual and cultural resources are rooted in the people's way of life, aesthetic tastes and values. Designers should extract these cultures and spirit and reproduce it in the creative product design, and find the emotional resonance with the masses. We should deeply understand the connotation of the Chinese traditional culture, the traditional thought and the furniture decoration, as well as the modeling elements, and comprehend the essence, only after infusing these elements into the designer's brain, taking root, and acting as a part of the inspiration of creation, can the design really step toward the realm of freedom from the realm of necessity, and can it inherit and carry forward the national style, shaping the design style with Chinese characteristics.

3.2. Changes in natural and social environment lead to changes in daily demand for storage furniture

With the increase of small-sized households, the multi-functional and user-friendly furniture are getting more and more popular. The urban living space becomes smaller increasingly, but people's demand on quality of life becomes ever-growing higher. In addition to meeting the basic storage function, consumers will also consider the use of space by furniture. The flexible, multi-functional and easy-to-assemble furniture is getting more and more popular with the people. Although the traditional cabinets can meet the basic storage function, most of them are large in size, inconvenient to use, take up space and not flexible enough.

People are facing more and more pressure and are more eager for emotional care in life. Furniture is the necessities of life, humanized, and is the most emotional care products. Therefore, we shall appropriately endow the cabinet with emotional meaning, increase the cultural connotation, induct the positive emotion, and bring the emotional satisfaction to the users. Its design sketch is shown in Figure 3. It can also be used to facilitate people's living habits and use habits, so as to make people feel comfortable and pleasant.

New products should be innovative, the production technology and materials of the original products can be retained, and the reproduction of traditional technology is the inheritance and development of traditional culture. On the basis of maintaining the traditional cultural symbols, the "old" collection cabinet is innovated in form, and the modern design technology and method are used to make observation and behavioral analysis on all kinds of scenes of life by combining with the people's demand of humanization and individualization of the collection product, and the structure of the original storage cabinet is broken. The pattern of the fretwork is integrated and reorganized and is shown through the structure and way of use of the cabinet.

3.3. The embodiment of the fretwork symbol in the design of the cabinet

The product innovation design is inseparable from the consumer group's recognition and support, and also, the product use and purchase are inseparable from the consumer group's appreciation and choice. Therefore, the appearance of design is influenced by the change of people's aesthetic psychology. Designers, however, cannot be influenced by the audiences' aesthetic, they need to lead the design trend, start from beautifying life and improving the public's aesthetic level, and be mutually promoted and driven together with the public's aesthetic, entering into a virtuous cycle, only in this way can they design the outstanding works and add fashion elements and interest for innovative design.

In the light of the aesthetics of modern people and the reference of the outstanding creative products at home and abroad, this paper uses the "brainstorming method" to design the combined collection storage cabinet based on the traditional fretwork symbol, including product planning, text analysis, as well as finalization, and moreover, the material objects are produced by working together with enterprises, promoting in an all-round way, focusing on the creative experience design of the products, and developing the modern collection furniture creative products with strong interactivity, personality as well as taste.

In the design, the composition is formed through the fretwork in a square, simple and repeated way, shaped like the character of the word "return" (HUI in Chinese character), the fretwork is three-dimensional, and slightly changed into four parts to be used for designing cabinets. The cabinet is divided into four divisible cabinets which can be reduced in proportion. Users can combine the cabinets according to their own hobbies and residential space. There are horizontal and vertical combination forms, and even three-dimensional combination forms, so as to meet people's various needs and to bring people the fun in the use process. In the shape, the simple and orderly design aesthetics same as the fretwork are inherited, being simple and generous, and when the storage cabinet are completely combined together we can see a simple fretwork pattern, and moreover, the increase of the storage space can be realized through the different forms of combination. The design applications are Figure 4-1 and Figure 4-2. At the same time, it can be used as a table or chair, meeting the needs of contemporary people for flexible collection, and it is convenient for small residential population to arrange the collection of items in the room according to the actual situation, and there are many ways to use it.

The solid wood material is applied for the following reasons: on the one hand, the solid wood material is environmentally friendly, which is non-pollution to the indoor environment and is durable; on the other hand, wood is a warm material, which can make people think of nature thus to make people feel relaxed. The storage cabinet is collocated with white and burlywood, the white gives the person bright and clean, smooth, simple, elegant and tranquil feeling, and forms the strong contrast with the dull burlywood, adding the joyful vigor breath to the storage cabinet. The rendering renderings are shown in Figure 5.

IOP Conf. Series: Materials Science and Engineering **825** (2020) 012024 doi:10.1088/1757-899X/825/1/012024

Figure 3. Design Sketch

Figure 4-1. Design Application

Figure 4-2. Design Application

Figure 5. Rendering renderings

The design uses deductive method to deconstruct and refine the fretwork symbol, and makes the application related to modern life. In this process, the greatest difficulty is how to embody the spiritual connotation of the fretwork through the external form, the proper means of expression and the selection and use of the symbols. As an exploratory design practice, the "way of life" is chosen as the centre to express the form of the connotation of the traditional fretwork symbol in the modern life. This process leads the author to have a deeper understanding of the design concept of the redesign of the fretwork symbol. Its appearance has the vivid graphic symbol element, with wider functional extension, and moreover, it contains the practicality, the ornamental, the material is also more

IOP Conf. Series: Materials Science and Engineering **825** (2020) 012024 doi:10.1088/1757-899X/825/1/012024

diversified, and moreover, the design values the users' sensory enjoyment, the interactive experience and the emotion resonance, thus to endow the users with the good feeling.

4. Summary

The Chinese nation has a long-standing history, the Chinese traditional culture is broad and profound enough to make the Chinese proud, and it is the important cohesion of the Chinese nation. China's famous master of art design, Mr. Han Meilin, believes that the success of art lies in the "root".[7-8] If we liken China's art to a wide-spreading ancient tree, the reason why it is durable and luxuriant is that it has root, and the root is exactly the traditional culture spread far and wide in China. This design highlights the use of the traditional fretwork elements, rather than a single pursuit for the imitation of the ancient, it aims at discovering the valuable and classic traditional symbols and making the symbols integrate into modern life, so as to constantly innovate the modern design in the inheritance, and take the responsibility of national cultural inheritance.

Acknowledgments

Authors wishing to acknowledge assistance from Guangdong Planning Office of Philosophy and Social Science (GD16XYS23) and Department of Education of Guangdong Province (530002001064).

References

[1] Yao S and Kai H 2014 *Journal of Anhui Polytechnic University (Social Science Edition)* (Huainan: Anhui University of Science and Technology Press) p105-108
[2] Min P 2014 *Packaging Engineering* (Chongqing: Southwest Technology and Engineering Research Institute) p 55-58
[3] Juan P and Jianxiong Z 2016 *Furniture and Interior Design* (Changsha: Central South University of Forestry and Technology Press) p 54-55
[4] Yang Y 2013 *Shandong Textile Economy* (Weifang: Shandong Chemical Fiber Research Institute) p 77-78
[5] Qianru L and Yantao M 2018 *Green Environment Protection Building Materials* (Beijing: Economic Daily Press) p 256
[6] Jiaojiao L and Jiqing L 2014 *Journal of Minjiang University* (Fuzhou: Minjiang University Press) p 120-123
[7] Yaming L 2013 *Packaging Engineering* (Chongqing: Southwest Technology and Engineering Research Institute) p 8-11
[8] Daiqiang J 2011 *Decoration* (Beijing: Tsinghua University Press) p 10-11

IOP Conf. Series: Materials Science and Engineering **825** (2020) 012025 doi:10.1088/1757-899X/825/1/012025

Appling TRIZ Theory in Ship Overall EMC Design

Xiangxiang Cheng

China Ship Development and Design Centre, Wuhan, China

e-mail: alice_andrea@126.com

Abstract. Same frequency interference between electronic equipment still cannot be wiped out, which leads electromagnetic compatibility (EMC) system engineers to the one and only way by making an EMC rule to manage and control all the equipment onboard. However, while the sensors are following the EMC rule in battle, the weapons' maximal combat capability is usually not able to be attained. Therefore, EMC management and control widely used in today's ship engineering design is actually a compromise. This paper presents an unconventional design method combining the TRIZ methodology into traditional EMC design methods. The proposed method used TRIZ tools such as function analysis, causal analysis, ideal solution analysis and resources analysis to make deep analysis of the same frequency interference problem on ship. The final solution is obtained through contradiction resolution theory, Su-Field analysis and 76 standard solutions in this paper. At the end, a few enlightening ideas are harvested for the engineering innovative design problem.

1. Background

The mission of ship overall electromagnetic compatibility (EMC) system is to ensure that equipment can work well compatibly on board under the ship electromagnetic environment (EME), which is the combined results of both wanted signals and unwanted emissions, containing a wide range of waveforms with differing characteristics of frequency, time, amplitude and energy that can disturb the normal functioning of a ship platform and its associated systems. Major electromagnetic interference (EMI) problems and considerable additional costs are likely to arise if EMC requirements and ideas are not taken into account at all stages of projects for the design and development of equipment and systems. It is therefore essential and important to implement suitable EMC designs and measures from the beginning and during ship's whole life cycle.

Experience has shown that the ideal EMC status is unrealistic for a ship, especially for those with a big mass of antennas. In practical engineering, ship overall EMC always compromises with other performances of the ship. In order to achieve good EMC status for a ship, that is workable condition of a set of systems and equipment on board simultaneously when their components do not generate degradation in performance, overall EMC design must be carried out by a series of measures and technological process [1–2], including antenna arrangement and optimization, shielding, filtering, ground connection, cable classification and optimized distribution, etc., which are all simple, economical and well-effective. Besides, EMC management and control is a significant step in ship overall EMC system, which is designed based on the optimization of the whole ship EMC status, for example, antenna arrangement is totally confirmed, but causing operational effectiveness loss to the confined equipment in frequency, time, space and energy domain, that directly impact on the general combat performance of the whole ship.

IOP Conf. Series: Materials Science and Engineering **825** (2020) 012025 doi:10.1088/1757-899X/825/1/012025

2. Problem Description

Electromagnetic degradation of ship platform systems and weapon systems is a complex and often misunderstood area of ship overall EMC design. Electro technical systems and equipment are commonly selected on their individual merit with little of no thought of the problems of integration into the complex electromagnetic environment of a modern ship platform. With the rapidly growth of electrical and electronic equipment and systems on platforms, the number of EMI sources has increased rapidly. In ship overall EMC management and control strategy, many equipment are specifically fitted to generate electromagnetic radiation or receive electromagnetic signals under controlled and limited conditions because transmitters produce unintended and unwanted emissions that are harmful to susceptive receivers. The majority of EMI problems encountered on ships are self-generated mutual interference with both the source and victim working on the same frequency, which is also the most difficult to deal with for the reason that EMI is usually exaggerated when the equipment are operating at the same time and on the same frequency.

When the same frequency interference problem on ship cannot be settled by means of conventional EMI control methods, such as, overall antenna optimized arrangement, construction technological process, and so on, they can only turn to EMC management and control for help, in which way the combat equipment are forced to lower their combat capabilities, due to the restrictions on their operational frequency, time, space and energy. As almost every design engineer encounters, system incompatibility and design conflict are the two most common problems, which seems more prominent in ship overall EMC design.

Under the situation, this paper introduces a method named TRIZ (Theory of inventive problem solving, in Russian), which has been widely used in mechanical engineering but rarely in EMC design. The TRIZ method was developed in the former Soviet Union by Altshuller, who had analysed over 400,000 patents in designing 40 inventive principles and developed a contradiction matrix to improve innovative design method [3]. In the innovative design problem solving process, the TRIZ method [3–7] is a tool for the designer to handle these design contradictions just as what we are facing.

The ultimate goal is a well-designed ship with good electromagnetic compatibility, which can get rid of EMC manage and control measures, or at least making them invisible, that is the EMC manage and control measures will not bring evident loss to the whole combat ability of a warship. In this paper, we are attempting to incorporate the TRIZ method into our traditionally overall EMC design, in order to get some new ideals and constructive enlightenment.

3. Problem Analysis

3.1. Function Analysis

The function model for a ship overall electromagnetic compatibility (EMC) system is illustrated in figure 1, which shows clearly the output product of ship overall EMC system is a function that can be described as ensuring equipment working well compatibly on board under the condition of all possible EMI. The components of ship overall EMC system are electrical and electronic equipment, including those with transmitters and receivers and EMC management and control equipment, which are unnecessary to decompose further since every equipment onboard should meet its own self-EMC requirement. Supra-system contains ship surface environment, people on board, and ocean environment, which also affect with components of ship overall EMC system like they do to each other. F_{EM} is the most typical effect in EMC system, we call harmful F_{EM} EMI which is denoted by wavy arrow line; another effect is electrical signal control F_E, imaginary lines mean not effective.

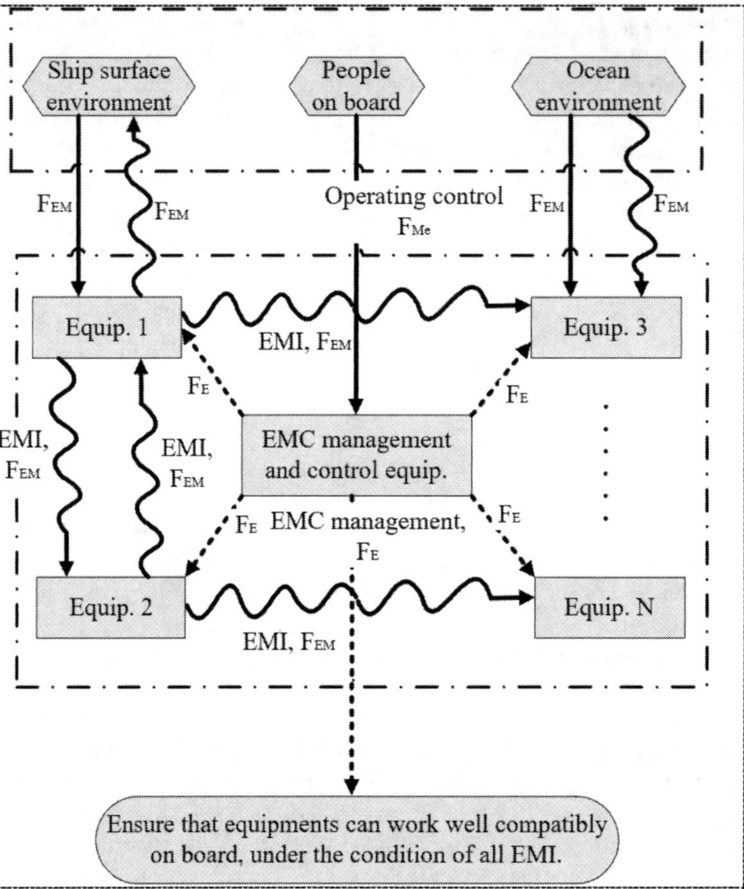

Figure 1. Function model figure of a ship overall electromagnetic compatibility (EMC) system.

3.2. Causal Analysis

Figure 2 shows root cause analysis searching for root causes, if the root cause is controllable then it can be the right direction to solution. In the first causal link path, all three root causes are uncontrollable, so this causal link path is out of the discussion of this paper.

Analysis of the second causal link path is based on the interference equation as follow:

$$\frac{I}{N} = P_t + G_t - L_p + G_r - R_s - \phi(\beta) - \phi(\Delta f) \tag{1}$$

where the only factor that can be controlled is transmission path loss Lp, which is expressed as:

$$L_p = 20 \log d + 20 \log f - 26.8 + k \tag{2}$$

Therefore, short distance between two antennas *d* is determined as a root cause in the second causal link path, while the other is low isolation degree between same frequency antennas. As indicated in figure 2, these two root causes are both controllable, but the first one is not in the range of this paper since antenna arrangement is already established, which only left the second root cause to be a investigation direction.

For the third causal link path, although only three uncontrollable root causes are obtained, but analysis process guides us a way, which is, in order to minimize the operational effectiveness loss, working principle and style involving electromagnetic wave emission and reception of related equipments must be intensively studied during the process of working out the EMC management and control strategy.

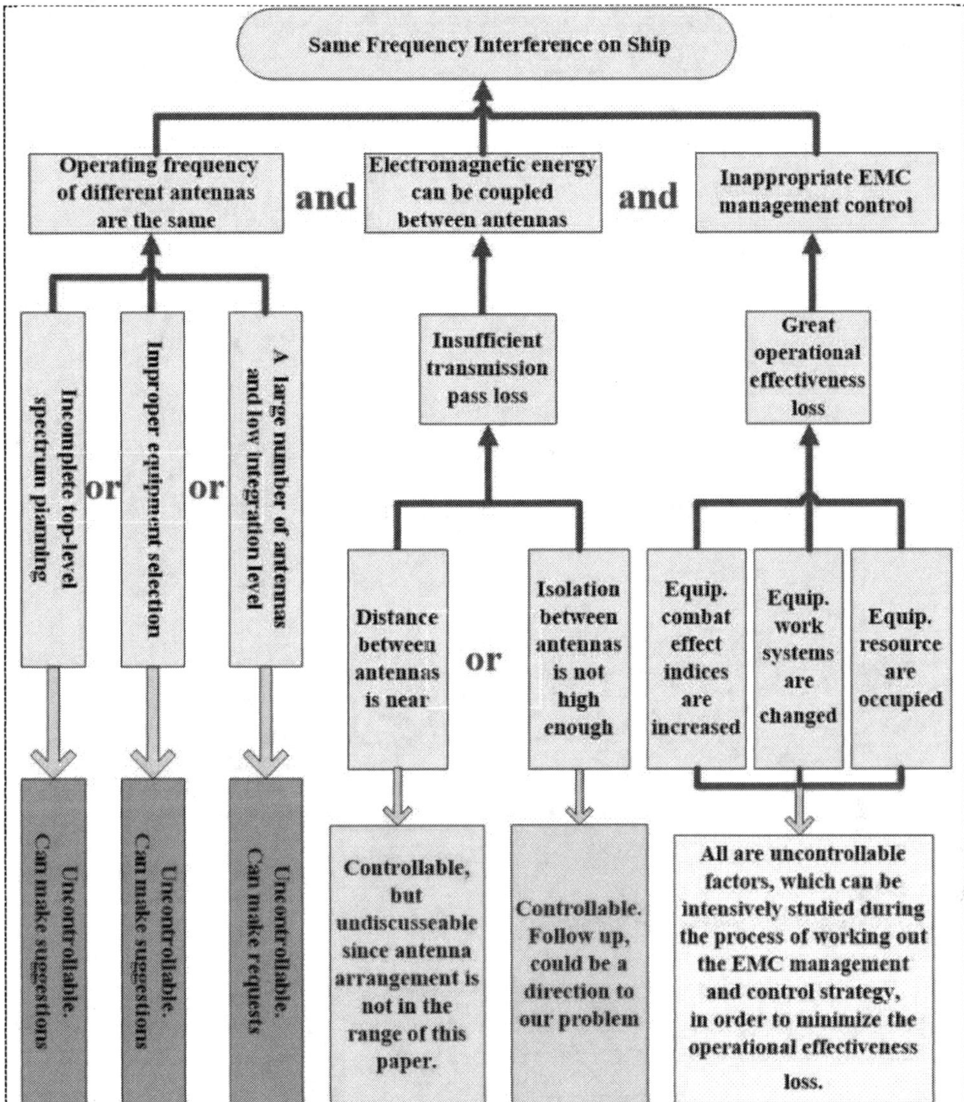

Figure 2. Root cause analysis figure of electromagnetic interference (EMI) between same frequency antennas on ship.

3.3. Contradiction Zone Determination

According to root cause analysis result, this paper determines its research scope on the following two contradiction zones:

- How to increase the isolation degree of same frequency antennas whose location is fixed on a ship.

The model figure of this problem is shown in figure 3, where the problem areas are coupling paths between antenna.

a) Key point 1: the distance between same frequency antennas is short due to limited space resourse on ship.

b) Key point 2: same frequency antennas can look straight to each other without obstacles on transmitting path.

- How to design and optimize EMC management and control strategy for each same frequency interference pairs.

a) Key point 3: space resourse contradiction.
b) Key point 4: frequency resourse contradiction.
c) Key point 5: time resourse contradiction.

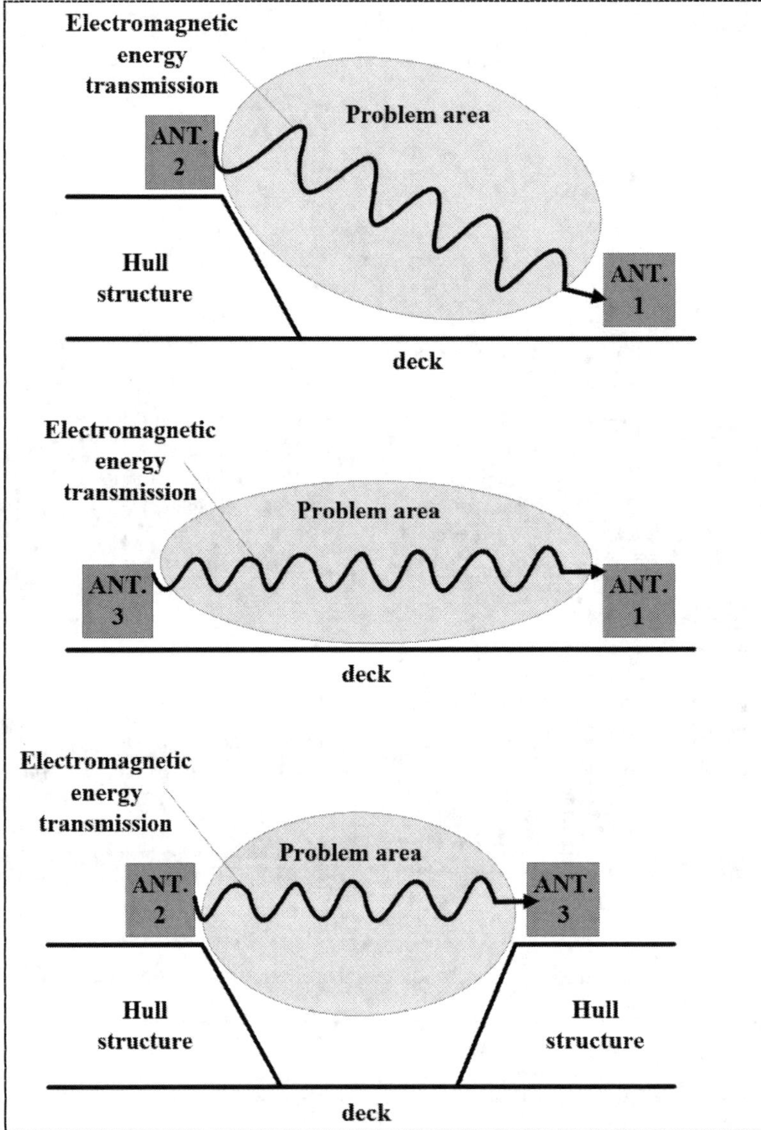

Figure 3. Model figure for the problem of how to increse isolation between two fixed antennas working on the same frequency.

3.4. Ideal Solution Analysis

Ideal solution can be divided into two types as final ideal solution and ideal solution. As the final ideal solution is usually hard to achieve, we need to define ideal solution to indicate direction following these steps:

1) What is the final goal? All electrical and electronic equipments on ship realize EMC.

2) What is the ideal solution? Causing no operational effectiveness loss.
3) What is the obstacle to achieve ideal solution? There exits EMI on ship between same frequency antennas.
4) What is the results when this obstacle occurs? Equipments are forced to lower their combat capabilities due to EMC management and control.
5) What is the condition of avoiding this obstacle? No EMI between same frequency antennas, or EMC management causing no operational effectiveness loss.
6) What are the available resources of creating this condition? Electromagnetic characteristic, working principle and style of related equipments are different from each other, in other words, equipments themselves are available resources. More detailed available resources are listed follow on.

3.5. Available Resources Analysis

System internal resources contains matter resource, field resource, space resource, time resource, function resource, frequency resource, information resource, which are all belong to direct application resources, while function resource and information resource are belong to differential resources at the same time. Information resource can be equipments' electromagnetic characteristic including pulse (pulse width, pulse repetition frequency, pulse compression ratio, etc.), frequency agility, frequency hopping, and so on, which are also export resources for ship overall EMC system. System external resources contains matter resource (e.g., air, metal, composite material), field resource, etc.

4. Problem Solving

4.1. Contradiction Resolution Theory

When using the TRIZ method in innovative design problem solving, the designer needs to first find the corresponding contradictions for the problem at hand, as we have done in section III.C. Next, the designer matches the meaning of each contradiction with two appropriate parameters from 39 engineering parameters that have been defined in the TRIZ contradiction matrix [3–7], that is what we will do in the following part. Using this contradiction matrix, the designer then can locate 3–4 suitable principles for solving that particular engineering innovative design problem.

4.1.1. Describe key point.

a) Key point 1: Ship EMC sacrifices for the good of ship overall performance, which results in the fact that the distance between same frequency antennas can not be increased.

b) Key points 2: The reason why there is no obstacle on EM wave transmitting path is because the obstacle is usually metal, which would bring negative influence to ship's Radar Cross-Section (RCS), aesthetic property, and practicabilities.

c) Key point 3: If space resourse contradiction happens, look angle of certain antenna is limited, whose detecting and tracking range would shrink, leading to space operational effectiveness loss.

d) Key point 4: If frequency resourse contradiction happens, filters must be added to susceptive receiver, which would narrow down the detecting frequency range of the receiver, leading to frequency operational effectiveness loss.

e) Key point 5: If time resourse contradiction happens, certain receiver must shut down to avoid the same frequency signal on board, whose detecting time would reduce, leading to time operational effectiveness loss.

4.1.2. Turn key point contradiction into TRIZ standard contradiction pair [3–6] as listed in table 1.

4.1.3. Find inventive principle in the contradiction matrix for each contradiction pair as listed in table 1.

4.1.4. Get the solution to problem on the basis of applicable inventive principle.

a) Scheme 1: According to Principle 35. Parameter Changes, ship overall EMC system can require equipments to make appropriate changes on their working style or electromagnetic property, in order to adapt to ship's particularity.
b) Scheme 2: According to Principle 31. Porous Materials, wave-absorbing material or EM composite material can be put into use.
c) Scheme 3: Combining Principle 22. "Blessing in Disguise" with Principle 10. Preliminary Action guided us to the way of how to making refined EMC management and control strategy, using harmful factors to achieve positive effects, which is management combination of time management, space management, frequency management, etc.

Table 1. Contradiction Pair and Inventive Principle Table

No.	Improving Feature	Worsening Feature	Inventive principle
1	Length of stationary object	Reliability	No.15 Dynamics No.28 Mechanics Substitution No.29 Pneumatics and Hydraulics
2	Length of stationary object	Object-affected harmful factors	No.1 Segmentation No.18 Mechanical Vibration
3	Length of stationary object	Adaptability or versatility	No.1 Segmentation No.35 Parameter Changes
4	Object-affected harmful factors	Adaptability or versatility	No.35 Parameter Changes No.11 Beforehand Cushioning No.22 "Blessing in Disguise" No.31 Porous Materials
5	Object-affected harmful factors	Loss of information	No.22 "Blessing in Disguise" No.10 Preliminary Action No.2 Taking Out
6	Reliability	Loss of information	No.10 Preliminary Action No.28 Mechanics Substitution
7	Reliability	Loss of time	No.10 Preliminary Action No.30 Flexible Shells and Thin Films No.4 Asymmetry

4.2. Su-Field Analysis and 76 Standard Solution

4.2.1. Build Su-Field model as shown in figure 4.

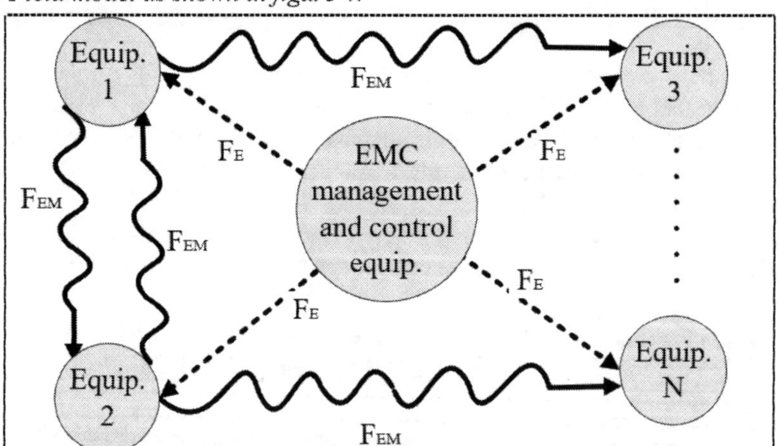

Figure 4. Su-Field model figure of electromagnetic interference (EMI) between same frequency antennas.

4.2.2. Apply resolution process of Standard Solution [3–6] to determine the general solution.

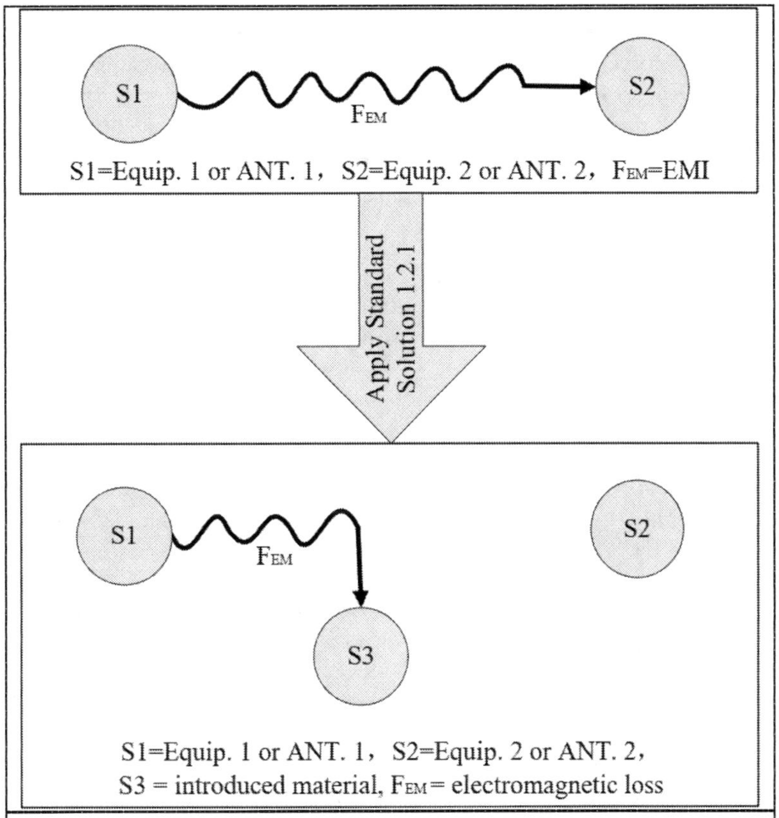

Figure 5. Su-Field model figure of applying Standard Solution 1.2.1 to slove the electromagnetic interference (EMI) problems between same frequency antennas.

- Because same frequency interference FEM in figure 4 belongs to totally harmful function, follow Standard Solution 1.2.1 (NO.9) in 1.2 Class, introduce S3 to eliminate harmful function, as shown in figure 5.
- Another Standard Solution 1.2.2 (NO.10) in 1.2 Class it to change S1 or S2 to eliminate harmful function.
- For insufficiency of function FE, in figure 4, Standard Solution 2.2.2 (NO.17) can be chosen, which tells major trends of the evolution, enlightening us to increasing the degree of fragmentation of time, space, and frequency for the propose of operational effectiveness loss reduction.

4.2.3. Get the solution to problem on the basis of obtained Standard Solution.

- Scheme 1: According to No.9 Standard Solution, as shown in figure 5, the solution is to introduce new substance (S3) on the path where unwanted EM waves transmit.
- Scheme 2: According to No.10 Standard Solution, the solution is to design constraints for S1 and S2 in ship EMC system, by modifying the parameters of whose working system or antenna's electromagnetic properties.
- Scheme 3: According to No.17 And No.42 Standard Solution, the solution is to subdivide the time, space and frequency resources, for the purpose of a delicacy EMC management rules.

5. Final Solution

Summarizing the above schemes solved with different inventive principles of TRIZ tools, we identify the final solution as listed in table 2.

Table 2. Final Solution and Usability Evaluation Table

No.	Scheme	Used inventive principles	Usability evaluation
1	Demand equipment to modify parameter	Principle 35 (Parameter Changes) in contradiction matrix; Principle 10 (change S1 or S2 to eliminate harmful function) in 76 Standard Solution.	Ship overall EMC design system has certain ability to restrain equipments onboard, which requires a lot of coordination and communication.
2	Add wave-absorbing material between antennas	Principle 31 (Porous Materials) in contradiction matrix; Principle 9 (introduce S3 to eliminate harmful function) in 76 Standard Solution.	This solution can reduce energy transfer between antenna with the same working frequency, which has no influences on ship's stealth and aesthetic. But the effect depends heavly on the electromagnetic property of used composite EM materials or coatings.
3	Combine several kinds of EMC managing means and pick the optional	Principle 22 (Blessing in Disguise) and Principle 10 (Preliminary Action) in contradiction matrix.	Based on in-depth study into operating principle and working manner of relevant electric equipments and combined with operational flow of a warship, customized EMC managing and controlling stratage can be worked out for each EMI pair, by applying TRIZ methodology, which aims at different EMI pairs and makes use of work characteristics and operating feature of both the interference source and the victim.
4	Subdivide the time, space and frequency resources for delicacy manage	Principle 17 (turn S2 from macroscopic to microscopic) and Principle 42 (system transform(d): to microscopic level) in 76 Standard Solution.	

6. Conclusion

Based on the TRIZ theory, this paper developed a complete set of ship EMC design method for a specific problem known as the same frequency interference on ship. The frameworks of this method are illustrated that integrate with selected TRIZ tools and corresponding inventive principles. This new method provides the designer with a supporting tool to develop new design inspiration with less limitation. More concrete EMC engineering problems are to be solved by further research of the method discussed in the paper.

References

[1] Zhou K J 2003 *Electromagnetic Compatibility Theory* (Harbin Engineering University press)
[2] Jimenez Gonzalez V H 2013 Electromagnetic compatibility in ships design and construction *IEEE ESTS 2013 NY USA* pp 333-336
[3] Ahshuller G 1996 And Suddenly the Inventor Appeared: TRIZ, The Theory of Inventive Problem Solving (Worcester, MA: Technical Innovation Center)
[4] Savransky S D 2000 Engineering of Creativity: Introduction to TRIZ Methodology of Inventive Problem Solving (Boca Raton: CRC Press)
[5] Orloff M A 2003 Inventive Thinking Through TRIZ: A Practical Introduction (Springer Verlag)
[6] Clausing D and Fey V R 2004 *Effective Innovation* (ASME press)
[7] Chen J L and Liu C-C 2001 *The Journal of Sustainable Product Design* **1** pp 263–272

IOP Conf. Series: Materials Science and Engineering **825** (2020) 012026 doi:10.1088/1757-899X/825/1/012026

Design of Pre-stressed Cable Rod Radar Array Structure Based on Finite Element Analysis

JinWei Wang[1], BaoFu Tang[1,*], JianGuo Cai[2] and HongHu Jiang[2],

[1]Nanjing Research Institute of Electronics Technology, Nanjing 210039, China
[2]National Prestress Engineering Research Center, Southeast University, Nanjing 210096, China

*tangbaofu@126.com

Abstract. In recent years, with the increasingly high requirements of radar thinness, the radar array appears thinner and thinner, which is easy to make the stiffness of the radar array is not enough, leading to serious deformation of the radar array, and thus affecting the work of the radar and the whole machine life. As a new type of space structure with light weight and high strength, the pre-stressed cable-stayed structure can play a good role in replacing the traditional front structure. According to the overall design requirements of a radar and the RMSE requirements for the plane flatness, the pre-stressed cable-stayed structure of the radar array is preliminarily designed, and the finite element analysis results are used to optimize the design and finally meet the design requirements.

1. Introduction

Radar array is one of the most important and complex components of phased array radar. Due to the increasingly high requirements on radar thinness in recent years, the radar array appears to be thinner and thinner, which often leads to the insufficient stiffness of the radar array. The bending and deflection of the radar array seriously affect the work and overall life of the radar [1]. Pre-stressed cable-stayed structure is a new type of space structure, through the adoption of cable-strut structure system to achieve to support the back of the radar antenna, the poles against the side panel for direct support to resist deformation to panel under load conditions, with lightweight, high strength characteristics [2-3], has become a traditional frame type phased array radar antenna structure one good alternative forms.

According to the overall design requirement of a radar and the RMSE requirement of flatness required for electronic communication of the front panel, the preliminary design of the pre-stressed cable- stayed front structure of the radar front is carried out, and the simulation calculation of the radar structure front is carried out by using ABAQUS software to verify the rationality of the structure design。 Then the model layout is optimized and adjusted, and the fish-belly layout is considered. The calculation results show that the fish-belly layout of the pre-stressed cable- stayed radar front structure can meet the design requirements, which can provide a reference for the radar front design of the pre-stressed cable- stayed structure.

2. Pre-stressed cable bar array structure design

2.1. Preliminary design of pre-stressed cable strut structure

In order to meet the requirements of radar electronic communication, it is required to ensure that the RMSE value of the panel flatness is less than 0.5 under various working conditions. Considering that the working condition of the radar front is mainly wind load working condition, the 8 wind load in the direction of the vertical panel is analyzed and calculated during design, and the wind speed of the 8 wind is about 17.2-20.7 m/s. . The basic wind pressure and wind load of the 8th grade wind are as shown in equation (1) and formula (2).

$$w_0 = \frac{1}{2} \times 0.001 \times 20.7^2 = 0.214\mathrm{kN/m^2} \tag{1}$$

$$w_k = \beta_z \mu_s \mu_z w_0 = 1.0 \times 1.0 \times 1.4 \times 0.214 = 0.23\mathrm{kN/m^2} \tag{2}$$

According to the structure of a whole radar system design requirements, the radar antenna using pre-stressed cable-stayed structure of front of parallel design, as shown in figure 1, the structure is mainly composed of skeleton line-rod and the front panel, cable-pole skeleton adopts parallel line layout, by the steel bar 1, 2, 3, line 4, 5, support bar support bar and support panel 6 through ball ground connection. The steel rod 1 has a diameter of 40 mm, the steel rods 2 and 3 have a diameter of 25 mm, the cable 4 has a diameter of 5 mm, the support rod 5 has a diameter of 10 mm and a length of 100 mm, and the support panel 6 has a size of 1500 × 1500 × 10 mm.

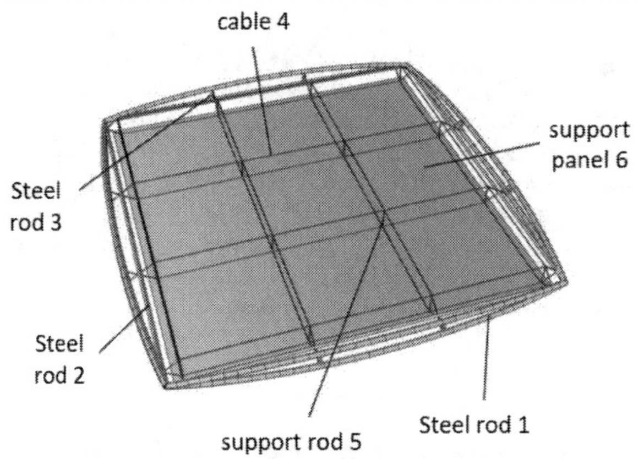

Figure 1. Model 1 structure layout

2.2. *Establishment of finite element model*

In order to improve the efficiency and accuracy of analysis, the surface model is simplified as follows:

1) the intersecting parts are connected through a common point, and the intersecting between the two cables is regarded as independent and independent;

2) the mass of TR components and antenna units fixed before and after the support panel 4 in the real radar is loaded by increasing the density of panel 4.

3) due to the large stiffness of the side beam, simplified analysis was carried out in the preliminary design. In figure 1, steel rod 1, 2 and 3 were omitted and the cable end was hinged for analysis. Strut bar 5 density =7850 kg/m^3, elastic modulus =2.06×1011N/m^2, Beam188 beam element is adopted. Cable 4 density =7300 kg/m^3, elastic modulus =1.9×1011N/m^2, linear expansion coefficient =1.1× 10-5/°C, using rod unit Link10; Support panel 6 density rho =8100 kg/m^3, elastic modulus EX=7× 1011N/m^2, shell unit Shell93. The boundary condition is that the four vertices of the cable-arch skeleton are completely constrained.

The pre-stress applied by the "cooling method" has the same effect as the natural frequency of the pre-stressed string calculated by the finite element vibration mechanics theory [4-5]. In this analysis, the pre-stress of the cable-stay structure is applied through the temperature field. The temperature was lowered by 100 degrees, the pre-stress was applied 209 MPa, and the analysis results were shown by magnification 50x.

2.3. Analysis of simulation results

The pre-stress condition of the whole front is pre-stressed, and the pre-stress + eight-level wind load condition is simulated. The vertical displacement cloud map of the obtained array is shown in figure 2 and figure 3 respectively. The RMSE of the plane flatness is shown in figure 4.

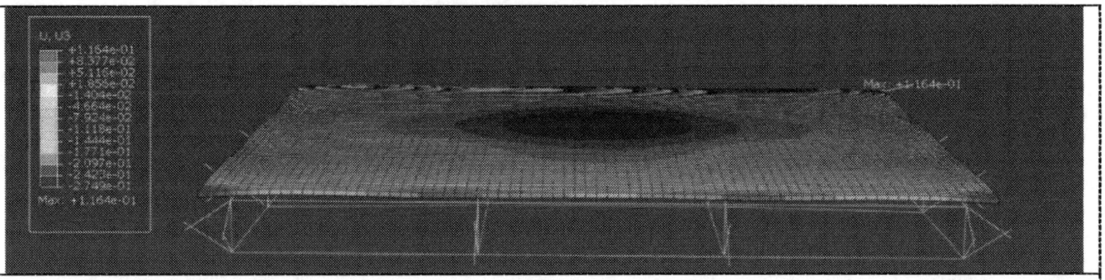

Figure 2. Vertical displacement cloud map under pure pre-stressed conditions (50x)

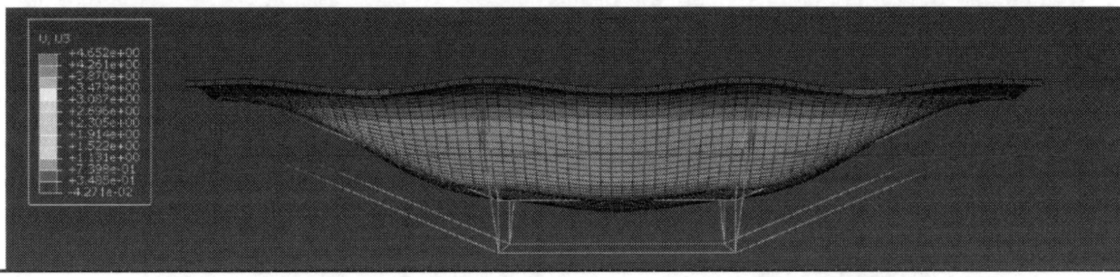

Figure 3. Vertical displacement cloud map under prestress + load conditions (50x)

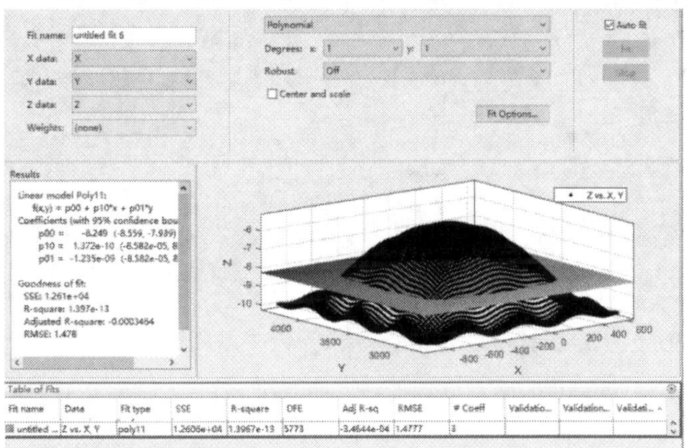

Figure 4. RMSE under pre-stress + load condition

It can be seen from figure 2, figure 3 and figure 4 that under the condition that only the pre-stressing force is applied to the front surface, the displacement of the front panel at the middle is the largest, and the displacement is 0.275 mm. Under the condition of pre-stressing + eight-stage wind load on the front, the displacement at the center of the panel is the largest, the displacement is 4.652mm, and the RMSE of the front panel is 1.477mm, which cannot meet the design requirements.

According to the theoretical analysis, the stiffness of the lattice structure can be increased by increasing the length of the struts. Therefore, the length of the struts 5 in the array is sequentially increased from the original 120 mm to 140 mm and 160 mm, respectively, and the simulation calculations are performed respectively. The results obtained from the simulation of the working conditions were counted as shown in table 1.

Table 1. Comparison of maximum displacement under various working conditions

category	Max displacement / mm	RMSE/mm
Panel 10mm, struts 120mm. No wind load	0.275	\
Panel 10mm, struts 120mm. Eight wind load	4.652	1.4777
Panel 10mm, struts 140mm. Eight wind load	4.590	1.4618
Panel 10mm, struts 160mm. Eight wind load	4.512	1.4434

It can be seen from Table 1 that the entire array has almost no effect on the stiffness of the entire array structure by increasing the length of the struts under the eight-stage wind load condition, and the RMSE of the front panel cannot meet the design requirements. The design of the cable-stay structure is invalid, and the failure analysis and optimization of the entire structure are required.

3. Failure analysis of pre-stressed cable rod front structure

According to the above simulation calculation results, the deformation side view of the structure under the eight-level wind load is selected for analysis, as shown in figure 5. According to figure 5, it can be seen that the outer ring of the outer surface of the cable is short due to the cable of the outer ring, and the load of the edge equivalent to the strut is small, so the displacement of the outermost ring is small, and there is almost no deformation. The outer ring cable bar structure is more reasonable. There is almost no relative deformation of the middle span cable, and the deformation is concentrated on the side spanning cable bar, showing a parallelogram deformation state.

Figure 5. Deformation side view of parallel cable-stay model

Since the equivalent load on the edge portion of the outer ring cable rod is small and the deformation amount is relatively small, it can be concluded that the force analysis is shown in figure 6. The outer ring cable rod can be equivalent to the support, and the rest of the cable rods are at the junction. Equivalent to the articulation, and then the force of each span cable is approximated to the force model shown in figure 7 below for simulation analysis.

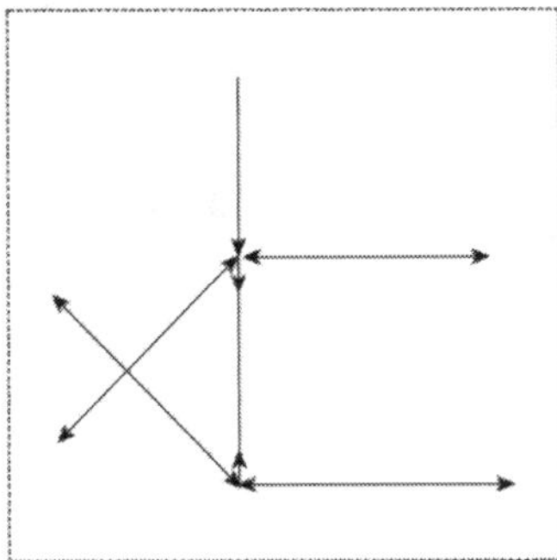

Figure 6. Force analysis of the edge cable to save space. Justify the caption.

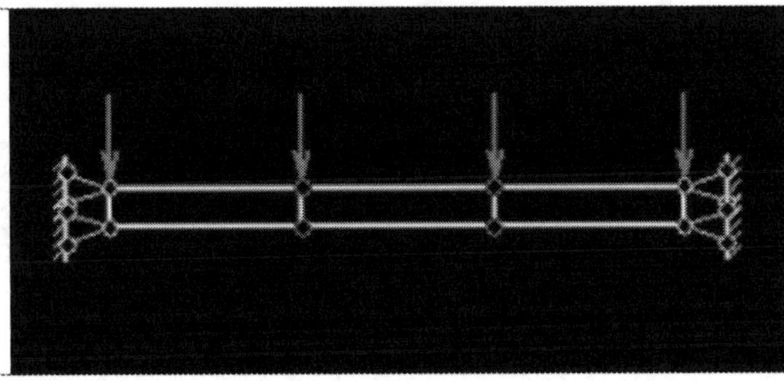

Figure 7. Simplified force model

As can be seen from figure 7, because of the unbalanced force, the cable-stay structure of the front is a transient system, and the transient system is extremely unstable when subjected to external loads, and the increase of the length of the brace will not change the force mode of the whole structure, so the stiffness of the cable-strut front structure cannot be increased by changing the length of the brace 5.

Further demonstration and analysis were carried out to extract the single trusses for analysis. With double cable design and unit load, the pre-stress is 209 MPa. The single truss structure is uniformly modeled as shown in figure 8 below.

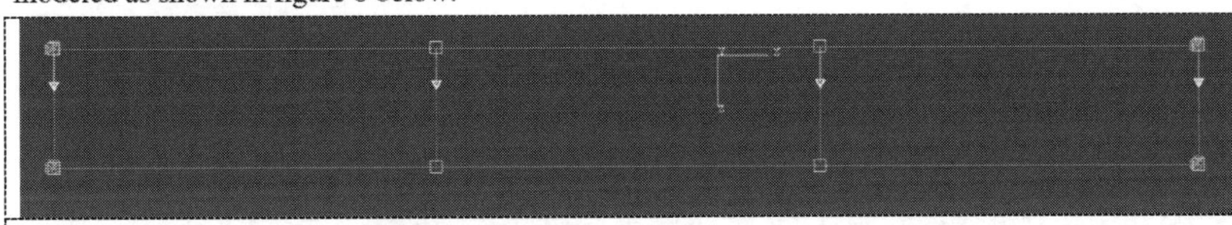

Figure 8. Single trussed parallel cable layout

In addition, on the basis of the original single truss structure with the strut length of 120mm, the strut length was set to 140mm and 160mm for simulation analysis, and the resulting displacement cloud map (all magnified 1000 times) was shown in figure 9, 10 and 11 respectively.

Figure 9. 120mm displacement of a single trussed parallel cable + strut FIG. U3 (1000x)

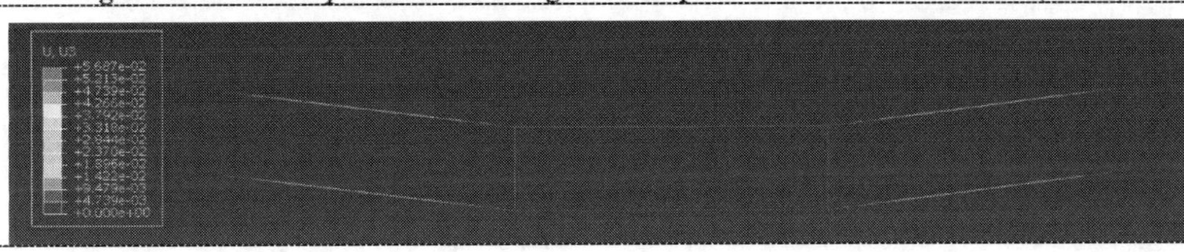

Figure 10. 140mm displacement of a single trussed parallel cable + strut FIG. U3 (1000x)

IOP Conf. Series: Materials Science and Engineering **825** (2020) 012026 doi:10.1088/1757-899X/825/1/012026

Figure 11. 160mm displacement of a single trussed parallel cable + strut FIG. U3 (1000x)

As can be seen from figure 9, 10 and 11, when the length of the bracing rod is 120mm, 140mm and 160mm respectively, the maximum displacement of the single trusses parallel cable structure is located in the middle span, and the maximum deformation is 0.0569mm. Therefore, it is further explained that the layout of the rod-rod structure of the array is designed as a transient system. By increasing the length of the strut, the stress mode of the overall structure will not be changed and the stiffness of the rod-rod array structure cannot be increased. Therefore, it is necessary to optimize and improve the layout of the rod-rod array structure.

4. Optimization design of pre-stressed cable strut structure

In order to avoid the pre-stressed cable-stay front structure system of transient condition affects the whole stiffness matrix surface, integrated the advantages of the stomach type design [6], against the cable-stay structure optimization, using the stomach type is shown in figure 12 single cross structure modeling, applying unified unit load at the same time, the pre-stress is 209 MPa, at the same time, the length of the strut is respectively 120 mm, 140 mm, 160 mm under the conditions of the simulation analysis, the displacement nephogram obtained as shown in figure 13,14,15 respectively.

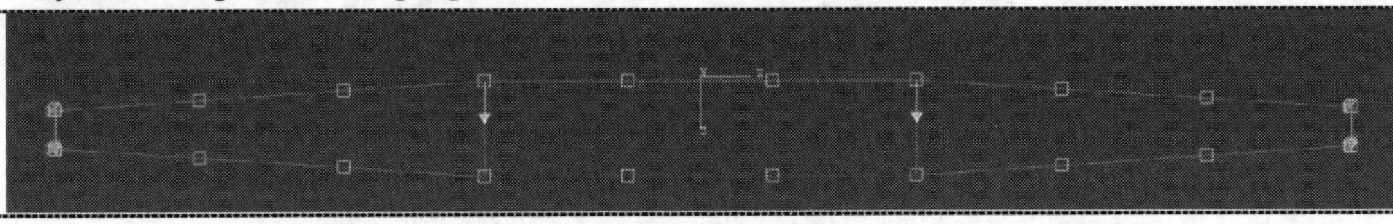

Figure 12. Layout of single trusses of fish belly structure

Figure 13. 120mm displacement of a single truss-belt-type + intermediate rod FIG. U3 (1000x)

Figure 14. 140mm displacement of a single truss-belt-type + intermediate rod FIG. U3 (1000x)

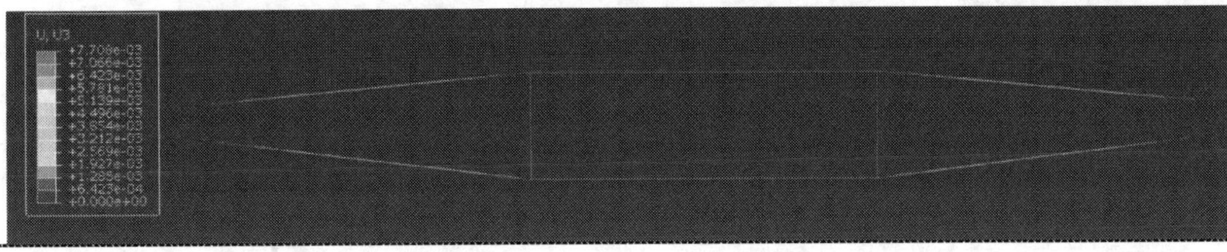

Figure 15. 160mm displacement of a single truss-belt-type + intermediate rod FIG. U3 (1000x)

As can be seen from figure 13, 14 and 15, when the length of the strut is 120mm, 140mm and 160mm respectively, the maximum displacement of the single trusses parallel cable structure is located in the middle span, but the maximum deformation is 0.02301mm, 0.01084mm and 0.007708mm respectively. This indicates that with the increase of the strut length, the rigidity of the single trussed parallel cable structure is larger, which can ensure the rigidity requirement of the whole pre-stressed cable-stay array structure, thus controlling the deformation of the array panel.

According to the calculation results of a single trussing fish-bellied parallel cable structure, the structural layout of model 1 (figure 1) was optimized, the cable layout was adjusted and the fish-bellied layout was adopted. The specific model layout is shown in figure 16.

Figure 16. Rendering of fish belly model

After the layout of the model is determined, the specific material parameters of the model are adjusted, and the design is carried out according to the requirements of the overall design of the radar and the corresponding simulation analysis. The specific parameters are defined as follows:

- The cable adopts stainless steel wire rope, with a diameter of 5mm, a minimum breaking tension of 14.7 kN, and an elastic modulus of 110GPa.
- The support rod adopts stainless steel S316, the diameter is 10mm and the thickness is 2mm round rod, the elastic modulus is 195GPa.
- The panel adopts aluminum plate with thickness of 10mm, 1.5m x 1.5m, and elastic modulus of 70GPa.
- The rest are made of Q345b steel.

IOP Conf. Series: Materials Science and Engineering **825** (2020) 012026 doi:10.1088/1757-899X/825/1/012026

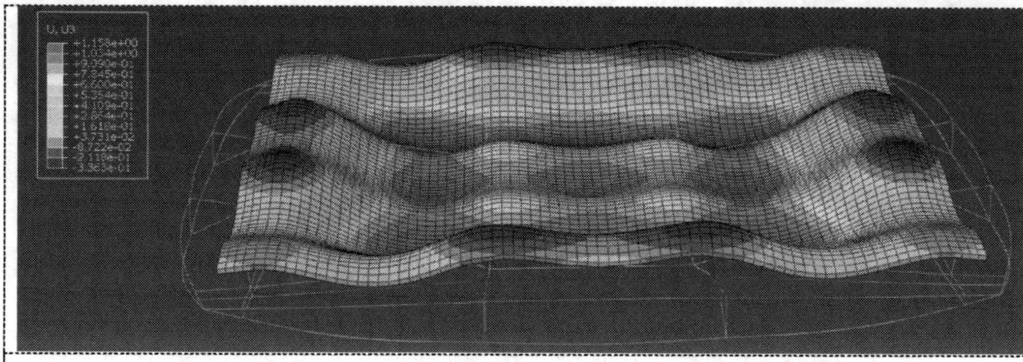

Figure 17. Displacement top view u3 of fish belly model

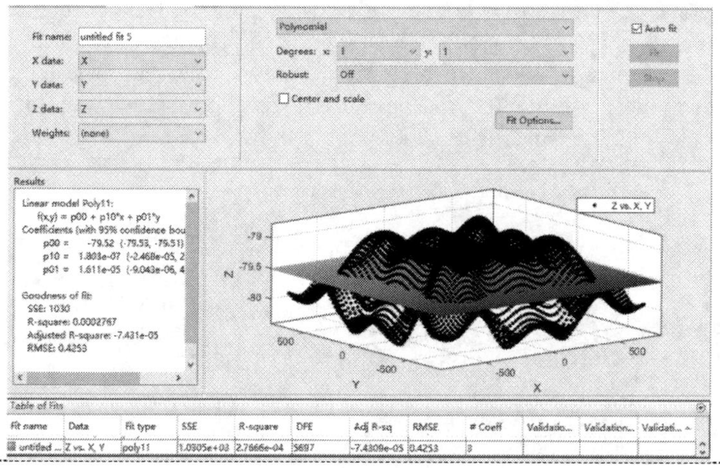

Figure 18. RMSE under load condition of fish belly model

Under the eighth wind load condition, 117MPa is applied to the pre-stressed element. The finite element analysis results are shown in figure 17 and 18. The maximum displacement is in the middle span of the front panel, the maximum displacement is 1.158mm, and the RMSE is 0.4253, which meets the design requirements of the front panel.

5. Conclusion

In this paper, according to the request of the overall design of a radar and electronic communications required against surface panel flatness RMSE requirements of the radar antenna using pre-stressed cable-stay front structure to the preliminary design, and through the application of ANSYS software to the simulation calculation of the radar antenna structure, verify the rationality of the structure design, the results show that the member structure exists transient system, it is difficult to guarantee the integral structure stiffness and RMSE of front panel flatness requirement.

To avoid transient system, adopting the new belly type structure of prestressed cable-strut radar antenna structure optimization design and simulation calculation verification, the results of the surface of pre-stressed cable-stay radar antenna structure using the stomach type decorates can ensure the stiffness of the front structure and make the front panel flatness RMSE value less than 0.5, can meet the design requirements, can be designed for radar antenna pre-stressed cable-stay structures provide a reference basis.

References

[1] Tang Baofu, Gu Yeqing, Wang Chao, et al. Application of Intelligent Structure in Phased Array Antenna Array[J]. Modern Atlas, 2014, 36(11): 8-10.

IOP Conf. Series: Materials Science and Engineering **825** (2020) 012026 doi:10.1088/1757-899X/825/1/012026

[2] Wang Hong, Guo Yanlin, Cui Xiaoqiang. Cable－arch hybrid structure dynamic response analysis［J. Engineering Mechanics,2003（1）:144~148.
[3] T Shen Shizhao, Xu Chongbao, Zhao Chen. Design of suspension structures [M]. Beijing, China, 1997.
[4] Yang Fei, Yang Wei, Xu Lei, et al. Application of Prestressing Effect Method in Various Unit Types[J]. Journal of Highway and Transportation Research and Development, 2012(8): 257-260.
[5] Liu Wei, Zhao Jianchang. Finite Element Analysis of Cooling Method for a Prestressed Concrete Simply Supported Beam[J]. Shanxi Architecture, 2008, 34(5): 92-93.
[6] Ma Yan. Design and Analysis of Fish-bellied Truss Support System for Glass Curtain Wall[J]. Silicon Valley, 2009(16):85.

IOP Conf. Series: Materials Science and Engineering **825** (2020) 012027 doi:10.1088/1757-899X/825/1/012027

Research on Artificial Intelligence in the Field of Art Design Under the Background of Convergence Media

Yan Yu, Zhan Binghong*, Guo Fei and Tian Jiaxin

Beijing institute of fashion technology

*Corresponding author: 646292998@qq.com

Abstract. With the continuous development of artificial intelligence, the application of artificial intelligence no longer repeats the "science and technology", "medical research and development", "computer" and other engineering disciplines. In recent years, artificial intelligence has also been fully developed and innovated in the "art" field. . This paper will focus on the research of artificial intelligence in the field of art in the context of convergence media, and hope to provide new ideas and new dimensions for the future of art design.

1.Introduction

On January 25, 2019, the 12th collective study of the political bureau of the communist party of China (CPC) central committee put "classroom" at the forefront of media integration. General Secretary Xi Jinping emphasized in the presidency that it has become an urgent task for us to promote the integration and development of the media and build the entire media. At the same time, artificial intelligence has officially entered a period of vigorous development in China from 2019, and a wide range of artistic works in the art design industry are also emerging. Under this background, how artificial intelligence will play its unique advantages and attractiveness in the field of art and design, which will be the key to this discussion and research papers.

2.Convergence media and artificial intelligence in the field of art design

Since the 18th National Congress of the Communist Party of China was held, "Convergence Media" has entered people's field of vision with its "advanced" and "unique" attitude. Now we are in the context of the era of media integration, and artificial intelligence is gradually involved in various industries. Among them, the application and function of artificial intelligence in the field of art design are becoming more and more extensive.

2.1. The meaning and development of the convergence media

With the continuous development of the times and the vigorous development of technology, big data is also applied to our lives, and the form of media is constantly evolving. From traditional media to new media, and then to today's convergence media. " convergence media" is a carrier that takes full advantage of the media and combines the plastic characteristics of traditional media such as radio, newspapers, television and new media (including digital technology, computer networks and wireless communication networks, as well as computer and mobile phone terminal communications form). In summary, convergence media will combine the advantages of traditional media and new media to create an integrated emerging media that is "flexible in resources, complete in content, and integrated in publicity" [1].

One of the innovations of convergence media is to blur the boundaries of various media and merge the advantages of different media. (Figure 1). For example, the convenience of broadcasting (which has been retro in the art field), the intuitiveness of television, the high loyalty of users, the unbounded nature of the Internet, and the media can be well integrated with these media. Integrated media is the product of the times. In addition to the innovation of ideas and models, it also combines the development of artificial intelligence. Establishing an "inclusive" database to reconstruct the original print media content is of great significance for preserving historical culture and better developing future culture. In the era of integrated media, innovation is eternal life. We need to better integrate traditional media with new media, and at the same time adhere to the core of "artificial intelligence" to achieve the integration of "users" and "services."

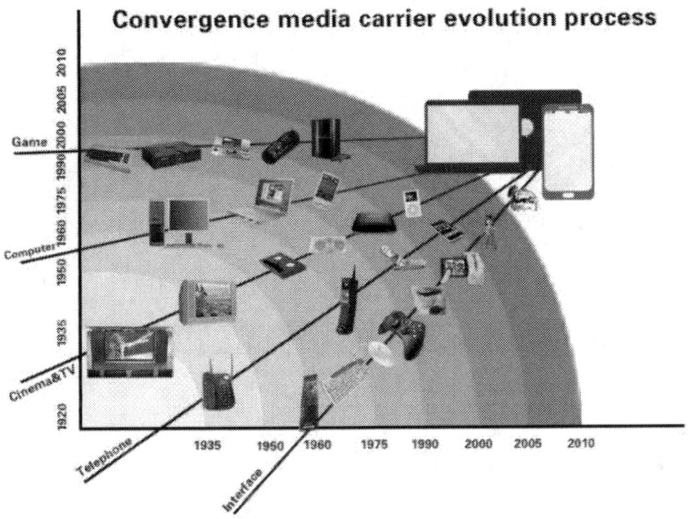

Figure1. Convergence media carrier evolution process

2.2. Artificial intelligence enters the field of art design

Over the years, the emergence and development of artificial intelligence have been accompanied by controversial voices, such as "replacing humans", "non-emotional" and "harm to humans". However, with the advent of the era of science and technology, we have to admit that artificial intelligence has not only not brought too many negative effects, but also promoted the accelerated progress of science and technology. Artificial intelligence has gradually entered the field of art design. As we all know, the reason why works of art are precious is because the human brain constantly receives new things while combining the author's own inspiration and ability to create. From the beginning, artificial intelligence only assisted humans in providing technology for design, and now it has begun to create works independently. Artificial intelligence is gradually entering the field of art design.

3.The performance of artificial intelligence in the field of art design

At the present stage, people's diversity and acceptance of artistic works are increasing. Under this premise, artificial intelligence provides a new creative idea and form for the creation of art works. For example, intelligent data-based devices, art interactive video works, chip technology and future clothing[2]. These products are the new era design works created by artificial intelligence technology. "Perhaps in the near future, art will no longer be exclusive to humans, and artificial intelligence has become a trend to intervene in art." Lu xiaobo, (dean of the academy of fine arts at Tsinghua university)[3].With the continuous maturity of artificial intelligence, emerging artificial intelligence

IOP Conf. Series: Materials Science and Engineering **825** (2020) 012027 doi:10.1088/1757-899X/825/1/012027

independently created paintings, text content generation design, poetry collections, trailer clips and other constantly attract people's attention.

3.1. Device space design of artificial intelligence technology

In May 2018, an exhibition called "Poetic AI" was launched by digital creation studio OUCHHH (Figure 2). The exhibition uses learning machines and artificial intelligence algorithms to study more than 20 million lines of text, including books, articles and theories of important scholars from ancient times to the present. When words and images are converted in a 3,300 square meter space, with sound, light, and shadow, the audience is immersed in the impact of the combination of history and modernity. This work is a typical creation of artificial intelligence under the background of convergence media. It perfectly combines the paper media of traditional media and the new media device through artificial intelligence to achieve the symbiotic creation of interaction with the audience.

Figure 2. "Poetic AI" exhibition organized by digital creative studio OUCHHH

In 2019, the Ultravioletto studio created an AI neural mirror art installation called "Neural Mirror" (Figure 3)[4]. This work uses artificial intelligence and facial recognition to build human data images, enabling viewers to think about the development of artificial intelligence in the artistic environment. This work fully demonstrates the design and application of artificial intelligence technology in today's art installation space. It can be seen that artificial intelligence can be used by artists to combine with art to create.

Figure 3. Ultravioletto studio installation "Neural Mirror"

3.2. Art paintings created by artificial intelligence technology

As early as 2015, Google launched the Deep Dream system using artificial intelligence technology for painting creation. It redesigns images by recognizing them and drawing[5].In 2017, Google again launched a tool called Autodraw, which uses algorithms to help users draw simple numbers and effectively help users optimize. Google has been working to combine artificial intelligence with the art of painting.

In 2016, the artificial intelligence work "Next Rembrandt" came out (Figure 4), which caused a lot of discussion in the art world. This work is actually a portrait drawn by artificial intelligence, which using data analysis and algorithms to mimic the painting style of famous Netherlands painter Rembrandt. The computer analyzed the data from Rembrandt's 346 works, identified more than 60 points that could be aligned on the face in the picture, and fully reproduced this work. It can be seen that artificial intelligence technology can already be used for painting creation, and more realistic and valuable works will be created in the future.

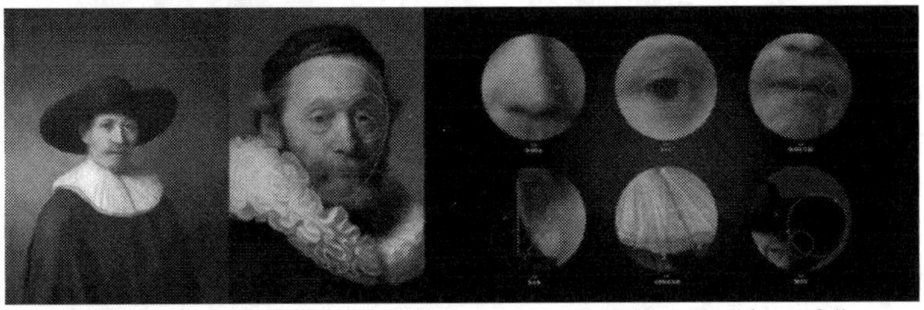

Figure 4. Artificial intelligence work "Next Rembrandt"

In addition to these works, there are many design works in which artificial intelligence technology is applied to painting. For example, the artist Gene Kogan from New York also used machine learning algorithms to create many works, such as "Alice's 17 Famous Paintings", "painting on glass" and so on. At the Christie's auction in 2018, a portrait painted by artificial intelligence, "Portrait of Edmund Bellamy" (Figure 5), was auctioned, which eventually sold for $ 432,500. This indicates that artificial intelligence has begun to enter our field of vision at a rapid speed, and people have begun to question whether the paintings created by artificial intelligence are real works of art? At present, even algorithms are designed by human-led ideas. People just use cutting-edge technologies to create newer and more singular works. In the prospect that the media is gradually coming into our sight, we should not exclude the application of new technologies creation.

Figure 5. "Edmund Bellamy's Portrait" exhibition renderings

3.3. Creating Movie Images with Artificial Intelligence

In 2016, IBM's artificial intelligence Watson completed the creation of the movie "Morgan" trailer (Figure 6). Engineers showed Watson trailers for hundreds of horror movies. Watson analyzes the images, sounds, intonations, and music of these trailers to determine emotional relationships. The trailer edited by artificial intelligence not only greatly shortens the working time of the staff, but also on this basis, the trailer edited by artificial intelligence is very creative and promising.

Figure 6. Screenshot of "Morgan" trailer

In the same year, the world's first artificial intelligence film "Sunspring" (Figure 7) also appeared in the 48-hour challenge unit of the London Science Fiction Film Festival. Benjamin, the screenwriter of this movie, has read thousands of science fiction movies. Although the filmed lines, pictures, and plots have many problems, this movie is a landmark work of artificial intelligence moving into the field of art design and image creation.

Figure 7. Screenshot of the "Sunspring" movie

4. The artistic creation of artificial intelligence in the context of convergence media

Through analysis, we can see that artificial intelligence is now gradually entering the art field and showing the power of technology with its innovative attitude. Nowadays, in the context of convergence media, how to use artificial intelligence for better artistic creation will be a major trend that cannot be ignored in future art design.

4.1. The promoting function of artificial intelligence in art teaching

In many years of art teaching, humans have experienced both paper and face-to-face teaching. Until now, we have used Internet e-learning and technology to lead students to recognize the charm of art. Under the guidance of this idea, we can gain insight into the problem from a new perspective by understanding and reconstructing the nature of the problem[6].This is the stage of research and exploration from traditional media to new media. Furthermore, setting up majors such as new media research, artificial intelligence technology, etc. in many universities is enough to prove that people attach great importance to innovative education and teaching.

Now it is in the age of media convergence, taking advantage of the possibility of media breakage, and combining the comprehensiveness, diversity, technology and innovation of both traditional media and new media to better conduct art teaching. At the same time, we attach importance to the application and involvement of artificial intelligence in art teaching, to improve students' creative awareness, accept the rapid development of the times, and accept the convenience, efficiency, and innovation of artificial intelligence. In addition to retaining its own artistic expression in the creation of artificial intelligence technology, it will create more avant-garde and forward-looking artistic works. Making good use of artificial intelligence technology will play an important role in promoting art teaching in universities.

4.2. Artistic creation of artificial intelligence through media convergence

At present, the prominent advantages of artificial intelligence in the field of art design are algorithms, data processing, and so on. Relying on big data platforms, artificial intelligence has demonstrated technological strength beyond human beings. Many artists' artificial intelligence art works usually take advantage of this outstanding advantage. However, in works related to thinking, emotion and innovation, artificial intelligence is still insufficient. It is believed that with the continuous development of science and technology, designs like this will also be overcome by artificial intelligence.

On this basis, making good use of the universality and inclusiveness of media convergence will enable artificial intelligence to improve the creation of future artistic works from multiple levels and perspectives. Take advantage of big data to make the most of the advantages of artificial intelligence at this stage. It can not only reduce the burden on the staff, effectively reduce unnecessary time for creation, but also integrate the power of technology into the work. With the advent of the 5G network and financial media era, more and more art works full of artificial intelligence technology will appear.

4.3. Emotional reflection of artificial intelligence works of art

Based on the background of the media era, the artworks created by artificial intelligence show a strong level of algorithmic application and technology. At this stage, when people use artificial intelligence technology to create artworks, the core themes are produced by human beings, and creative and imaginative works are produced by what the human brain sees, hears, and thinks about. From this we can see that the current independent works of artificial intelligence are just statistics and stacking of big data. There is a great lack of the most critical emotional expression in art. At the same time, this is why the academic world is full of controversial artificial intelligence works.

As we all know, emotional integration is very important in the creation of artistic works, so that the works can be reflected by the audience and users. Tolerance to society and empathy are important criteria for evaluating a work. Therefore, in the future, the use of artificial intelligence technology in the creation of artists must complement their emotional expressions into the works, and combined with the advantages and characteristics of the media will be able to create excellent works of art.

5. Conclusion

In summary, we can see that artificial intelligence has gradually entered the art field in recent years. Through the analysis and interpretation of works, in the context of today's convergence media, artificial intelligence art works have shown a trend of diversity, technology and independence. But it is still lacking in emotion and reflection. More and more artists are paying attention to the artistic design of artificial intelligence. From the perspective of future trends, whether it is the “convergence media” core development strategy advocated by the General Secretary or the 5G era that we are all embracing, we must believe that the pace of artificial intelligence exploration will not stop. Accompanying it is the development in the art field will definitely flourish. Artists will create more outstanding, independent and innovative artificial intelligence works of art, adding glory to the new art era.

Acknowledgments

Thanks to the guidance of professor Zhan Binghong and Guo Fei of Beijing institute of fashion technology and the help of Tian Jiaxin, I completed this research. I hope I can have more ideas and innovations in the field of artificial intelligence and design in the future. Thank you again.

IOP Conf. Series: Materials Science and Engineering **825** (2020) 012027 doi:10.1088/1757-899X/825/1/012027

Reference

[1] Huang Dan. Talking about the "Media convergence": A Historical Perspective [J]. Journalists, 2019 (03): 20-26.
[2] Ren Nan. Research on the application of artificial intelligence in art design [J] .Art Science and Technology, 2019,32 (13): 149-150.
[3] Zhang Tingting. Art is more integrated into reality in the era of artificial intelligence [N]. China Art News, 2019-11-11 (002).
[4] Du Chengwei, Zhou Acheng. Research on the Application of Interactive Installation Art in Urban Commercial Space [J]. Design, 2019, 32 (07): 56-57.
[5] Tao Feng. Research on Artificial Intelligence Visual Art [J]. Literary Controversy, 2019 (07): 73-81.
[6] Liao Wei. Liao Wei: Design Thinking + Education = Creative Education [J / OL]. Design, 2019 (18): 44-48 [2019-11-23].

IOP Conf. Series: Materials Science and Engineering **825** (2020) 012028 doi:10.1088/1757-899X/825/1/012028

Park Accessibility Analysis Based on Location Information and GIS: Take Shanghai Hongkou District As An Example

Ziran Zhang[1, 2, 3,4] , Shengxi Fan[2,3, a] and Huayan Zhang[1]

[1] Shanghai University of Engineering Science, School of Art and Design, 201620 Shanghai, P.R. China
[2] Tongji University, Shanghai Institute of Design and Innovation, 200092 Shanghai, P.R. China
[3] Tongji University, College of Design and Innovation, 200092 Shanghai, P.R. China
[4] Tongji University, College of Architecture and Urban Planning, 200092 Shanghai, P.R. China

[a] Corresponding author's E-mail: 123952771@qq.com

Abstract. Spatial accessibility is one of the methods to study the rationality of park design. This paper is taking Hongkou District Park in Shanghai as an example, by using the ARCGis application as analysis tool to analyze that the relation between the location of Park entrance and location of the community, and to calculate the different time cost of people who is under the walking and cycling mode to put forward the distribution of park entrance location which is based on the accessibility of riding and walking.

1. Introduction

Parks and other green environments in a city are public green spaces open to the general public with integrative functions of ecological maintenance, recreation, disaster shelter, etc.[1][2][3]. The public often consider the rationality of the distribution of urban public green spaces in a city based on data collection such as Green Space Ratio of City and/or Green Land Area Per Capita. The application of accessibility analysis in related public green space has led to the emergence of some proper nouns, i.e. in this case, accessibility of public green space. Accessibility, originally proposed and defined by Hansen as the probability of nodes in a transportation network interacting with each other, is now considered to be the ability of people get access to a certain destination[4][5].

At present, most researches on accessibility of parks design accessibility study based on the object of the area of residential zone, and make the proportion of such area the criteria for accessibility evaluation. It wrongly considered as accessible of all the park bounds, which in reality is partially accessible, and as a result is easy to overestimate the accessibility of the park. To replenish the gap like this, this paper enrolled the entrances and exits of residential areas and parks in Hongkou District, shanghai for a point-to-point accessibility test followed by comparative analysis.

"Accessibility" and "Park accessibility" were used as the keywords of searching on CNKI, of which Fig. 1 and 2 showed the metrological and visual analysis results.

IOP Conf. Series: Materials Science and Engineering **825** (2020) 012028 doi:10.1088/1757-899X/825/1/012028

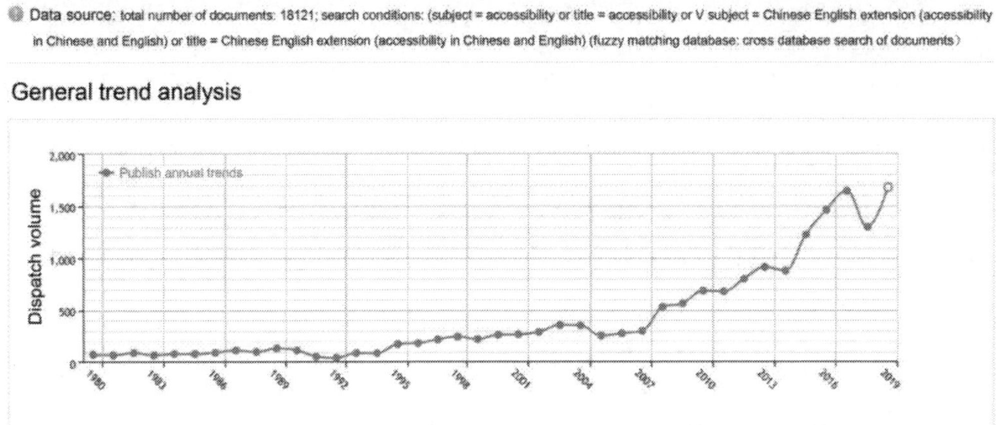

Figure 1. Metrological and visual analysis with keyword "accessibility"

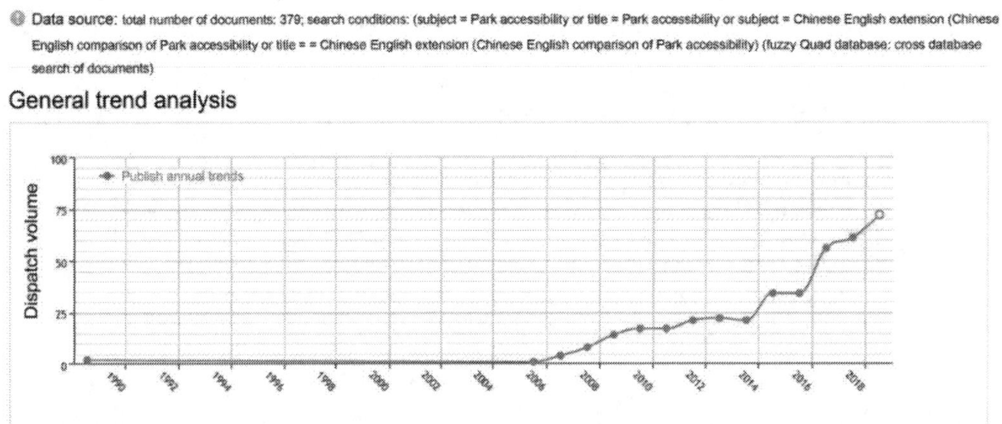

Figure 2. Metrological and visual analysis with keyword "park accessibility"

It can be seen that the quantity of papers involving "accessibility" and "GIS accessibility" had a huge growth around 2007, while those involving "park accessibility" increased obviously until 2017 around. Besides, the earliest literature records were ten years later than that of "accessibility". It suggested more attention were paid to subjects on accessibility in the past two years and the hot discussion has been staying at a high level. Now, GIS-based road network analysis serves as the principal methods for studying in accessibility of urban park. These include buffer analysis, network analysis, gravity model, and cost weighted distance, etc. [6]-[10].

2. Data research

2.1. Data collection

The location information of the entrances and exits of parks and communities in, as well as data of road network of, Hongkou District, Shanghai were required for the research in this paper. During early surveys[11], latitude and longitude information of entrances and exits of those parks were recorded. Park addresses were collected using Python crawler technology in the later data analysis to ensure the accuracy of information before they were converted into latitude and longitude information under different coordinate systems by "Address-GPS coordinate converter". Community addresses were captured using Python crawler technology from "58.com", "FangDD", and "fang.com" websites to increase the accuracy of service data. Road network data of the Hongkou District came from Google Maps in Bigemap's map library. Given the accessibility test was built on walking and cycling, accuracy of road network was at nine-grade.

2.2. Data cleansing and conversion

The optimal Data Cleaning method was determined after geographical information processing with three different methods and comparison to the feasibility in GIS accessibility analysis. On the basis of importing in GIS of location data of communities in Hongkou District, all coordinates of communities were corrected to two decimal places (i.e. hectometre) for proofreading, where non-Hongkou District data were deleted.

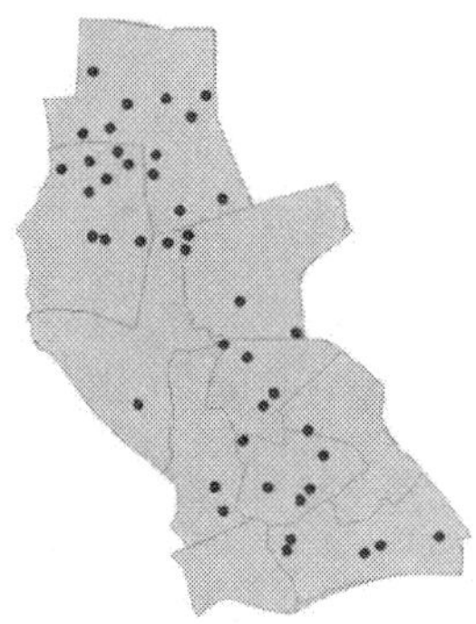

Figure 3 Method III for address translation and GPS data processing and verification

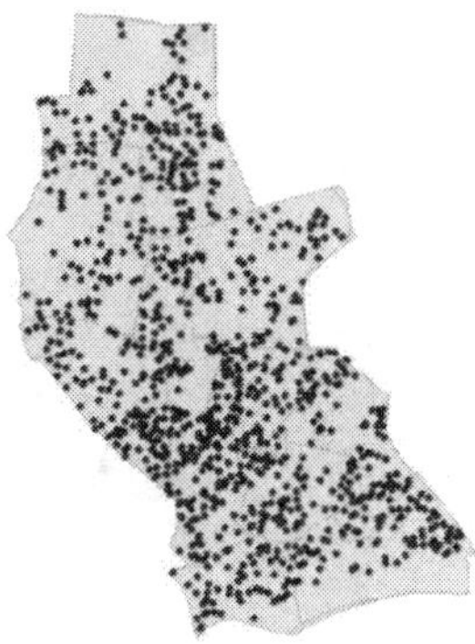

Figure 4. Addresses of communities in Hongkou District

The first 45 lines of Hongkou District screened out using the method following served as experimental data. Conformably before each community name it was added with "Hongkou District, Shanghai" before imported in "Address-GPS Coordinate Converter". From the detection, the derived data and the pickup system were 100% consistent; no consistency between the keyword column and the address column was observed. These were imported into GIS, as shown in Fig. 3. No significant deviation can be seen.

Data of communities in Hongkou District were processed using the Method and imported into GIS, as shown in Fig. 4. All of them fell in the administrative division of Hongkou District without obvious deviation. Method III was considered feasible, reproducible, and mechanically applicable in processing such data.

From the overall importing in GIS and the consistency of translated coordinates with Baidu coordinate pickup system, the conformable addition of "Hongkou District, Shanghai" before each community name produced the most accurate results. No obvious deviation would occur.

Table 1. Temporal information list

	Walking	**Bicycles Primary routes**	**Bicycles Non-primary routes**	**e-bicycles Primary routes**	**e-bicycles Non-primary routes**
Speed per hour (km/h)	5	15	12	20	15
Time (min)	12	4	5	3	4
No Date(min)	15	3	3	2.4	2.4

3. Analysis of accessibility in two traveling modes

3.1. Data importing and preliminary processing

For data processing in GIS, previously obtained geographic location information must be first imported into it and transformed as SHP files. SHP is one of the commonly used file formats for GIS software. It stores a part of geographic data, such as streets, points of interest, and postal code boundaries. From the road network, non-required provincial highway, national highway and other useless data were deleted. Only nine-grade roads, pedestrian paths and other demanded roads should be reserved. For the sake of accessibility test subsequent, in road attribute list the Value field of speed per hour was added according to Table 1. The Values were assigned in accordance with walking or cycling and then saved.

3.2. Accessibility test in walking mode

3.2.1. Accessibility test. The data of road network after processed is rasterized, since the information of the road is much too fine, the pixel accuracy is adjusted to be10^{-4}. The rasterized data is re-classified according to Table 2. With the cost-distance tool, the park entrance as the source data and the reclassified rasterized data as the cost data, the cost distance from each pixel to the nearest source in the Hongkou area is analyzed. The time domain after analysis is re-divided according to Table 2.

Table 2. Walking Accessibility Level

Walking time(min)	Level
>10	Level I
10-15	Level II
15-20	Level III
20-30	Level IV
30-60	Level V
60<	Level VI

3.2.2. Test result. Beautify the analytical drawing and export the map as shown in Figure 5. From Table 2, it can be seen that the accessibility of Quyang and Pinggong Park is lower, while the accessibility of Jiangwan Park, Ace Children Park and Huoshan Park is higher. Besides, it can be seen that the park is densely distributed and the district distribution is also denser.

3.3 Accessibility test in riding mode

3.3.1.Bicycle-based accessibility test. The bicycle-based accessibility is also analyzed with the walking method. The time domain after analysis is divided according to Table 3.

Table 3. Bicycle-based Accessibility Level

Riding time (min)	Level
>10	Level I
10-20	Level II
20-30	Level III
30-45	Level IV
45-60	Level V
60<	Level VI

The diagram is exported as shown in Figure 6. Combining the results with Table 3, it can be seen that the accessibility of Liangcheng Park and Kunshan Park is lower, and the accessibility of Jiangwan Park, Peace Park and Lu Xun Park Park is higher under the condition of bicycle. It can be assumed primarily that the accessibility of the large-scale parks is higher in the bicycle-based model.

IOP Conf. Series: Materials Science and Engineering **825** (2020) 012028 doi:10.1088/1757-899X/825/1/012028

3.3.2.E-bicycle Accessibility Test. Analyze the accessibility of e-bicycles with same method. It is exported as shown in Figure 7. In combination with Table 3, it can be seen that the accessibility of Liangcheng Park and Kunshan Park is lower, and that of Ace Children Park, Peace Park and Lu Xun Park is higher under the condition of bicycles. It is found that the coincidence rate of the accessible result for e-bicycles and bicycles is higher.

3.3.3.Comparison of Two Riding Models. It can be seen from Figure 6 and 7, in the same time domain, the coverage of e-bicycles increases more significantly compared to bicycles, but there is no significant change in the accessibility of parks.

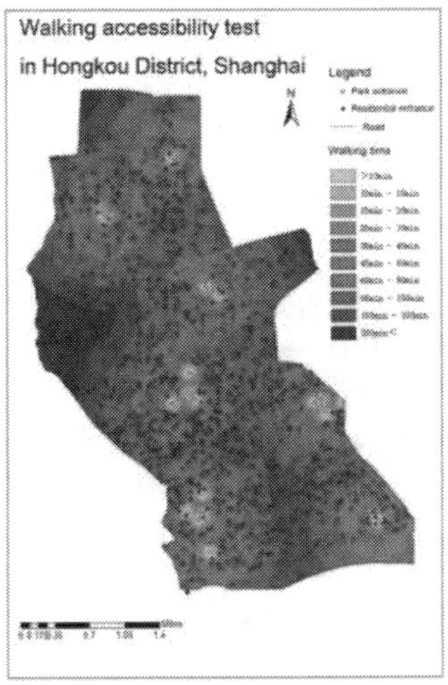

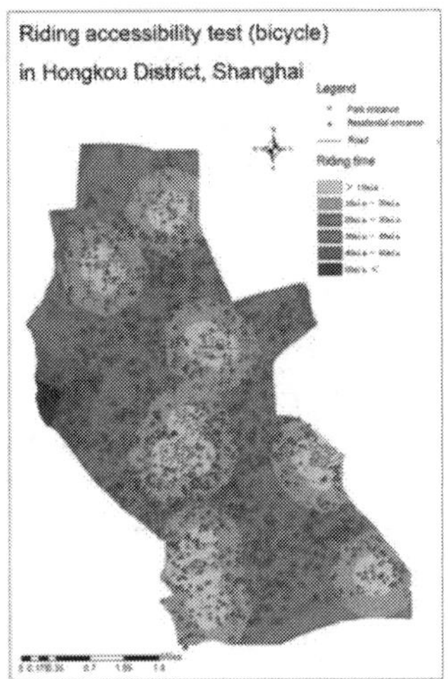

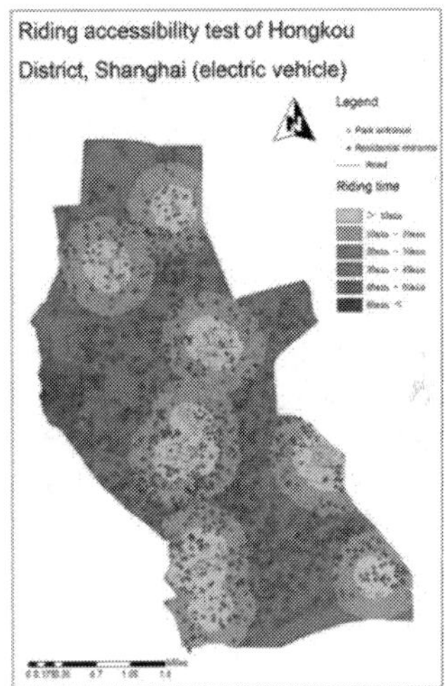

Figure 5. Walking Accessibility Test

Figure 6. Riding Accessibility Test (Bicycle)

Figure 7. Riding Accessibility Test (e-bicycles)

3.4 Comparison with walking and riding

It can be seen from Figures 5, 6, and 7, that at the same time domain, there is a significant increase in the accessibility of the park compared to that of walking. Accessibility of park also changes, and that of the smaller parks is higher.

4. Conclusion

Based on different travel modes, the accessibility of Hongkou Park in Shanghai is analyzed in this paper and the following conclusion is drawn:

Under the walking mode, the comfort accessibility of park is smaller. Smaller parks with only one or two entrances, such as River Bay Park, Ace Children Park and Huoshan Park, have the higher accessibility. Large parks with large entrances and exits, such as peace park, has less accessibility. It can be seen that in this travel mode, the relevance of the park's entrance and exit and community is not higher.

In the riding mode, the range of park's comfort increase significantly. In this case, large parks with more entrances, such as Peace Park and Lu Xun Park, have higher accessibility, while smaller parks have slightly lower accessibility. It can be seen that in the riding mode, multiple entrances and exits enables relatively improve the park's accessibility range.

In order to further improve the quality of urban residents' life, the independent road system can be established for larger ranger of riding accessibility which enhances the safety and convenience for urban residents accessible to the park green space in a short time. According to the distribution density of residential region, the number of the park's entrance and exit can be increased or deduced, which can shorten the time for residents to reach the park and improve the accessibility of the park.

Acknowledgments
Project Research Team of Intelligence Sustainable Package Design Support by Shanghai Summit Discipline in Design; Foundation item: Chenguang Plan (13CG74) sponsor; Shanghai Style Fashion Design &Value Creation Collaborative Innovation Center Support by Shanghai Summit Discipline in Design (DB18212).

References
[1] Xiaopan W, Yan H and Jian L 2017 *J. Beijing mapping.* Analysis on Urban Park Green. **05** 80
[2] Yuenwen T, Yi D, Yan H and Songjiang D 2019 *J. Beijing mapping.* **09** 1072
[3] Haiwei Y, Fanhua K and Yueguang Z 2008 *J. Ecology Journal.* **28** 3375
[4] Hansen W G 1959 *J. Jam Inst Planners.* **15** 3
[5] Yun Q and Shibo L 2019 Research on Progress of Urban Park Green Land Accessibility in China *J. Western leather.* 10-0047-01
[6] Xiaoma L and Changfu L 2009 *J. Ecology Journal.* **29** 1554
[7] Linbing M and Xiaoshu C 2006 *J. Journal of National Sun Yat-sen University* **45** 111
[8] Kongjian Y, Tiewu D, Dihua L and Jinfu P 1999 *J. Urban Planning.* **23** 8
[9] Yi H, Zhaoling H and Jian W 2009 *J. Journal of Xuzhou Normal University.* **27** 72
[10] Zhibin H, Xingyuan H and Qingxuan H 2005 *J. Journal of Shenyang Jianzhu University.* **21** 671
[11] Ziran Z, Yujin T, Yin Y and Shenxi F 2019 *4th Int. Conf. on Design Engineering and Science* (England: IOP Publishing). 10.1088/1757-899X/520/1/012018

IOP Conf. Series: Materials Science and Engineering **825** (2020) 012029 doi:10.1088/1757-899X/825/1/012029

Study on the Brine Storage Design During the Engine Running in the Climatic Facility

Zhang Hui[1,*], Wu Jingtao[1] and Cheng Zhu[1]

[1]Aircraft Strength Research Institute, Xi'an, Shan Xi, China

[*] Email: zhanghui07282006@163.com

Abstract. The engine will consume a large amount of air in the climatic environment facility when the engine is running in the test. The pressure in the chamber may decrease and the temperature in the chamber will change. In order to ensure the continuity and safety of the climatic test, the jet engine make-up air unit (JMAU) will compensate the cold/heat to guarantee the continuous and effective performance of the engine start-up test. Therefore, we should design the brine storage to ensure the supply of the brine.

1. Introduction

The engine is the heat of the airplane, the weather ability of the engine will directly affect the normal flight of the airplane and the normal work of the other systems. Therefore, before the airplane is on service, it is necessary to test the working conditions of the engine based on the full condition aircraft in the various harsh environments. The engine performance、the failures of the engine and the cooperative work of the engine system with other systems are assessed in the test, which will provide favorable basis for the optimal design of the airplane and safe flight.

The engine will consume a large amount of air in the climatic environment facility when the engine is running in the climatic environment facility. In order to ensure the stability of the temperature and the pressure and guarantee the safety, we should compensate the cool/heat. However, whether the cool/heat can be released immediately and continuously will influence the duration of the engine running. Therefore, the investigation on the cooling storage and released will ensure the engine is operated in the facility effectively.

2. The storage system in the climatic facility

As the engine needs to discharge a large amount of high-temperature gas during the start-up, the temperature in the laboratory will rise up. So it is necessary to supply cold air to the laboratory in time. Direct refrigeration cannot meet the cold demand of the laboratory in a short time but it is not easy to control. Therefore, the laboratory adopts indirect refrigeration and stores the cold energy in the coolant storage tank in advance. The secondary refrigeration exchanges heat with the air through the heat exchanger of the JMAU. The cold storage system of the climate laboratory includes cold source of refrigerating unit, secondary refrigeration and brine storage tank.

2.1. Refrigerating unit

The climatic environment laboratory adopts cascade refrigeration, using the 4 high temperature compressor units, the high temperature refrigerant is R507, using the 3 low temperature compressor

IOP Conf. Series: Materials Science and Engineering **825** (2020) 012029 doi:10.1088/1757-899X/825/1/012029

units, the low temperature refrigerant is R23, when the test temperature is higher than -25°C, the cooling from the R507 is transferred to the AS-6 to store when the test temperature is lower than -25°C, the cooling is transferred to the AS-6 and dichloromethane to store[1-3].

2.2. Choose of the secondary secondary refrigerant

The climatic environment facility using two secondary refrigerants -the high secondary refrigerant and the low secondary refrigerant[4-7]. The freezing point is lower than -70°C, the freezing point of the high secondary refrigerant is lower than -45°C. According to the analysis and investigate of the market[8], we decide to use the AS-6 as the high secondary refrigerant and use the dichloromethane as the low secondary refrigerant. The physical property parameter is shown in the table 1. The freezing point of the AS-6 is -55°C. It hasn't flash point, meet the requirement of the high secondary refrigerant. The low secondary refrigerant of the climatic facility using the dichloromethane. The physical property parameter of the dichloromethane is shown in the table 2.

Table 1 .The physical property parameter of AS-6

Temperature (°C)	**Viscosity (cp)**	**density (g/cm^3)**	**Heat conductivity Coefficient (W/m/k)**	**Specific heat (KJ/kg/K)**
-50	54.7	1.388	0.38	2.55
-40	25.7	1.383	0.39	2.57
-30	11.96	1.378	0.41	2.58
-20	8.70	1.370	0.42	2.59
-10	5.31	1.368	0.43	2.60
0	3.87	1.366	0.44	2.61
10	3.53	1.362	0.45	2.61
20	3.20	1.358	0.47	2.62
40	2.3	1.338	0.53	2.74
80	1.4	1.316	0.57	2.82

Table 2. The physical property parameter of the dichloromethane

Temperature (°C)	**Viscosity (cp)**	**density (g/cm^3)**	**Heat conductivity Coefficient (W/m/k)**	**Specific heat (KJ/kg/K)**
-70	0.8064	1.480	0.1892	1.2339
-60	0.7545	1.464	0.1858	1.1242
-50	0.7023	1.448	0.1825	1.2392
-40	0.6589	1.431	0.1791	1.1288
-30	0.6153	1.414	0.1757	1.1317

When the test temperature in the laboratory is above -25 °C, the secondary refrigerants of medium or high temperature exchanges cooling with the refrigerants of high temperature-R507. When the test temperature in the laboratory is below -25 °C, the secondary refrigerants of medium and high temperature exchanges cooling with the refrigerants of low temperature- R23 in the refrigeration system.

3. The design of the secondary refrigerant storages

After the secondary refrigerants are determined, the number and size of the secondary refrigerant storages should be designed according to the cooling requirement and the physical property parameters of the secondary refrigerants.

3.1. The flow of the secondary refrigerant AS-6

AS-6 treats the moist air from 35 °C to - 25 °C, AS-6 flowing out of the brine storage is -35°C, AS-6 flowing into the brine storage is -7.07°C. The relevant parameters to be used for calculation are as follows:

The dry bulb temperature of the handling moist air is 35 °C, the wet bulb temperature is 25.8 °C. According to the table, the moisture content of the moist air is $d = 17g/kg$, the moisture content of the saturated air at - 25 °C is $d_1 = 0.38g/kg$. The latent heat is $Q_0 = 2257KJ/s$ when the water is changed into vapor. The melting heat of the ice is Q_{melt}=334.72KJ/s. The air compensation is 400kg / s, there are 5 FMAUs, the air flow rate of each FMAU is 80kg / s, and the specific heat capacity of the air is $C_0 = 1.005KJ/s$, the moisture content of 80 kg / s moist air is $d_0 = 16.7g/kg$.

The calculation of the flow is as follows.

The water content in the moist air can be obtained as follows:

$$M_{water}=m\times(d_0-d_1)=80\times(16.7-0.38)\times10^{-3}=1.306kg \quad (1)$$

The dry air content in the moist air can be obtained as follows:

$$m=80-1.306=78.694kg \quad (2)$$

The calculation of the cooling needed when the vapor changed from 35°C to 0°C can be expressed as:

$$Q_1 = m \cdot c \cdot \Delta T = 1.306 \times 4.2 \times 35 = 191.9 \text{ KJ/s} \quad (3)$$

The calculation of the cooling needed when the vapor changed to the water can be expressed as:

$$Q_2 = m \cdot Q_0 = 1.306 \times 2257 = 2947.6 \text{ KJ/s} \quad (4)$$

The latent heat when the water changed to the ice can be calculated as:

$$Q3=m\times Q_{melt}=1.306\times334.72=437.14KJ/s \quad (5)$$

The cooling when the ice at 0°C changed to the ice at -25°C can be calculated as:

$$Q_4 = m \cdot c \cdot \Delta T = 1.306 \times 2.1 \times 25 = 68.54 \text{ KJ/s} \quad (6)$$

The cooling when the dry air at 35°C changed to the dry air at -25°Ccan be calculated as:

$$Q_4 = m \cdot c \cdot \Delta T = 1.005 \times 60 \times 78.694 = 4745.24 \text{ KJ/s} \quad (7)$$

The cooling when the compensation air changed from35°Cto-25°Ccan be calculated as:

$$\begin{aligned} Q &= Q_1 + Q_2 + Q_3 + Q_4 + Q_5 \\ &= 191.9 + 2947.6 + 437.14 + 68.54 + 4745.24 \\ &= 8390.42KJ/s \end{aligned} \quad (8)$$

The density of the AS-6 at -35°C is $\rho = 1378\,kg/m^3$, the specific heat is $C_0 = 2.57\,J/(g \cdot K)$

The temperature difference of the AS-6 between the inlet and outlet is:

$$\Delta T = 35 - 7.07 = 27.93℃ \quad (9)$$

The volume flow rate of AS-6 is:

$$V = \frac{Q}{(\rho \cdot C_0 \cdot \Delta T)} = \frac{8390.42}{1378 \times 2.57 \times 27.93} = 0.082\,m^3/s = 82\,L/s \quad (10)$$

3.2. The flow of the secondary refrigerant---dichloromethane

Dichloromethane treats the Saturated air from -25 °C to - 55 °C, Dichloromethane flowing out of the brine storage is -63°C, Dichloromethane flowing into the brine storage is -41.96°C.

The flow of the air is 80kg/s, the temperature difference of the air is:

$$\Delta T = 55 - 25 = 30℃ \tag{11}$$

The cooling needed when the air is handled can be calculated as follows:

$$Q = m \cdot c \cdot \Delta T = 1.005 \times 80 \times 30 = 2412\ \mathrm{KJ/s} \tag{12}$$

The density of dichloromethane is $\rho = 1487\, kg/m^3$.

The specific heat of dichloromethane at -63°C is c_0=0.903KJ/(kg·K)

The temperature difference of dichloromethane between the inlet and outlet can be obtained as follows:

$$\Delta T_0 = T_2 - T_1 = 21.04\mathrm{K} \tag{13}$$

The volume flow rate of dichloromethane is:

$$V = \frac{Q}{(\rho \cdot C_0 \cdot \Delta T)} = \frac{2412}{1487 \times 0.903 \times 21.04} = 0.0853\, m^3/s = 85.3\, L/s \tag{14}$$

3.3. The number and the size of the brine storage

The cooling storage capacity can provide 20 minutes of the test demand every 4 hours.

Due to incomplete stratification and interference of the inlet and outlet, it is assumed that only 90% of brine in the tank can be used for the experiment. The other 10% offset the energy loss caused by the start, pump, fan and pipeline accessories.

The design flow of AS-6 is 82L/s, and the volume of AS-6 which is required is:

$$82\mathrm{L/s} \times 20\mathrm{min} \times 1.2(\mathrm{comsume}) = 118.08\mathrm{m}^3/\mathrm{Unit} \tag{15}$$

There are 5 tanks, each one has a volume of 118.08m^3.

The design flow of dichloromethane is 213.4L/s, and the volume of dichloromethane which is required is:

$$85.3\mathrm{L/s} \times 20\mathrm{min} \times 1.2(\mathrm{comsume}) = 99.809\mathrm{m}^3/\mathrm{Unit} \tag{16}$$

There are 5 tanks, each one has a volume of 99.809m^3.

3.4. The design of the brine storage

The design of the brine storage tanks is mainly used for the cooling / heat storage for the secondary refrigerant in the engine running in the laboratory. As the total air compensation of the laboratory is 400kg / s, the storage tanks of the medium or high temperature refrigerant AS-6 are designed to be 5, each volume is 118.08m^3.The storage tanks of low temperature refrigerant dichloromethane are designed to be 5, each volume is 99.809m^3.

4. Conclusions

In order to maintain the normal operation of the engine starting in the laboratory, the air first passes through the cooling coil with refrigerant AS-6, then passes through the cooling coil with refrigerant dichloromethane, and finally enters the engine test room. The higher the engine thrust, the more air is consumed. Therefore, the faster the salt water is consumed, the shorter the duration time is.

The cooling capacity stored in the refrigerant storage tank can continuously provide about 80 kg / s of air at - 55 °C for 20 minutes, which can meet the requirement of running an engine with an exhaust speed of 204.3 kg / s for 20 minutes.

The cold storage system designed in this paper can meet the cooling demand of the engine start-up test in the laboratory, and provide technical support for the later test capacity expansion of the aircraft in the climate environment laboratory.

Acknowledgments

Authors wishing to acknowledge assistance or encouragement from colleagues, special work by technical staff or financial support from organizations should do so in an unnumbered Acknowledgments section immediately following the last numbered section of the paper.

IOP Conf. Series: Materials Science and Engineering **825** (2020) 012029 doi:10.1088/1757-899X/825/1/012029

References

[1] Zhou Xia, Liu Li, Numerical study on precooling process of cryogenic liquid oxygen tank, Cryogenic engineering, 2018,3(223), 36-40.
[2] Tang Hu, Li Ximing, Climatic Test of Aircraft, Equipment Environment Engineering, 2012, 9 (1), 60-65.
[3] Sun Laiyan, Huang Bengcheng, Cheng Zhixiang, et al. Cryonetics in Large Space Simulators, Cryogenics, 1999, 4, 218-223.
[4] Xu Zeqi, Zhang Shiyan, Xuan Weifang, Environmental Worthiness Evaluation of Equipment, Equipment Environment Engineering, 2012, 9(1), 54-59.
[5] Tang Hu, Liu Haiyan, Engineering Proposal for the Design of an Environmental Test Facility, Xi'an:Aircraft Strength Institute.
[6] Wang Jun, Technical Environmental Simulating, Beijing, National Defense Polytechnic Press, 1996:18-19.
[7] Cross M FMajumdar A K, Bennett J C, et al. Modeling of chill down in cryogenic transfer lines, Journal of Spacecraft and Rockets,2002. 39 (2): 284-289.
[8] Sun, Yang, Research on an intelligent behavior evaluation system for unmanned ground vehicles, Energies, 2018, 11, pp.1-22.

IOP Conf. Series: Materials Science and Engineering **825** (2020) 012030 doi:10.1088/1757-899X/825/1/012030

Titanium Coatings Plasma Sprayed with/without a Shroud

Hong Zhou[1,*], Cheng Peng[2], Zhi Liu[3], Mohammad Al-Rawi[1] and Jai Kanna[1]

[1] Centre for Engineering and Industrial Design, Waikato Institute of Technology, Hamilton, 3240, New Zealand;

[2] Technical Centre, GAC Motor Co. Ltd., Guangzhou 511434, China

[3] Science and Technology ltd., Hamilton 3210, New Zealand.

[*] zhouhong.nz@gmail.com

Abstract. Titanium coatings were deposited by plasma spraying with and without a shroud. The as-sprayed titanium coatings were then microstructurally examined. A comparison in microstructure between titanium coatings with and without the shroud was carried out. Based on the analytical results, it showed that the shroud played a critical role in protecting the titanium particles from oxidation. The presence of the shroud in the plasma spraying brought about a better heating of the particles in the plasma jet due to a mitigation of air entrainment with the shroud, thus a reduction in coating porosity was obtained. An enhanced microstructure in the shrouded titanium coatings were observed compared to the air plasma sprayed counterpart.

1. Introduction

Titanium and its alloys have been extensively used due to their good corrosion resistance and high specific strength. This illustrates their early success in the aerospace and the chemical processing markets [1-2]. Other industries like architecture, medicine, power generation, marine, sports and leisure, and transportation are also seeing an increased application of titanium [1-4]. As we know, a self-sealing oxide layer can be formed on the surface of titanium and its alloys when exposed to oxygen. The dense film provides a strong resistance to various corrosive chemical media [5- 6]. However, the metal is expensive because of the difficulty in processing. Processing for titanium must be carefully carried out in an inert atmosphere.

Thermal spray technologies have been developed throughout the major sectors of engineering industry for component protection and reclamation. These technologies allow combining a thin layer of high value material onto cheap structural materials. Plasma spraying is considered as the most flexible coating technique among them since there are plenty of materials for coating to meet various different requirement [7-8]. But titanium and its alloys are not very suitable for plasma spraying because they become reactive at high temperatures and have a strong affinity with oxygen titanium, nitrogen and hydrogen. Consequently, plasma spraying of titanium coatings in ambient air becomes challenging. Meanwhile, coating porosity may be increased due to oxidation of these metals and mechanical properties of coating would be degraded [9-10]. The spraying process for titanium should be carried out carefully in an inert atmosphere or vacuum to prevent from oxidation. Within the area of thermal spray, controlled atmosphere plasma spray, low pressure plasma spray, or vacuum plasma spray can offer an inert or vacuum atmosphere. They have been the potemtial techniques applied for titanium fabrication since they can form dense coatings with low oxygen content [11-14]. Some researchers reported that a good corrosion resistance like that of bulk titanium can be reached if proper spraying parameters were optimized [12, 14]. However, these processes are very expensive, and limited only to the small components.

Shrouded plasma spray can potentially produce titanium coatings at a low oxide containing level and be applicable for large components and when spraying under field conditions. The plasma spray torch is modified by using a gas-shrouded attachment [15-22]. The shroud attachment physically shields the plasma jet. A gas shroud surrounds the plasma jet after exiting the shroud attachment. Kinos et al [23] fabricated both titanium and tantalum coatings by shrouded plasma spraying. In his study, oxidation during plasma spraying was reduced with a shrouding system. Porosity and oxide content were minimized by optimizing the spraying parameters.

To avoid oxidation in plasma-sprayed coatings, feedstock should be melted, impacted onto substrate and solidified without contacting with appreciable amounts of oxygen. In the present study, a solid conical shroud was used to plasma spray titanium powders to reduce the oxide content in the coatings [24, 25]. The low-cost shrouding system can be operated in an open-air situation. Air plasma sprayed titanium coatings without the shroud were also deposited under the same plasma spraying parameters. The difference in microstructure between the two types of titanium coating has been investigated in terms of microstructure.

2. Experimental procedure

2.1. Materials and methods

Titanium powder was commercially available from Xi'an Lilin company, China, which was used to fabricate titanium coating. The powder's particle size has a range of 20 - 90 μm. The morphology of the titanium powders is shown in figure 1, which illustrates that the titanium powders presents irregular shapes.

Figure 1. The particle morphology of the titanium powders

A shroud has been used for the fabrication of titanium coatings by plasma spray [24,25]. A plasma gun SG-100 from Praxair surface technologies, USA, with and without the shroud, was taken to plasma spray the titanium powder. The plasma gun on a 6-axis robot moved in a raster pattern to spray the samples. The substrates were some rectangular plates of mild steel with size of 100 ×25 ×3 mm^3. They were degreased, and sand blasted before plasma spraying. Table 1 presents the parameters for plasma spraying with and without the shroud. The spraying parameters were the same except for a shroud gas flow in the plasma spray process with the shroud. For plasma spraying with the shroud, argon was used as an external shrouding gas at the exit, and the flow rate of argon is 300 slpm.

2.2. Characterization

Samples of cross sections of the titanium coatings sprayed with and without the shroud were metallographically prepared for microstructural analysis. Scanning electron microscope (SEM, Hitachi S4700, Japan and Zeiss EVO 18, German) with second electron and back scatter electron was used to observe the microstructure and morphology of the specimens.

Table 1. The parameters for plasma spraying with and without the shroud

Spray parameter	setting
Voltage, V	80
Current, A	800
Argon, Primary gas, slpm	85
Helium, Auxiliary gas, slpm	18
Feed rate of powder, g/min	30
Passes of spray	10
Distance of spray, mm	100
Transverse moving speed, mm/s	500
Nozzle	"Mach II"–forward injection anode nozzle

3. Results and discussion

The titanium coatings with and without the shroud were microstructurally analysed on the polished cross sections, as presented in figure 2. The SEM images reveals that a dense cross section at a low level of porosity in the shrouded titanium coating has been obtained; the pores in the titanium coating with the shroud are small, sparse and scattered. Whereas the titanium coating plasma sprayed without the shroud possesses a lot of porosity; the pores are relatively large and gathered into strips. This demonstrates that the presence of the shroud attachment resulted in a reduction in the porosity of titanium coating. In principle, there is a shielding effect of the shroud and the external shrouding gas, which would significantly mitigate the contact of in-flight molten titanium particles with the ambient air. This enabled the plasma spraying with the shroud to delay the corresponding drop in temperature. On the other hand, the reduction in air entrainment with the shroud could extend hot core of the plasma jet, and then increased the dwell time of particles in the plasma [20]. Thus, a better particle heating during the in-flight time was expected for the plasma spraying with the shroud, which led to more titanium particles fully molten at higher temperatures and a good splat spreading on the substrates. This helped to generate a dense microstructure and decrease the porosity in the shrouded plasma sprayed titanium coatings.

a) b)

Figure 2. SEM images showing the polished cross sections for a) plasma sprayed Ti coating with the shroud and (b) air plasma sprayed Ti coating without the shroud.

The back-scattered electron (BSE) images of the titanium coatings with and without the shroud are presented in figure 3. Backscattered electrons are normally sensitive to the atomic mass of the nuclei from which they scatter. Consequently, heavier element that backscatters more efficiently appears brighter than lighter element in a BSE image. Therefore, the BSE images contain compositional information. In figure 3, the lightest contrast features were identified as the titanium splats, the grey features were titanium oxides and the darkest contrast features were identified as porosity. The images show that there exist some titanium oxides in both the titanium coatings. However, the shrouded titanium coating presents much fewer grey areas than the unshrouded counterpart, which implies less oxidation took place during the shrouded plasma spraying. Most of the titanium oxides and porosity in both titanium coatings are located at the pass boundaries, which can be verified by many large grey or the titanium oxide stringers with some dark spots distributed along the pass boundaries in the images. This suggests that the dominating oxidation in the plasma spraying with and without the shroud might be the post deposition oxidation, which associated with the build-up of titanium coating when the spray gun moved away from the deposited material at each pass. As the plasma gases exited the torch or the shroud, they mixed with air to generate a high temperature, oxygen rich environment, which gave rise to the post oxidation. The shrouded titanium coating presented a much lower level of titanium oxides at the pass boundaries, which is clearly shown in figure 3a. It is postulated that the low oxygen in the plasma jet and a high external shroud gas flow which prevented the surrounding air ingress around the periphery of the shroud that was near the surface, could reduce the extent of titanium oxide formation at the pass boundaries during the shrouded plasma spraying. It was also observed that several thin submicron size (thickness) titanium oxide stringers homogeneously and broadly distributed within both types of titanium coating. The submicron titanium oxide stringers should be associated with splat formation, and more densely distributed in the titanium coatings without the shroud. During the plasma spraying without the shroud, air entrainment into the plasma resulted in molten or semi-molten particles being easily oxidized on the particle surface; the particles then impacted onto the substrate and flattened during splat spreading. Finally, those submicron titanium oxide stringers were formed at splat boundaries. In the shrouded titanium coatings, only a few of submicron titanium oxide stringers were observed. This can be explained that the shroud attachment and external shroud gas were able to create an effective shielding, which allowed the splats to form in a low oxygen environment. Thus, there were less submicron titanium oxides generated during the shrouded plasma spraying.

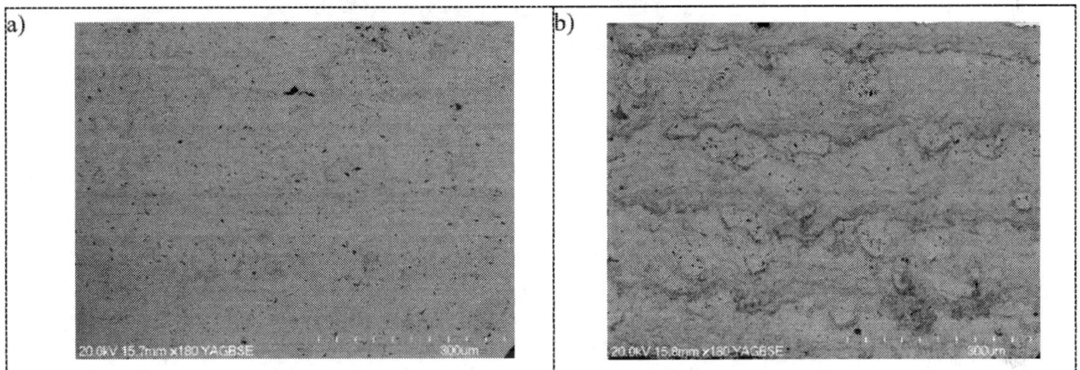

Figure 3. BSE images showing the polished cross sections for a) plasma sprayed Ti coating with the shroud and (b) plasma sprayed Ti coating without the shroud.

Figure 4 shows the SEM images of fractured cross-section for the plasma sprayed titanium coatings with and without the shroud. It is clear to observe that there exists a lamellar structure in the shrouded titanium coating, as shown in figure 4a. This is a typical structure for plasma sprayed coatings. Some voids in the titanium splats and some porosity in between the titanium splats could be observed. The splats with thin thickness were closely contacted within the titanium coating. None of unmelt particle was found in the titanium coating plasma sprayed with the shroud. In the spraying process, a plasma flame was produced by the ionization process of an argon gas, and the feedstock powders were injected into the plasma flame. The plasma could heat the titanium particles to a molten state. The titanium particles then impacted on the substrate, spread and quickly solidified to build up the titanium coating and the lamellar structure was then formed in the coatings. The presence of the shroud led to a better heating of particle during the plasma spraying. A higher temperature resulted in a better molten state of the particles. Therefore, the splats could spread fully when impacting upon the substrate and formed a dense lamellar structure with low porosity. Figure 4b presents the laminas in the shrouded titanium coating with thickness of about 1-2 μm, and the laminas have a nanosized columnar microstructure. The rapid heterogeneous nucleation should take the responsibility for the generation of the columnar nanostructure. The nucleation normally occurs at the cooler boundaries of the flattened droplets at large undercooling. The titanium coating without the shroud showed a different structure as seen in figure 4c. The lamellar structure is not clearly observed, whereas some solid particles with pores inside indicated that the molten state of splat during spraying was not well achieved. Some micrometre-sized particle could also be seen in the figure, which resulted from the droplets from well molten splats at impact or at the end of flattening.

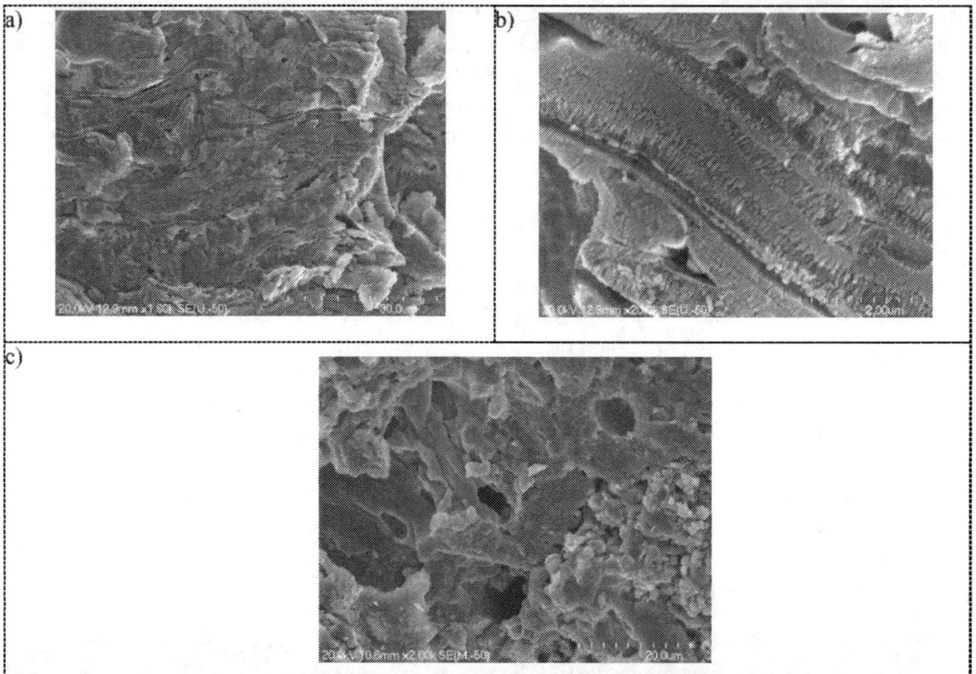

Figure 4. SEM images of fractured cross sections. a & b) plasma sprayed Ti coatings with the shroud at different magnification; c) plasma sprayed Ti coatings without the shroud.

4. Conclusion
This work presents the feasibility of using the shroud to fabricate titanium coatings by plasma spraying, and air plasma sprayed titanium coatings were also deposited under the same conditions. Microstructural characterisation was carried out for both titanium coatings. The physical isolating of the shroud attachment and the external shroud gas resulted in a better heating of in-flight titanium particles and resulted in a reduction in terms of coating porosity. The titanium coating with the shroud had a dense and reinforced microstructure; meanwhile, the titanium coating plasma sprayed without the shroud possessed a high porosity.

References
[1] Leyens C, Peters M. Titanium and titanium alloys. Weinheim (Germany): Wiley-VCH GmbH & Co. KGaA; 2003; 6-18.
[2] Inagaki I, Takechi, T, Shirai Y. Application and features of titanium for aerospace industry. Nippon Steel and Sumitomo Metal Technical Report. 2014; 106: 22-27.
[3] Gangwar K, Ramulu M. Friction stir welding of titanium alloys: A review. Materials and Design. 2018; 141:230-255.
[4] Zhou H, Gabbitas B, et al. Titanium and titanium alloy coatings for corrosion protection, In: Zhou L, Chang H, editors. Proceedings of the 12th World Conference on Titanium, V3; 2011 June 19-24; Beijing, China, Science press; 2012 : 1287-1291.
[5] Kitashima T, Kawamura T. Prediction of oxidation behavior of near-α titanium alloys. Scripta Materialia. 2016; 124: 56-58.
[6] Rajabi A, Mashreghi A, Hasani S. Non-isothermal kinetic analysis of high temperature oxidation of Ti–6Al–4V alloy. J. alloys and Compounds. 2020; 815: 151948.
[7] Fauchais P, Vardelle M, Goutier S. Latest Researches Advances of Plasma Spraying: From Splat to Coating Formation. Journal of Thermal Spray Technology. 2016; 25(8): 1534-1553.
[8] Fauchais P, Montavon G, Bertrand G. From Powders to Thermally Sprayed Coatings. Journal of Thermal Spray Technology. 2010; 19(1-2): 56-80.
[9] Matthews S. Shrouded plasma spray of Ni-20Cr coatings utilizing internal shroud film cooling. Surface and coating technology. 2014; 249:56-74.
[10] Gan J, Berndt C. Review on the Oxidation of Metallic Thermal Sprayed Coatings: A Case Study with Reference to Rare-Earth Permanent Magnetic Coatings. Journal of Thermal Spray Technology. 2013;

22(7):1069–1091.
[11] Kawakita J, Kuroda S, Fukushima T, et al. Dense titanium coatings by modified HVOF spraying. Surf. Coat Technol. 2006; 201:1250–1255.
[12] Steffens H, Erturk E, Busse K. A comparison of low-pressure arc and low-pressure plasma sprayed titanium coatings. J. Vac. Sci. Technol. 1985; A3:2459-2463.
[13] Gan J, Berndt C. Titanium Powder Metallurgy: Science, Technology and Applications. Amsterdam (Netherlands): Elsevier Academic Press; 2015. Chapter 23, Thermal spray forming of titanium and its alloys; p. 425-446.
[14] Lugscheider E, Lu P, Haeuser B, et al. Optimized vacuum plasma-sprayed titanium coatings. Surf. Coat. Technol. 1987; 32:215-226.
[15] Liu T, Zheng L, Zhang H. Effect of Solid Shield on Coating Properties in Atmospheric Plasma Spray Process. Journal of Thermal Spray Technology. 2016; 25(8):1502-1515.
[16] Mukherjee B, Kumar R, Islam A, et al. Evaluation of strength-ductility combination by in-situ tensile testing of graphene nano platelets reinforced shroud plasma sprayed Nickel-Aluminium coating. J. alloys and Compounds. 2018; 765: 1082-1089.
[17] Kim J, Kang K, Yoon S.et al. Enhancement of metallic glass properties of Cu-based BMG coating by shroud plasma spraying. Surface and Coating Technology. 2011; 205(8-9):3020-3026.
[18] Chen Q, Peng X, Yang G, et al. Characterization of Plasma Jet in Plasma Spray-Physical Vapor Deposition of YSZ Using a <80 kW Shrouded Torch Based on Optical Emission Spectroscopy. J. Thermal Spray Technology. 2015; 24: 1038-1045.
[19] Gawne D, Zhang T, Liu B. Computational analysis of the influence of a substrate, solid shield and gas shroud on the flow field of a plasma jet. Surface and Coating Technology. 2002; 153: 138-147.
[20] Chen H, Pfender E, Heberlein J. Improvement of plasma spraying efficiency and coating quality. Plasma Chemestry and Plasma Processing. 1997; 17: 93-105.
[21] Li J, Li C, Yang G, Li, C. Effect of vapor deposition in shrouded plasma spraying on morphology and wettability of the metallic Ni20Cr coating surface. Journal of Alloys and Compounds. 2018; 735:430-440.
[22] Vetrivendan E, Jayaraj J, Ningshen S, et al. Argon Shrouded Plasma Spraying of Tantalum over Titanium for Corrosion Protection in Fluorinated Nitric Acid Media. Journal of Thermal Spray Technology. 2018; 27(3):512-523.
[23] Kinos T, Chen S, Siitonen P, et al. Densification of plasma-sprayed titanium and tantalum coatings. J. Thermal Spray Technol. 1996; 5:439-444.
[24] Zhou H, Liu Z, Luo L. Microstructure and phase composition of titanium coatings plasma sprayed with a shroud, in Proceedings of the MATEC Web of Conferences, Qingdao, China, 25-27 August 2017.
[25] Zhou H, Liu Z, Luo L. Microstructural characterization of shrouded plasma sprayed titanium coatings. J. Manuf. Mater. Process. 2019; 3(4): jmmp3010004.

IOP Conf. Series: Materials Science and Engineering **825** (2020) 012031 doi:10.1088/1757-899X/825/1/012031

A Welding Unit Partition Method of Complex Weld Parts

Yunfeng Hu, Bo Wu, Jie Lu*, Jinli Xu and Xiaogang Liu

Wuhan University of Technology, School of Mechanical and Electronic Engineering, 430070 Wuhan, P.R. China

*lujiehotmail@163.com

Abstract. Welding modularization can make the process planning and manufacturing of complex welding products easier, convenient and synergy. The appropriate units partition is critical for welding modularization. This paper proposes a method of welding units partition for complex welding parts. Firstly, the basic parts and sub parts of welding units are defined. Then, the basic parts are evaluated by using the fuzzy analytical hierarchy process (FAHP) method and the number of basic parts is determined by the improved units partition formula. Furthermore, the welding units are classified by the minimum spanning tree (MST) algorithm and the interference matrix is used to evaluate partition results. Finally, a rear sub frame of automobile is taken as an example to validate this method.

1. Introduction

Welding and mechanical assembly are the two most important processes to assemble the automobile as a whole. It is estimated that almost 50% of the production cost is due to the assembly planning costs [1]. As welding parameters such as welding sequence, welding time, impact the final quality of the automobile, various methods have been used by researches to optimise welding sequence and to reduce welding time in the past study. The first research on welding sequences optimisation began in the 1940s [2]. With the development of artificial intelligence technology, different algorithms, such as neural network method [3], knowledge-based method [4], genetic algorithm [5], particle swarm optimisation algorithm, ant colony optimisation algorithm [6], have been used to solve the welding sequence optimisation problem with different fitness functions. However, these studies mainly aim to optimise robot spot welding sequence with fixed location. No previous study has addressed welding sequence optimisation for complex welding parts, which would be assembled at different work stations.

For complex welding parts with multiple welding processes, welding sequence planning is complex and time-consuming task. To improve welding sequence planning efficiency, an effective way is to divide the complex welding parts into several welding units. The unit partition as an effective method has always been applied in complex assembly planning. Liu Bo et al [7], used fuzzy clustering method to classify assembly units, which improves the intelligent level and efficiency of ship block assembly process design. Liu Ying [8], used the previous design results of assembly unit partition scheme to assist in the design of new assembly unit partition scheme. Appropriate units partition is critical for achieving collaborative design. However, the unit partition method used in complex welding parts has not been found in past studies.

Therefore, this paper proposes a method of welding unit partition for complex welding parts. FAHP is used as the calculation method of the basic parts, and the number of basic parts is determined by the improved unit partition formula. Then the welding units are classified by the minimum spanning tree (MST) algorithm and the interference matrix is used to evaluate partition results. Finally, a rear sub frame of automobile is taken as an example to validate this method.

2. Definition of welding units

A welding assembly P needs n parts to be welded as a whole. Under certain constraints, P can be partitioned into n welding units $U=\{u_1,u_2,...,u_n\}$, and a welding unit includes a welding basic part and a certain number of welding sub parts. The structure of welding assembly is as shown in Figure. 1.

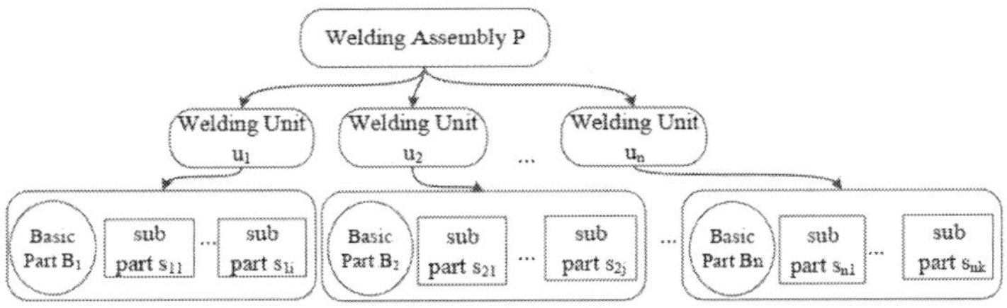

Figure 1. The structure of welding assembly

The welding basic part is the centre connecting carrier of various welding sub parts, which should have the following features:

1) Uniqueness: One welding unit has and only one basic part, and can have no sub parts.

2) Connectivity: The base part is directly connected to sub parts in a welding unit.

3) Convenience: The basic part need withstand the welding deformation caused by the heat during welding. Therefore, the basic part is usually with the largest mass and volume in a welding unit.

4) Similarity: Symmetrical welding is used in the welding process, and similar structures can be welded at the same time to improve the quality of welding [9]. The similarity is calculated by the Euclidean Distance formula 1:

$$S=\left[\frac{(w_i-w_j)^2}{w_{max}}+\frac{(a_i-a_j)^2}{a_{max}}+\frac{(b_i-b_j)^2}{b_{max}}\right]^{1/3} \quad (1)$$

Where w is the weight of welding part, a is the area of welding part, b is the centroid of welding part.

5) Inclusion: For a shell structure welded by a plurality of stamped metal sheets, if there is a support or a functional weld parts inside the shell, these parts should be partitioned into a welding unit.

3. Determination of basic parts

The analytic hierarchy process (AHP) is a multi-objective decision analysis method combining quantitative and qualitative analysis. However, this method has some problems, such as the difference between judgment consistency, difficulties in consistency testing and the lack of scientificity [10]. To solve the problem of consistency, the FAHP method is used to determine the basic parts, which uses the fuzzy complementary judgment matrix to evaluate the weights of parts. The steps of FAHP method are as follows:

Step1: Building the hierarchical structure model, shown in Figure. 2.

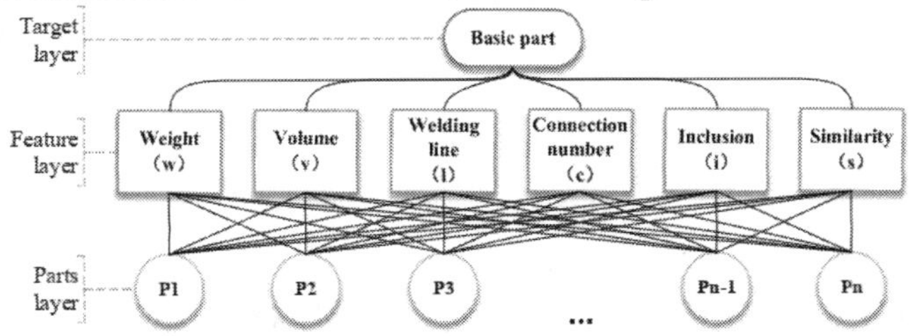

Figure 2. Hierarchical structure model

Step2: According to the Table 1, the fuzzy judgment matrix A is obtained by comparing the target layer with the feature index layer.

Table 1. Pair-wise comparison scale.

Score	Comparison index
0.5	Equal
0.6	Moderately preferred
0.7	Strongly preferred
0.8	Very strongly preferred
0.9	Extremely preferred
0.1,0.2 0.3,0.4	In contrast, if the judgment matrix of C_m to C_n is $R_{mn,}$ the judgment matrix of C_n to C_m is R_{nm}=1-R_{mn}

Step3: The data of welding parts layer need to be normalized to the range of the fuzzy score before comparing the welding parts layer with the feature index layer. The normalization formula is as shown in equation 2.

$$A_{ij} = \frac{P_i - P_j}{\max(P_i, P_j)} \times 0.4 + 0.5 \tag{2}$$

Where P_i is ith welding parts; A_{ij} is fuzzy scale value of P_i versus P_j.

Step4: The fuzzy complementary consistent matrix R can be obtained by transforming matrix A. The transformation formula is as shown in equation 3.

$$R = \begin{pmatrix} r_{11} & r_{12} & \cdots & r_{1n} \\ r_{21} & r_{22} & \cdots & r_{2n} \\ \vdots & & \ddots & \vdots \\ r_{n1} & r_{n2} & \cdots & r_{nn} \end{pmatrix}, \quad r_{ij} = \frac{r_i - r_j}{2(n-1)} + 0.5 \tag{3}$$

Where r_i and r_j is the sum of the elements in ith and jth row of the matrix A respectively; r_{ij} is the element of the ith row and jth column in the matrix R; n is the order of the matrix R;

Step5: The weight values of elements in each layer can be calculated, as shown in equation 4.

$$\omega_i = \frac{1}{n} - \frac{1}{2\alpha} + \frac{1}{n\alpha}\sum_{j=1}^{n} r_{ij} \tag{4}$$

Where α is *(n-1)/2*.

4. Welding units partition and interference check method

4.1. Partition of welding units

Since the number of welding units and basic parts is the same, it is feasible to limit the number of assembly units. Ericsson et al [11] proposed the optimal unit number formula 5:

$$1 < N_P \le \sqrt{N} \tag{5}$$

Where N_P is the number of welding units; N is the number of welding parts.

Although formula 5 determines the range of unit number, it's not suitable for the partition of welding units. It pays more attention to the heat deformation during welding, which will affect the quality of products. Thus, the formula is improved as shown in 6.

$$1 < N_P \le (N-N_S)^{1/2} + (N-N_S)^{1/3} \tag{6}$$

Where N_S is a special welding unit.

After determining the basic parts and the weights between the basic parts, the partition of the welding units is performed by the MST algorithm. It starts from the welding basic part to the sub parts.

4.2. Interference check

When any part is assembled in any direction of the coordinate system, the interference of the part with another part is represented in a matrix, which is called an interference matrix [12] .

In the welding process, the welding unit is usually welded to another unit in six direction (±x, ±y, ±z). In the welding assembly U composed of n welding units, the orthogonal interference matrix of the welding units in the direction k (k=x, -x, y, -y, z, -z) is I_y, as shown in equation 7.

$$I_y = \begin{matrix} & U_1 & U_2 & \cdots & U_n \\ U_1 & i_{11y} & i_{12y} & \cdots & i_{1ny} \\ U_2 & i_{21y} & i_{22y} & \cdots & i_{2ny} \\ \vdots & \vdots & \vdots & & \vdots \\ U_n & i_{n1y} & i_{n2y} & \cdots & i_{nny} \end{matrix} \quad (7)$$

Where I_{ijy} is the interference between the ith part and the jth part when the ith part is assembled in the direction of y. The value of I_{ijy} is 0 (interference)or 1 (no interference).

When the welding unit U_1 moves in the +y direction, as shown in Figure. 3(a), it interferes with U_2, I_{12y}=0; when U_1 moves in the -z direction, it does not interfere with U_2, I_{21z}=1, The interference matrix is shown in Figure. 3(b).

$$I_x = \begin{matrix} & U1 & U2 \\ U1 & 0 & 0 \\ U2 & 1 & 0 \end{matrix} \quad I_y = \begin{matrix} & U1 & U2 \\ U1 & 0 & 1 \\ U2 & 0 & 0 \end{matrix}$$

$$I_z = \begin{matrix} & U1 & U2 \\ U1 & 0 & 0 \\ U2 & 1 & 0 \end{matrix}$$

(a) Interference instance explosion picture (b) Interference matrix

Figure 3. Interference between two welding units

5. Case study

A rear sub frame of automobile shown in Figure. 4 is taken as an example to verify the method.

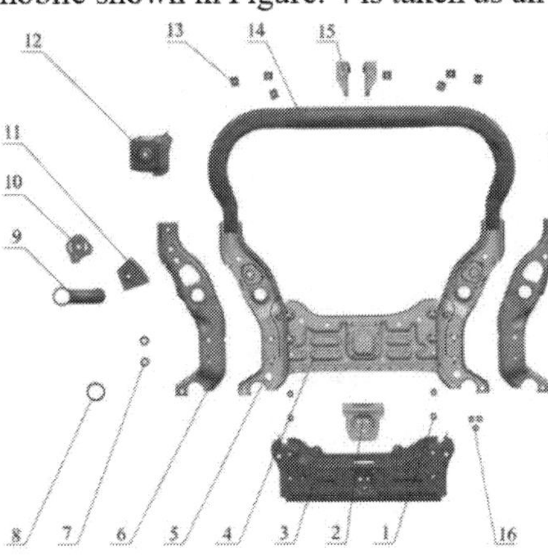

P1:Steering gear mounting sleeve P2:Power suspension reinforcement P3:Upper piece of sub-frame P4:Lower piece of sub-frame P5:Lower longitudinal beams P6:Upper longitudinal beams P7:Swing arm mounting sleeve P8:Drivepipe P9:Elbow pipe P10:Reinforcing plate under longitudinal beam P11:Reinforcing plate upper longitudinal beam P12:Welded fitting supports P13:Nut plate P14:Front sub-frame beam P15:Left bracket P16:Nut

Figure 4. The three-dimensional explosion view of the rear sub frame

5.1. Building a hierarchical model

As shown in Figure. 4, the structure of rear sub frame is welded by large volume and mass parts. As P_{12} with large quality is welded in advance, if it is classified into a welding unit at the same time with P_{14}, It is easy to make P_{14} produce large welding deformation. In this case, the indexes of connections need to be reduced. The relative weights of the basic parts in the rear sub frame are shown in Table 2.

Table 2. The fuzzy consistent judgment matrix and weights of the target layer.

	w	v	l	c	i	s	Weight
w	0.5	0.55	0.6	0.9	0.6	0.85	0.2067
v	0.45	0.5	0.5	0.8	0.55	0.7	0.1867
l	0.4	0.5	0.5	0.65	0.5	0.75	0.1787
c	0.1	0.2	0.35	0.5	0.3	0.7	0.1327
i	0.4	0.45	0.5	0.7	0.5	0.8	0.1806
s	0.15	0.3	0.25	0.3	0.2	0.5	0.1146

5.2. Units partition and interference check

The rear sub frame contains a special welding part P_{12}, $N_S=1$. According to formula 6 and the weight calculation of each part, as shown in Table 3, therefore, P_{12}, P_5, P_4, P_6, P_3, P_{14} are basic parts. Then using MST method shown in Figure. 5(a), which starts with the basic parts to sub parts based on the weight values shown in Table 3, welding units are determined as shown in Figure. 5(b). U1 (P_1, P_4), U2(P_2, P_3, P_{16}), U3(P_5, P_7, P_8, P_{10}), U4(P_6, P_9, P_{11}), U5(P_{13}, P_{14}, P_{15}), U6(P_{12}). The rationality of partition result is further verified by interference check. According to a certain order U1→U3→U5→U4→U2→U6, it is found that there is no interference, and the welding unit is classified reasonably.

Table 3. The weights of basic parts.

Part	Weight	Part	Weight	Part	Weight	Part	Weight
P_1	0.05066224	P_5	0.07444707	P_9	0.06747178	P_{13}	0.04557972
P_2	0.05919667	P_6	0.07610272	P_{10}	0.04958779	P_{14}	0.07886231
P_3	0.0770381	P_7	0.05671893	P_{11}	0.07886231	P_{15}	0.05898788
P_4	0.07522823	P_8	0.06156757	P_{12}	0.06858044	P_{16}	0.04466278

$$
I_x=\begin{array}{c} \\ U1\\U2\\U3\\U4\\U5\\U6\end{array}\begin{array}{c}U1\ U2\ U3\ U4\ U5\ U6\\ \begin{bmatrix}0&1&1&0&0&0\\1&0&1&1&0&0\\1&1&0&1&0&0\\0&0&1&0&0&0\\0&0&1&1&0&1\\0&0&0&0&1&0\end{bmatrix}\end{array}\quad
I_y=\begin{array}{c} \\ U1\\U2\\U3\\U4\\U5\\U6\end{array}\begin{array}{c}U1\ U2\ U3\ U4\ U5\ U6\\ \begin{bmatrix}0&1&1&1&0&0\\1&0&1&1&0&0\\1&1&0&1&1&0\\1&1&1&0&1&0\\0&0&1&1&0&1\\0&0&0&0&1&0\end{bmatrix}\end{array}\quad
I_z=\begin{array}{c} \\ U1\\U2\\U3\\U4\\U5\\U6\end{array}\begin{array}{c}U1\ U2\ U3\ U4\ U5\ U6\\ \begin{bmatrix}0&1&1&1&0&0\\0&0&0&0&0&0\\0&1&0&1&1&0\\0&1&0&0&0&0\\0&0&0&1&0&1\\0&0&0&0&0&0\end{bmatrix}\end{array}\tag{8}
$$

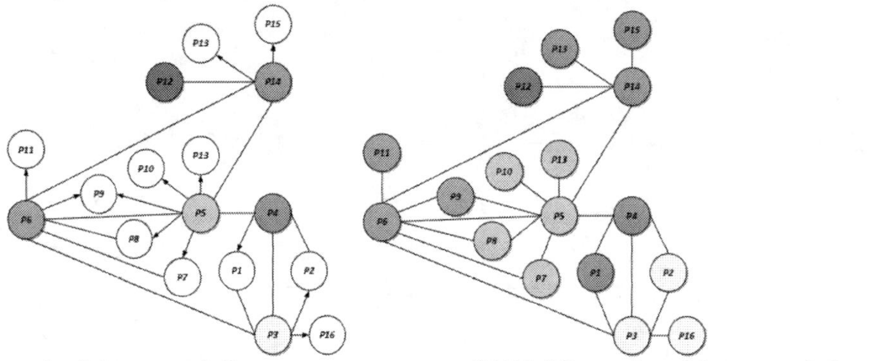

(a) Vector graph of the rear sub frame (b) Welding units of the rear sub frame

Figure 5. Welding units partition of the rear sub frame

6. Conclusion

A welding unit partition method is proposed in this paper, which can be applied to welding sequences planning to improve the efficiency of process planning. In order to verify the method, the rear sub frame of automobile is taken as an example. The results show that:

1) The method of determining basic parts by FAHP based on the features of the welding parts can solve the problem of consistency, and the units partitioned are more independent and convenient.

2) For the partition of basic parts with different welding structure features, the comparison between feature index layer and target layer is evaluated subjectively; the comparison between parts is evaluated objectively. It makes the method to be more applicable.

3) The result is determined by the improved unit partition formula and the interference matrix, and is verified to be feasible by the case study.

Acknowledgments
The authors would like to acknowledge the financial support from National Ministry of Industry and Information (Grant no. 2159999) and Key Science and Technology Program of Liuzhou (Grant no. 2018A0201e002).

References

[1] Beik Varshan, Marzbani Hormoz, Jazar Reza 2019 Welding sequence optimisation in the automotive industry: A review Proceedings of the Institution of Mechanical Engineers *Part C: Journal of Mechanical Engineers Science* 223: 5945-5952.
[2] Meriam JL, DeGarmo EP and Jonassen F 1946 A method for the measurement of residual welding stresses *Welding Research Supplement*; 340s–343s.
[3] Fukuda S and Yoshikawa K 1990 Determination of welding sequence: a neural net approach *Eng Anal Boundary Elem* 7: 78–82.
[4] Lian J, Lai XM, Lin ZQ 2002 Application of data mining and process knowledge discovery in sheet metal assembly dimensional variation diagnosis *J Mater Process Technol* 129: 315–320.
[5] Gonc¸ alves JF and Resende MGC 2016 Random-key genetic algorithms In: Martı´ R, Panos P and Resende MGC (eds) Handbook of heuristics *Cham: Springer International Publishing* 1–13.
[6] Tabar RS, Wa¨ rmefjord K and So¨ derberg R 2018 Evaluating evolutionary algorithms on spot welding sequence opti-mization with respect to geometrical variation *Procedia CIRP* 75: 421–426.
[7] Liu Bo, Li Rui, Li Sheng and Wang Ji 2019 Assembly unit partition method for ship block based on assembly relationship *J. Huazhong Univ. of Sci. & Tech. (Natural Science Edition)* 47(07):118-123.
[8] Liu Ying, Sun Yunyan and Zhang Genbao 2017 Error Propagation Model and Calculating Method of Effective Transfer Path for Meta-action Assembly Unit *Journal of Chongqing University* 40(3):1-11.
[9] Fahlström K, Andersson O, A. Melander, Karlsson L and Svensson L.-E 2016 Correlation between laser welding sequence and distortions for thin sheet structures *Science and Technology of Welding and Joining* 1-7.
[10] LAN Jibin, Xu Yang, Huo Liang'an and Liu Jiazhong 2006 Research on the weight of Fuzzy Analytic Hierarchy Process *Systems Engineering-Theory & Practice* (09):107-112.
[11] Ericsson A, Erixon C 1999 Controlling Design Variants: Modular Product Platforms New York, USA: ASME Press.
[12] Dini G and Santochi M 1992 Automated sequencing and subassembly detection in assembly planning *CIRP Annals - Manufacturing Technology* 41(1), 1-4.

IOP Conf. Series: Materials Science and Engineering **825** (2020) 012032 doi:10.1088/1757-899X/825/1/012032

Study on Influence of Micro-tool Tip Arc Waviness on Cutting Micro Aspheric Surface

Yi Liu[1], Chunyu Zhang[1], Guo li[1,a], Junjie zhang[2,b], Haijun Zhang[1] and Chengwei song[1]

[1]Research Center of Laser Fusion, China Academy of Engineering Physics, Mianyang 621900, China
[2]Center of Precision Engineering, Harbin Institute of Technology, Harbin 150001, China

[a] liguo51404@163.com; [b] junjie.zhang@hit.edu.cn

Abstract. In this paper, the influence of the waviness of tip arc profile on the surface quality of micro transition arc workpiece is studied. Firstly, the contour curve of the surface of the micro tool was measured by using the radius amplitude measuring instrument, and the characteristic parameters of the waviness in the contour curve were found. A simulation study of the effect of feed rate on the quality of the machined surface of a sine wave model with different characteristics was performed by Matlab. It can be known from the simulation results that when a sine wave of 0.05sin(10t) exists in the contour of the tool tip arc, when the feed rate is small, the impact on the surface accuracy of the workpiece is greater, and a sin wave of 0.02sin(30t) locally affects the regional surface roughness is greater. With the increase of feed rate, a sin wave of 0.02sin(30t) has no effect on the surface roughness of the partial area, while a sin wave of 0.05sin(10t) still has a greater effect on the surface roughness of the partial area.

1. Introduction

Micro aspheric element has been widely used in optics, aerospace, and laser fusion experiments [1-2]. The surface roughness of nanometer scale and the surface shape precision of submicron scale can be obtained by means of ultra-precision turning. However, when machining some aspheric array microstructures or modulation structures, it is necessary to select micro-arc single-point diamond turning tool for cutting to avoid interference. Micro-arc tool refers to the tip of the arc radius of 0.1~0.001mm (even smaller) of the micro tool, at present micro-arc tool tip of the arc waviness can achieve less than 0.1μm accuracy with long production cycle and so many production difficulties. Therefore, Xia et al proposed to obtain a high-precision micro-arc tool by mechanically grinding the tool with a large radius of arc at the tip of the tool, and finally obtained a high-precision micro-arc tool with R8.2μm and a waviness less than 0.1μm by grinding the arc tool with R0.2mm [3]. In the cutting process, if the cutting position of the tool always keeps the curvature radius unchanged, the workpiece will obtain excellent surface shape accuracy, while in the cutting process, the profile error of the tool tip arc will "copy" on the workpiece surface, thus affecting the surface quality of the processed workpiece [4]. Therefore, the accurate measurement of the profile of the micro arc tool is very important for the error compensation of the actual machining. It is a mature and universal method in industrial production to measure the radius of the tool tip by optical microscope and obtain the roundness error and waviness error of the tool tip by processing the image data [5]. He C et al. established a mathematical model for predicting the surface contour of turning plane workpiece by measuring the profile of tool tip and combining machining parameters and material parameters of the workpiece, verified the correctness of the model through experiments, and analyzed the influence of tool profile error on the surface quality of processing workpiece [6-7]. Zhang R used sinusoidal wave to simulate the waviness of tool tip arc to theoretically deduce the influence of waviness on the surface shape accuracy of machining workpiece, and experimentally studied the influence of different tool waviness on the surface shape accuracy of convex circular workpiece with machining radius of 50mm and diameter of 45mm [8].

At present, there are few reports about the influence of tool waviness on the surface quality of the workpiece with small arc shape. In this paper, the characteristic parameters of the waviness in the profile curve of micro arc tool are measured, and the influence of the waviness of the profile of micro arc tool tip on the surface shape accuracy and surface roughness of micro aspheric workpiece is studied by the simulation of Matlab.

2. Numerical simulation model

As shown in figure 1, the profile error curve of tool tip arc measured by the radius amplitude measuring instrument. In this paper, the sine wave (It can be expressed like this $A\sin(2\pi / \lambda \cdot t)$) is used to simulate the ripple existing in the tool profile, which is the wavelength of the whole period of the sine wave. Through the analysis of the corrugation error curve in figure 1, there are two kinds of typical corrugations. One is the wavelength $\lambda = 2\pi / 10$, the amplitude is 0.05μm,it is a medium frequency and large amplitude waviness, the other is the wavelength $\lambda = 2\pi / 30$, the amplitude is 0.02μm, it is a medium frequency and small amplitude waviness, and the sine wave can be expressed as respectively 0.05sin(10t) and 0.02sin(30t).

As shown in figure 2, it is the theoretical derivation diagram of micro transition arc turning. In the turning process, only the tool arc profile "copy" on the workpiece to generate the workpiece surface profile, without considering other factors. In the figure 2, the blue line segment represents the surface profile generated by the ideal tool turning, and the red line segment represents the surface profile generated by the turning when there is a certain waviness on the arc profile of the tool. The rectangular coordinate system is established with the center of the workpiece arc as the origin, R1 as the radius of the workpiece arc, the cutter feeds along the workpiece arc contour and O1, O2, O3, O4…It is the center of the tool tip arc for each feeding of the tool, a is the length of half of the center distance, Rt is the radius of the tool tip arc, and α is the rotation angle of the tool for each feeding relative to the workpiece center.

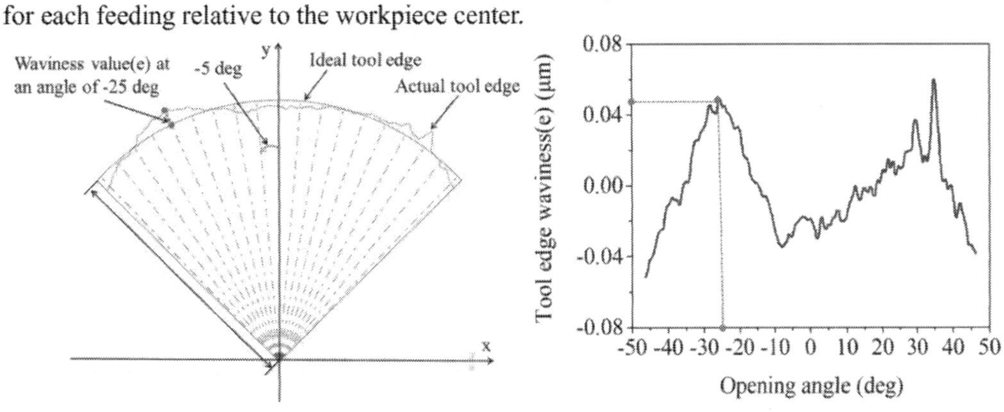

Figure 1.Profile of tool cutting edge and tool edge waviness.

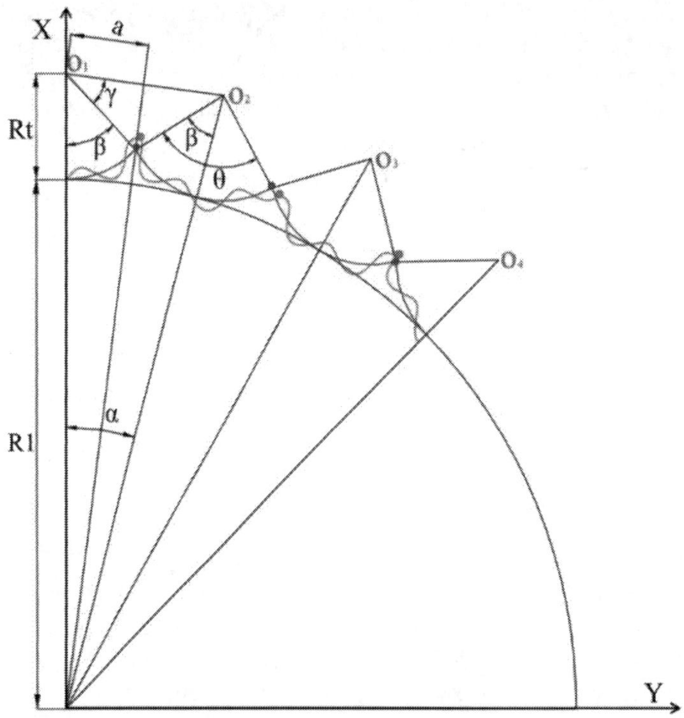

Figure 2. Schematic diagram of theoretical derivation.

$$\alpha = \frac{f}{2\pi(R1+Rt)} \times 360° \tag{1}$$

where the f is feed rate.

$$a = 2(R1+Rt)\sin\frac{\alpha}{2} \approx \frac{f}{2} \tag{2}$$

So in this coordinate system, the center of tool arc (X_i, Y_i) can be expressed as follows:

$$\begin{cases} X_i = (R1+Rt)\sin(i\alpha) \\ Y_i = (R1+Rt)\cos(i\alpha) \end{cases} \tag{3}$$

Where i is the natural number starting from zero.

$$\gamma = ar\cos\frac{f}{2Rt} \tag{4}$$

$$\beta = 90° - \frac{\alpha}{2} - \gamma \tag{5}$$

It can be seen from the geometric analysis that the contour of the ideal tool "copy" on the workpiece surface is the same radian θ every time it is feed, and β is the radian corresponding to half of the contour. So θ can be expressed as follows:

$$\theta = 2\beta = 180° - \frac{f}{2\pi(R1+Rt)} \times 360° - 2\arccos\frac{f}{2Rt} \tag{6}$$

Under the rectangular coordinate system established with the workpiece center as the origin, if the second coordinate system is established with the cutter center, the cutter outline of each section "copy" on the workpiece surface shall be in the third and fourth quadrants of the second coordinate system, then the value range of θ_i shall be:

$$\theta_i = \left(270^\circ - i\alpha - \beta\right) \sim \left(270^\circ - i\alpha + \beta\right) \tag{7}$$

The contour of each feeding tool "copy" on the workpiece can be expressed as

$$\begin{cases} x_i = (R1 + Rt)\sin(i\alpha) + Rt \cdot \cos\theta_i \\ y_i = (R1 + Rt)\cos(i\alpha) + Rt \cdot \sin\theta_i \end{cases} \tag{8}$$

As shown in figure 2, the blue mark points are the intersection point of the "copy" contour of the ideal tool in two feeds, and the red mark points are the intersection point of the "copy" contour in two feeds when the tool has certain corrugations. The blue mark point will be regularly displayed in the arc position with radius R, and the value expression of R is equation.9, while the red mark point will not be regularly displayed each time.

$$R = \sqrt{(Rt + R1)^2 - \left(\frac{f}{2}\right)^2} - \sqrt{Rt^2 - \left(\frac{f}{2}\right)^2} \tag{9}$$

The coordinates of the blue mark point (R_x, R_y) can be expressed as

$$\begin{cases} R_x = \left[\sqrt{(Rt + R1)^2 - \left(\frac{f}{2}\right)^2} - \sqrt{Rt^2 - \left(\frac{f}{2}\right)^2}\right]\sin\left(i\alpha - \frac{\alpha}{2}\right) \\ R_y = \left[\sqrt{(Rt + R1)^2 - \left(\frac{f}{2}\right)^2} - \sqrt{Rt^2 - \left(\frac{f}{2}\right)^2}\right]\cos\left(i\alpha - \frac{\alpha}{2}\right) \end{cases} \tag{10}$$

The surface profile generated by the ideal tool cutting, and the blue mark point shall be the highest point of the generated surface profile. After the profile generated by the ideal tool is flattened along the arc radius, the PV value of the generated surface profile can be expressed as follows:

$$R_{th} = \sqrt{(Rt + R1)^2 - \left(\frac{f}{2}\right)^2} - \sqrt{Rt^2 - \left(\frac{f}{2}\right)^2} - R1 \tag{11}$$

Through the above derivation, each formula can be input into Matlab for numerical calculation. Since the calculated values in Matlab are all discrete points, the intersection points in Matlab need to be obtained by solving the equation to get the minimum value, then the value of θ_i increased appropriately ϕ, and the value of ϕ only needs to ensure that the simulation can calculate the intersection point of the contour left by the two feed "copy". φ_i can be expressed as follows:

$$\varphi_i = \left(270^\circ - i\alpha - \beta - \phi\right) \sim \left(270^\circ - i\alpha + \beta + \phi\right) \tag{12}$$

When there is a sine wave on the tool profile, the coordinate (x_{1i}, y_{1i}) value of part of the tool profile in Matlab can be expressed as follows:

$$\begin{cases} x_{1i} = (R1+Rt)\sin(i\alpha) + \left[Rt + A\sin\left(\dfrac{2\pi}{\lambda}\varphi_i\right)\right]\cos\varphi_i \\ y_{1i} = (R1+Rt)\cos(i\alpha) + \left[Rt + A\sin\left(\dfrac{2\pi}{\lambda}\varphi_i\right)\right]\sin\varphi_i \end{cases} \tag{13}$$

$$\sqrt{\left(x_{1(i+1)} - x_{1i}\right)^2 + \left(y_{1(i+1)} - y_{1i}\right)^2} \le \varepsilon \tag{14}$$

where the ε is error factor.

According to the criterion of Equation.14, two groups of coordinate points which satisfy the error factor ε in the profile of two feed tools can be obtained, they are regarded as intersection points. In order to ensure the calculation efficiency ε is taken as 0.0005μm, the coordinate value between the intersection points is the surface profile of the workpiece obtained by "copy" each feed tool on the workpiece.

Equation.13 is used as the unified expression formula of tool profile in Matlab program for simulation calculation, and the unit is μm. When $R1$ = 1000, Rt = 10, f = 5, A = 0, the simulation results are shown in figure 3(a) after the removal of the workpiece radius from the contour surface in the range of $i\alpha$= 20-70 degrees and the angle flattening, and the simulation results of partial44-45 degrees are shown in figure 3(b).

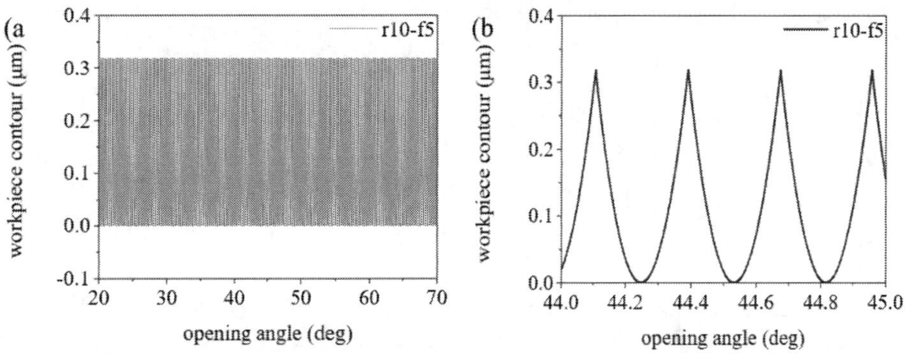

Figure 3. Simulation results of ideal tool.

The surface roughness Ra is used to evaluate the surface quality. Since the data calculated by Matlab are all discrete points, the calculation formula of surface roughness is expressed as follows:

$$Ra = \frac{1}{n}\sum_{n=1}^{n}|Z_n| \tag{15}$$

According to the theoretical derivation, the surface roughness and PV value of the simulation results of the ideal tool in a large area range of 20-70 degrees are consistent with those of the simulation results of the partial area. The surface roughness Ra of partial area is 80.6nm, and the PV value of partial area is 318.87nm

3. Results and discussion

3.1. Influence of medium frequency and small amplitude waviness on workpiece surface

When there is a sine wave with an amplitude of 0.02μm and a wavelength λ=2π/30, that can be expressed as 0.02sin(30t), and the radius of the tool tip arc are 10μm, the simulation is carried out at different feed rates of 2, 3, 4 and 5μm/r. The simulation results obtained from the range of $i\alpha$=30-50 degrees are shown in figure 4.

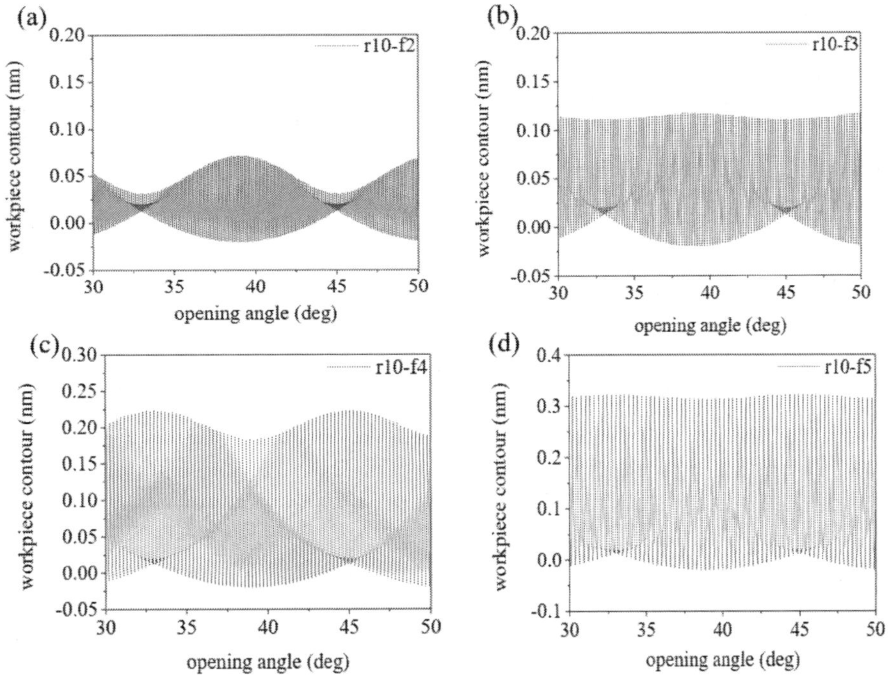

Figure 4. Simulation results of tool with 0.02sin(30t) under different feed rates.

As shown in figure 4, the radius of the tool tip arc is 10μm, and the feed rate is 2, 3, 4 and 5μm/r respectively. The simulation results are obtained. It can be seen from the figure 4 that with the increase of the feed rate, the trend of the upper edge curve is in the range of 33-45 degrees, from the trend of first increasing and then decreasing when the feed rate is 2μm/r to the trend of first decreasing and then increasing when the feed rate is 4μm/r to the trend of flattening when the feed rate is 5μm/r.

The main reason for the change trend of the contour curve of the upper surface is that with the increase of the feed rate, the depth of the tool contour under the "copy" of the single feed on the workpiece increases, as shown in figure 5. As shown in Figure 5(a)(b) are the comparison between the contour generated by the partial area and the contour generated by the ideal tool when the feed rate is 2μm/r, and figure 5(c)(d) are 5μm/r respectively .When the feed rate is 2μm/r, the depth of the contour curve generated by the minimum partial area is obviously smaller than that generated by the ideal tool, while that generated by the maximum partial area is significantly smaller than that generated by the wheel. When the feed rate is 5μm/r, it can be seen that the contour generated by the partial area almost coincides with the contour generated by the ideal tool, and the depth difference is very small. This shows that with the increase of the feed rate, the influence of the corrugation on the surface roughness is smaller and smaller, so the curve trend of the upper surface profile tends to be flat. Since the contour curve of the lower surface always keeps decreasing first and then increasing, and the values of discrete points in the curve will not change, the difference between the PV value of the contour formed in the partial area and the PV value of the contour formed by the ideal tool will be smaller and smaller with the increase of the feed rate, as shown in figure 6.

As shown in figure 7, the suffixes of Max and min respectively represent the maximum and minimum surface roughness values of partial areas, the suffixes of whole curve represent the surface roughness values within the range of 30-50 degrees, and the suffixes of ideal represent the surface roughness values of ideal tools. It can be seen from the figure.7 that when the feed rate is less than 4μm/r, the surface roughness values of whole > Max > ideal > min, and the whole roughness values are greater than others The reason is that when the feed rate is too small, the surface profile of the workpiece is greatly affected by the amplitude of sine wave on the tool, so it

IOP Conf. Series: Materials Science and Engineering **825** (2020) 012032 doi:10.1088/1757-899X/825/1/012032

will lead to the calculation of the roughness too large. when the feed rate is greater than or equal to 4μm/r the whole ≈ Max ≈ ideal ≈ min, which can be seen from Figure.8, at this time, the difference of contour PV value obtained is very small, and the influence brought by sine wave on the tool is very small, so when calculating the roughness, the difference between them will be very small.

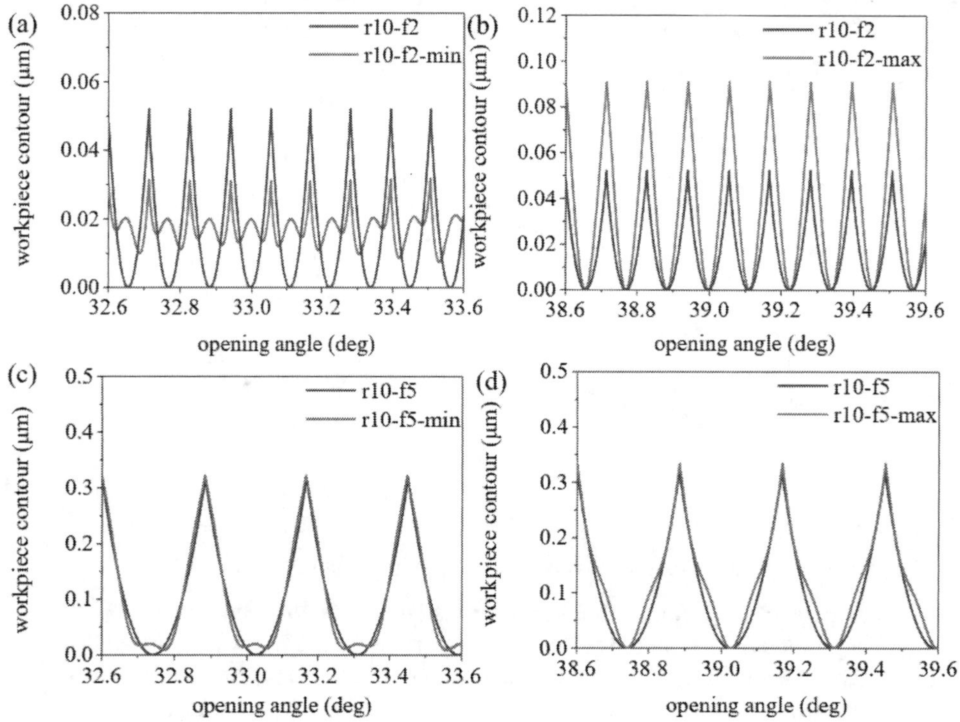

Figure 5. Simulation results of contour curves.

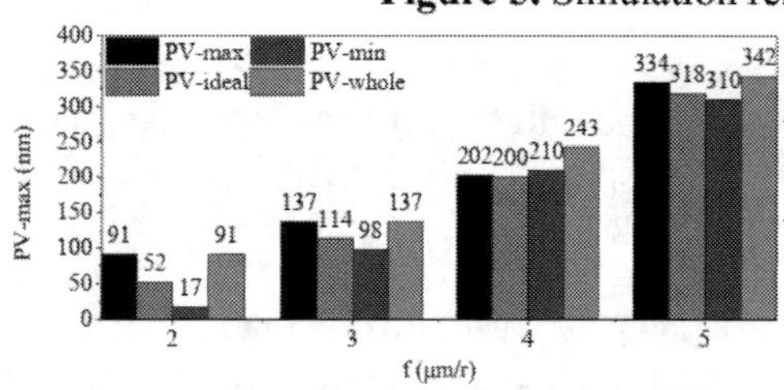

Figure 6. Comparison of PV.

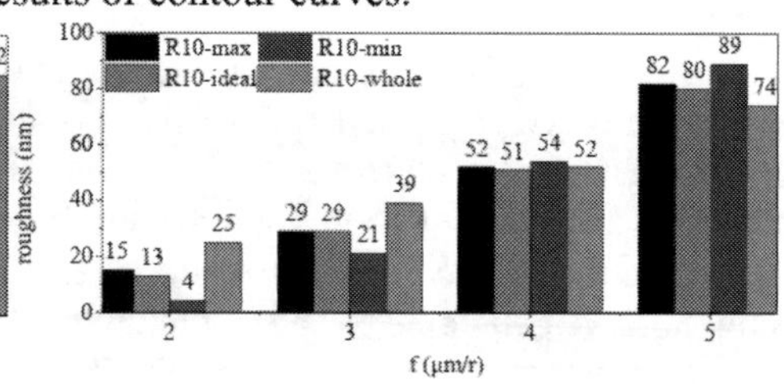

Figure 7. Comparison of Ra.

Through the above analysis, it can be seen that when there is a sine wave with an amplitude of 0.02μm and a wavelength $\lambda=2\pi/30$ on the tool, such a sine wave will have a greater impact on the surface shape accuracy of the generated workpiece surface when the feed rate is small, and at the same time, it will affect the surface profile generated in the partial area. In some areas, the surface quality can be better or even better than that of the ideal tool, but in some areas, the surface quality of the tool is relatively poor. When the feed rate is large, the influence of such sine wave on the surface roughness of workpiece is small.

3.2. Influence of medium frequency and large amplitude waviness on workpiece surface

When there is a sine wave with an amplitude of 0.05μm and a wavelength $\lambda=2\pi/10$, that can be expressed as 0.05sin(10t), and the radius of the tool tip arc are 10μm, the simulation is carried out at different feed rates of 2, 3, 4 and 5μm/r. The simulation results obtained from the range of $i\alpha$=20-70 degrees are shown in figure 8.

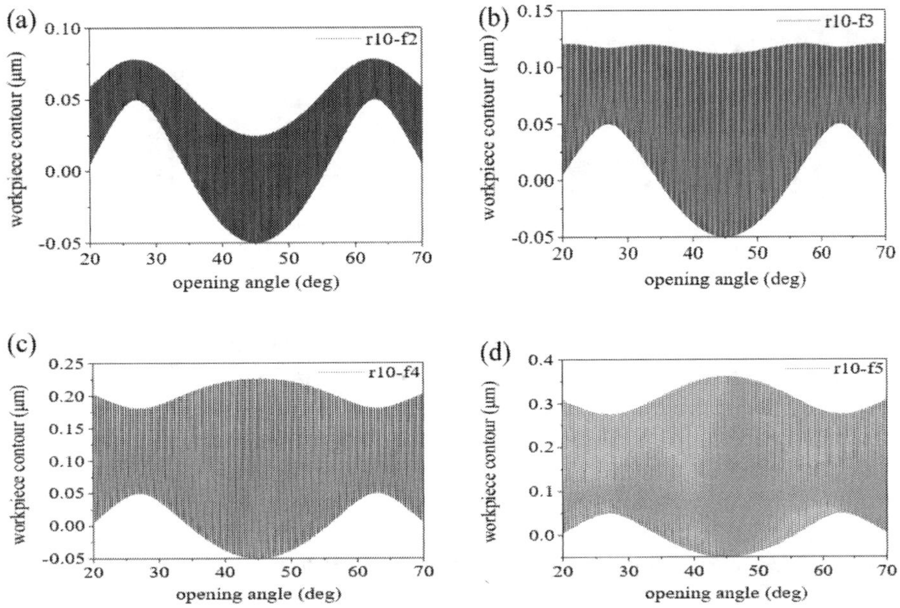

Figure 8. Simulation results of tool with 0.02sin(30t) under different feed rates.

As shown in figure 8, it can be seen that the trend of the upper edge curve in the range of 27-63 degrees decreases first and then increases with the increase of feed rate at 2μm/r, increases first and then decreases at 3μm/r, increases first and then decreases at 3μm/r, increases first and then decreases at 4μm/r, and decreases at 5μm/r, but the slope increases. The lower edge curve is formed at the lowest point of each feed, and these points are approximately equal to the value of the corresponding cutting angle on the sine wave 0.05sin (10t). Therefore, the main reason for the change of the trend of the upper edge curve is that with the increase of the feed rate, the depth of the "copy" contour on the workpiece surface will be different in different areas due to the change of the cutter contour participating in the cutting position in the partial area, and there is a maximum and minimum value of the contour depth in the partial area, so PV is carried out for the maximum and minimum value areas of 44.6-45.6 degrees and 26.6-27.6 degrees respectively Calculation of values and surface roughness.

As shown in figure 9, that is the comparison of PV values in different regions. The suffix is ideal, which indicates the PV value of the workpiece surface profile generated by the ideal tool, and the whole, which indicates the PV value in the range of 20-70 degrees. It can be seen in the figure 9 that there is always the maximum and minimum value of the partial area, and there is a certain difference. At the same time, it can be seen that with the increase of the feed rate, the maximum value of PV value of the partial area is the PV value of the whole area. The difference between the maximum value of PV value in the partial area and the contour PV value formed by the ideal tool is from + 42%, + 41%, 37.5% to + 27.5%; the difference between the minimum value of PV value in the partial area and the contour PV value formed by the ideal tool is from - 48%, - 50%, - 35%, to - 29%; the difference between the PV value of the contour formed by the overall area and the ideal tool is from + 146%, + 49%, + 37.5 to + 27.5%. The percentage of difference decreases with the increase of the feed rate.

As shown in figure 10, that is the comparison of surface roughness in different areas, and the suffix of the curve in the figure is the same as that in figure 9. It can be seen from the figure 10 that when the feed rate is 2μm/r, the surface roughness is whole > Max > ideal > min, when the feed rate is 3μm/r, the surface roughness is Max > whole > ideal > min, and when the feed rate is 3μm/r, the surface roughness is Max > whole ≈ ideal > min.

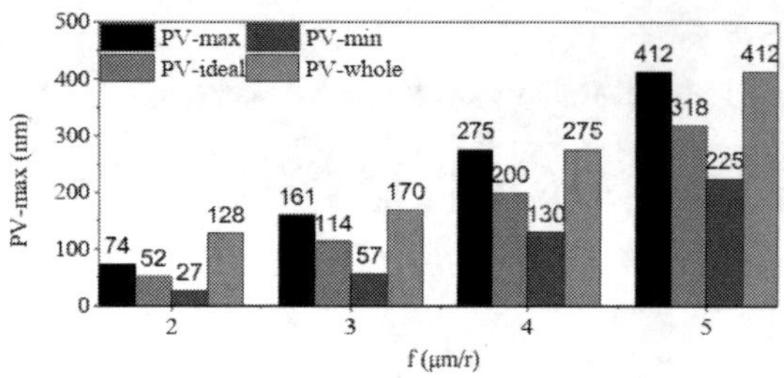

Figure 9. Comparison of PV.

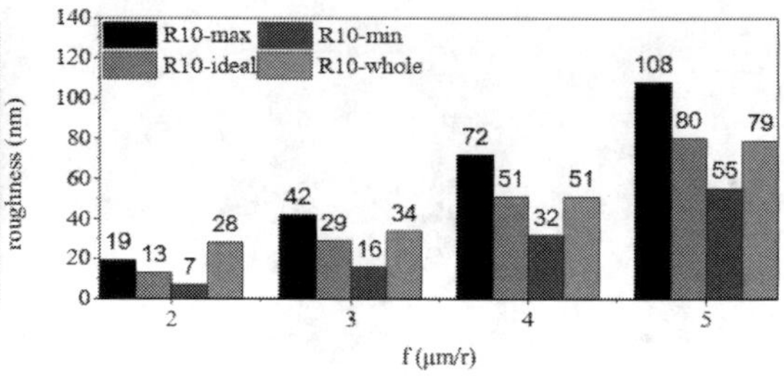

Figure 10. Comparison of Ra.

Through the above analysis, we can know that when there is a sine wave with a wavelength $\lambda=2\pi/10$ and an amplitude of 0.05μm on the tool tip arc, when the feed rate is small, such sine wave will seriously affect the surface accuracy of the workpiece, and at the same time, it will cause a large difference in the surface quality of the partial area. With the increase of the feed rate, the surface quality of the partial area is different greatly, and the surface quality of the partial area is still better than that of the ideal tool.

4. Conclusion

In this paper, the sine wave is used to simulate the corrugation error in the circular arc profile of the micro tool tip. Two kinds of corrugations with typical characteristics on the corrugation error curve are simulated and studied. The two corrugation parameters are amplitude 0.02μm, wavelength $\lambda=2\pi/30$, and the other is amplitude 0.05μm, wavelength $\lambda=2\pi/10$,they can be expressed as 0.02sin(30t) and 0.05sin(10t) respectively. Through the theoretical formula derivation, and then the derivation process and formula using Matlab software to write a simulation program for numerical simulation calculation. It can be seen from the simulation results that when there is a sine wave of 0.05sin(10t) in the arc profile of the tool tip, the influence on the surface accuracy of the workpiece is greater when the feed rate is smaller, while the influence of a sin wave of 0.02sin(30t) on the surface roughness of the partial area of the workpiece surface is greater. With the increase of feed rate, a sin wane of 0.02sin(30t) has no effect on the surface roughness of partial area, while a sin wave of 0.05sin(10t) still has great effect on the surface roughness of partial area.

Acknowledgment

The authors would like to thank the Science Challenge Project (No.TZ2018006) and the Laboratory Fund Project of Ultra-Precision Manufacturing Technology (No. ZD18007) for the support of this work.

References

[1] Fang F Z, Zhang X D and Hu X T 2008 *Cylindrical coordinate machining of optical freeform surfaces* Optical Express 16 7323-7329.
[2] Lee W B, Kong L B and Zhang H 2013 *An overview of ultra-precision diamond machining of micro structured freeform surfaces* Journal of mechanical Engineering 49 144-155.
[3] Xia Z, Jia L, Zhang X and CHU C 2017 *Research on dressing and truing of micro-arc diamond tool* Diamond& Abrasives Engineering 3 69-73 .
[4] Lei D, Yue X, Cui H and Zhang X 2017 *Measurement and evaluation of tool tip arc waviness of diamond tool* Optics and Precision Engineering 25 2697-2705.
[5] Zhou J, Li Z, Wang Y, Sun T and Zong W 2013 *Tool Tip Arc Measurement and Characterization of Micro-Arc Diamond Tools* Nanotechnology and Precision Engineering 11 334-340.
[6] He C, Zong W and Sun T 2016 *Origins for the size effect of surface roughness in diamond turning* International Journal of Machine Tools and Manufacture 106 22-42.
[7] He C, Zong W and Sun T and Xue C 2018 *An accurate 3D surface topography model for single-point diamond turning* International Journal of Machine Tools and Manufacture 134 42-68.
[8] Zhang R 2017 *Studies on ultra-precision optical fabrication technology with the round nose diamond tool*

IOP Conf. Series: Materials Science and Engineering **825** (2020) 012033 doi:10.1088/1757-899X/825/1/012033

Numerical Smulation of Radial Compression and High-temperature Creep of Metal Elastic Elements

J Chang[1,3], Y Cai[1], L Peng[2], L Yang[1], H Liu[1] and H Li[1,*]

[1] State Key Laboratory of Metal Matrix Composites, School of Materials Science and Engineering, Shanghai Jiao Tong University, Dongchuan Road No.800, Shanghai 200240, China.
[2] School of Mechanical Engineering, Shanghai Jiao Tong University, Dongchuan Road No.800, Shanghai 200240, China.
[3] China Academy of Launch Vehicle Technology, South Dahongmen Road No.1, Fengtai District, Beijing 100076, China.

[*]E-mail: lih@sjtu.edu.cn

Abstract: With the help of the numerical calculation method—Finite Element Method, the numerical simulation and result analysis of a kind of metal elastic element-knitted spring are carried out, mainly focusing on the radial compression and high temperature creep behaviours of the spring. The purpose of this study is to analyse the previous with numerical simulation and explain the influence of configuration parameters on its mechanical properties and to provide data support for design, manufacture and application of the prepared material as an elastic component.

1. Introduction

With the development of high-speed aircraft, it is vital to study the high-temperature sealing technology and high-temperature metal elastic elements. Since the shape reinforced skeleton is often used with other high-performance materials to prepare various seals for specific technological functions [[1]-[2]]. Therefore, to achieve these particular requirements and understand the mechanical behaviour of elastic material and prevent seal leakage [3], several actual configurations and material tests were performed[4]-[5], especially the high-temperature performance of elements[6][7]. However, mechanical modelling, performance prediction, and parametric design of components employing numerical analysis can replace expensive material tests in a certain extent and make it cost-effective. The research content includes the influence of geometric configuration parameters, boundary constraints, and high temperature creep on the resilience performance of metal elastic elements. The adjustment in a geometric configuration, boundary constraints, and working environment led to the optimized value. However, the listing of commonly used high-temperature metal flexible elements is as follows: inclined coil spring, knitted spring, helical spring, and cutting spring (Figure 1). Each configuration has multiple configuration parameters, which determine the mechanical behaviour of the elastic elements.

The main problems discussed in this work are: (1) combining the material constitutive model and the finite element model (for metal elastic element). (2) To establish a relationship between the main mechanical performance parameters i.e. force-displacement relationship, high temperature creep

IOP Conf. Series: Materials Science and Engineering **825** (2020) 012033 doi:10.1088/1757-899X/825/1/012033

characteristics, and resilience energy, and the geometric configuration characteristic parameters of elastic elements, obtained through different geometric configurations of numerical simulation. (3) The testing and analysis of typical structural samples, design, manufacture, and application of high-temperature metal elastic elements data. (4) While for knitted winding spring, ABAQUS is used to carry out numerical simulation and describe the finite element modeling process.

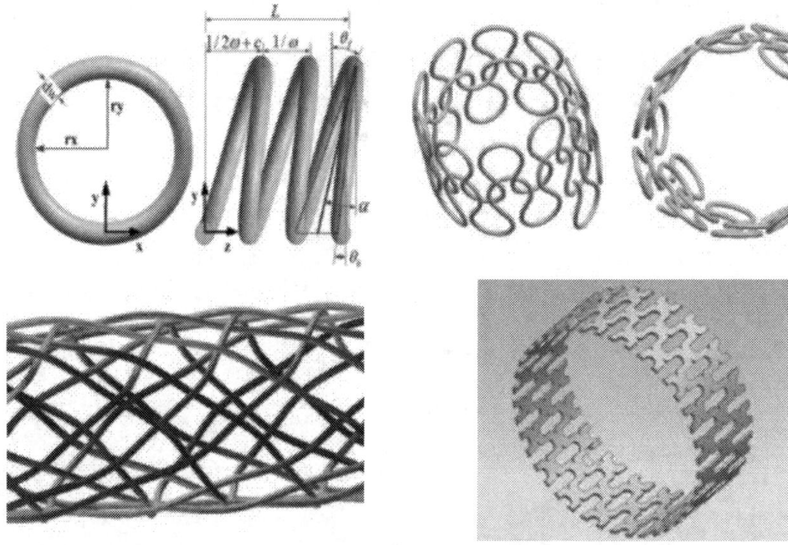

Figure 1 .configuration of common elastic elements

2. Finite element calculation model and material parameters

Knitted spring is a new type of spring (Figure 2), which is widely used by NASA and other departments in the United States. Through flexible winding, design the spring can be made relatively superior in rebound performance. The same feature unit spirally wounds the spring, with various geometric parameters, as shown in Figure 3.

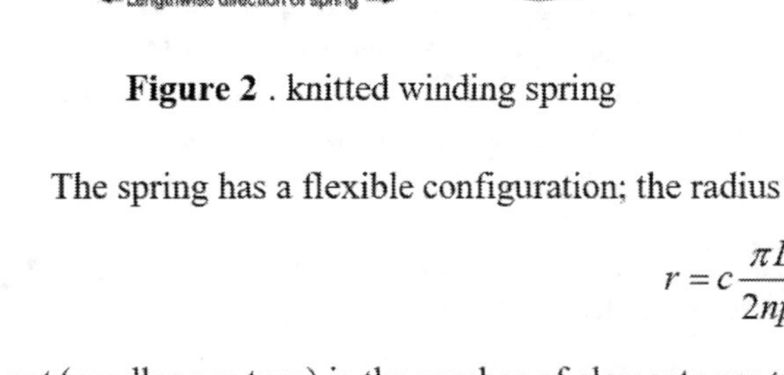

Figure 2 . knitted winding spring

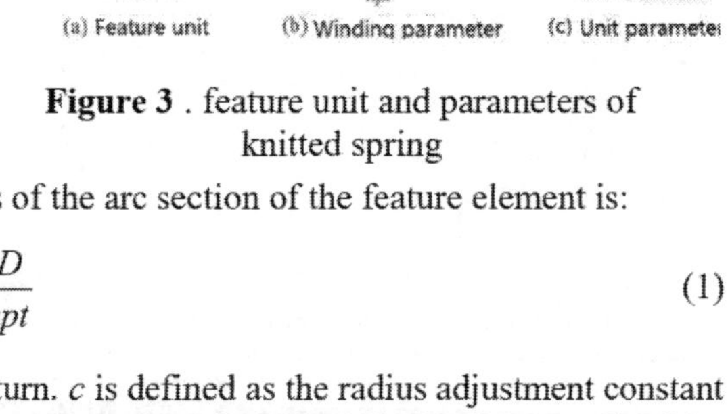

Figure 3 . feature unit and parameters of knitted spring

The spring has a flexible configuration; the radius of the arc section of the feature element is:

$$r = c\frac{\pi D}{2npt} \tag{1}$$

npt (needles per turn) is the number of elements per turn. c is defined as the radius adjustment constant of the feature element which in the range of $0 < c < 1$. Its value changes with the diameter D of the knitted spring tube. However, we selected the constant value for c is equal to 0.8, here in this work.

Besides, when modeling with CAD, the feature units are offset correspondingly in the radial direction to avoid the mutual penetration of the adjacent coil (Figure 4a). Then the CAD model is imported into the finite element software ABAQUS, and the simulation analysis of the radial compression and resilience process can be carried out after meshing, setting the boundary and loading conditions. The final finite element simulation model is shown in Figure 4b. The lower rigid plate is fixed, and the stiff upper plate is pressed down to an absolute displacement for radial loading. Because friction has little influence on the displacement-load characteristics of the radial compression, therefore the self-contact and the friction between the rigid plate and the spring are ignored.

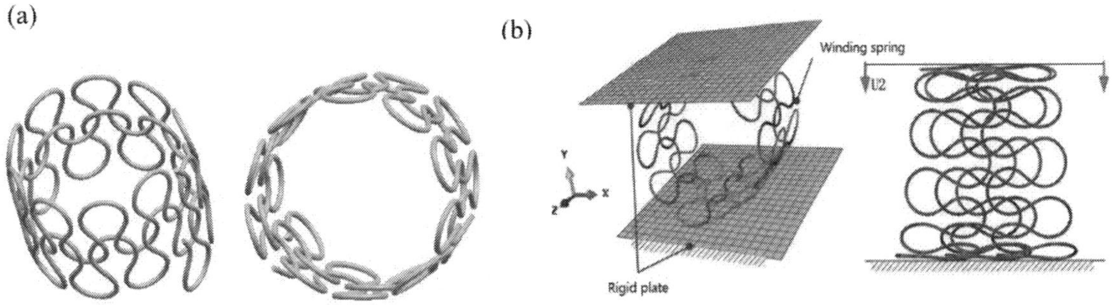

Figure 4 . finite element simulation model for radial compression of knitting spring

The Superalloy K4169 is used in the design of spring while setting the material parameters at 25°C, 600°C, 700°C in Table 1.

Table 1 . Parameters of Superalloy K4169 elastic plastic material

T^a (°C)	E^b (GPa)	μ^c	σ_y ^{d}(MPa)	σ_b ^{e}(MPa)	δ ^{f}(%)
25	199.6	0.2903	1123.9	1365.2	21
600	166.4	0.2768	989.6	1142.6	15.44
700	159.0	0.2913	933.6	1014.4	8.56

[a] T is temperature.
[b] E is Young's modulus.
[c] μ is Poisson's ratio.
[d] σ_y is Yield strength.
[e] σ_b is Tensile strength.
[f] δ is Elongation.

3. Calculation results of radial compression process

3.1. Influence of length and end effect

The listing for various designs of the springs is in Table 2, and the result of the simulation is shown in Figure 5, and Figure 6. These results are consistent with the experimental and simulation results obtained by NASA [3] and other relevant researchers. The load increases linearly with the downforce displacement in a broad range, and the spring stiffness increases significantly when the movement of downforce exceeds about 45%. The whole displacement-load characteristic curve is bilinear. Figure 5 shows the maximum stress of different length springs under the same pressure displacement, so the end effect has little effect on the stress inside the spring. When the pressure displacement reaches 51%, the maximum Mises stress in the spring is close to the yield strength of 1460 MPa.

Compared with the displacement-load curve in Figure 6, it can be seen that the length and end effect have little effect on the stiffness of the knitted spring. Therefore, we can obtain the exact solution just by simulating two coil length springs, not to consider the actual length of the springs in the simulation analysis.

Table 2 . simulation parameters of knitted spring (different lengths)

No.	L^a (mm)	p^b (mm)	d^c (mm)	$d_w{}^d$ (mm)	npt^e	LD^f (loops/mm^2)	T^g (°C)
L4.2	4.2	2.1	10.5	0.3	15	0.217	25
L6.3	6.3	2.1	10.5	0.3	15	0.217	25
L10.5	10.5	2.1	10.5	0.3	15	0.217	25

[a] *L* is the length of spring.
[b] *p* is the screw-pitch.
[c] *d* is Diameter.
[d] d_w is Wire diameter.
[e] *npt* is Needles per turn.
[f] *LD* is Loop Density.
[g] *T* is Temperature.

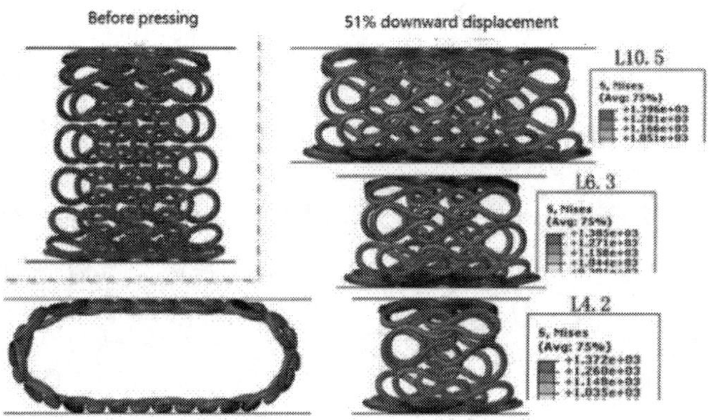

Figure 5 . simulation results with different length of knitted spring

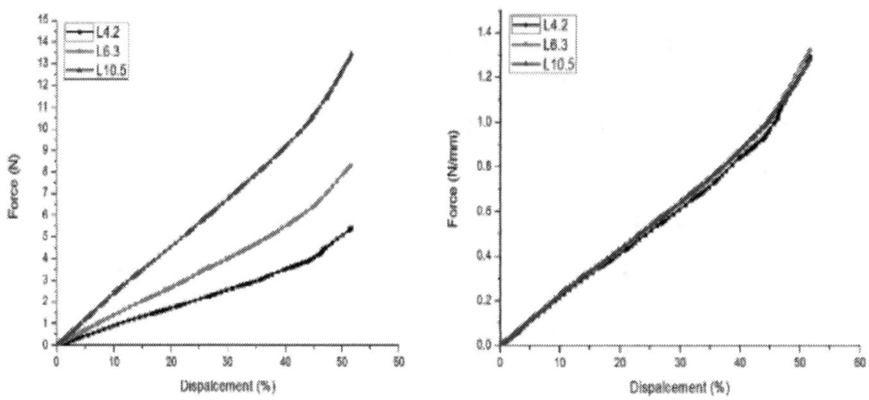

Figure 6 . radial compression displacement-load curve with different length of knitted spring

3.2. Influence of wire diameter

The wire diameter is 0.2, 0.3, and 0.4mm, respectively, and the length is 8.4mm. Other parameters, such as the placement-load curve, are the same as in Table 1 and Figure 7. The wire diameter has a significant influence on the stiffness of the knitted spring. The larger the wire diameter is, the higher is the spring stiffness.

IOP Conf. Series: Materials Science and Engineering **825** (2020) 012033 doi:10.1088/1757-899X/825/1/012033

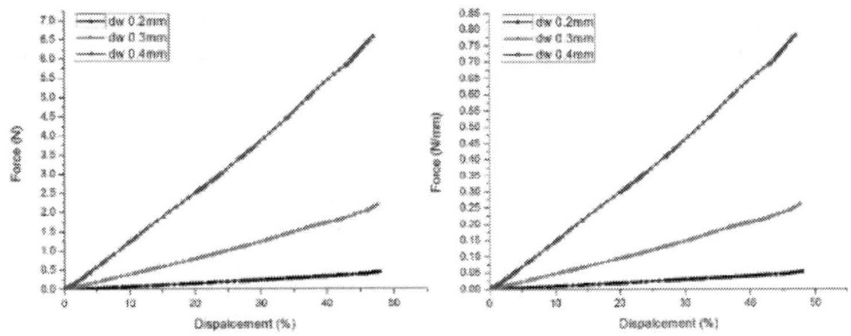

Figure 7 . displacement-load curve of knitted spring with different wire diameters

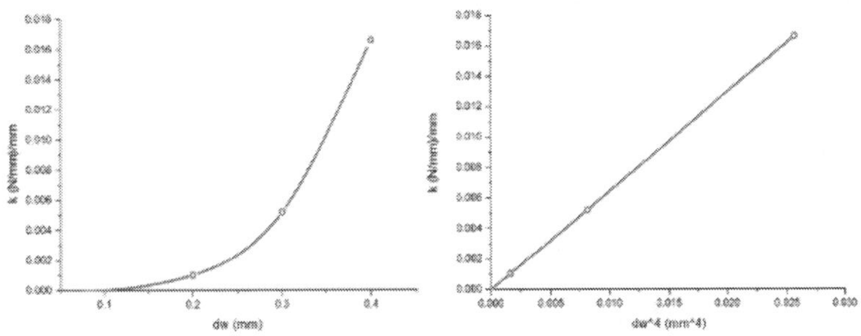

Figure 8 . Influence of wire diameter on the stiffness of knitted spring

A linearly fitted curve is drawn in Figure 7, taking the slope value of the fitting line as the stiffness K of the knitted spring. The drawing of the wire diameter-stiffness curve is shown in Figure 8, shows that the stiffness of the knitted spring is directly proportional to the fourth power of the wire diameter, that is:

$$k = ad_w^4 \tag{2}$$

where a is a constant number. It can be seen that the spring with different stiffness can be obtained by changing the wire diameter in a small range, which is of great reference value for the design of knitted spring.

4. Effect of high temperature creep on resilience performance

The simulation calculation and analysis of high-temperature creep are carried out according to the creep data of superalloy K4169 at 600°C and 700°C, which are fitted with time hardening model as follows:

$$\dot{\bar{\varepsilon}}^{cr} = A\tilde{q}^n t^m \tag{3}$$

Where $\dot{\bar{\varepsilon}}^{cr}$ is equivalent to creep strain rate, $\tilde{q}^n$ is equivalent to deviator stress and t is the time. Since the curve obtained from the creep test is the creep deformation-time curve, therefore the above formula expressed by the strain rate must be integrated. The integration result is as follows:

$$\bar{\varepsilon}^{cr} = \frac{A}{1+m}\tilde{q}^n t^{m+1} \tag{4}$$

Where $n > 0$, $0 \geq m > -1$. Then, the test data can be fitted with the above-mentioned integral formula to obtain the appropriate three parameters A, n and m, as shown in Table 3.

Table 3 . fitting parameters of K4169 creep time hardening model

T	A	n	m
600°C	4.569E-19	3.608	-0.3361
700°C	3.647E-36	10.381	-0.24347

The calculation results of high temperature creep at different temperatures are shown in Figure 9 and Figure 10. It can be seen that at 600°C, the spring can rebound after 10 hours creep at 50% compression, and the rebound rate is close to 100%. At 700 °C, the rebound performance decreases obviously after 1 hour creeps at 50% compression, and the spring tube height decreases compared with the full rebound height after complete unloading. The ratio between the reduction height and the full rebound height is defined as the height reduction, as shown in Figure 11, to quantitatively measure the impact of high temperature creep on the radial compression rebound performance of spring tube.

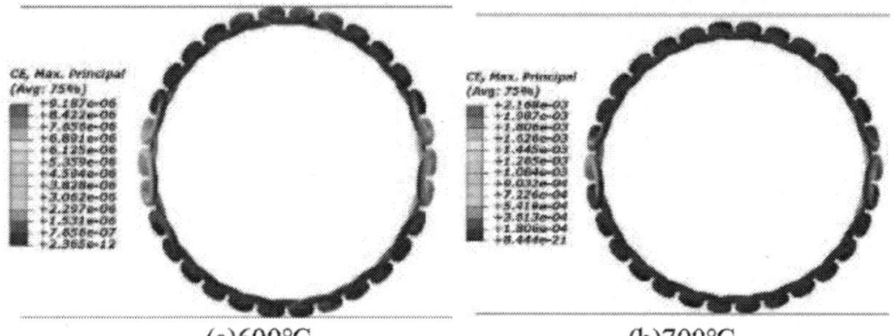

(a)600°C (b)700°C

Figure 9 . creep results of knitted spring at 600/700 °C (50% compression, 3600s creep)

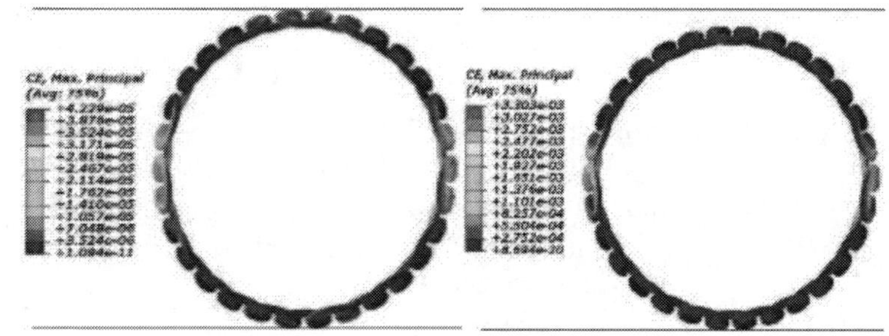

(a)600°C (b)700°C

Figure 10 . creep results of knitted spring at 600/700 °C (50% compression, 36000s creep)

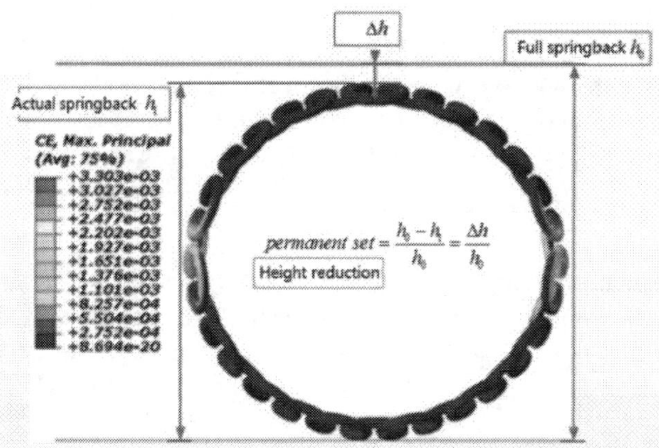

Figure 11 . Definition of rebound height reduction of knitted spring

Figure 12 shows the comparison of the height reduction after creep at 600 °C and 700 °C, from which it can be seen that the height reduction after creep at 600 °C for 1 hour and 10 hours is basically zero, while at 700 °C, the height reduction is as high as 2% and 5% for 1 hour and 10 hours respectively. It can be seen that if K4169 alloy is used to make knitting spring, effective heat insulation measures should be taken to avoid the spring tube directly bearing the high temperature above 600 °C.

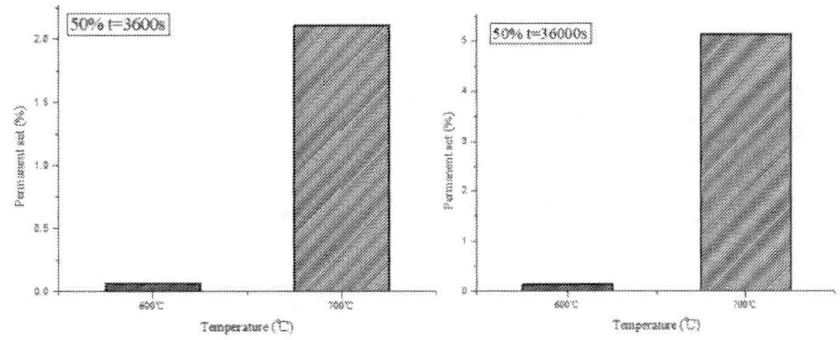

Figure 12 . the height reduction of knitted spring after creep at 600 °C and 700 °C

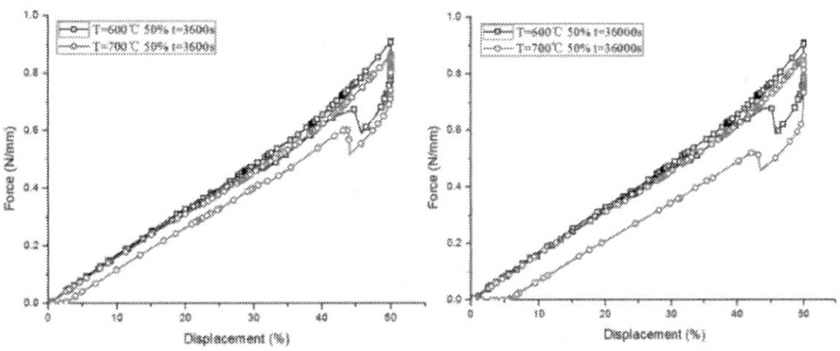

Figure 13 . the resilience-the radial compression displacement curve of knitted spring under high temperature creep at 600 °C and 700 °C

Figure 13 shows the influence of high temperature creep at 600 °C and 700 °C on the radial compression and resilience characteristics of knitted winding spring. It can be seen that the resilience has no visible attenuation after creeping at 600 °C for 1 hour or 10 hours, the loading and unloading lines almost coincide. While creep at 700 °C for 1 hour, there has apparent attenuation that the resilience force becomes to zero when the rebound reaches about 2% compression displacement. After 10 hours of creep, the mitigation of resilience is more serious, and the horizontal intercept is about 5%.

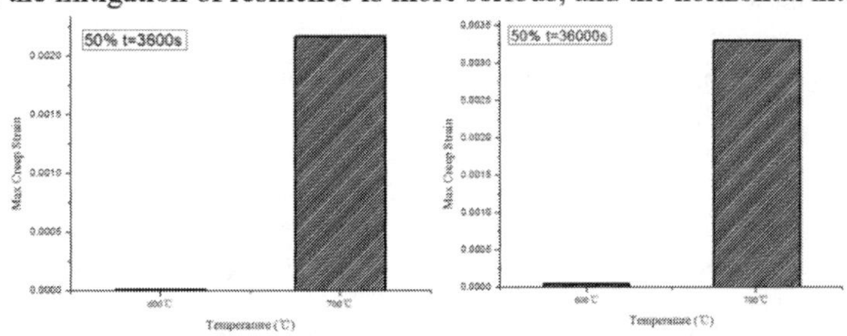

Figure 14 . maximum creep strain at 600 °C and 700 °C

It is evident that the maximum creep strain (Figure 14) at 600 °C is much lower than that at 700°C, the magnitude increases from 1e-6 for 1h to 1e-4 for 10h. Compared with that at 700 °C for 1h i.e., 1E-3, the creep rate at 600 °C is relatively slow. The creep strain is the fundamental reason that influences the resilience performance of elastic elements on high-temperature creep, along with that, the main factors include temperature, time, and stress. In other words, different temperatures, different creep time, and different radial compression displacements (different stress) will affect the resilience of elastic elements. The above content has made a detailed analysis of the influence of temperature. It can be seen that the creep effect at 600 °C in a short time (less than 10 hours) has no apparent impact on the resilience of the knitted spring, while the creep effect at 700 °C can significantly change the strength of the knitted spring within 1 hour.

The influence of creep time on the resilience performance of knitted spring is analyzed, and the simulation results are shown in Figure 15, and in Figure 16a with the reduction of resilience height. It can be seen that under 50% compression and half an hour of creep at 700°C, there is a noticeable

IOP Conf. Series: Materials Science and Engineering **825** (2020) 012033 doi:10.1088/1757-899X/825/1/012033

influence on the resilience performance, where the maximum creep strain reached 1E-3 from Figure 16b. While in Figure 17, the effect of creep time on the characteristic curve on the resilience and the radial compression displacement is mentioned in detail.

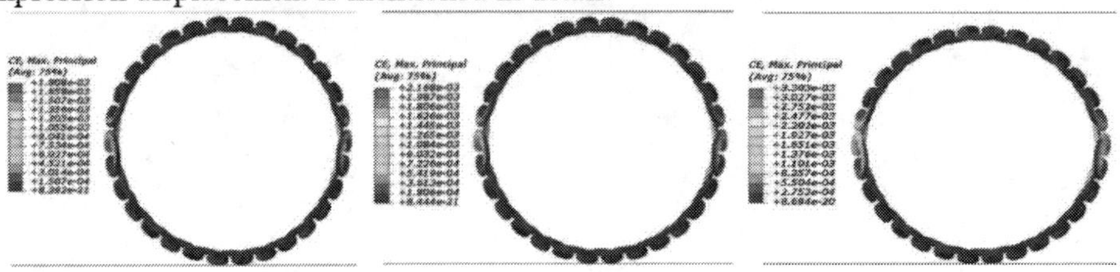

(a) 1800s (b) 3600s (c) 36000s

Figure 15 . the reduction of resilience height of knitted spring under different creep time

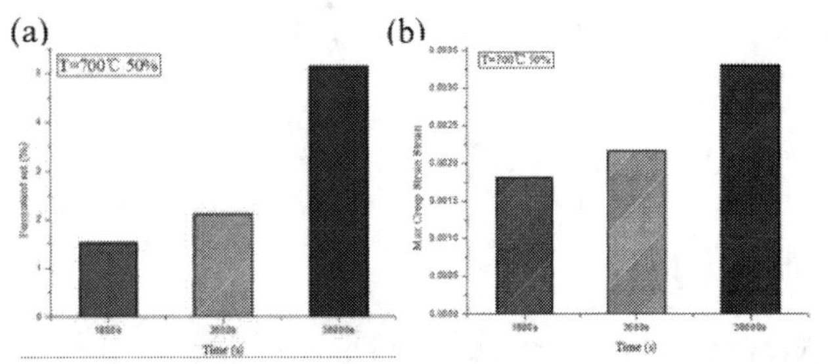

Figure 16 . the reduction of resilience height and maximum creep strain under different creep time

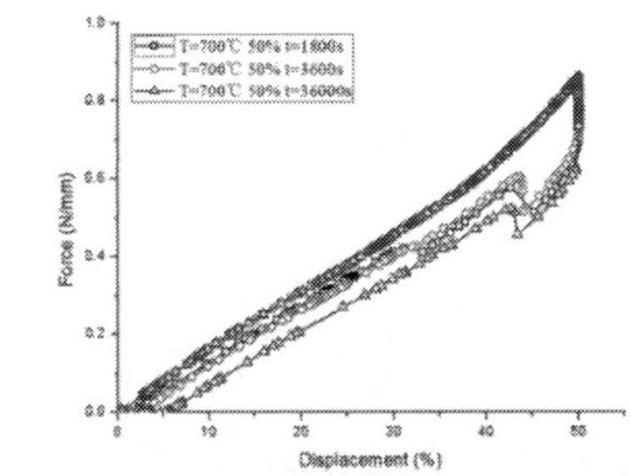

Figure 17 . the resilience-the radial compression curve of knitted spring under different creep time

In Figure 18, Figure 19 and Figure 20, the influence of different radial compression displacement under high temperature creep on the resilience performance of knitted spring is alnalyzed and compared the results under 50% and 40% compression. It has little effect on rebound performance when creeping one hour under 40% compression, but it has a noticeable impact on rebound performance after 10 hours of creep. Under 50% compression, it has an evident effect on rebound performance after 10 hours of creep. Under 50% compression, just one hour of creep has an evident impact on resilience, more significant after 10 hours, especially.

It can be seen that no matter whether the variable is temperature, creep time, or radial compression when the maximum creep strain reaches the order of 1E-3, the apparent decrease of rebound height and resilience can be observed. That is, when the maximum creep strain reaches the order of 1E-3, the effect of high temperature creep on the resilience performance of knitted spring becomes clear.

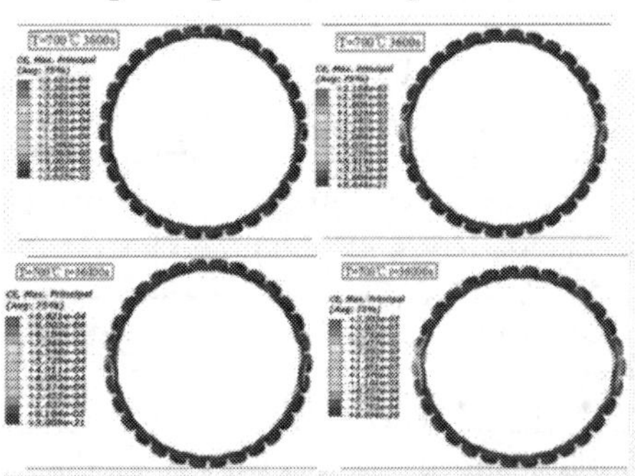

(a) 40% (b) 50%

Figure 18 . resilience height of knitted spring under different radial compression (creep 1h and 10h)

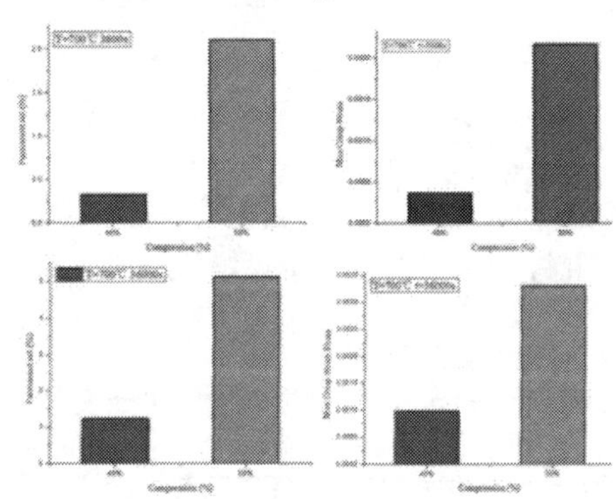

Figure 19 . the reduction of resilience height and maximum creep strain under different radial compression (creep 1h and 10h)

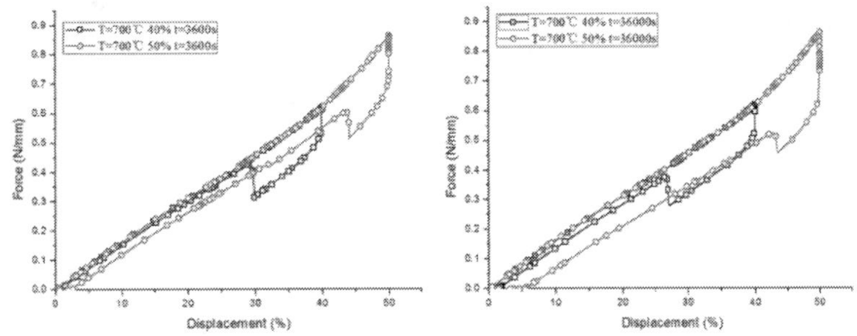

Figure 20 . the resilience-the radial compression curve of knitted spring under different radial compression

5. Summary

Based on the above simulation results, the following are the main conclusions:

1) compared with other types of springs, although the structural characteristics of knitted springs are more complicated. There are many variable characteristic parameters, the displacement-load features of radial compression are relatively simple, and the load increases linearly with the increase of compression.

2) The total length of the spring has little effect on the stiffness of the winding spring per unit length, so the influence of the end effect can be ignored.

3) The wire diameter has a significant impact on the stiffness of the knitted spring. The larger the wire diameter is, the higher the stiffness is, and the stiffness is proportional to the fourth power of the wire diameter. The radial compression stiffness required by design can be easily obtained by changing the wire diameter.

4) Determining the effect of high temperature creep on the resilience properties of elastic elements lies in the magnitude of creep strain, and the main factors affecting creep strain are temperature, time, and compression. Different temperatures, different creep time, and different compression (different stress) will change the rebound performance of the elastic element.

5) it can be seen that if K4169 alloy is used to make knitted spring, effective thermal insulation measures should be taken to prevent the spring tube from directly bearing the high temperature above 600°C.

Acknowledgment

There are no conflicts to declare. This work is supported by the National Natural Science Foundation of China (No.11704244 and U1733130); CALT Foundation (No.201707); Shanghai Natural Science Funding (Grant No. 17ZR1441000); Basic Research Field of Shanghai Science and Technology Innovation Program (No.16JC1401500); Science and Technology Innovation Special Zone Program (No.18-163-13-ZT-008-003-06); Cross Research Fund of Biomedical Engineering of Shanghai Jiao Tong University (YG2016MS70). The Instrumental Analysis Center of Shanghai Jiao Tong University and Xi'an Respro Applied Materials Science and Technology Co., Ltd. are sincerely acknowledged for assisting with the relevant analyses.

References:

[1] Demange J, Dunlap P, Steinetz B and Drlik G 2007 An Evaluation of High Temperature Airframe Seals for Advanced Hypersonic Vehicles 43rd AIAA/ASME/SAE/ASEE Joint Propulsion Conf. & Exhibit (July 8–11 Ohio: Cincinnati/AIAA) 10.2514/6.2007-5743

[2] Zhang J, Jiang J, Ren Q and Li X 2015 The Review of Test Technology for the Ameriacan High Temperature Structural Seals Hydraulic pneumatic & seals 10 p13-15+45.

[3] Dunlap P H, Steinetz B M, Daniels C, Wasowski J, Robbie M, Erker A, Drlik G and Mayer J 2010 Full-Scale System for Quantifying Loads and Leak Rates of Seals for Space Applications

46th AIAA/ASME/SAE/ASEE Joint Propulsion Conf. & Exhibit (July 25-28 Tennessee: Nashville/AIAA) 10.2514/6.2010-6987

[4] Dunlap P H and Steinetz B M 2012 A Comparison of Candidate Seal Designs for Future Docking Systems 48th AIAA/ASME/SAE/ASEE Joint Propulsion Conf. & Exhibit (July 30-Aug 01 Georgia: Atlanta/AIAA) 10.2514/6.2012-4075

[5] Steinetz B M and Hendricks R C 2001 Overview of NASA Glenn Seal Development Program NASA Seal/Secondary Air System Workshop (Oct 01 Ohio: Cleveland) 1 p1-21

[6] Demange J, Bott R and Dunlap P 2011 Investigation of the Mechanical Performance of Compliant Thermal Barriers 17th Int. Space Planes and Hypersonic Systems and Technologies Conf. (Apr 11-14 CA: San Francisco/AIAA) 10.2514/6.2011-2388

[7] Garafolo N and Daniels C. 2013 An Experimental Investigation of Leak Rate Performance of a Subscale Candidate Elastomer Docking Space Seal Journal of Spacecraft and Rockets 50(3) p709-714

IOP Conf. Series: Materials Science and Engineering **825** (2020) 012034 doi:10.1088/1757-899X/825/1/012034

Investigation on Drill Wear and Micro Hole Quality in High Speed Drilling of High Frequency Printed Circuit Board

Xianwen Liu, Hongyan Shi*, Guang Huang, Sha Tao and Zhisen Gao

College of Mechatronics and Control Engineering, Shenzhen University, Shenzhen 518060, People's Republic of China

*Corresponding author's email: shy-no.1@163.com

Abstract. Due to the severe wear of micro drill and the difficulty in controlling of hole quality, high speed drilling process of high frequency printed circuit board (PCB) was attracted comprehensive attention. In this investigation, the wear width of drills' primary face was measured, and the wear morphologies of micro drills were studied at once. Furthermore, regularity of micro hole quality concerning drilling burr and nail heading while drilling with different types of drills were analyzed. The research results indicated that the wear width of diamond coating drill reduced slightly. Abrasive wear and adhesive wear were mainly occurred on uncoated drill and diamond like carbon coating drill, but some micro breaches and pits appeared on uncoated drill. Meanwhile, drilling with diamond like carbon coating drill could obtain better hole quality than drilling with uncoated ones. However, diamond coating drill performed well in terms of ability of wear resistance, as well as micro hole quality.

1. Introduction

Transmission of high frequency signal required the high frequency PCB with the properties of lower dissipation factor (Df), lower dielectric constant (Dk) and lower coefficient of thermal expansion (CTE). Yet materials in high frequency PCBs, such as polytetrafluoretyhylene (PTFE) and ceramic filler, which usually cause micro drill wear severely and bring the difficulty in controlling of hole quality during the high speed drilling.

There were many studies regarding the micro hole drilling of PCBs. With respect to drill wear, Rawat et al. demonstrated the wear mechanisms of micro drills, and they found that the friction between drill and materials raised the drilling temperature, and aggravated drill wear. Furthermore, abrasive wear mainly occurred on primary cutting edge of drill[1]. Iliescu et al. presented an effective method to predict tool wear, which was validated via the drilling experiments, so that it provided a reference to investigate the drill wear in the high speed drilling process of high frequency PCBs [2]. In order to study the wear mechanisms of micro drill, the morphologies of drills were observed, and the flank wear of the micro drills was measured, and the factors affected flank wear were studied by Zheng et al. [3]. Imran et al. created a characteristic wear map of cutting conditions which aimed at helping to identify the zones of lowest wear rate, so it provided a valid method for selecting suitable cutting parameters for micro drilling [4].

In aspect of micro hole quality, the effect of drilling parameters (spindle speed and feed rate) to micro hole quality was studied [5-9]. Guo et al. found that with the decrease of chisel edge length, drilling with thinned micro drill could get smaller entrance burrs, less roundness errors, and regular shapes. Meanwhile, drilling with thinned chisel edge generated more small-size chips, as well as improved chip breaking [10]. Nakagawa et al. depicted that it was of great importance to reduce the workload for the sake of excellent hole quality[11]. Shi et al. maintained that coating technique in micro drill was the trend while considered to improve the ability of wear resistance for micro drill, and they reviewed the major techniques which regarding micro hole drilling of high speed PCB and high frequency PCB [12].

Micro hole drilling was of significance to the fabrication of high frequency PCB, in view of the challenges in micro drilling of high frequency PCB, investigation concerning drill wear and micro hole quality were meaningful and necessary. In this paper, wear width of the uncoated drill and coating drills were measured after a given number of holes were finished, and the wear morphologies were observed simultaneously. Additionally, micro hole quality, including drilling burr and nail heading, while drilling with different types of drills were investigated.

2. Experiment and method

2.1 Materials

A kind of high frequency PCB was treated as the process object in this experiment, which model was RO3003, mainly consisted of copper foil, PTFE and ceramic filler. The combination of aluminum sheet and cold punching plate were used to be the entry board, the phenol aldehyde board was used for back board. The laminated construction of those materials, therefore, was shown as figure 1.

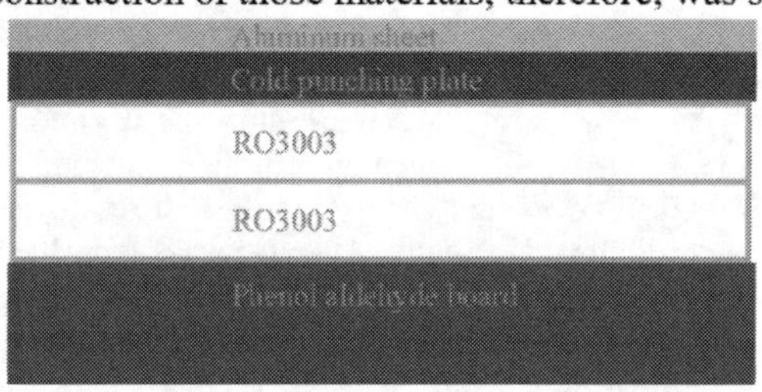

Figure 1. Laminated construction of materials.

Three types of drills which with the same geometry structure were prepared for the drilling process, and they were named uncoated drill, diamond like carbon coating drill and diamond coating drill, respectively. Therein, diamond like carbon coating drill was fabricated through physical vapor deposition, and this kind of coating with low friction coefficient. Diamond coating drill was fabricated via chemical vapor deposition, which was processed under hydrocarbon atmosphere.

2.2 Experiment method

Firstly, the drilling parameters, such as spindle speed, feed rate and retract rate were selected through the consideration of drill diameters and the drilling experiments, they were listed as, spindle speed: 155 krpm; feed rate: 40 mm/s; retract rate: 300 mm/s.

Additionally, in order to find out the influence of high frequency PCB to the drill wear, the width of drill's primary face would be measured via microscope (OLYMPUS, model: SZ61TRC) after drilling a certain holes for each kind of drill. The wear width (w) was calculated through the following equation (1):

$$w = L_0 + L_0' - (L_n + L_n') \quad (1)$$

Where, L_0 and L_0' were the width of primary face of new tool, L_n and L_n' were the width of primary face of the drill which drilled a certain number of holes, the subscript n identified the sequences of measurement (n=1, 2, 3, 4, 5, 6, 7). Consequently, the measuring principle as shown in figure 2. Also, the wear rate (R) was defined by the following equation (2):

$$R = \frac{w}{L_0 + L_0^{'}} \times 100\% \qquad (2)$$

The wear morphologies of three types of drills were going to be observed through scanning electron microscope (FEI company, model: Quanta FEG 450). Micro hole quality of high frequency PCB, including drilling burr and nail heading would be compared while using different drills. Therein, the height of drilling burr and nail heading were measured through microscope. Moreover, the height of drilling burr and the value of nail heading were the mean value in each measurement period. The method of measurement was shown as figure 3, H_1, H_2 represented the height of drill burr (figure 3a), and the nail heading was expressed by the value of B divided by A (figure 3b).

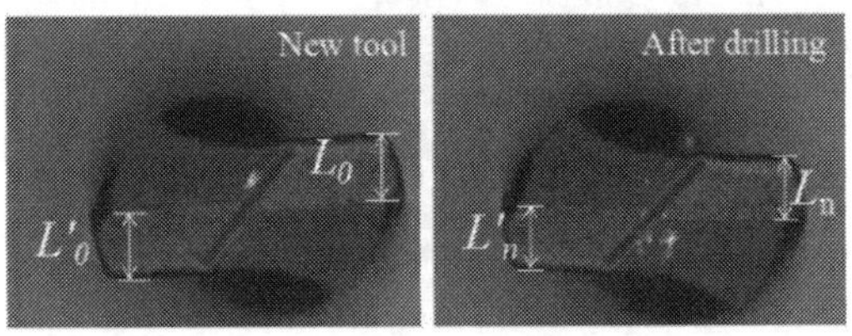

Figure 2. Measurement of wear width

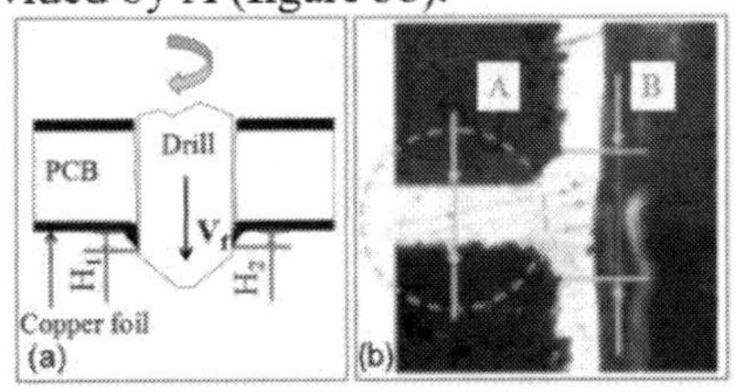

Figure 3. Measurement of drilling burr (a) and nail heading (b) [13]

3. Results and discussions

3.1 Drill wear

3.1.1 Wear width of drill's primary face. Comparisons of primary face between the new drills and the drills finished 1200 holes were shown as figure 4, figure 5, and figure 6. These figures showed that the primary face of uncoated drill and diamond like carbon coating drill were worn severely when they drilled 1200 holes. The wear width for uncoated drill and diamond like carbon coating drill were 77.9 μm and 70.1 μm. Furthermore, there were some resins adhered on the secondary face where near chisel edge for each kind of drill, and the most serious adhesion occurred on uncoated drill, it might because the roughness of that positions was worse.

The details of wear width variations were shown as figure 7, which indicated that the wear width of diamond like carbon coating drill was less than uncoated drill after drilling same number of holes, but the diamond coating drill showed the best performance while considered the wear width of primary face. Meanwhile, it was clear that the wear width was increased tremendously while finished 100 holes for uncoated drill and diamond like carbon coating drill, and then increased at slower rate in the following process. The reason to explained this phenomenon was that the primary cutting edges of uncoated drill and diamond like carbon coating drill were loaded heavy pressure at that period. The wear width of diamond coating drill, however, rose slightly and steadily during the whole drilling period, it was only 11.63 μm after drilling 1200 holes.

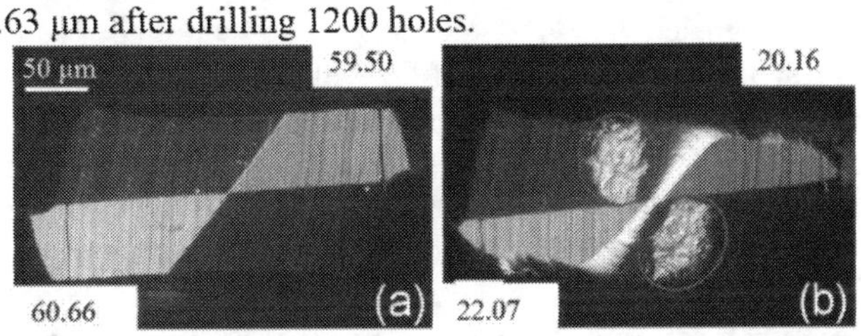

Figure 4. Comparison of uncoated drill before and after drilling

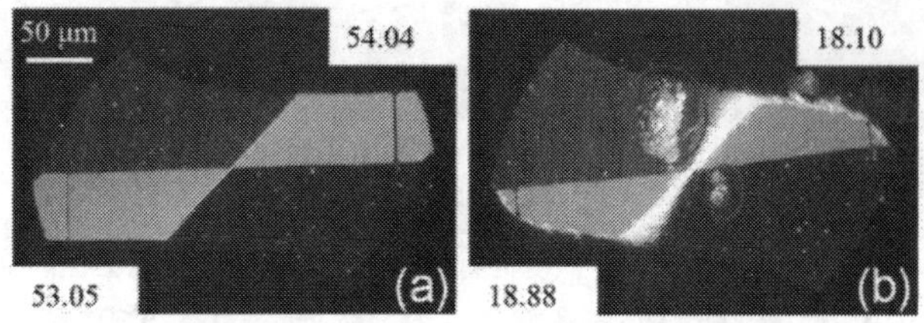

Figure 5. Comparison of diamond like carbon coating drill before and after drilling

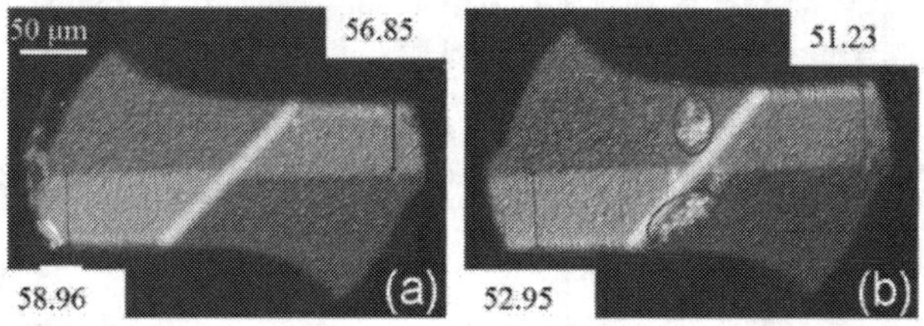

Figure 6. Comparison of diamond coating drill before and after drilling

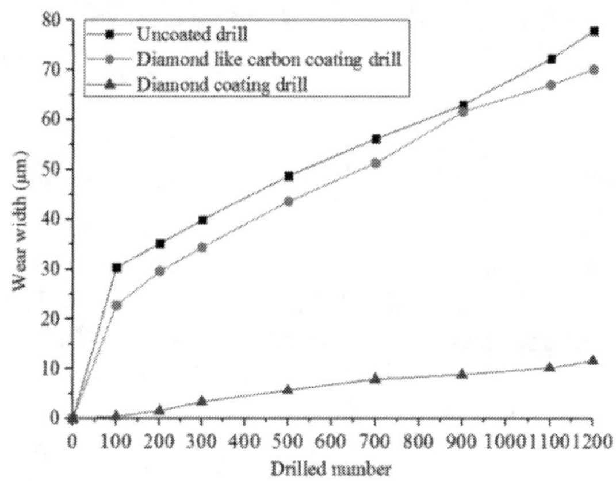

Figure 7. Wear width variations of primary face

In terms of wear rate, there were 64.9%, 65.4%, 10.0% for uncoated drill, diamond like carbon coating drill and diamond coating drill, respectively. These results illuminated that diamond coating drill had excellent ability of wear resistance. Moreover, these two kinds of coated drills were propitious to reduce the wear width of micro drill in the micro hole drilling of high frequency PCB.

3.1.2 Wear morphology. In this part, each figure were divided into part (a) and part (b), and part (b) was the partial enlargement of part (a). Wear morphologies of uncoated drill were shown in figure 8, figure 9 and figure 10. The breach occurred on the primary cutting edge (figure 8), it might be explained that the wallop while drilling ceramic fillers made the primary cutting edge arose brittle rupture, especially when the cutting edge was sharp. This form of wear would directly decrease the drill's cutting ability. Some pits appeared near chisel edge (figure 9), that was mainly because chisel edge and drilling object engendered strong squeezing action, which caused the temperature around this position went up quickly, so that the cohesive action from cobalt decreased and the micro grains that consisted of drill fell off easily. In addition, the fracture on land (figure 10) was caused by the friction from the hole wall and extrusion pressure from chips, particularly the chips were made of copper foil and ceramic filler.

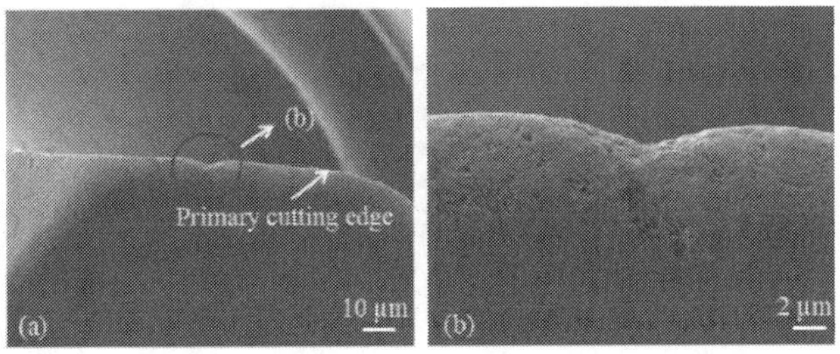

Figure 8. Wear morphology around primary cutting edge of uncoated drill

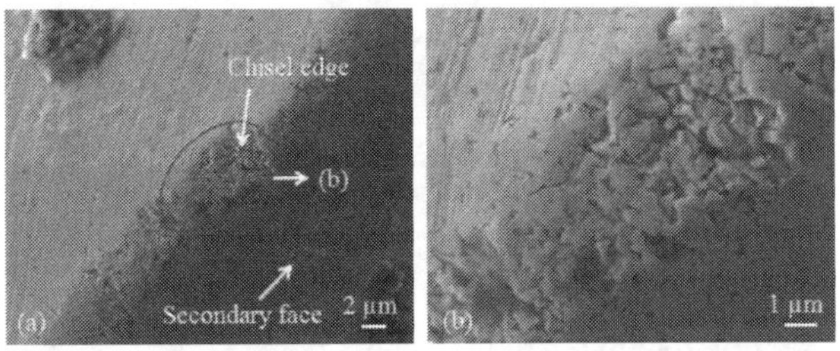

Figure 9. Wear morphology around chisel edge of uncoated drill

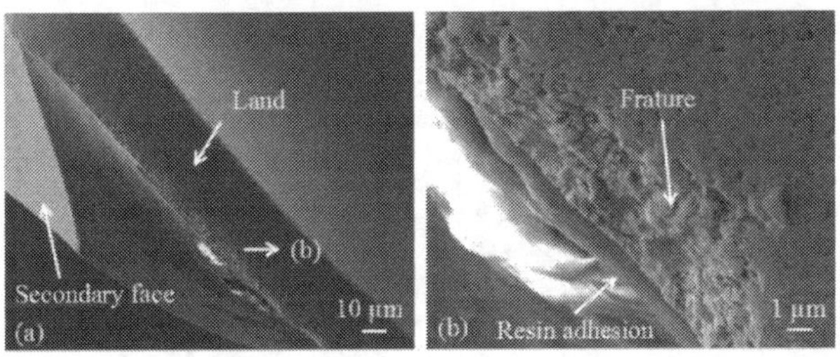

Figure 10. Wear morphology around land of uncoated drill

Wear morphologies of diamond like carbon coating drill were shown in figure 11, figure 12 and figure 13. The wear on primary cutting edge and chisel edge were similar as uncoated drill, yet the boundary between coating and raw materials was palpable and clear in these two areas, and the phenomenon of coating thinned was observed on land of drill (figure 13). The abrasive wear and adhesion wear were the main forms on primary cutting edge and chisel edge, and coating shedding was occurred in these two areas. Additionally, because the land was a part of cylinder, the friction action on land would be more balanced, so the coating on land took on thinned trend with increment of drilled hole. Moreover, no breach or pit were found around primary cutting edge, chisel edge and land.

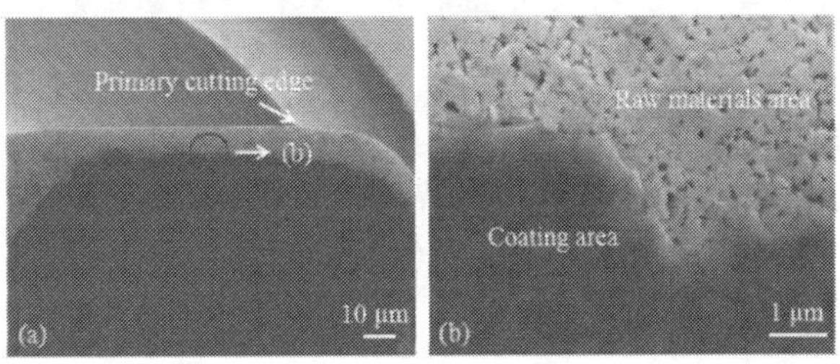

Figure 11. Wear morphology around primary cutting edge of diamond like carbon coating drill

IOP Conf. Series: Materials Science and Engineering **825** (2020) 012034 doi:10.1088/1757-899X/825/1/012034

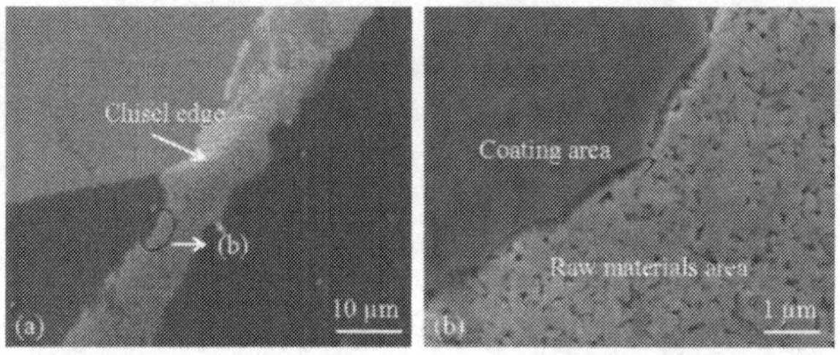

Figure 12. Wear morphology around chisel edge of diamond like carbon coating drill

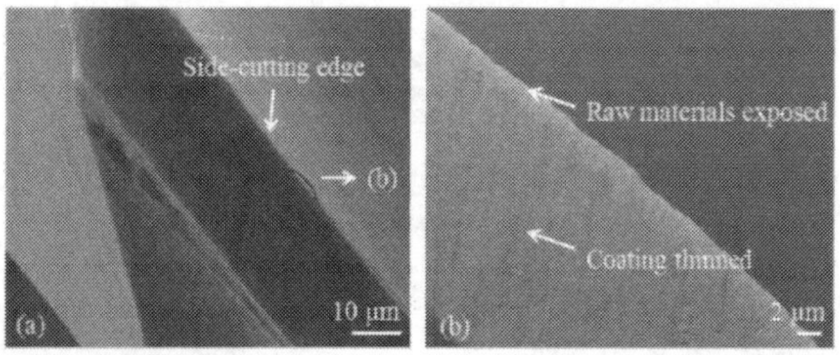

Figure 13. Wear morphology around land of diamond like carbon coating drill

Wear morphologies of diamond coating drill were shown in figure 14, figure 15. Firstly, some copper chips adhered near the primary cutting edge, and the coating on primary cutting edge and side-cutting edge caused them not as sharp as the uncoated drill and diamond like carbon coating drill. But the outlines of primary cutting edge and side-cutting edge were kept well as the new tool, no coating shedding or breach occurred on all places. This kind of coating, therefore, displayed excellent ability of wear resistance and predominant adhesion with micro drill matrix.

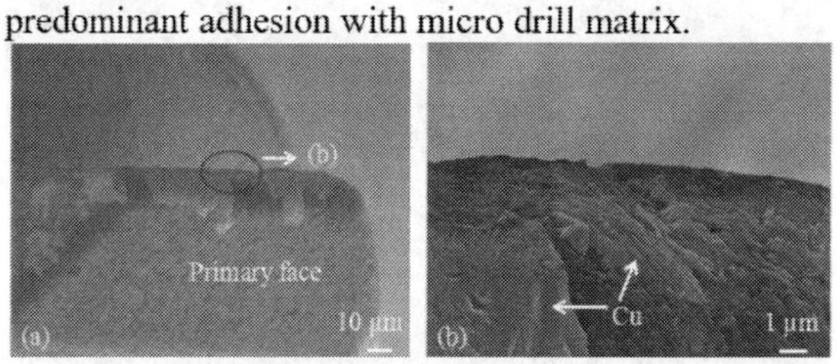

Figure 14. Wear morphology around primary cutting edge of diamond coating drill

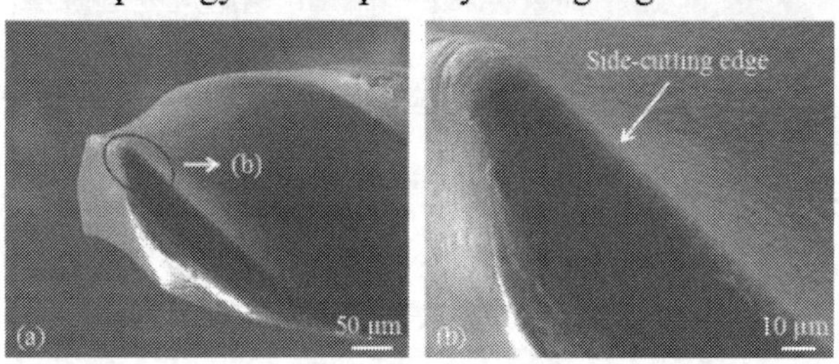

Figure 15. Wear morphology around side-cutting edge of diamond coating drill

3.2 Micro hole quality

3.2.1 Drilling burr. The variations of drilling burr after a given number of drilled holes were shown as figure 16. The result indicated that the height of drilling burr was really short before drilling 200 holes, especially drilling with uncoated drill and diamond like carbon coating drill. That was because the cutting edges of drills still kept sharp. But drilling with diamond like carbon coating drill could obtained shorter drilling burr than drilling with uncoated drill in every measurement stage. The height of drilling burr was increased gradually after drilling 200 holes, from 2.69 μm to 21.61 μm while drilling with uncoated drill, and from 0.34 μm to 19.69 μm while drilling with diamond like carbon coating drill. For diamond coating drill, the height of drilling burr held a slight fluctuation in the three measurement stages. Besides, the height of drilling burr kept less than 7 μm steadily in each measurement stage.

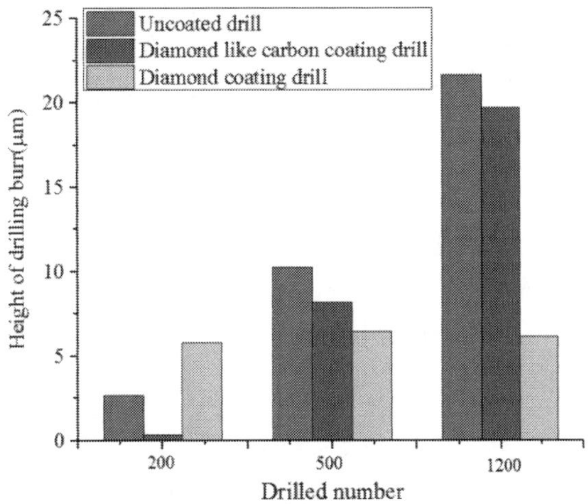

Figure 16. Variations of drilling burr in different drilling stages

With regard to the variation of drilling burr while drilling with uncoated drill and diamond like carbon coating drill, since the wear width increased and the primary cutting edges rusted with the increment of drilled number, the drilling performance became weaker, and drilling burr increased gradually. Furthermore, the height of drilling burr was stable and short while drilling with diamond coating drill, this superiority on drilling burr may be explained by the excellent ability of wear resistance of diamond coating drill, and this ability kept the cutting edge sharp for a long term.

3.2.2 Nail heading. As figure 17 shown, regularity on nail heading was similar with the trend of drilling burr. The nail heading increased with the increment of drilled number while drilling with uncoated drill and diamond like carbon coating drill, meanwhile the nail heading reached to 1.89 and 1.71, respectively, while the uncoated drill and diamond like carbon coating drill finished 1200 holes. But for the process using diamond coating drill, the nail heading became smaller with the increase of drilled number.

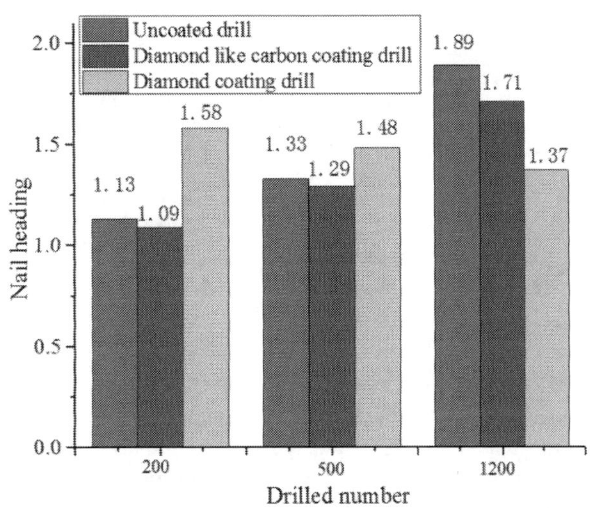

Figure 17. Variations of nail heading in different drilling stages

According to the results we could know that the wear width of primary face was the most significant factor when considering the variation of nail heading, because the changes of wear width reflected the develop trend of nail heading, such as the nail heading's variations while drilling with uncoated drill and diamond like carbon coating drill. But the biggest nail heading occurred when drilling 200 holes, that might because the cutting edge of diamond coating drill was blunt at that period. Meanwhile, the variation trend of nail heading was a little different to the drilling burr while drilling with diamond coating drill, that meant the sharpness of primary cutting edge, side-cutting edge and other factors ought to be considered when we desired to find out the influence factors of nail heading and drilling burr more synthetically.

4. Conclusion

Drill wear, drilling burr and nail heading in micro hole drilling of high frequency PCB were investigated in this study. In aspect of drill wear, the wear width for uncoated drill, diamond like carbon coating drill and diamond coating drill were 77.9 μm, 70.1 μm, and 11.6 μm, respectively, after drilling 1200 holes, furthermore, their wear rates were 64.9%, 65.4%, 10.0%, respectively. Thus, the coating drills indicated the better performance in terms of wear width, especially for the diamond coating drill. Abrasive wear and adhesive wear were mainly occurred for uncoated drill and diamond like carbon coating drill, and some pits and breaches appeared on uncoated drill. The boundary between coating and raw materials was palpable and clear on diamond like carbon coating drill, but the phenomenon of coating thinned was still observed on land. The outlines of primary cutting edge and side-cutting edge maintained well as the new tool, no coating shedding and breach were found on every position of diamond coating drill.

In addition, diamond like carbon coating drill performed better than the uncoated drill on drilling burr and nail heading. Because of the excellent ability of wear resistance of diamond coating drill, the height values of drilling burr was steady and short, but the nail heading became smaller while its cutting edge became sharper. However, further study regarding drilling with diamond coating drill was encouraged simultaneously.

Consequently, study on drill wear revealed the wear mechanism of the micro drills, it would provide a guideline for improving the ability of wear resistance of micro drill. Moreover, these two types of coating drills were beneficial to raise the hole quality in high speed drilling of high frequency PCBs.

Acknowledgements
This research is supported by the National Natural Science Foundation of China (Nos. 51971150 and 51605304), the Natural Science Foundation of Guangdong Province (No. 2016A030310036), the Science and Technology Innovation Commission Shenzhen (Nos. JCYJ20170817101008562 and JSGG20190219152602381), and the Natural Science Foundation of Shenzhen University (No. 2017034).

References

[1] Rawat S, Attia H (2009) Wear mechanisms and tool life management of wc-co drills during dry high speed drilling of woven carbon fibre composites *Wear* **267**(5-8):1022-1030
[2] Iliescu D, Gehin D, Gutierrez ME, Girot F (2010) Modeling and tool wear in drilling of CFRP *Int J Mach Tools Manuf* **50**:204-213
[3] Zheng LJ, Wang CY, Fu LY, Yang LP, Qu YP, Song YX (2012) Wear mechanisms of micro-drills during dry high speed drilling of PCB *J Mater Process Technol* **212**(10):1989-1997
[4] Imran M, Mativenga PT, Withers PJ (2012) Assessment of machining performance using the wear map approach in micro-drilling *Int J Adv Manuf Technol* **59**(1-4):119-126
[5] Zheng LJ, Wang CY, Zhang X, Huang X, Song YX, Wang K, Zhang LQ (2016) The tool-wear characteristics of flexible printed circuit board micro-drilling and its influence on micro-hole quality *Circuit World* **42**(4):162-169
[6] Sahoo S, Thakur A, Gangopadhyay S (2016) Application of analytical simulation on various

IOP Conf. Series: Materials Science and Engineering **825** (2020) 012034 doi:10.1088/1757-899X/825/1/012034

characteristics of hole quality during micro-drilling of printed circuit board *Adv Manuf Process* **31**(14): 1927-1934

[7] Bhandari B, Hong YS, Yoon HS, Moon JS, Pham MQ, Lee GB, Huang YC, Linke BS, Dornfeld DA, Ahna SH (2014) Development of a micro-drilling burr-control chart for PCB drilling *Precis Eng* **38**(1):221-229

[8] Zheng XH, Dong DP, Huang LX, An QL, Wang XB, Chen M (2013) Research on fixture hole drilling quality of printed circuit board *Int J Preci Eng Manuf* **14**(4): 525-534

[9] Huang, LX (2015) Investigation of chip formation and burr control in pcb fixture hole *Circuit World* **41**(2):61-69

[10] Guo H, Liang Z, Wang X (2018) Influence of chisel edge thinning on helical point micro-drilling performance *Int J Adv Manuf Technol* **99**: 9-12

[11] Nakagawa H, Ogawa K, Kihar A, Hirogaki T (2007) Improvement of micro-drilled hole quality for printed wiring boards *J Mater Process Technol* **191**:293-296

[12] Shi HY, Liu XW, Lou Y (2019) Materials and micro drilling of high frequency and high speed printed circuit board: a review *Int J Adv Manuf Technol* **100**:827-841

[13] http://www.uniontool.co.jp/english/tech_02-06.html

IOP Conf. Series: Materials Science and Engineering **825** (2020) 012036 doi:10.1088/1757-899X/825/1/012036

Continuous Forming System for Forming 3D Surface Parts and Its Finite Element Model

Mi Wang [a], **Shuchen Yang** [a,*] **and Guolong Lu** [b]

[a] Engineering College, Changchun Normal University, Changchun, Jilin Province, 130022, P. R. China.
[b] Key Laboratory of Bionic Engineering (Ministry of Education), College of Biological and Agricultural Engineering, Jilin University, Changchun, Jilin Province, 130022, P. R. China.

*Ysc2017@mail.cncnc.edu.cn

Abstract. Continuous forming system (CRS) is a novel technique for manufacturing 3D surface parts. CRF employs two flexible rolls to generate deformation by controlling roll gap between two flexible rolls. Finite element model is set up to investigate the CRF process. The influence of maximum height difference and maximum compression ratio difference in cross-section has been investigated. The accuracy of the longitudinal was demonstrated through numerical simulations. CRF experiments were performed and the results show the shape of the formed part is in good agreement with the shape of the desired part.

1. Introduction

3D surface parts are widely used in modern architectures. For many years, tamping, stretch-bending, hydroforming have been used to form three-dimensional surface parts. Nevertheless, large initial investments and long setup time are needed in these forming processes, making them only profitable for mass production. Yoon and Yang [1,2] explored an incremental roll forming design using one centre roll and two pairs of support rolls to manufacture 3D surface parts, which is very suitable for single and small batch products. However, it is low in productivity and the formed part has large thickness reduction. Shim et al. [3-5] proposed discrete-rolls forming which is time-saving for small batch production. The discrete-roll is composed of small rolls arranged in a linear array. Each small roll can rotate around its curved axis, and the shape of forming roll can be adjusted by changing the height of each small roll. Li et al. [6,7] proposed flexible roll forming based on three roll-bending process, and the sheet metal can be bent simultaneously in transverse and longitudinal directions. Cai et al. [8,9] proposed the CRF process using two roll for manufacturing 3D surface parts. In the present paper, the finite element model is established with the aim of investigating the forming accuracy of CRF numerically.

2. Continuous roll forming system and Finite element model

2.1 Principle of continuous roll forming

Continuous roll forming (CRF) system employs two flexible roll to generate deformation in transverse and longitudinal directions. Figure 1 shows a schematic diagram of CRF equipment. The height of control unit on the flexible roll can be adjusted, resulting in curvature and roll gap changing [10,11]. In forming process, the flexible roll rotate around its curved axis and a flat sheet metal is manufactured into a desired part [12].

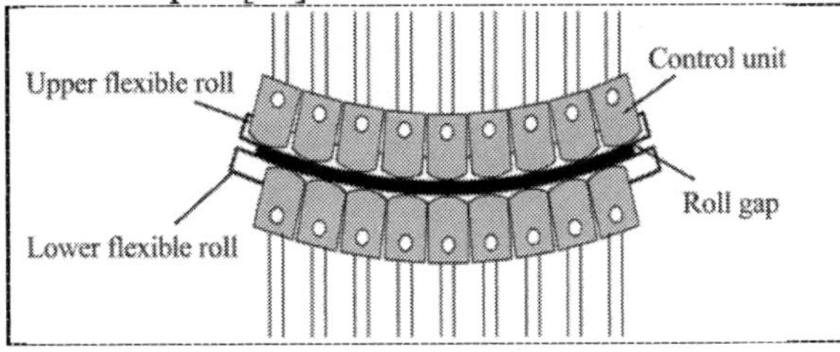

Figure 1. Principle of Continuous Roll Forming

2.2 Finite element model of CRF

Finite element model of CRF process was performed based on dynamic explicit FEM under ABAQUS/Explicit environment [13-16]. The sheet metal was meshed by C3D8R element. Relevant material properties are: Young's modulus E = 207 GPa, yield stress σ_s = 91 MPa, Poisson's coefficient μ= 0.29 and density ρ = 7845 kg/m^3. The flexible roll was established using R3D4 elements. Figure 2 shows the schematic diagram of 1/2 finite element model from the symmetrical axis (xx') to the outer edge.

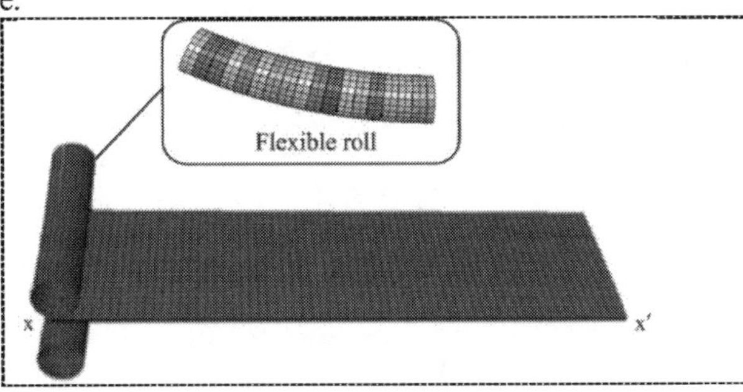

Figure 2. 1/2 Finite element model of blank sheet and flexible rolls

A typical CRF process is shown in figure 3. At the first step, the upper flexible moves downward, and sheet metal is pressed by the flexible rolls. At the second step, flexible roll rotates on its curved axis and sheet metal is bent in two directions. With the rotation of flexible rolls, 3D surface parts are formed eventually.

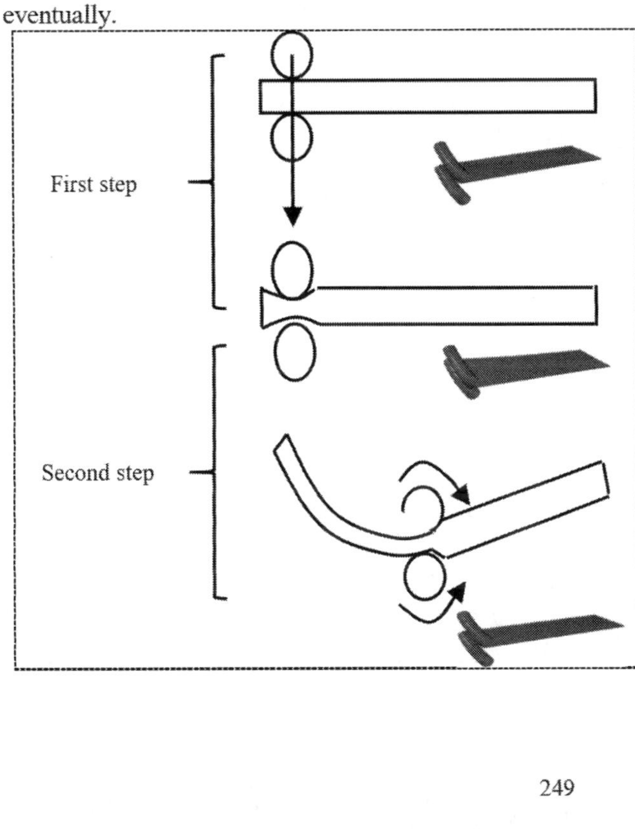

Figure 3. Typical CRF numerical simulation process

2.3 Results of numerical simulation analysis

Plentiful simulated results were presented to compare with the obtained results based on the theoretical calculation. The central function of the flexible roll is calculated according to the transverse curvature of the desired part.

$$z(y) = \rho_T\left[1-\sqrt{1-\left(y\rho_T^{-1}\right)^2}\right] \tag{1}$$

In which, ρ_T^{-1} is the transverse curvature, y is the coordinate in transverse direction.

The distributions of the roll gap are different for saddle parts and torus parts. For a saddle part, the distribution of the roll gap is wide in middle section, for a torus part, the distribution of the roll gap is narrow in middle section, respectively. In order to form a saddle part, the height of the control unit is minor adjusted based on by equal 2. A torus part is formed when the distribution of the roll gap is given as equal 3.

$$h(y) = \left\{1-\rho_T\rho_L^{-1}\left[1-\sqrt{1-\left(y\rho_T^{-1}\right)^2}\right]\right\}H \tag{2}$$

$$h(y) = \left\{1-\rho_T\rho_L^{-1}\left[\sqrt{1-\left(y\rho_T^{-1}\right)^2}-\sqrt{1-\left(y_1\rho_T^{-1}\right)^2}\right]\right\}H \tag{3}$$

In which, H is the thickness of the sheet metal.

To evaluate the accuracy of relationship between calculation curvature ρ_L and simulation curvature ρ_L^*, shape error ratio e_L is defined as shape difference between calculation curvature ρ_L and simulation curvature ρ_L^*.

$$e_L(y) = \left|\frac{\rho_L^* - \rho_L}{\rho_L}\right| \tag{4}$$

2.3.1. Case 1: saddle part. Finite element model of CRF is established, in order to form a saddle part, the position of short rigid roll is minor adjusted based on by equal 2. Figure 4 presents calculation curvatures and simulation curvatures of saddle part under a series of maximum height difference γ and maximum compression ratio difference δ. Figure 4(a) shows calculation curvatures and simulation curvatures of saddle parts under same γ, which is 13.864 mm. Shape error ratio is small, the maximum value of e_L is 9.64%, the average of e_L is 5.43%.

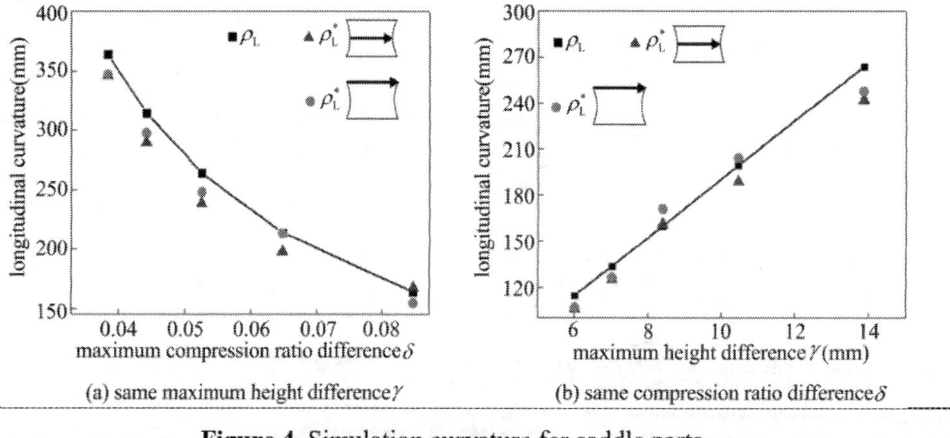

(a) same maximum height difference γ (b) same compression ratio difference δ

Figure 4. Simulation curvature for saddle parts.

Figure 4(b) shows calculation curvatures and simulation curvatures of saddle parts under same δ, which is 0.0525. Shape error ratio is small, the maximum value of e_L is 8.26%, and the average of e_L is 5.62%. Dates in figure 4 clearly demonstrate that CRF can be controlled to form saddle part based on equal 2, and prove the relationship between longitudinal curvature of saddle part and γ/δ.

2.3.2 Case 2: torus part. A torus part is formed by minor adjusting the position of short rigid roll based on by equal 3. Simulation results of CRF for torus part under a series of maximum height difference γ and maximum compression ratio difference δ are shown in figure 5. Figure 5(a) shows calculation curvatures and simulation curvatures of torus parts under same γ, which is 13.864 mm. Shape error ratio is small, the maximum value of e_L is 9.49%, the average of e_L is 5.36%.

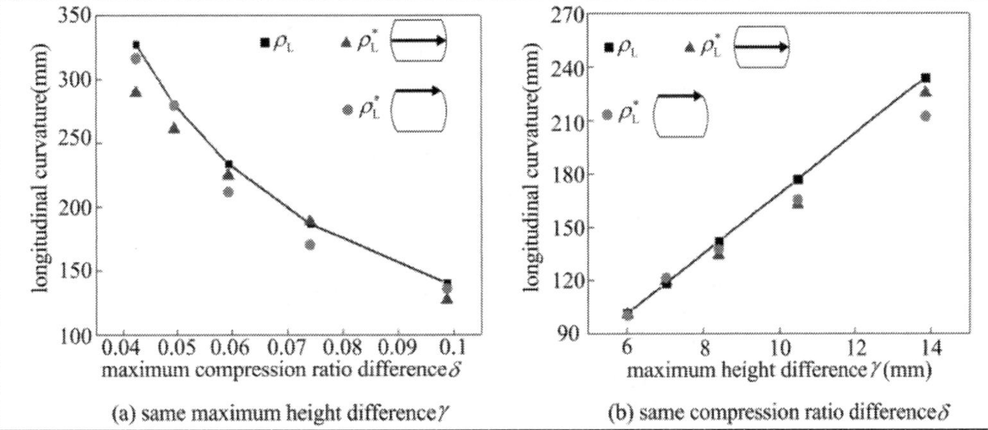

Figure 5. Simulation curvature for torus parts.

Figure 5(b) shows calculation curvatures and simulation curvatures of torus parts under same δ (0.0593). Shape error ratio is small, the maximum value of e_L is 9.39%, and the average value of e_L is 4.32%. Dates in figure 4 clearly demonstrate that CRF can be controlled to form torus part based on equal 3, and prove the relationship between the longitudinal curvature of torus part and γ/δ.

3. Conclusions

Continuous roll forming has been developed to manufacture 3D surface part. The longitudinal curvature can be controlled by the gap between upper and lower flexible rolls. The influence of maximum height difference γ maximum compression ratio difference δ in cross-section has been investigated. According to γ and δ, longitudinal curvature of formed part can be calculated. Extensive numerical simulations of CRF processes had been performed by dynamic explicit finite element methods, a series of numerical results were obtained. These results reveal the relationship between longitudinal curvature of formed part and γ/δ.

References

[1] Yoon SJ and Yang DY. Development of a highly flexible incremental roll forming process for the manufacture of a doubly curved sheet metal. Cirp Ann-Manuf Technol 2003; 52(1): 201-4.
[2] Yoon SJ and Yang DY. An incremental roll forming process for manufacturing doubly curved sheets from general quadrilateral sheet blanks with enhanced process features. Cirp Ann-Manuf Technol 2005; 54(1): 221-4.
[3] Shim DS, Yang DY, Han MS, Chung SW, Kim KH, and Roh HJ. Experimental study on manufacturing doubly curved plates using incremental rolling process. ICTP 2008 (The 9th International Conference on Technology of Plasticity),Korea 2008; 2378-83.

[4] Shim DS, Yang DY, Kim KH, Han MS, and Chung, SW. Numerical and experimental investigation into cold incremental rolling of doubly curved plates for process design of a new LARS (line array roll set) rolling process. Cirp Ann-Manuf Technol 2009; 58(1): 239-42.
[5] Shim DS, Yang DY, Kim KH, Chung SW, and Han MS. Investigation into forming sequences for the incremental forming of doubly curved plates using the line array roll set (LARS) process. Int J of Mach Tool Manuf 2010; 50: 214-8.
[6] Cai ZY, Li MZ. The principle and theoretical analysis of continuous roll forming for three-dimensional surface parts. Sci China Technol Sci 2013; 56(2): 351-8.
[7] Cai ZY, Li MZ, Lan YW. Three-dimensional sheet metal continuous forming process based on flexible roll bending: the principle and experiments. J Mater Process Technol 2012; 212(1): 120-7.
[8] Li MZ, Cai ZY, Li RJ, Lan LW, Qiu NJ. Continuous forming method for three-dimensional surface parts based on the rolling process using bended roll. J Mech Eng 2012; 48(14): 44-9.
[9] Cai ZY, Li MZ. Mechanical mechanism of continuous roll forming for three-dimensional surface parts and the calculation of bending deformation. J Mech Eng 2013; 49(2): 35-41.
[10] Li MZ, Hu ZQ, Cai ZY, Gong XP. Method of efficient continuous plastic forming for freeform surface part. J Jilin Univ (Eng Technol Ed) 2007; 37(3): 489-94.
[11] Li MZ, Cai ZY, Sui Z, Li XJ. Principle and applications of multi-point matched-die forming for sheet metal. Proc Inst Mech Eng B: J Eng Manuf 2008; 222(5): 581-9.
[12] Dong YG, Zhang WZ, Song JF. Theoretical and experimental research on the elongation law of the rail in rail rolling by a universal mill. J Mech Eng 2010; 46(6): 87-92.
[13] Cai ZY, Lan YW, Li MZ. Continuous sheet metal forming for doubly curved surface parts. Int J Precis Eng Man 2012; 13(11): 1997-2003.
[14] Zeng J, Liu ZH, Champliaud H. FEM dynamic simulation and analysis of the roll-bending process for forming a conical tube. J Mater Process Technol 2008; 198(1-3): 330-43.
[15] Feng GK, Champliaud H. Modeling and simulation of asymmetrical three-roll bending process. Simul Model Practr Th 2011; 19(9): 1913-7.
[16] Cai ZY, Wang SH, Hong XP, Li MZ. Numerical simulation for the multi-point stretch forming process of sheet metal. J Mater Process Technol 2009; 209(1): 396-407.

IOP Conf. Series: Materials Science and Engineering **825** (2020) 012037 doi:10.1088/1757-899X/825/1/012037

A Resonant Pressure Sensor based on Magnetostrictive/Piezoelectric Magnetoelectric Effect

Lunan Liu, Yikun Yang and Bintang Yang*

State Key Laboratory of Mechanical System and Vibration, Shanghai Jiao Tong University, Shanghai 200240, China

* btyang@sjtu.edu.cn

Abstract. This study proposed a resonant pressure sensor based on magnetoelectric effect induced in magnetostrictive/piezoelectric materials. Due to the magnetoelectric effect, the resonant frequency of the sensor varies as the external applied pressure changes. By studying the nonlinear constitutive parameters of the magnetostrictive material and adopting the equivalent circuit method, the sensor is theoretically analysed and the relationship between the resonant frequency and applied pressure is determined. The prototype of the sensor is optimized based on theoretical model. Optimal parameters of the sensor are determined to ensure high sensitivity. The measurement range of the sensor is 0-360 kPa and experimental studies show that the sensitivity reaches to 11.63 Hz/kPa. The experimental result is in good agreement with theoretical analysis. It can be concluded that the pressure measurement method based on magnetostrictive/piezoelectric magnetoelectric effect is of robustness and accuracy. Analysis shows the proposed mechanism also has the potential in mass measurement.

1. Introduction

The magnetoelectric (ME) effect is the electric polarization induced by a magnetic field or the magnetization induced by an electric field [1]. A strong extrinsic ME effect is found in magnetostrictive/piezoelectric composites [2,3]. The high magnetoelectric coefficient at room temperature makes the ME composites be practically applied in sensing technology. The ME sensors are developed to measure magnetic field [3], current [4], rotational speed [5], crankshaft position [6] and susceptibility [7]. Given the high-precision measurement achieved by the ME sensors, they are promising in pressure sensing.

Pressure sensors are essential and widely applied in consumer electronics, medical, industrial, automotive and structural health monitoring [8,9]. Among them piezoresistive pressure sensors are of high linearity but with temperature drift and relatively low sensitivity. The capacitance pressure sensors are of high precision and rapid dynamic response. But the applied range of capacitance sensors is limited by the nonlinearity outputs and the strict environment requirement. Compared with above pressure sensors, resonant pressure sensors are of high signal-noise ratio, broad frequency bandwidth and high sensitivity [10].

This study proposed a resonant pressure sensor based on magnetostrictive/piezoelectric magnetoelectric effect. The magnetostrictive/piezoelectric composite structure works as the resonator and the resonant frequency varies with the applied pressure due to magnetoelectric effect. In this paper, Section 2 presents the sensing principle and theoretical model of the sensor. Section 3 introduces the prototype and its optimization. Section 4 delivers experimental studies and determines the sensitivity of the sensor. Finally, conclusions are given in Section 5.

2. Modelling of the pressure sensor

2.1. Sensing principle

The configuration of the pressure sensor is shown in figure 1. The Terfenol-D bar is put inside the coil, which provides magnetic field. The magnetic yokes ensure the magnetic field a complete circuit. The stress generated by the magnetostriction of Terfenol-D applies on PZT-5 through mechanical coupling. An induced voltage is generated because of the piezoelectric effect in PZT-5. The measured pressure acts on the pressure applied plate. The resonant system consists of Terfenol-D, PZT-5, output rod and pressure applied plate. Due to the magnetoelectric effect, the resonant frequency of the system varies with the external pressure. Therefore the measurement of pressure is converted into measuring the resonant frequency of the system, which can be obtained by applying swept excitation to the coil.

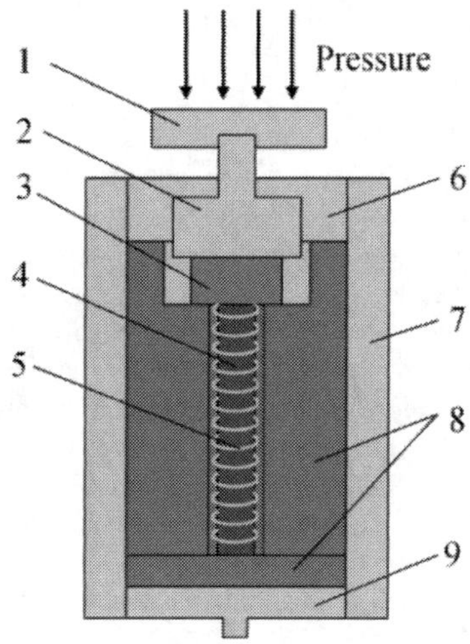

Figure 1. Configuration of the sensor. 1. Pressure applied plate; 2. Output rod; 3. PZT-5; 4. Terfenol-D; 5. Coil; 6. Front cover; 7. Housing; 8. Magnetic yokes; 9. Rear cover.

2.2. Constitutive parameters of magnetostrictive phase

The magnetostrictive constitutive equations can be expressed as

$$\begin{cases} {}^{\mathrm{m}}\varepsilon_3 = {}^{\mathrm{m}}s_{33\mathrm{e}}^{H}\,{}^{\mathrm{m}}\sigma_3 + {}^{\mathrm{m}}d_{33\mathrm{e}}\,{}^{\mathrm{m}}H_3 \\ {}^{\mathrm{m}}B_3 = {}^{\mathrm{m}}d_{33\mathrm{e}}\,{}^{\mathrm{m}}\sigma_3 + {}^{\mathrm{m}}\mu_{33\mathrm{e}}^{\sigma}\,{}^{\mathrm{m}}H_3 \end{cases} \tag{1}$$

where ${}^{\mathrm{m}}\varepsilon_3$ and ${}^{\mathrm{m}}\sigma_3$ are the longitudinal strain and stress, ${}^{\mathrm{m}}B_3$ and ${}^{\mathrm{m}}H_3$ are the magnetic flux density and magnetic field, ${}^{\mathrm{m}}s_{33\mathrm{e}}^{H}$, ${}^{\mathrm{m}}d_{33\mathrm{e}}$ and ${}^{\mathrm{m}}\mu_{33\mathrm{e}}^{\sigma}$ are the nonlinear elastic compliance at constant *H*, nonlinear piezomagnetic constant, and nonlinear magnetic permeability at constant stress. The subscript e denotes the parameters are derived from the nonlinear model of magn etostrictive phase and are equivalent to the parameters in linear constitutive equations. The nonlinear constitutive parameters are defined as

$$ {}^{\mathrm{m}}s_{33\mathrm{e}}^{H} = \frac{\partial\,{}^{\mathrm{m}}\varepsilon_3}{\partial\,{}^{\mathrm{m}}\sigma_3}, \qquad {}^{\mathrm{m}}d_{33\mathrm{e}} = \frac{\partial\,{}^{\mathrm{m}}\varepsilon_3}{\partial\,{}^{\mathrm{m}}H_3}, \qquad {}^{\mathrm{m}}\mu_{33\mathrm{e}}^{\sigma} = \frac{\partial\,{}^{\mathrm{m}}B_3}{\partial\,{}^{\mathrm{m}}H_3} \tag{2}$$

Zhou Haomiao *et al* [11] derived the nonlinear constitutive parameters based on the nonlinear model for Terfenol-D [12]. The model is expressed as follows:

$$\begin{cases} H_e = {}^{m}H_3 + \alpha M + \left[2\text{-tanh}\left(\dfrac{2\,{}^{m}\sigma_3}{\sigma_s}\right)\right]\dfrac{3\,{}^{m}\sigma_3\lambda_s M}{2\mu_0 M_s^2} \\ M = M_s\left(\coth\dfrac{H_e}{a} - \dfrac{a}{H_e}\right) \\ \lambda = \left[1\text{-}\dfrac{1}{2}\tanh\left(\dfrac{2\,{}^{m}\sigma_3}{\sigma_s}\right)\right]\dfrac{\lambda_s M^2}{M_s^2} \\ {}^{m}\varepsilon_3 = \dfrac{{}^{m}\sigma_3 - \sigma_0}{E_s} + \dfrac{\lambda_s}{2}\left[\tanh\left(\dfrac{2\,{}^{m}\sigma_3}{\sigma_s}\right)\text{-anh}\left(\dfrac{2\sigma_0}{\sigma_s}\right)\right] + \lambda \end{cases} \tag{3}$$

where H_e and M are the effective magnetic field and magnetization respectively, λ is the magnetostriction coefficient, α represents the effective field constant, a denotes the effective domain density, M_s, λ_s, σ_s and E_s are the saturation magnetization, saturation magnetostriction coefficient, saturation stress and saturation Young's modulus of the magnetostrictive material, and σ_0 is the prestress. The parameters are identified using hybrid niching coral reefs optimization algorithm (HNCRO) [13]. The results are shown in table 1.

Table 1. Model parameters identified by HNCRO.

Parameters	α	a (kA/m)	M_s (kA/m)	λ_s	σ_s (MPa)	E_s(GPa)
Values	0.031	40	915	0.0014	200	110

In equation (3), ${}^{m}\sigma_3$ is the pressure applied on Terfenol-D and can be expressed as

$${}^{m}\sigma_3 = \frac{\sigma A_a}{A_m} + \sigma_0 \tag{4}$$

where σ is the measured pressure acting on pressure applied plate, A_a and A_m are cross section areas of pressure applied plate and Terfenol-D.

2.3. Magnetoelectric equivalent circuit analysis

The equivalent circuit method is adopted to model magnetoelectric effect [14]. At high frequency, equations of motion are established to study the magneto-elasto-electric coupling of the system. Assume the applied magnetic field is sinusoidal so the vibration of the system along the longitudinal or z axis is also sinusoidal (harmonic motion). The equations of motion are expressed as

$$\begin{cases} \Delta m_m \dfrac{\partial^2 u}{\partial t^2} = \Delta\,{}^{m}\sigma_3 A_m & 0 \le z \le l_m \\ \Delta m_p \dfrac{\partial^2 u}{\partial t^2} = \Delta\,{}^{p}\sigma_3 A_p & l_m < z \le l_m + l_p \\ \Delta m_r \dfrac{\partial^2 u}{\partial t^2} = \Delta\,{}^{r}\sigma_3 A_r & l_m + l_p < z \le l_m + l_p + l_r \\ \Delta m_a \dfrac{\partial^2 u}{\partial t^2} = \Delta\,{}^{a}\sigma_3 A_a & l_m + l_p + l_r < z \le l_m + l_p + l_r + l_a \end{cases} \tag{5}$$

where u is the displacement, $\Delta m_m = \rho_m A_m \Delta z$, $\Delta m_p = \rho_p A_p \Delta z$, $\Delta m_r = \rho_r A_r \Delta z$, $\Delta m_a = \rho_a A_a \Delta z$. ρ_m and l_m are the density and length of Terfenol-D. ρ_p, A_p and l_p are the density, cross section area and length of PZT-5. ρ_r, A_r and l_r are the density, cross section area and length of output rod.

ρ_a and l_a are the density and length of pressure applied plate. The constitutive equations of PZT-5 are written as

$$\begin{cases} {}^{p}\varepsilon_3 = {}^{p}s_{33}^{E}\,{}^{p}\sigma_3 + {}^{p}d_{33}\,{}^{p}E_3 \\ {}^{p}D_3 = {}^{p}d_{33}\,{}^{p}\sigma_3 + {}^{p}\varepsilon_{33}^{\sigma}\,{}^{p}E_3 \end{cases} \tag{6}$$

where ${}^{p}\varepsilon_3$ and ${}^{p}\sigma_3$ are the longitudinal strain and stress, ${}^{p}D_3$ and ${}^{p}E_3$ are the dielectric displacement and electric field, ${}^{p}s_{33}^{E}$, ${}^{p}d_{33}$ and ${}^{p}\varepsilon_{33}^{\sigma}$ are the elastic compliance at constant E, the piezoelectric constant, and the dielectric stiffness at constant stress. The impedance equations considering magneto-elasto-electric coupling are obtained as

$$\begin{aligned} &F_{1n} = Z_{1n}\dot{u}_1 + Z_{2n}(\dot{u}_1 - \dot{u}_2) + \varphi_{mn} H_3, \\ &F_{2n} = -Z_{1n}\dot{u}_2 + Z_{2n}(\dot{u}_1 - \dot{u}_2) + \varphi_{mn} H_3 = Z_3\dot{u}_2 + Z_4(\dot{u}_2 - \dot{u}_3) + \varphi_p V, \\ &F_3 = -Z_3\dot{u}_3 + Z_4(\dot{u}_2 - \dot{u}_3) + \varphi_p V = Z_5\dot{u}_3 + Z_6(\dot{u}_3 - \dot{u}_4), \\ &F_4 = -Z_5\dot{u}_4 + Z_6(\dot{u}_3 - \dot{u}_4) = Z_7\dot{u}_4 + Z_8(\dot{u}_4 - \dot{u}_5), \\ &F_5 = -Z_7\dot{u}_5 + Z_8(\dot{u}_4 - \dot{u}_5) \end{aligned} \tag{7}$$

where

$$Z_{1n} = j\rho_m v_{mn} A_m \tan\frac{k_{mn} l_m}{2}, \quad Z_{2n} = \frac{\rho_m v_{mn} A_m}{j\sin k_{mn} l_m}, \quad Z_3 = j\rho_p v_p A_p \tan\frac{k_p l_p}{2},$$

$$Z_4 = \frac{\rho_p v_p A_p}{j\sin k_p l_p}, \quad Z_5 = j\rho_r v_r A_r \tan\frac{k_r l_r}{2}, \quad Z_6 = \frac{\rho_r v_r A_r}{j\sin k_r l_r}, \quad Z_7 = j\rho_a v_a A_a \tan\frac{k_a l_a}{2},$$

$$Z_8 = \frac{\rho_a v_a A_a}{j\sin k_a l_a}, \quad \varphi_p = \frac{A_p\,{}^{p}d_{33}}{{}^{p}s_{33}^{E} l_p}, \quad \varphi_{mn} = \frac{A_m\,{}^{m}d_{33e}}{{}^{m}s_{33e}^{H}}$$

F_{1n}, F_{2n}, F_3, F_4 and F_5 are forces at $z = 0$, $z = l_m$, $z = l_m + l_p$, $z = l_m + l_p + l_r$ and $z = l_m + l_p + l_r + l_a$ respectively. The subscript n denotes the forces at $z = 0$ and $z = l_m$ are derived considering the nonlinearity of the magnetostrictive phase. V is the induced voltage. The wave velocities are defined as $v_{mn} = \sqrt{1/\rho_m\,{}^{m}s_{33e}^{H}}$, $v_p = \sqrt{1/\rho_p\,{}^{p}s_{33}^{E}}$, $v_r = \sqrt{1/\rho_r\,{}^{r}s_{33}}$, $v_a = \sqrt{1/\rho_a\,{}^{a}s_{33}}$, where ${}^{r}s_{33}$ is the elastic compliance of output rod and ${}^{a}s_{33}$ is the elastic compliance of pressure applied plate. The wave numbers are defined as $k_{mn} = \omega/v_{mn}$, $k_p = \omega/v_p$, $k_r = \omega/v_r$, $k_a = \omega/v_a$, where ω is the angular frequency. The capacitance C_0 is defined as $C_0 = A_p/\bar{\varepsilon}_{33}$, where $\bar{\varepsilon}_{33} = {}^{p}\varepsilon_{33}^{\sigma}(1 - k_{33}^2)$. The electromechanical coupling factor $k_{33} = {}^{p}d_{33}/\sqrt{{}^{p}s_{33}^{E}\,{}^{p}\varepsilon_{33}^{\sigma}}$.

Considering the boundary conditions, one end of the system is clamped and the other end is applied by the external pressure σ. The equivalent circuit of the system is shown in figure 2.

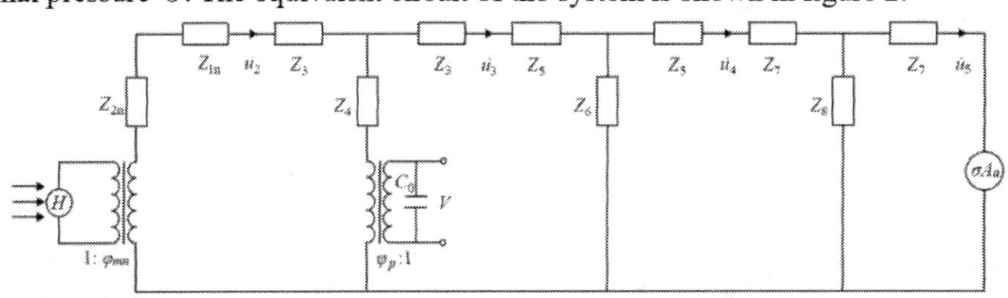

Figure 2. Magneto-elasto-electric equivalent circuit of the system

According to the Kirchhoff's voltage law and current law, the equivalent circuit can be expressed as

$$\begin{cases} \varphi_{\text{mn}}{}^{\text{m}}H_3 - (Z_{1\text{n}} + Z_{2\text{n}} + Z_3)\dot{u}_2 - \left(Z_4 + \dfrac{\varphi_{\text{p}}^2}{j\omega C_0}\right)(\dot{u}_2 - \dot{u}_3) = 0 \\ \left(Z_4 + \dfrac{\varphi_{\text{p}}^2}{j\omega C_0}\right)(\dot{u}_2 - \dot{u}_3) - (Z_3 + Z_5)\dot{u}_3 - Z_6(\dot{u}_3 - \dot{u}_4) = 0 \\ Z_6(\dot{u}_3 - \dot{u}_4) - (Z_5 + Z_7)\dot{u}_4 - Z_8(\dot{u}_4 - \dot{u}_5) = 0 \\ \sigma A_{\text{a}} + Z_8(\dot{u}_4 - \dot{u}_5) - Z_7\dot{u}_5 = 0 \\ (\dot{u}_2 - \dot{u}_3) - \dfrac{\varphi_{\text{p}} V}{\varphi_{\text{p}}^2 / j\omega C_0} = 0 \end{cases} \tag{8}$$

The output voltage V of the system is calculated as

$$V(\sigma) = \frac{\varphi_{\text{mn}}{}^{\text{m}}H_3 - \dfrac{(Z_{1\text{n}} + Z_{2\text{n}} + Z_3)Z_6 Z_8 \sigma A_{\text{a}}}{\left[(Z_7 + Z_8)(Z_5 + Z_6 + Z_7 + Z_8) - Z_8^2\right]Z_{\text{int}}}}{\dfrac{j\omega C_0 Z_4}{\varphi_{\text{p}}} + \varphi_{\text{p}} + (Z_{1\text{n}} + Z_{2\text{n}} + Z_3)\left(\dfrac{j\omega C_0}{\varphi_{\text{p}}} + \dfrac{\dfrac{j\omega C_0 Z_4}{\varphi_{\text{p}}} + \varphi_{\text{p}}}{Z_{\text{int}}}\right)} \tag{9}$$

where

$$Z_{\text{int}} = Z_3 + Z_5 + Z_6 - \frac{Z_6^2}{Z_5 + Z_6 + Z_7 + Z_8 - \dfrac{Z_8^2}{Z_7 + Z_8}}$$

According to equation (9), the resonant angular frequency of the system ω_S under different σ can be derived as

$$\omega_{\text{s}}(\sigma) = \omega\left[V_{\max}(\sigma)\right] \tag{10}$$

where $V_{\max}(\sigma)$ is the maximum of V (resonant voltage) under σ and $\omega_S(\sigma)$ is the corresponding angular frequency.

3. Prototype of the sensor and optimization

The prototype of the sensor is shown in figure 3. The PZT (PA4FKW) is supplied by Thorlabs China with dimensions of $5 \times 5 \times 3$ mm^3 and the Terfenol-D is supplied by Gansu Tianxing Co. Ltd with dimensions of $14 \times 2.5 \times 2.5$ mm^3. The parameter values of the prototype are listed in table 2. As part of the resonant system, the pressure applied plate largely affects the resonant frequency and performance of the sensor. By optimizing the design of the pressure applied plate, high sensitivity of the sensor can be achieved and corresponding optimal resonant frequency is determined.

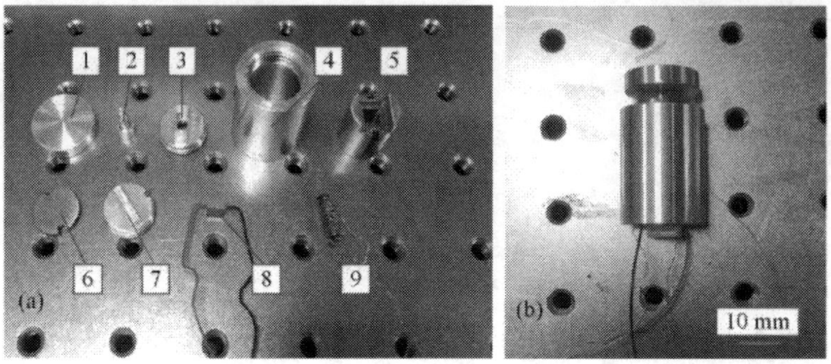

Figure 3. Prototype of the sensor. (a) 1. Pressure applied plate; 2. Output rod; 3. Front cover; 4. Housing; 5 and 6. Magnetic yokes; 7. Rear cover; 8. PZT-5; 9. Terfenol-D and coil. (b) Assembly.

Table 2. Parameter values of the prototype.

Parameters	Values	Parameters	Values
ρ_m	9230 kg/m^3	l_p	3 mm
ρ_p	7600 kg/m^3	l_r	10 mm
ρ_r, ρ_a	2750 kg/m^3	${}^{p}s_{33}^{E}$	20.7×10^{-12} m^2/N
A_m	6.25 mm^2	${}^{p}d_{33}$	400 pC/N
A_p	25 mm^2	${}^{p}\varepsilon_{33}/\varepsilon_0$	1750
A_r	28.3 mm^2	${}^{r}s_{33}$, ${}^{a}s_{33}$	14.5×10^{-12} m^2/N
l_m	14 mm		

Pressure applied plates with different dimensions are studied and the parameters are listed in table 3. Based on the theoretical model, the resonant frequency f_s under different σ can be obtained. For sensors with pressure applied plates of different dimensions, the output curves can be calculated and the results are shown in figure 4.

It is found as the diameter of the pressure applied plate increases, the variation of f_s increases. It is because σA_a, i.e. the force applied on the sensor increases, thus the corresponding variation in the equivalent circuit increases. In addition, as the height of pressure applied plate increases, the variation of f_s decreases. The reason is that the mass of the resonant system increases so f_s decreases and its variation decreases as well. To ensure high sensitivity, considering the overall size of the sensor and the mechanical strength of the pressure applied plate, the optimal size is determined to be $\varphi 20 \times 5$ mm^2.

Table 3. Pressure applied plates with different dimensions.

Number	Dimensions (mm^2)	Mass (g)	Number	Dimensions (mm^2)	Mass (g)
1	$\varphi 10 \times 5$	1.08	7	$\varphi 15 \times 15$	7.29
2	$\varphi 10 \times 10$	2.16	8	$\varphi 15 \times 20$	9.72
3	$\varphi 10 \times 15$	3.24	9	$\varphi 20 \times 5$	4.32
4	$\varphi 10 \times 20$	4.32	10	$\varphi 20 \times 10$	8.64
5	$\varphi 15 \times 5$	2.43	11	$\varphi 20 \times 15$	12.96
6	$\varphi 15 \times 10$	4.86	12	$\varphi 20 \times 20$	17.28

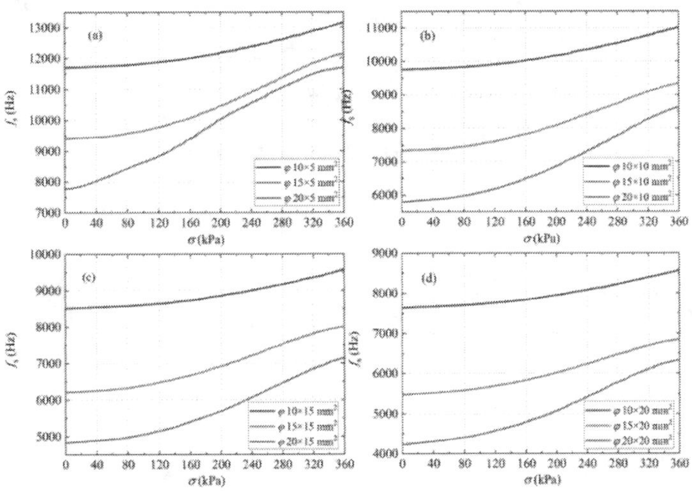

Figure 4. Output curves of sensors with pressure applied plates of different dimensions.

Based on the above results, it is found the resonant frequency of the sensor decreases as the mass of the plate increases. The resonant frequencies at $\sigma = 0$ with pressure applied plates of different masses are shown in figure 5. It is indicated that the proposed sensor has the potential to be applied in mass measurement.

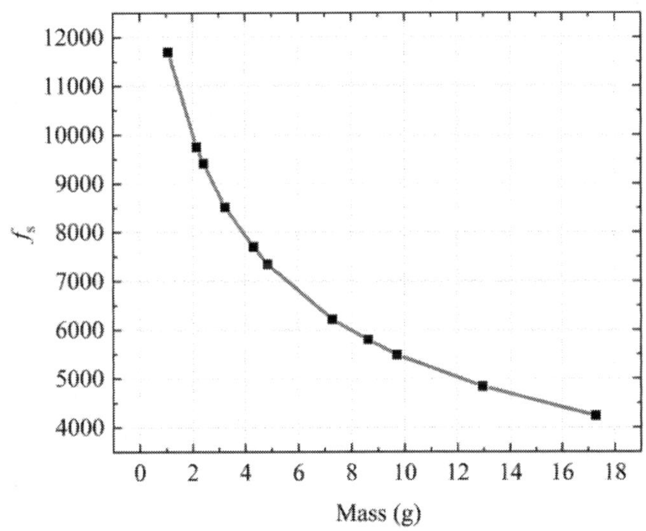

Figure 5. Resonant frequency of the sensor as the mass of the pressure applied plate changes.

4. Experimental studies

With the configurations and parameters determined in Section 3, the performance of the sensor is tested by experiments. The experimental platform is shown in figure 6. The pressure sensor is clamped with a jig. A force gauge (HANDPI Instruments: HP-500) applies and measures the pressure σ acting on the sensor. The input current is sent to the coil after being amplified by the power amplifier (NF: BP4610). The output signal V_{out} is acquired using a lock-in amplifier (Zurich Instruments: MFLI) and sent to PC.

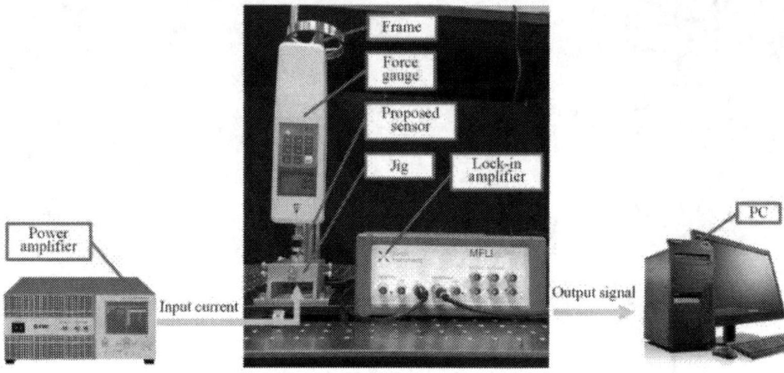

Figure 6. Experimental platform

The range of the applied pressure is 0-360 kPa where the output of the sensor is optimal. Based on theoretical analysis, the output tends to be saturated above 360 kPa. Under different σ, V_{out} of the sensor is acquired by MFLI in sweep frequency mode and corresponding f_s is obtained. The results of sweep frequency are shown in figure 7. Based on above results, the σ vs. f_s curve of the sensor is shown in figure 8.

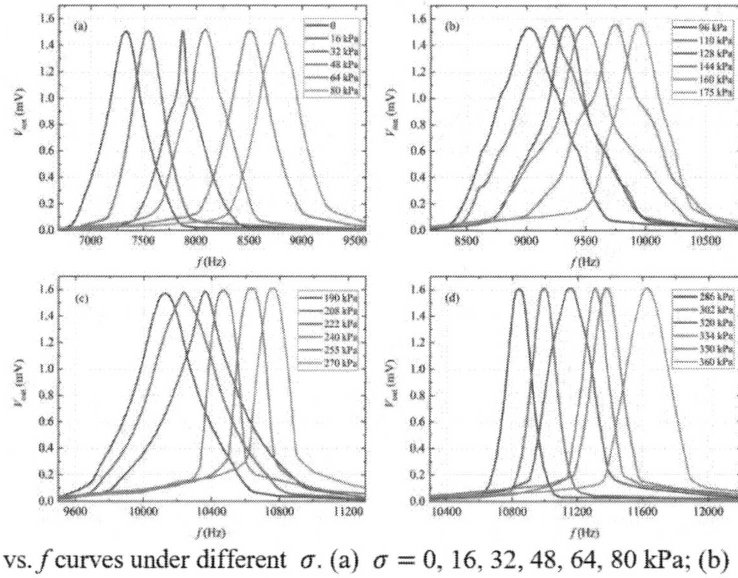

Figure 7. V_{out} vs. f curves under different σ. (a) $\sigma = 0$, 16, 32, 48, 64, 80 kPa; (b) $\sigma = 96$, 110, 128, 144, 160, 175 kPa; (c) $\sigma = 190$, 208, 222, 240, 255, 270 kPa; and (d) $\sigma = 286$, 302, 320, 334, 350, 360 kPa.

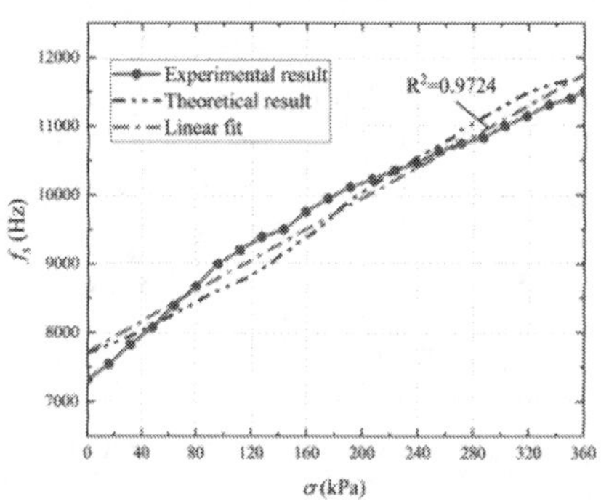

Figure 8. σ vs. f_s curve of the sensor

The linear fit of the experimental curve is shown in figure 8 as well and the R-square value is 0.9724. The sensitivity of the sensor is determined to be 11.63 Hz/kPa. The experimental result shows good agreement with the theoretical one. The deviation might be caused by the assembly error of the sensor and the extra prestress applied by the front cover.

5. Conclusions

In this study, a resonant pressure sensor based on magnetostrictive/piezoelectric magnetoelectric effect is designed, fabricated and tested. The resonant system consists of Terfenol-D, PZT-5, output rod and pressure applied plate. The external pressure acting on the pressure applied plate changes the resonant frequency of the system. Nonlinear model of the magnetostrictive phase is established to evaluate the effect of pressure to the response of the system. Combining with the equivalent circuit method, the resonant frequency of the sensor under different pressure is obtained. The optimal parameters of pressure applied plate are determined to be $\varphi 20 \times 5\ \text{mm}^2$ to improve the performance of the sensor. In the measurement range of 0-360 kPa, the sensitivity of the sensor is 11.63 Hz/kPa and the experimental result is close to the theoretical result. Future work is focused on further studying the performance of the sensor, reducing the dimensions and exploring the applications in Non-destructive Testing, mass measurement and other fields.

Acknowledgement

This work is supported by National Natural Science Foundation of China (51775349) and National Key R&D Program of China (2017YFF0108000), for which the authors are most grateful.

References

[1] Yang, Y, Yang, B and Niu, M (2018). Dynamic/static displacement sensor based on magnetoelectric composites. *Applied Physics Letters*, 113(3), 032903.

[2] Nan, C. W, Bichurin, M. I, Dong, S, Viehland, D and Srinivasan, G (2008). Multiferroic magnetoelectric composites: historical perspective, status, and future directions. *Journal of Applied Physics*, 103(3), 31101.

[3] Wang, Y, Li, J and Viehland, D (2014). Magnetoelectrics for magnetic sensor applications: status, challenges and perspectives. *Materials Today*, 17(6), 269-275.

[4] Leung, C. M, Or, S. W, Zhang, S and Ho, S. L (2010). Ring-type electric current sensor based on ring-shaped magnetoelectric laminate of epoxy-bonded $Tb_{0.3}Dy_{0.7}Fe_{1.92}$ short-fiber/NdFeB magnet magnetostrictive composite and $Pb(Zr,Ti)O_3$ piezoelectric ceramic. *Journal of Applied Physics*, 107(9), 149.

[5] Myers, R, Islam, R. A, Karmarkar, M and Priya, S (2007). Magnetoelectric laminate composite based tachometer for harsh environment applications. *Applied Physics Letters*, 91(12), 122904.

[6] Petrov, R. V, Bichurin, M. I and Leontiev, V.S (2016) The crankshaft position sensor based on magnetoelectric materials. *Proceedings of the Power Electronics and Motion Control Conference*, 848-853.

[7] Xi, H, Qian, X, Lu, M. C, Mei, L, Rupprecht, S and Yang, Q. X, et al (2016). A room temperature ultrasensitive magnetoelectric susceptometer for quantitative tissue iron detection. *Scientific Reports*, 6, 29740.

[8] Borole, U. P, Sasikala, S, Kulkarni, I. R, Saravanan, P, Barshilia, H. C and Chowdhury, P (2018). Highly sensitive giant magnetoresistance (GMR) based ultra low differential pressure sensor. *Sensors and Actuators A: Physical*, 280, 125-131.

[9] Ko, J. M and Ni, Y. Q (2005). Technology developments in structural health monitoring of large-scale bridges. *Steel Construction*, 27(12), 1715-1725.

[10] Tian, B, Zhao, Y and Jiang, Z (2010). The novel structural design for pressure sensors. *Sensor Review*, 30(4), 305-313.

IOP Conf. Series: Materials Science and Engineering **825** (2020) 012037 doi:10.1088/1757-899X/825/1/012037

[11] Zhou, H. M, Ou, X. W, Xiao, Y, Qu, S. X and Wu, H. P (2013). An analytical nonlinear magnetoelectric coupling model of laminated composites under combined pre-stress and magnetic bias loadings. *Smart Materials and Structures*, 22(3), 035018.

[12] Zheng, X. J and Liu, X. E (2005). A nonlinear constitutive model for Terfenol-D rods. *Journal of Applied Physics*, 97(5), 053901.

[13] Yang, Y, Yang, B and Niu, M (2017). Parameter identification of Jiles–Atherton model for magnetostrictive actuator using hybrid niching coral reefs optimization algorithm. *Sensors and Actuators A: Physical*, 261, 184-195.

[14] Dong, S, Li, J. F and Viehland, D (2003). Longitudinal and transverse magnetoelectric voltage coefficients of magnetostrictive/piezoelectric laminate composite: theory. *IEEE Transactions on Ultrasonics, Ferroelectrics and Frequency Control*, 50(10), 1253-1261.

IOP Conf. Series: Materials Science and Engineering **825** (2020) 012038 doi:10.1088/1757-899X/825/1/012038

Output Force Control of a Pneumatic Soft Gripper with a Jointed Endoskeleton Structure

Zhaoping Wu[*]**, Xiaoning Li and Zhonghua Guo**

Nanjing University of Science and Technology, School of Mechanical Engineering, 210094 Nanjing P.R. China

[*]895349885@qq.com

Abstract. Soft grippers are suitable for gripping various objects without damage owing to their high flexibility. But current soft grippers have a general problem of insufficient gripping force. To grip objects with different mass and sizes on assembly line safely and reliably, a novel pneumatic soft gripper with a jointed endoskeleton structure was developed and fabricated successfully. The actuation and force bearing functions of the gripper were separated to increase the gripping force. Moreover, to control the output force of the finger tip of the soft gripper, a fuzzy auto-tuning PID controller was designed and the relevant parameters of the controller were determined. Furthermore, a test platform with the designed fuzzy auto-tuning PID controller was developed to control the output force of the finger tip. The test results showed that the controller was able to maintain the desired target output force with a maximum deviation of 0.6 N. At last, the soft gripper was used to grip material objects for testing. From the experiment, it indicated that the output force of the finger tip can reach up to the desired values accurately and various objects can be gripped safely and reliably.

1. Introduction

Grippers act as an important part of modern industrial manufacturing lines to grip and transport workpieces automatically. With the improvement of productivity level and labor efficiency, new manufacturing technologies such as mixed flow production put forward higher requirements to the functional diversity of grippers, especially to grip objects with different mass and sizes safely and reliably. The soft robotics has become a current research focus, in which pneumatic soft gripper has been applied widely owing to its high flexibility, low fabrication process, rapid response and excellent environment adaptability [1-2].

A series of pneumatic soft actuators have been developed at home and abroad. The concept of soft robotics was proposed firstly by Harvard University. Ilievski et al. developed a fully soft starfish-like gripper fabricated by PneuNets technology, which was used for gripping small objects such as eggs and guinea pigs without damage [3]. A rapidly bending pneumatic actuator was designed by Mosadegh. The inner chamber was divided into two parts to expand and extrude each other, which caused the bending motion [4]. Such fully soft grippers act as both actuators and executors, which lead to the low clamping force and load capacity. Meanwhile, the soft material lacks stiffness which will generate undesired deformation such as radial expansion and irregular stretch. Further, a composite fluidic actuator with a circumferential fiber-reinforced structure has been studied by Polygerinos et al. With the circumferential fiber-reinforced structure, radial expansion could be effectively reduced such that the rigidity and gripping force could be increased [5]. But local unexpected deformations still exist, which consumes the energy and the gripping force is insufficient.

Furthermore, gripping the target object safely and reliably is the most basic function of a gripper so that the output force control of the gripper is necessary. Guoliang Zhong of Central South University proposed a novel soft pneumatic dexterous gripper that consisted of four soft fingers and a movable sucker. The gripper had four grasping modes namely as perpendicular mode, parallel mode, two-finger mode, and sucker mode [6]. John Morrow achieved low steady-state error and overshoot in position and force of a soft pneumatic actuator with rigid fingernails using feed-forward models that related pressure, force, and curvature along with a PID controller [7]. Useok Jeong proposed the use of a position-based impedance controller to control a soft wearable robotic hand with a slack enabling tendon actuator [8]. Minou Kouh Soltani introduced a nonlinear hybrid force/position control strategy for soft tendon driven catheters and the catheter was capable of regulating the force of 4.9 mN [9]. At present, the research of the control of pneumatic soft grippers, especially the control of force is comparatively little and the control method should be improved.

Therefore, it is crucial to increase the gripping ability, decrease or even limit the unnecessary deformation of the gripper and to control the output force accurately. To solve the above-mentioned problems, a novel pneumatic soft gripper with a jointed endoskeleton structure which separated the actuation and force bearing function has been proposed and fabricated successfully. Besides, a fuzzy auto-tuning PID controller is applied on the control of the output force of the finger tip to increase the force accuracy and to grip the target object safely, reliably and precisely.

2. Structure

A jointed endoskeleton structure is proposed in this paper. The design principle is that: an endoskeleton consist of several jointed skeleton units acts as the force bearing mechanism; the outer actuation part made of silicon rubber with an embedded radial-restrained fiber achieves the bending motion by pressurized. Thus the gripping force can then be increased and high flexibility can be maintained.

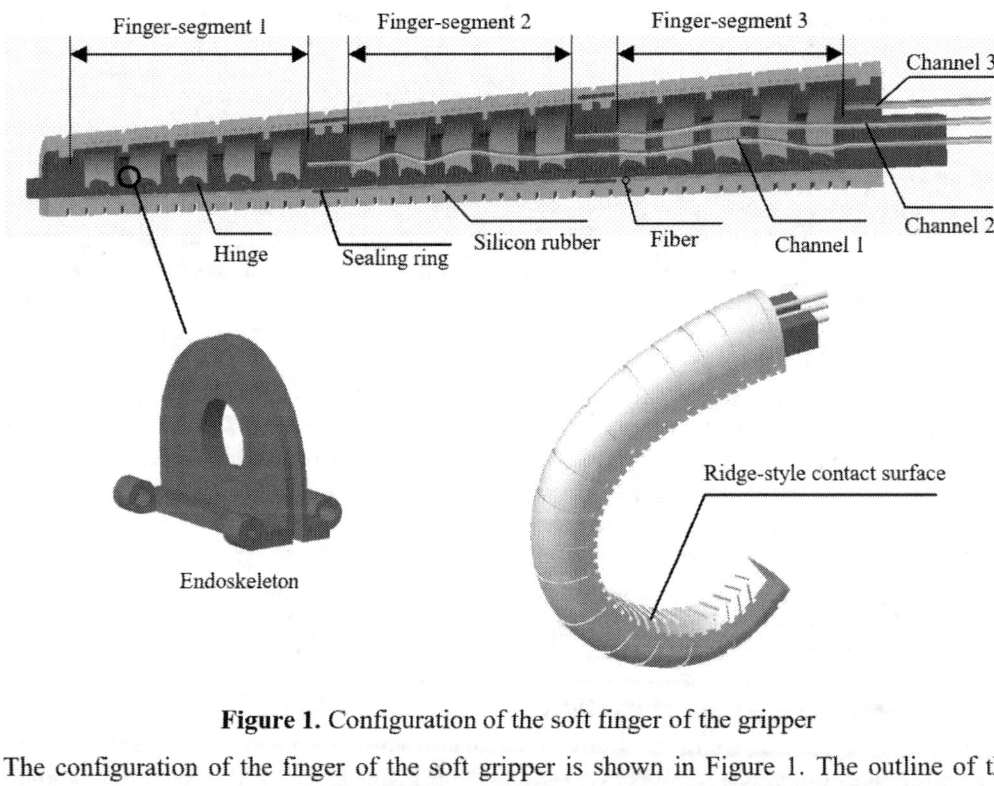

Figure 1. Configuration of the soft finger of the gripper

The configuration of the finger of the soft gripper is shown in Figure 1. The outline of the soft finger imitates the octopus tentacles. A smooth taper shape is formed from the fingertip to the finger root with a size that increases gradually. The shape of the contact surface of the finger is the ridge style which can increase the friction when gripping. The endoskeleton consists of a serious of skeleton units that can rotate from − 45° to 180° at the hinge joints. The outer silicon rubber layer is bonded with the endoskeleton by super soft glue, which constitutes a series of inner chambers. To increase the operational efficiency and realize a greater number of postures, the entire finger is divided into three independent finger segments that can be independently pressurized using different air pressures to control the bending angle. Each finger segment consists of 5 finger joints that work together to acquire the bending motion of one finger segment.

When the soft finger works, compressed air flows through the channels from the finger root to control three finger segments, and the air pressure is marked as p1, p2, and p3, respectively, for the three finger segments. Bending deformation occurs at each finger joint under the air pressure and the tension of the silicon rubber. The expected output bending angle and torque could be obtained. The total bending angle of the finger is the sum of each finger joint's bending angle, and the total output torque of the finger is the sum of each finger joint's torque.

The soft gripper is set up by three soft fingers through the palm. The three fingers are evenly distributed on the circumference of the palm. The angle between the two adjacent fingers is 120°, and the angle between the axis of a finger and that of the palm is 35°. The palm comprises flow channels passing to the fingers. The operation of the soft gripper is convenient and both enveloping gripping mode and fingertip gripping mode can be chosen.

3. Fabrication

The soft gripper is made of several materials. Light and rigid material should be used for the endoskeleton so that fatigue damage is not apt to occur after repeated bending. Then PolyPlusTM (a polylactic acid material produced by Polymaker) is used for making the endoskeleton by 3D printer. The actuating layer is made of Smooth-on Dragon Skin silicon rubber with the Shore-hardness of 10 A. The palm is also fabricated by 3D printed so that its streamline surface can fix the finger well. Fibers are used for the embedded reinforced layer to limit the radial deformation.

The soft gripper is fabricated through modular components. The fabrication process is shown in Figure 2. First, two skeleton units are linked by two cylindrical pins and a spring [Figure 2(a)]. A series of skeleton unties composes the endoskeleton [Figure 2(b)]. Then three air tubes are embedded into the endoskeleton so as to control different finger-segment independently [Figure 2(c)]. The soft actuating layer is cast through three steps. In the first step, one silicon rubber layer is glued surrounding the endoskeleton [Figure 2(d)]. In the second step, circumferential fiber is evenly winded on the surface [Figure 2(e)]. In the last step, the outer silicon rubber layer is glued to form the finger [Figure 2(f)]. At last three fingers are connected with the palm to complete the fabrication of the soft gripper [Figure 2(g)]. Figure 2(h) shows the bending state of the soft gripper.

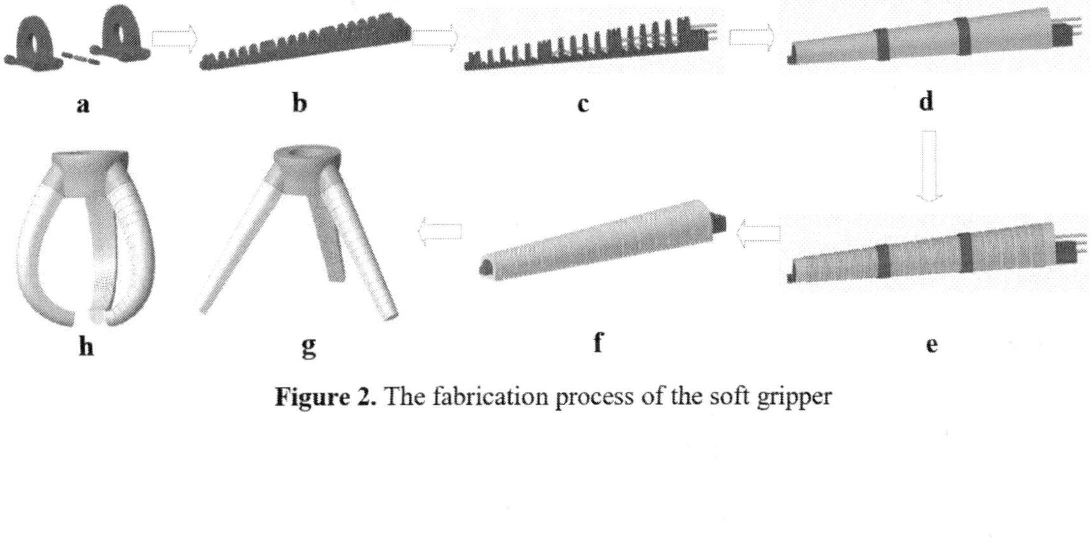

Figure 2. The fabrication process of the soft gripper

IOP Conf. Series: Materials Science and Engineering **825** (2020) 012038 doi:10.1088/1757-899X/825/1/012038

4. Control algorithm

The pneumatic soft gripper with jointed endoskeleton can be used to grip multiple objects with different masses and shapes safely and reliably. Gripping methods include enveloping gripping and fingertip gripping. Fingertip gripping is also known as accurate gripping. The fingertip touches the surface of the object and the object is clamped. The object relies on the fingertip to realize a stable state. High dexterity, reliable accuracy and more operation styles will be acquired under fingertip gripping. It requires accurate control of the output force of the finger tip. Because the soft gripper is a coupled pneumatic system which consists of large nonlinear deformation and multiple materials, the mathematical model is complex and difficult to achieve the desired control effect. Therefore, an algorithm to design the controller without a mathematical model of the system is needed.

The PID controller has been used in the industrial automation field because of the advantages of the simple principle, convenient application and strong adaptability. The control variables of PID consist of proportion (P), integration (I) and differentiation (D) of the control deviation linearly. The control law can be described as

$$u(t) = K_p \left\{ e(t) + \frac{1}{T_i} \int_0^t e(t)\mathrm{d}t + \frac{T_D \mathrm{d}e(t)}{\mathrm{d}t} \right\} \quad (1)$$

where K_p is the coefficient of proportion; T_i is the coefficient of integral time; T_D is the coefficient of differential time.

The tuning of PID parameters is the core content of the control system design. The regulation quality and robustness of the controller are influenced by the quality of parameters directly. A typical PID controller can only be controlled by a group of fixed parameters, which can not take consideration of the contradiction between the dynamic and static performance of the system, the set value and disturbance suppression. Therefore, an intelligent control system should be introduced into the controller to improve the dynamic and static performance of the system by adding the tuning of parameters based on the initial PID values. The fuzzy control algorithm is robust and insensitive to the change of process parameters, which is suitable for the nonlinear time-varying system with large delay nonlinear, In this paper, a fuzzy auto-tuning algorithm is adopted to realize the automatic online optimal adjustment of PID parameters [10-11]. The control scheme is shown in Figure 3.

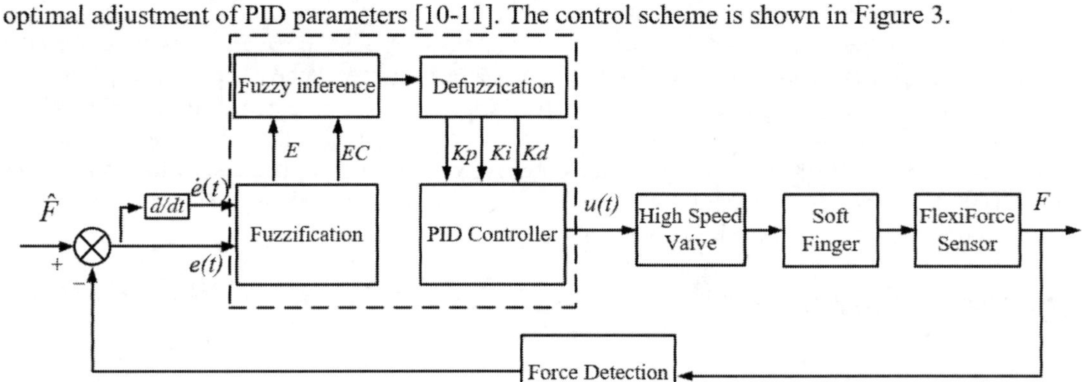

Figure 3. Scheme of the fuzzy auto-tuning PID controller

The fuzzy auto-tuning PID controller is double input and three output system. The force error *e(t)* and the change rate of force error $\dot{e}(t)$ is the input language variable. The PID parameters *Kp, Ki, Kd* act as the output language variable, which is adjusted in real-time according to the fuzzy inference rules. The output of the PID controller *u* is the duty cycle of control frequency of the high speed on-off valve, which is used to control the air inflow of the soft finger.

In order to reflect the dynamic characteristics of the output variables of the control system better, according to the measurement range of the flexible pressure sensor and the actual opening and closing characteristics of the high speed on-off valve, etc., the fuzzy set of *e(t)* is defined to be {NB, NM, NS, ZO, PS, PM, PB}, the value range is [-5 5], the fuzzy set of $\dot{e}(t)$ is defined to be {NB, NS, ZO, PS, PB}, the value range is [-2 2], the fuzzy set of K_p, K_i, K_d is {NB, NM, NS, ZO, PS, PM, PB}, the value range of K_p is [-10 -3 -0.5 0 0.5 3 10], the value range of K_i is [-50 -15 -1 0 1 15 50], the value range of K_d is [-1 -0.2 -0.01 0 0.01 0.2 1]. Considering the stability, overshoot, response speed and steady-state error of the algorithm synthetically, the tuning principle of the PID parameters is as follows:

(1). When $|e(t)|$ is large, the value of K_p is larger, the value of K_d is smaller, thus the better tracking performance of the algorithm can be achieved; meanwhile, the value of K_i is taken 0 to avoid large overshoot.

(2). When $|e(t)|$ is middle, the value of K_p should be smaller to control the overshoot. Then the value of K_d has large influence on the response speed of the system, the value of K_i should be proper.

(3). When $|e(t)|$ is small, K_p and K_i should be larger to achieve better system stability. The value of K_d is determined by $|\dot{e}(t)|$.

According to the above principles, the fuzzy inference rules are built as shown in Table 1. Fuzzy control rules of Kp, Ki, Kd lay from left to right in each cell.

Table 1. The fuzzy inference rules

$\dot{e}(t)$	*e(t)*						
	NB	NM	NS	ZO	PS	PM	PB
NB	PM/ZO/PS	PS/NS/PM	PM/NM/PM	PB/NB/ZO	NM/PM/ZO	NM/PS/PM	NM/ZO/PS
NS	PB/ZO/NB	PM/ZO/NB	PB/NB/NS	PB/NB/NS	NB/PB/NS	NM/ZO/PB	NB/ZO/PB
ZO	PB/ZO/NB	PM/ZO/NB	PB/NB/NS	PB/NB/ZO	NB/PB/NS	NM/ZO/PB	NB/ZO/PB
PS	PB/ZO/NB	PM/ZO/NB	PB/PB/NS	PB/NB/NS	NB/PB/NS	NM/ZO/PB	NB/ZO/PB
PB	PM/ZO/PS	PS/NS/PM	PM/PM/ZO	PM/NB/ZO	NM/PM/ZO	NS/PS/PM	NM/ZO/PS

The membership functions of each fuzzy state are generally symmetric triangle, symmetric trapezoid and normal membership functions. The shape of the triangle membership function is only related to the slope of its line, the operation is simple and the memory space is small, thus it is suitable for on-line tuning fuzzy control with membership function. Here the triangle style is selected as the membership function of the language variable.

The coefficient weight average method is used to solve fuzzification. The formula is as follows:

$$u(t) = K\frac{\sum_{i=0}^{m}\sum_{j=0}^{n} c_{ij}u_i[e(t)]u_j[\dot{e}(t)]}{\sum_{i=0}^{m}\sum_{j=0}^{n} u_i[e(t)]u_j[\dot{e}(t)]} \tag{2}$$

where *u(t)* is the control signal of the high speed on-off valve; c_{ij} is the weight; $u_i[e(t)]$ and $u_j[\dot{e}(t)]$ are the values of membership functions of force error and its change rate respectively.

IOP Conf. Series: Materials Science and Engineering **825** (2020) 012038 doi:10.1088/1757-899X/825/1/012038

5. Experiment

5.1 Control test of the output force of fingertip

The prototype of the soft gripper was fabricated according to the previous design structure. After the successful design of the controller, the test platform was set up to control the precision of the fingertip output force. The flexible force sensor (FlexiForce a201-25) was selected to collect the force signal of fingertip. Different from the general force sensor, the flexible force sensor can deform with the soft actuator and output the force value on the basis of not damaging its electronic performance, which is very suitable for the soft gripper with high flexibility. Three pressure sensors (SMC ISE4-01-26) were used to collect the pressure of the air chamber of three finger segments. The high speed on-off valve (Festo P7944411) was used to control the air inflow of the soft finger.

The soft finger was placed horizontally on the test platform and the flexible force sensor was bonded on the finger tip. The bending deformation of the finger tip was limit to zero and the output force of the fingertip was controlled in this condition. Matlab Simulink software was used to build the control system. The signals of the flexible force sensor collected through the integrated circuit board of the industrial control computer in real-time were input to fuzzy auto-tuning PID controller after modulation, amplification and filtering. The controller output duty cycle signal of PWM to control the high speed on-off valve. Meanwhile, the pressure signals of each finger segment were collected. The schematic diagram of the test control circuit is shown in Figure 4.

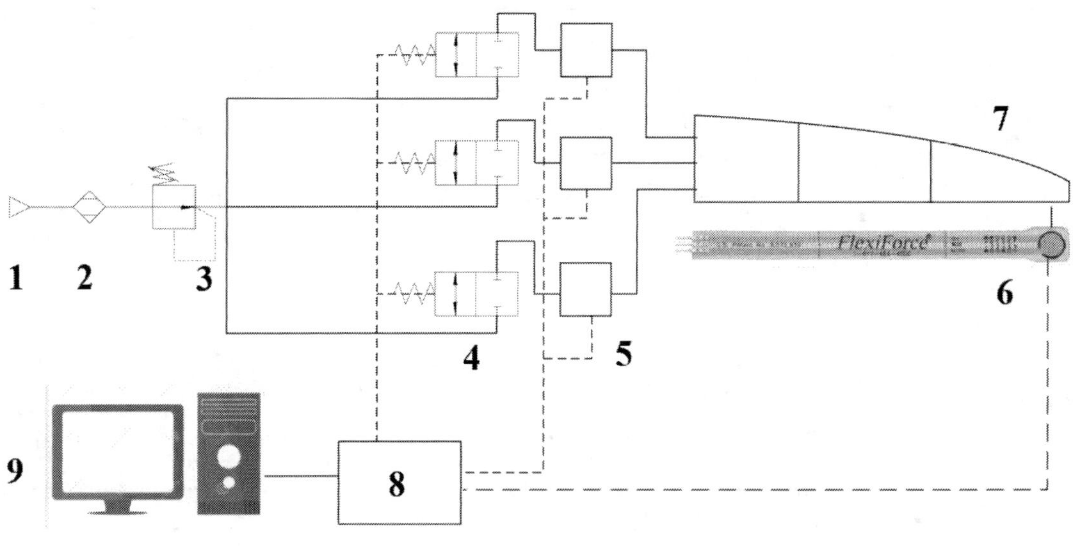

Figure 4. Circuit diagram of the control test for output force. 1.Air source; 2.Dryer; 3. Reducing valve; 4. High speed on-off valve; 5. Pressure sensor; 6.Flexible force sensor; 7. Soft finger; 8. Integrated circuit board; 9. IPC.

The output force values of 5 N, 10 N and 15 N are chosen to test. The curves of the response of the output force were acquired and drawn through Simulink, as shown in Figure 5. The stability time of the system is about 2s which is not influenced by the increase of the output force greatly. The initial overshoot is relatively large to increase the response speed. The maximum deviation final of the final output force is maintained within 0.6 N. It indicates that the system is satisfied with the actual control requirements.

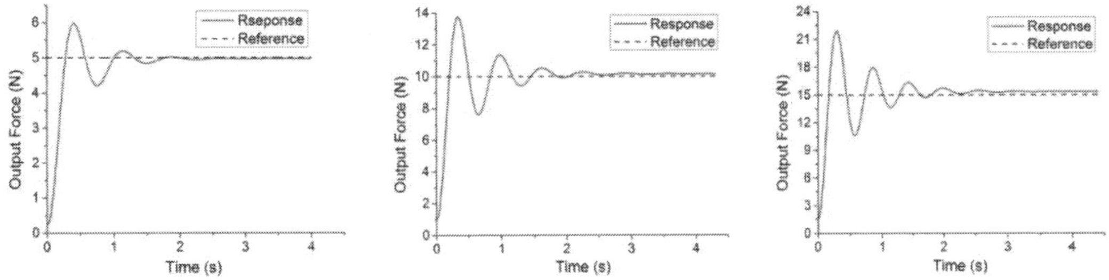

Figure 5. Response of the output force of finger tip

5.2 Objects gripping demonstration

Gripping objects is the most basic performance of the gripper. In order to test the gripping performance of the soft gripper, fingertip gripping experiments were carried out on objects with different shapes, sizes and weights. The output force of the three fingers of the soft gripper was set to the same value and the output force values were planned and determined beforehand. The parameters of the test objects and corresponding output force values of the fingertip are shown in Table 2.

Table 2. The parameters of the test objects and corresponding output force

Name	Size (mm)	Mass (g)	Material	Output force (N)
Golf ball	D42	46	Rubber	3
Cube	56*56*56	80	Plastic	6.5
Bulb	About D130	98	Glass	8
Spirit	About D56*164	196	Plastic	14

Photographs of the soft gripper when gripping test objects are shown in Figure 6. Every test object was gripped by the soft gripper successfully according to the planned output force of fingertip. The gripping was stable and reliable which proved the correctness and effectiveness of the control algorithm of the output force.

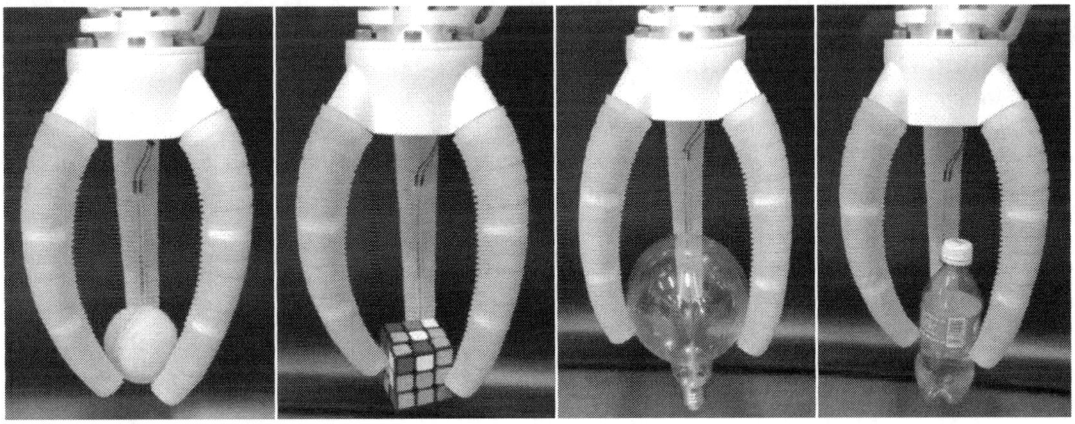

a. Golf ball **b. Cube** **c. Bulb** **d. Spirit**

Figure 6. Photographs of the gripping experiment

6. Conclusion

A novel pneumatic soft gripper with a jointed endoskeleton structure is proposed and fabricated successfully, in which, the actuation and force bearing functions can be separated. The bending action is performed by the soft rubber with an embedded radial-restrained fiber through pneumatic actuation, while the gripping force is borne and transferred by the jointed endoskeleton. The gripping force can then be increased, and high flexibility can be maintained.

A fuzzy auto-tuning PID controller is proposed and used to control the output force of the finger tip. Every parameter is debugged and determined. Through the fuzzy auto-tuning PID controller, the overshoot of the system is lower, the loading response time is shorter and the precision of the output force is higher.

A test platform is built and the closed-loop control of output force is conducted with the designed fuzzy auto-tuning PID controller. Through experiments, the maximum deviation is maintained within 0.6N. Moreover, the gripping test of several objects is carried out with the planned output force, which indicates that the soft gripper can complete the gripping operation safely and reliably.

Acknowledgements

This research was supported by National Natural Science Foundation of China (Grant No. 51305202) and Jiangsu Provincial Natural Science Foundation of China (Grant No. BK20130764).

References

[1] Kim S, C Laschi and B Trimmer 2013 soft robotics: a bioinspired evolution in robotics *Trends Biotech.* **31** 287–294
[2] Cao Y, Shang J and Liang K 2012 review of soft-bodied robots *J. Mech. Eng.* **48** 25–33
[3] Ilievski F, Mazzeo A D and Shepherd R F 2015 soft robotics for chemists *Angew. Chem.* **123** 1930–35
[4] Mosadegh B, Polygerinos P and Keplinger C 2014 soft robotics: pneumatic networks for soft robotics that actuate rapidly *Adv. Funct. Mater* , **24** 2109
[5] Polygerinos P, Wang Z and Overvelde J T B 2015 modeling of soft fiber-reinforced bending actuators *IEEE. T. Robot* **31** 778–789
[6] Zhong G, Hou Y and Dou W 2019 a soft pneumatic dexterous gripper with convertible grasping modes. *Int. J. Mech. Sci.* **153–154** 445–456
[7] John M, Shin H S, Calder P G, Jang S H, Jacob T, Riley L, Steven D, Park Y L and Dmitry B 2016 improving soft pneumatic actuator fingers through integration of soft sensors, position and force control, and rigid fingernails. 2016 *Int. Conf. on Robotics and Automation (ICRA)* (Stockholm: IEEE) p 16–21
[8] Useok J, Hyunki I, Haemin L, Brian B K and Cho K J 2015 investigation on the control strategy of soft wearable robotic hand with slack enabling tendon actuator *Int. Conf. on Robotics and Automation (ICRA)* (Seattle: IEEE) p 5004–09
[9] Soltani M K, Sohrab K, Farzan G and Farrokh J 2017 a soft robotics nonlinear hybrid position/force control for tendon driven catheters *Int. J. Control. Autom. Syst* **15** 54–63
[10] Wang S Y, Shi Y and Feng Z X 2011 a method for controlling a loading system based on a fuzzy PID controller *Mech. Sci. Tec. Aero. Eng* **30** 166–172
[11] Zhou Y F and Zhang Q C 2011 experimental study on self-adjusting fuzzy-PID control ofsynchro force for pneumatic servo system Chin. Hydr. Pneu. 9 19–22

IOP Conf. Series: Materials Science and Engineering **825** (2020) 012039 doi:10.1088/1757-899X/825/1/012039

A New Method to Identify the Diesel Engine Modal Parameters based on Self-Excitation

Biao Hou[1], Xing Liu[1,*], Jiheng chen[2], Feng Liu[1], Ye Tian[1] and Jingsong Tan[2]

[1] Diesel engine power plant teaching and research office, Naval Petty Officer Academy, Bengbu 230000, PR China
[2] Mechanical and electrical equipment maintenance teaching and research office, Naval Petty Officer Academy, Bengbu 230000, PR China

[*]E-mail address: liuxing.3n8b@163.com

Abstract. At present, the methods to study the dynamic characteristics of diesel engine structure include finite element analysis method and experimental analysis method. Among , 60% of the structural stiffness and 90% of the damping ratio of the whole diesel engine come from the joint part, and the dynamic characteristics of the joint part of the diesel engine are very complex, so it is difficult to establish an accurate finite element model, and the results of finite element analysis methods are often not accurate. For the experimental analysis method, it is mainly divided into the traditional Experimental Modal Analysis method (EMA) and the Operational Modal Analysis (OMA). The EMA it alays needs to measure the excitation force of the diesel engine and the vibration response of the structure, simultaneously. And the influence of motion factors on the structure dynamics cannot be considered. The structure is excitated by the working environment, and only analyzes the modal parameters of the structure with the vibration response.Few scholars analyze the OMA for diesel engines.This paper proposes a new method to identify the modal parameters of diesel engine based on dry run. In this paper, the design and research of the motion law of the air-running self-excitation is carried out, and the validity of the method is verified.

1. Introduction

Diesel engines are developing towards the direction of high power, lightweight, reducing fuel consumption and emission, etc., but the vibration and noise problems are increasingly prominent[1, 2]. It mainly show in the components wear increasely, reduce the service life, higher noise, and deterioration of working conditions.Under the periodic change of dynamic loading, the main parts of diesel engine can cause a strong resonance,and the dynamic stress and dynamic displacement will be several times or even dozens of times as much as static load,and it seriously affected the reliability of the diesel engine.Since it is great significance to study the diesel engine dynamics between the external exciting force and the structure itself inherent characteristics[3].And it plays a very important role,to improve the performance of diesel engine.

In 1982, walterOtt et al. established a finite element model of a four-cylinder in-line engine, respectively established four models for different elements, and conducted free mode analysis. Through modal test comparison, it was concluded that the model of combining plate between the shell elements and solid elements was reasonable.Anderton, D.Ghazy et al. studied the dynamic response of heavy machine with finite element software.J.p.randeis conducted finite element analysis on the diesel

IOP Conf. Series: Materials Science and Engineering **825** (2020) 012039 doi:10.1088/1757-899X/825/1/012039

engine body through finite element method, predicted the vibration of the diesel engine, and pointed out that the weak links of the engine body existed in the skirt and oil sump[4] .D. anderton and t. priede studied the noise radiation control of the diesel engine[5] .Lalor N and Birth M analyzed the finite element dynamic characteristics of the diesel engine body, calculated the natural frequency and modal shape, analyzed and summarized each mode, and had a relatively directional understanding of the vibration mode[6] [7] .In addition, many diesel engine manufacturers in the world, such as MTU, TOYOTA and MSSSAN, have also carried out extensive finite element analysis on many parts of diesel engine (body, cylinder head, bearing, connecting rod, sealing ring, etc.).In 1990, Shung. H Sung et al. used the finite element analysis software Nastran to study the vibration and noise of a diesel engine body[8] .The ISVR laser room has developed a laser velocimeter that can be used to monitor the running state and performance of an engine, as well as whether the engine can reduce noise.Through its strong experimental capability, LMS can better determine the incentive force of the body, so that the study of dynamic characteristics can be closer to the actual situation.Austria AVL consider the internal combustion engine has a good NVH(noise, vibration, ride comfort) performance as the main standard of a product can occupy a place in the market . A.F.Seybert, D.A.Hamilton used finite element software to conduct finite element sonic-solid coupling analysis aim at many parts of diesel engine[8] .

In 1997, Shung.H.sung,Donald.J.Nefske et al. carried out finite element analysis of vibration and noise prediction to the diesel engines [9].The university of Kentucky A.F.Seyber did a lot of finite element analysis to the various parts of the engine and completed the acoustic vibration coupling analysis[10] .In 1997, Robert Sou Thai J et al. developed Perkins V6HSDI (High Speed Direct Injection) diesel engine that met European standards by using finite element analysis technology[11].AVL (Austria) and LMS (Belgium) have made great achievements in determining the excitation force of cylinder block by virtue of their excellent experimental capabilities. Finite element analysis technology has been applied in practical engineering design, which makes the dynamic response results obtained by computer simulation closer to the actual situation[12] .In 2002, Baverstock L studied the structural performance of diesel engine body by means of experiment, and pointed out that the maximum stress appeared on the joint surface of main bearing cover [13] .

At present, most domestic and foreign researchers studied the dynamic characteristics of diesel engine are mainly focus on showing the working stability of diesel engine through simulation software, so as to guide the structural optimization of diesel engine. Occasionally, some scholars conduct experiments under the static condition of diesel engine with LMS or other test instruments, so as to input the obtained dynamic parameters into the simulation model or verify simulation results.

Therefore, this paper proposes a new method to identify the diesel engine modal parameters based on its self-excitation

2. Design of self-excitation motion law

2.1. The basic principle of self-excitation modal analysis method

The method of self-excitation analysis for diesel engine is mainly used the exciting force which generated by the deflagration of the combustion chamber against the engine body ,and then used the vibration response of the structure to identify the modal parameters.The schematic diagram of the self-excitation modal analysis method as the figure 1. shown . Under the action of the excitation force, the structure of the diesel engine will produce the corresponding vibration response. Finally, the vibration response signal is analyzed by the response-based modal analysis algorithm, and the modal parameters of the diesel engine can be identified with the vibration response signal of the structure.

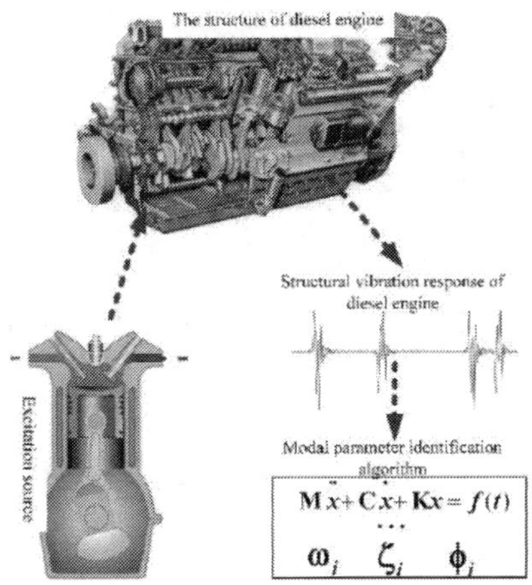

Figure 1. The schematic diagram of self-excitation modal analysis method

Airflow generated at the moment of deflagration combustion chamber (Turbulent) incentives the cylinder head, piston, cylinder liner, and then passed on to the diesel engine structure.The structure of the diesel engine can produce vibration response under the action of the exciting force.At the moment, the piston confined to a relatively small range, the small changes of piston connecting rod component will not affect the structural dynamic characteristic dynamic behavior of diesel engine.At the same time ,the crankshaft movement caused by the centrifugal force and coriolis acceleration can be ignored. Therefore, the structure of the diesel engine can be considered as linear time-invariant under this condition. Under the assumption of viscous damping, the dynamic Eq. of the diesel engine structure can be expressed as:

$$\mathbf{M}\ddot{\mathbf{X}}+\mathbf{C}\dot{\mathbf{X}}+\mathbf{K}\mathbf{X}=\mathbf{F} \tag{1}$$

Where M, C and K are the discrete physical mass, damping and stiffness matrices of the whole engine structure, X is the vibration displacement vector of the diesel engine structure, and F is the excitation force of the diesel engine structure generated by deflagration. The left side always represents the dynamic characteristics of the diesel engine structure, and the right side represents the inertial force generated by the self-excitation movement. If the inertial force can be measured, the Fourier transform of formula (1.1) can be obtained as follows:

According to the modal theory, the frequency response function matrix of diesel engine structure can be further expressed by the modal parameter model:

$$\mathbf{H}(\omega)=\sum_{r=1}^{N}\mathbf{Q}_r\frac{\psi_r\psi_r^{\mathrm{T}}}{j\omega-\lambda_r} \tag{2}$$

Where ψ_r is the r-order mode mode vector of diesel engine structure, the real part of λ_r is the r-order mode damping ratio of diesel engine, and the imaginary part is the natural frequency of the r-order mode of diesel engine. The Eq. (1.4) indicates that if the frequency-response function matrix of the diesel engine structure can be measured, the modal parameters of the diesel engine can be obtained through the corresponding experimental modal analysis algorithm.However, due to the complexity of the system composed of piston connecting rod, crankshaft, bearing, cylinder body and cylinder head, etc., the inertial excitation force acting on the structure of diesel engine is difficult to be measured ,so the frequency response function of the structure of diesel engine under the self-excitation of empty operation cannot be obtained.

2.2. Self-excitation motion characteristics analysis of diesel engine in dry run

(1) Free response signal extraction

When the excitation force meets the assumption of "white noise", the NEXT can extract the free response from the vibration response signal that can reflect the dynamic characteristics of the structure.

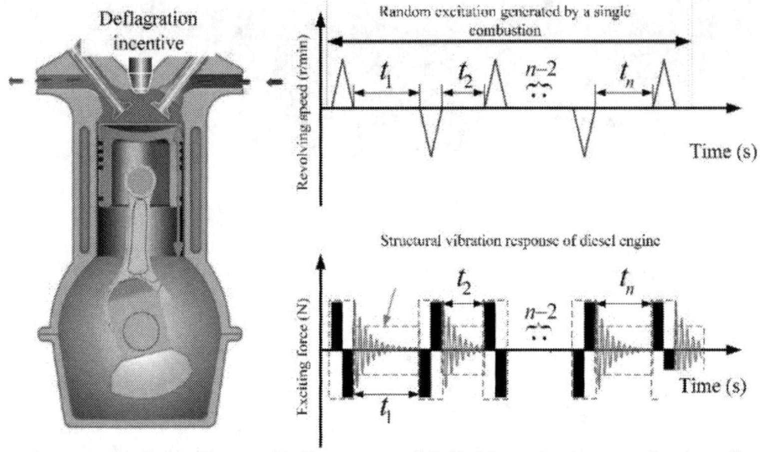

(a) Deflagration excitation schematic diagram (b) Self-excitation excitation force sequence

Figure 2. schematic diagram of self-excitation motion law of diesel engine in dry run

The figure 2 is a schematic diagram of the self-excitation motion law of the diesel engine in dry run.The Figure. (a) is the diagram of self-excitation: when the diesel engine in dry run, the structural change of the combustion chamber is limited to a small area during the moment of compression combustion, since the influence of the change of self-excitation movement position on the structure can be ignored in a small area.Figure (b). shows the one turbulence generated by deflagation during combustor combustion and the random pulse excitation striking the diesel engine.

Fast-burning stage in the process of diesel engine combustion time, the highest pressure appear.The growth rate of average pressure $\Delta p/\Delta\psi$is great, but unfavorable exceed 0.4 MPA / °CA. Therefore, when the gas hits the combustion chamber as shown in Figure b., the inertial impact sequence shown in Figure c. will be generated. Under the inertial impact sequence, the structure will produce corresponding vibration response. According to the RDT, the displacement response of structural vibration of diesel engine can be expressed as:

$$x(t) = x(0)D(t) + \dot{x}(0)v(t) + \int_{t_0}^{t} h(t-\tau)f(\tau)d\tau \tag{3}$$

Where $x(0)$ is the initial displacement of the vibration response, and $\dot{x}(0)$ is the initial velocity of the vibration response; $D(t)$ is the free displacement response of the structure only under the initial conditions $x(0)=1$, $\dot{x}(0)=0$. $v(t)$ is the free response of the structure only under the initial conditions $x(0)=0$, $\dot{x}(0)=1$. $h(t)$ is the unit impulse response function of the structure; The first two parts are deterministic and the second part is determined by the random inertial excitation sequence.

when the diesel engine is driven by dry run, the gas impact diesel engine should be as dense as possible in unit time, so as to obtain a higher capability level.The shorter the duration of gas impact sequence, the larger the energy spectrum bandwidth.

Therefore, when the research object is certain, it is necessary to improve the speed of the diesel engine as much as possible. However, if the speed is set too high, it may cause some damage to the diesel engine. Therefore, the principle of selecting the speed of the diesel engine that it is possible to select the speed under normal operation.

3. The validity of experimentally verified

In this paper, the self-excitation modal analysis experiment and the traditional percussion modal experiment are carried out.For the validity of the self-excitation modal analysis method is studied, and the experimental object of diesel engine is TBD234V12

3.1 Self - excited experimental platform
Diesel engine TBD234V12 is the experimental object of self-excited mode verification.The front end of the vibration signal acquisition of the self-excited mode experiment is dynamic acquisition equipment of Belgium LMS company. The vibration signal is measured by the 356A16 ICP three-phase acceleration sensor produced by PCB company.

3.2 Self-excitation experiment scheme
In order to fully study the whole mode of diesel engine, 66 measuring points were selected. Figure 3shows the measuring point layout model of the self-excitation mode experiment. Each node of the grid in the model represents the position of the sensor installed.

As shown in the figure 3(b), 8 measuring points are evenly arranged on the cylinder head, 28 measuring points are evenly arranged on the cylinder body, 14 measuring points are symmetrically arranged on the oil sump, and 16 measuring points are symmetrically arranged on the bracket.As the theory described in section 2 and the structural properties of TBD234V12, the engine speed is set to 1500r/min with no load.The sampling frequency set 1024Hz, and the sampling time set 3 minutes.As a comparison, the traditional experimental modal analysis (EMA) method is used to analyze the modal parameters of the diesel engine structure,and the knocking excitation is applied at the input end of the diesel engine.The location of the sensors in the knocking experiment is the same as the self-excited mode experiment.

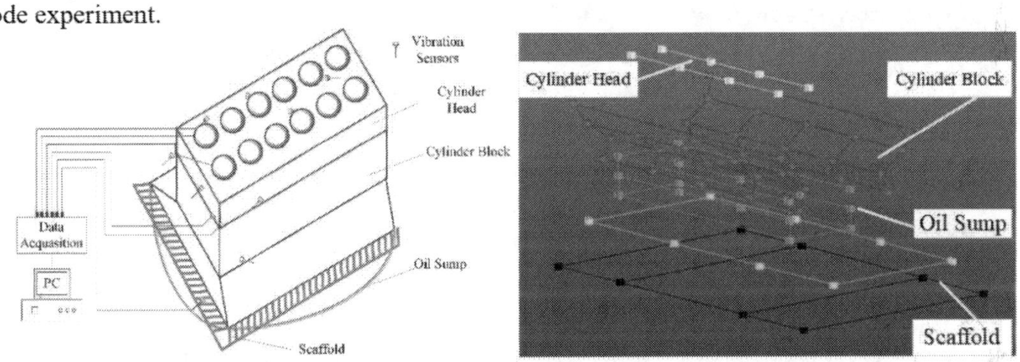

(a) (b)

Figure 3. Schematic diagram of self-excitation modal analysis experiment

3.3 Validity verification of identification modal parameters
Table 1 is the result of EMA and self-excitation experiment.

Table 1. The results of EMA and self-excitation experiment

		modal #1	modal #2	modal #3	modal #4	modal #5	modal #6
EMA	Natural Frequency	26.727Hz	35.352Hz	55.476 Hz	94.516Hz	120.085Hz	135.681Hz
	Damping ratio	3.25 %	1.53 %	1.17 %	3.89 %	1.62 %	1.57 %
OMA	Natural Frequency	24.507 Hz		54.524 Hz			147.260Hz
	Damping ratio	4.20 %		1.51 %			1.95 %

As the Table 1 shown, it can be seen that the natural frequency identified by the self-excitation modal analysis method is slightly less than that calculated by the EMA method. This is mainly due to the knocking process of diesel engine is in a state of rest. At work, the crankshaft of diesel engine piston component has been at the state of motion, the results of identification has a certain influence.

IOP Conf. Series: Materials Science and Engineering **825** (2020) 012039 doi:10.1088/1757-899X/825/1/012039

And the incentive to identify modal parameters less than the result of the EMA experiment, it is mainly because the incentive energy relative to knock experiment, result in some modal parameters are not excited.

4. Conclusions

In this paper, the basic concept of self-excitation modal analysis method for diesel engine is presented.Then the design method of the excitation law and the identification algorithm of the modal parameters are proposed for the modal analysis method of the diesel engine.At last, the modal analysis on TBD234V12 diesel engine, the traditional knocking experimental modal analysis and the selt-excitation modal experimental were carried out respectively, and the two results were analyzed and compared.The experimental results show that the self-excited modal analysis method can effectively identify the modal parameters of the diesel engine structure, and the motion state of the diesel engine will affect the modal parameters of the diesel engine structure to some extent.

References

[1] Anderton D, M T. Applications of engine noise control research engine noise,excitation,vibration and radiation[M]. 1982.
[2] Baverstock L. Investigation of diesel engine cylinder block structural performance using design of experiments[J]. Ford Motor Company Limited, Altair Engineering Ltd. 2002.
[3] Birth M, Papez S. Idealization, measurement and calculation of an engine block[R]. SAE Technical Paper, 1982.
[4] Brandeis J P. The use of finite element techniques to predict engine vibration[R]. SAE Technical Paper, 1982.
[5] F S A, A H D, A H P. Prediction of radiated noise from engine components using the bem and the rayleigh integral[J]. Noise Control Engineering Journal. 1997, 46(3): 77-82.
[6] Lalor N. Finite element optimisation techniques of diesel engine structures[R]. SAE Technical Paper, 1982.
[7] Priebsch H H, Affenzeller J, Kuipers G. Structure borne noise prediction techniques[J]. Sae Transactions. 1990: 42-52.
[8] Qian D. Noise optimization of the 493 diesel engine[J]. Journal of Hefei University of Technology(Natural Science). 2007, 30(12): 159-161.
[9] Southall R, Southall B, Trimm M. Noise and vibration technology for the perkins v6 hsdi demonstration engine[J]. Sae Transactions. 1997: 3191-3199.
[10] Sung S H, Nefske D J, Chen F H, Fannin M P. Development of an engine system model for predicting structural vibration and radiated noise of the running engine[J]. Sae Transactions. 1997: 3175-3180.
[11] Sung S H, Nefske D J, Chen F H, Fannin M P. Development of an engine system model for predicting structural vibration and radiated noise of the running engine[J]. Sae Transactions. 1997: 3175-3180.
[12] Tan D. Vibration control of internal combustion engine[M]. Southwest Jiaotong University Press, 1993.
[13] Zuan L, Liu X, Gao Z. Internal combustion engine [Z]. Beijing:China Machine Press, 1999.

IOP Conf. Series: Materials Science and Engineering **825** (2020) 012040 doi:10.1088/1757-899X/825/1/012040

Based on Vibration Signal to Identify the Diesel Engine Modal Parameters under Working Condition

Xing Liu[1,*], Jiheng chen[2], Biao Hou [1], Ye Tian[1], Feng Liu[1] and Jingsong Tan[2]

[1] Diesel engine power plant teaching and research office, Naval Petty Officer Academy, Bengbu 230000, PR China

[2] Mechanical and electrical equipment maintenance teaching and research office, Naval Petty Officer Academy, Bengbu 230000, PR China

[*]E-mail address: liuxing.3n8b@163.com

Abstract. The methods to study the dynamic characteristics of diesel engine structure include finite element analysis method and experimental analysis method. It is difficult to establish an accurate finite element model, since the results of finite element analysis methods are often not accurate. For the experimental analysis method, it is mainly divided into the traditional Experimental Modal Analysis (EMA) and the Operational Modal Analysis (OMA). The EMA method always needs to measure the excitation force of the diesel engine and the vibration response of the structure simultaneously. And the influence of motion factors on the structure dynamics cannot be considered. The structure is excitated by the working environment, and only analyzes the modal parameters of the structure with the vibration response. Few scholars analyze the OMA for diesel engines. This paper proposes a new method to identify the modal parameters of diesel engine based on self-excitation. In this paper, the theory of self-excitation mode parameter identification is demonstrated by derivation, and the validity of the method is verified.

1. Introduction

Traditionally, the EMA is used to obtain Modal properties of structures, through a hammer to excite the structure and obtain the modal parameters (natural frequency, modal damping and modal mode). The application of EMA has a big limitation, that needs obtain the structural excitation signal. Sometimes it's hard to get the excitation signal of the structure characterization [1].And the accuracy of Finite Element Modal(FEM) analysis is limited by the accuracy of the model. At this time, The OMA method was proposed to identify the Modal of the structure in the case of unknown input [2].

Guo et al. predicted the radiated noise from the oil pan of a diesel engine by coupling methodology of finite element analysis (FEA) and boundary element analysis (BEA), and optimized it using rib stiffeners. However, the optimized oil pan was excited by the same boundary conditions utilized for the original one, which would influence the accuracy of the structural responses after optimization.[3].Lalor N and Birth M analyzed the finite element dynamic characteristics of the diesel engine body, calculated the natural frequency and modal shape, analyzed and summarized each mode, and had a relatively directional understanding of the vibration mode[4] [5] .In addition, many diesel engine manufacturers in the world, such as MTU, TOYOTA , have also carried out extensive finite element analysis on many parts of diesel engine (body, cylinder head, bearing, connecting rod, sealing ring, etc.).In 1990, Shung. H Sung et al. used the finite element analysis software Nastran to study the vibration and noise of a diesel engine body[6] . Seybert A.F. et al.used finite element software to conduct finite element sonic-solid coupling analysis aim at many parts of diesel engine[7] .

In 1997, Southall R et al. developed Perkins V6HSDI (High Speed Direct Injection) diesel engine that met European standards by using finite element analysis technology[8].AVL (Austria) and LMS (Belgium) have made great achievements in determining the excitation force of cylinder block by virtue of their excellent experimental capabilities. Finite element analysis technology has been applied in practical engineering design, which makes the dynamic response results obtained by computer simulation closer to the actual situation .In 2002, Baverstock L studied the structural performance of diesel engine body by means of experiment, and pointed out that the maximum stress appeared on the joint surface of main bearing cover [9] .

At present, most researchers studied the dynamic characteristics of diesel engine are mainly focus on showing the working stability of diesel engine through simulation software, so as to guide the structural optimization of diesel engine. Occasionally, some scholars conduct experiments under the static condition of diesel engine with LMS or other test instruments, so as to input the obtained dynamic parameters into the simulation model or verify simulation results.

Therefore, this paper proposes a new method to identify the diesel engine modal parameters based on its self-excitation.

2. The theory of self - excitation mode parameter identification

The random subspace method is mainly based on the impulse response function of the system as the basic model. Through generalized Hankel matrix and singular value decomposition technology, the minimum realization of the system is obtained and the system matrix of the minimum order is obtained. On this basis, the modal parameters of the system are further identified[10] .

2.1. Structural modal parameter feature system implementation algorithm

According to modern control theory, the structural vibration equation of diesel engine can be further written in the form of state space:

$$\dot{y}(t) = Ay(t) + Bf(t)$$

$$z(t) = Gy(t) \tag{1}$$

Where A is the system matrix, B is the control matrix and G is the observation matrix. They jointly constitute an implementation of the system and are determined by the inherent characteristics of the system:

$$A = \begin{bmatrix} 0 & I \\ -M^{-1}K & -M^{-1}C \end{bmatrix} \tag{2}$$

$$B = \begin{bmatrix} 0 \\ M^{-1} \end{bmatrix} \tag{3}$$

According to the literature[11] , the modal parameters of the system can be identified through the system matrix A :

$$A = \Psi \Lambda \Psi^{-1} \tag{4}$$

$$\Lambda = \begin{bmatrix} \ddots & & \\ & \lambda_k & \\ & & \ddots \end{bmatrix}, \Theta = [\cdots \quad \phi_k \quad \cdots], k = 1, \ldots, N \tag{5}$$

Where, λ_k and the natural frequency ω_k and damping ratio ξ_k of diesel engine structure have the following relation:

$$\lambda_k = -\xi_k \omega_k + i\sqrt{1-\xi_k^{\ 2}}\,\omega_k \tag{6}$$

Where $i=\sqrt{-1}$, ϕ_k represents the mode vector of diesel engine structure. According to the above equation, if the system matrix of diesel engine structure can be obtained, the modal parameters of diesel engine can be identified by singular value decomposition of the matrix. However, the above is for a time continuous system. For actual modal analysis, since the data used is digital signal and the system is a time discrete system, it is assumed that $t=t_0$ when the initial condition is $y(t_0)$, the solution of the equation of state is:

$$y(t)=e^{A(t-t_0)}y(t_0)+\int_{t_0}^{t}e^{A(t-\tau)}Bf(\tau)d\tau \tag{7}$$

Set the discrete time points are $k=0$, 1,2, and the sampling time interval Δt, then $t=k\Delta t$,and the solution of the equation of state can be written as the discrete form:

$$y[(k+1)\Delta t]=e^{A\Delta t}y(k\Delta t)+e^{A\Delta t}Bf(k\Delta t)\Delta t \tag{8}$$

If $k\Delta t$ is simply denoted as k, then the discrete state equation of the system can be expressed as: $y(k+1)=A_1y(k)+B_1f(k)$

$$z(k+1)=Gy(k+1) \tag{9}$$

Among them,

$$A_1=e^{A\Delta t} \quad B_1=A_1B\Delta t \tag{10}$$

Substitute formula (4) into the above equation, and according to the inherent nature of the index, the following equation can be obtained:

$$A_1=e^{\Psi\Lambda\Psi^{-1}\Delta t}=\Psi e^{\Lambda\Delta t}\Psi^{-1} \tag{11}$$

Equation (11) can be further expressed as:

$$\Psi^{-1}A_1\Psi=e^{\Lambda\Delta t} \tag{12}$$

From the above equation, it can be known that the eigenvectors of A_1 and A are the same, the eigenroots μ_k of matrix A_1 and he eigenroots λ_k ofmatrix A satisfy: $\lambda_k=\dfrac{\ln\mu_k}{\Delta t}$

According to the above analysis, if the system matrix A_1 of discrete state equation of diesel engine structure can be obtained, the modal parameters of diesel engine structure can be obtained by singular value decomposition of the matrix.

2.2 The equation of state representation of the impulse response function

Z transformation of the transfer function of the system can be obtained as follows:

$$H(z)=\sum_{k=0}^{\infty}h_kz^{-k} \tag{13}$$

Where, $z=e^{s\Delta t}$ is the Z transformation factor; $s=\sigma+j\omega$ is the domain of complex Numbers; $H(z)$ is the transfer function matrix in the form of Z transformation; h_k is the impulse response function of the system.The discrete equation of state of the diesel engine structure is transformed by Z, as follows:

$$\begin{aligned} zY(z)&=A_1Y(z)+B_1F(z)\\ Z(z)&=GY(z) \end{aligned} \tag{14}$$

The above equation can be obtained:

$$Z(z)=G(zI-A_1)^{-1}B_1F(z) \tag{15}$$

Thus, the transfer function can be expressed as:

$$H(z)=z^{-1}G(I-z^{-1}A_1)^{-1}B_1 \tag{16}$$

As

$$(I-z^{-1}A_1)^{-1}=\sum_{k=0}^{\infty}(z^{-1}A_1)^k=\sum_{k=0}^{\infty}A_1^k z^{-k} \tag{17}$$

Substitute into equation (16), the transfer function can be expressed as:

$$H(z)=\sum_{k=1}^{\infty}GA_1^{k-1}B_1z^{-k} \tag{18}$$

By comparison with equation (14), the following equation can be obtained:

$$h(k)=GA_1^{k-1}B_1 \tag{19}$$

According to the above equation, the system matrix A_1 of the structural state equation can be constructed by using the impulse response signal $h(k)$ of the diesel engine structure, and then the system matrix can be used to identify the modal parameters of the diesel engine structure.

2.3Matrix construction of state equation system

System matrix A_1 can be obtained by constructing Hankel matrix.Hankel matrix is composed of Markov parameter matrix, and if the measured impulse response matrix is $M\times L$ -order matrix $h(k)$, Hankel matrix is composed as follows:

$$H(k-1)=\begin{bmatrix} h(k) & h(k+t_1) & \cdots & h(k+t_{\beta-1}) \\ h(k+s_1) & h(k+s_1+t_1) & \cdots & h(k+s_1+t_{\beta-1}) \\ \cdots & \cdots & \cdots & \cdots \\ h(k+s_{\alpha-1}) & h(k+s_{\alpha-1}+t_1) & \cdots & h(k+s_{\alpha-1}+t_{\beta-1}) \end{bmatrix} \tag{20}$$

Where, t_i, s_i are the integer corresponding to any discrete time. Since $h(k)$ is order $M\times L$, the $H(k-1)$ matrix is order $\alpha M\times\beta L$. However, due to noise pollution in measurement, $H(k-1)$ has rank deficit. Only when α , β increase to a certain degree, the rank of $H(k-1)$ is approximately equal to the order of the system. Usually, $t_i=s_i=i$, there is:

$$H(k-1)=\begin{bmatrix} h(k) & h(k+1) & \cdots & h(k+\beta-1) \\ h(k+1) & h(k+2) & \cdots & h(k+\beta) \\ \cdots & \cdots & \cdots & \cdots \\ h(k+\alpha-1) & h(k+\alpha) & \cdots & h(k+\alpha+\beta-2) \end{bmatrix} \tag{21}$$

$$H_{\alpha\beta}(k-1)=PA_1^{k-1}Q \tag{22}$$

Among $P=\begin{bmatrix} G \\ GA_1 \\ GA_1^2 \\ \vdots \\ GA_1^{\alpha-1} \end{bmatrix}$, $Q=\begin{bmatrix} B_1 & A_1B_1 & A_1^2B_1 & \cdots & A_1^{\beta-1}B_1 \end{bmatrix}$, α , β is, $\frac{2n}{M}\le\alpha\le 2n$ and $\frac{2n}{L}\le\beta\le 2n$, respectively.

Let $k=1$, then:

$$H_{\alpha M\times\beta L}(0)=PQ \tag{23}$$

Perform singular value decomposition for $H_{\alpha M\times\beta L}(0)$:

$$H_{\alpha M\times\beta L}(0)=U\sum V^T \tag{24}$$

$$U^TU=I\,,V^TV=1 \tag{25}$$

Where, U is the left singular vector matrix of order $\alpha M\times 2n$; V is the right singular vector matrix of order $\beta L\times 2n$; $\sum$ is the $2n\times 2n$ order diagonal matrix, and $\sum=diag\begin{bmatrix}\sigma_1 & \sigma_2 & \cdots & \sigma_{2n}\end{bmatrix},\sigma_i^2$ is the non-zero characteristic root of $H^T_{\alpha M\times\beta L}(0)H_{\alpha M\times\beta L}(0)$.

Introduce matrix $H^{\#}(0)$ such that

$$QH^{\#}P=I \tag{26}$$

There are:

$$H(0)H^{\#}(0)H(0)=PQH^{\#}PQ=PQ=H(0) \tag{27}$$

Formula (23)~(27) can be obtained as follows:

$$H^{\#}=V\sum{}^{-1}U^T \tag{28}$$

Set $E_M{}^T=\begin{bmatrix}I_M & 0_M & \cdots & 0_M\end{bmatrix},E_L{}^T=\begin{bmatrix}I_L & 0_L & \cdots & 0_L\end{bmatrix}$, Where $E_M{}^T$ is the matrix of order $M\times\alpha M$ and $E_L{}^T$ is the matrix of order $L\times\beta L$. According to equation (21), it can be known that:

$$h(k+1)=E_M{}^TH(k)E_L \tag{29}$$

Substitute equation (22) into the above equation:

$$h(k+1)=E_M{}^TPA_1{}^kQE_L \tag{30}$$

Insert equation (26) into the above equation for further arrangement:

$$h(k+1)=E_M{}^TPQH^{\#}PA_1{}^kQH^{\#}PQE_L=E_M{}^TH(0)H^{\#}PA_1{}^kQH^{\#}H(0)E_L \tag{31}$$

$H^{\#}PA_1{}^kQH^{\#}$ can be expanded as follows:

$$H^{\#}PA_1{}^kQH^{\#}=(H^{\#}PA_1Q)^k=\left[H^{\#}H(1)\right]^k \tag{32}$$

By substituting equation (28) into the above equation, the following equation can be obtained:

$$H^{\#}PA_1Q=\left[H^{\#}H(1)\right]^k=\left[V\sum{}^{-1}U^TH(1)\right]^k \tag{33}$$

Substitute the above equation and (24), (25) ,(26)into equations (31) ,then:

$$h(k+1)=E_M{}^TU\sum{}^{-\frac{1}{2}}(\sum{}^{-\frac{1}{2}}U^TH(1)V\sum{}^{-\frac{1}{2}})^k\sum{}^{-\frac{1}{2}}V^TE_L \tag{34}$$

By comparison with equation (19), the following equation can be obtained:

$$A_1=\sum{}^{-\frac{1}{2}}U^TH(1)V\sum{}^{-\frac{1}{2}} \tag{35}$$

$$B_1=\sum{}^{\frac{1}{2}}V^TE_L \tag{36}$$

$$G=E_M{}^TU\sum{}^{\frac{1}{2}} \tag{37}$$

According to the above analysis, the calculation of modal parameters of diesel engine structure includes: firstly, free response characteristics are extracted from vibration response signal of diesel engine structure; However, in the process of practical analysis, the vibration of the measured signal often contains a lot of noise, the results of the self-excited modal analysis algorithm include not only the modal parameters of the structure, but also the pseudo-modal introduced by noise and other components. it is necessary to further identify the modal parameters of the diesel engine structure from the calculated results.

3. Experimental verification

In this paper, the self-excitation modal analysis experiment and the traditional modal experiment are carried out. For the validity of the self-excitation modal analysis method is studied, and the experimental object of diesel engine is TBD234V12

3.1 Self - excited experimental platform

In order to fully study the whole mode of diesel engine, 66 measuring points were selected. Figure 1. shows the measuring point layout model of the self-excitation mode experiment. Each node of the grid in the model represents the position of the sensor installed. As shown in the figure 1, 8 measuring points are evenly arranged on the cylinder head, 28 measuring points are evenly arranged on the cylinder body, 14 measuring points are symmetrically arranged on the oil sump, and 16 measuring points are symmetrically arranged on the bracket. As the theory described in section 2 and the structural properties of TBD234V12, the engine speed is set to 1500r/min with no load. The sampling frequency set 1024Hz, and the sampling time set 3 minutes. As a comparison, the traditional experimental modal analysis (EMA) method is used to analyze the modal parameters of the diesel engine structure,and the knocking excitation is applied at the input end of the diesel engine. The location of the sensors in the knocking experiment is the same as the self-excited mode experiment.

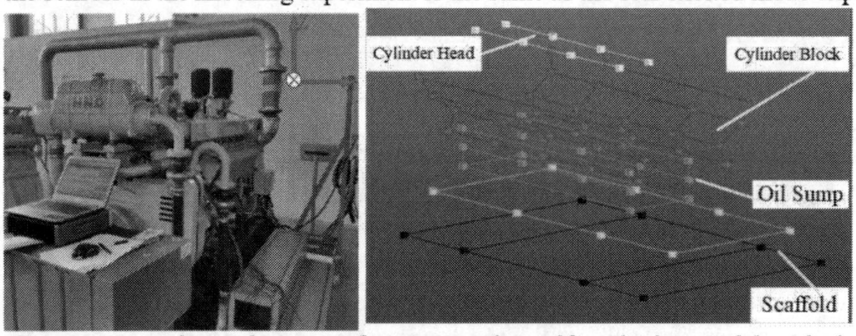

Figure 1. experimental system of empty running self-excitation modal analysis

3.2 Validity verification of identification modal parameters

Table 1. The results of EMA and self-excitation experiment

		modal #1	modal #2	modal #3	modal #4	modal #5	modal #6
EMA	Natural Frequency	26.727Hz	35.352Hz	55.476 Hz	94.516Hz	120.085Hz	135.681Hz
	Damping ratio	3.25 %	1.53 %	1.17 %	3.89 %	1.62 %	1.57 %
OMA	Natural Frequency	24.507 Hz		54.524 Hz			147.260Hz
	Damping ratio	4.20 %		1.51 %			1.95 %

As the Table 1 shown, the natural frequency of the modal parameters identified by EMA method is higher than the identified by OMA, except for the sixth modal. This is mainly because EMA is the result of diesel engine identification at rest, while OMA is the result of diesel engine at work The modal parameter orders identified by OMA is less than the identified by EMA, This is mainly because the incentive energy of OMA is less than the incentive energy of EMA.

4. Conclusions

In this paper, the basic concept of self-excitation modal analysis method for diesel engine is presented. Then the design method of the excitation law and the identification algorithm of the modal parameters are proposed for the modal analysis method of the diesel engine. At last, the two results are analyzed and compared. The experimental results show that the self-excited modal analysis method can effectively identify the modal parameters of the diesel engine structure, and the motion state of the diesel engine will affect the modal parameters of the diesel engine structure.

References

[1] Zaghbani I,Songmene V. Estimation of machine-tool dynamic parameters during machining operation through operational modal analysis [J]. International Journal of Machine Tools & Manufacture,2009,49(12): 947 −957.
[2] LIU Yufei,XIN Kegui,FAN Jiansheng,et al. A review of structure modal identification methods through ambient excitation [J]. Engineering Mechanics,2014,31(4): 46 −53.
[3] Guo L, Hao ZY, Xu HM. Integrated virtual-design methods for predicting radiated noise of cylinder block of a single cylinder diesel engine. Chin Inter Combust Engine Eng 2007;28:76–80 [in Chinese].
[4] Lalor N. Finite element optimisation techniques of diesel engine structures[R]. SAE Technical Paper, 1982.
[5] Birth M, Papez S. Idealization, Measurement and Calculation of an Engine Block[R]. SAE Technical Paper, 1982.
[6] Sung S H, Nefske D J, Chen F H, et al. Development of an engine system model for predicting structural vibration and radiated noise of the running engine[J]. SAE transactions. 1997: 3175-3180.
[7] Seybert A F , Hamilton D A , Hayes P A . Prediction of Radiated Noise from Engine Components Using the BEM and the Rayleigh Integral[J]. Noise Control Engineering Journal, 1998, 46(3).
[8] Southall R, Southall B, Trimm M. Noise and Vibration Technology for the Perkins V6 HSDI Demonstration Engine[J]. SAE transactions. 1997: 3191-3199.
[9] Baverstock L. Investigation of diesel engine cylinder block structural performance using design of experiments[J]. Ford Motor Company Limited, Altair Engineering Ltd. 2002.
[10] Mohanty P, Rixen D J. Modified ERA method for operational modal analysis in the presence of harmonic excitations[J]. Mechanical Systems and Signal Processing. 2006, 20(1): 114-130.
[11] Magalhães F, Cunha Á. Explaining operational modal analysis with data from an arch bridge[J]. Mechanical Systems and Signal Processing. 2011, 25(5): 1431-1450.

IOP Conf. Series: Materials Science and Engineering **825** (2020) 012041 doi:10.1088/1757-899X/825/1/012041

The Design of Hardware System of High Precision Dynamic Flow Controller

Jie Chen, Shuo Cheng, Bin Li* , Yang Gou and Xinghang Fan

School of Mechanical Engineering and Automation, Shanghai University, Shanghai 200444, PR China

*sulibin@shu.edu.cn

Abstract. At present, the dynamic flow controller produced in China cannot meet the demand of pharmaceutical, food, beverage and other industries in terms of its precision. Therefore, a high precision dynamic flow controller detection device is designed to improve the control accuracy of the dynamic flow controller. The device also has the function of flow calibration. The experimental results show that the high-precision dynamic flow controller detection device designed in this paper has a great improvement in the control accuracy of dynamic flow controller and can realize the high-precision calibration function of the tested flowmeter.

1. Introduction

As we all know, dynamic flow controller is widely used in industry, agriculture, medical industry and daily life. However, the domestic dynamic flow controller market is monopolized by foreign countries [1]. In recent years, with the rapid development of domestic pharmaceutical, food, beverage and other industries, medical agents, beverages and health products and other products on the dynamic flow controller filling control accuracy requirements are higher and higher. However, the current domestic production of electromagnetic flow controller, its control precision cannot meet the needs of the above industries. Therefore, the research and development of dynamic flow controller detection device with independent intellectual property rights, improve the control accuracy of dynamic flow controller, has become the current research priority [2].

With the continuous development of economy and technology, relevant technologies have also been improved to some extent. At present, China has certain economic and technical strength to develop high-precision dynamic flow detection device for filling with independent intellectual property rights. The high demand of the market and the continuous improvement of technology provide the corresponding conditions and power for the development of dynamic flow controller detection device. The research and development of dynamic flow controller detection device with independent intellectual property rights cannot only meet the increasing demand of various fields in China, improve the accuracy of dynamic flow detection device, but also break the situation that domestic dynamic flow detection device is monopolized by foreign countries, and fill the gap in related fields in China [3].

The dynamic flow controller detection device designed in this paper mainly realizes the following two functions: rapid filling by controlling the dynamic flow controller; high precision calibration of meter by standard meter. Therefore, in the design of the high-precision dynamic flow controller detection device, the related accuracy requirements of the dynamic flow detection device in the

IOP Conf. Series: Materials Science and Engineering **825** (2020) 012041 doi:10.1088/1757-899X/825/1/012041

documents "JJG 643-2003 standard table method flow standard device" and "gbt34268-2017" are mainly referred to.

2. Overall design and implementation of the device

The hardware system of this device uses the high precision (0.1%) Coriolis mass flowmeter of RHM06 for flow measurement, and the high precision electromagnetic filling flowmeter of E+H Dosimag 5BH04 for filling. The controller unit adopts the industrial control host of Advantech and the upper computer development software-LabWindows/CVI of NI. The software system of the device provides an easy-to-use program for calibration and filling. Input work orders and verification parameters through the software system can be stored in the MySQL database, and the traceable calibration data can be stored locally through the database and WORD file format. Figure 1, figure 2 and figure 3 are the schematic diagram of the system design.

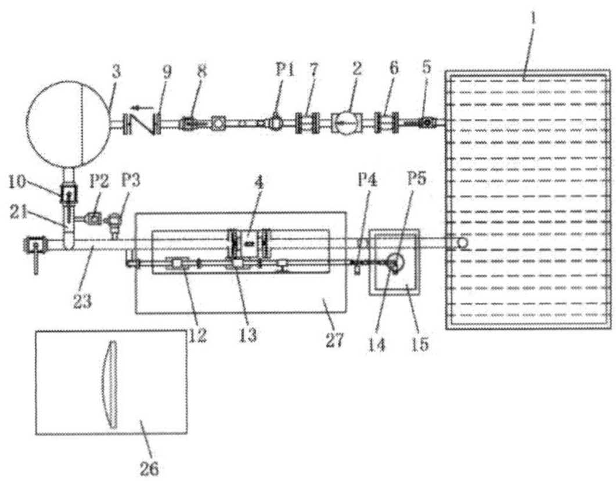

Figure 1. Device top view.

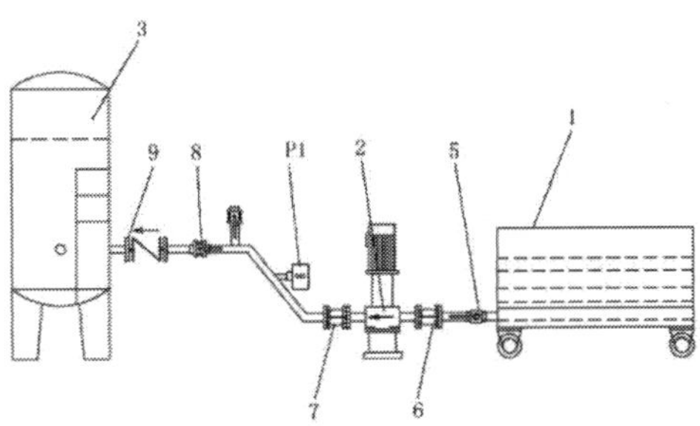

Figure 2. Part of the main view from the water tank to the inlet of the stabilizer tank in the main loop of the device

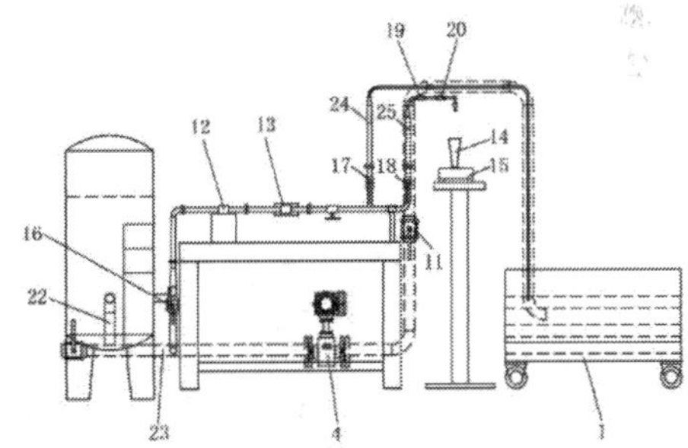

Figure 3. Main view of tank outlet to water tank and test circuit in main circuit of device

The part label in the figure is as follows:

1-Tank; 2-Variable frequency pump; 3-Pressure stabilizing tank; 4-Electromagnetic flowmeter; 5-The first valve; 6-The first bellows; 7-The second bellows; 8-The second valve; 9-Check valve; 10-The third valve; 11-The fourth valve; 12-Standard flowmeter; 13-Filling flowmeter; 14-Cup; 15-Electronic scales; 16The-fifth valve; 17-The sixth valve; 18-The seventh valve; 19-First solenoid valve; 20-The second solenoid valve; 21-The first horizontal segment;

22-Vertical section; 23-The second horizontal segment; 24-The first branch;
25-The second branch; 26-Control desk; 27-Calibration station; P1-The first pressure sensor;
P2-The second pressure sensor; P3-The third pressure sensor; P4-The fourth pressure sensor;
P5-The fifth pressure sensor.

The control unit of this device is mainly composed of IPC-940 of Advantech, PCI series board card and ADAM data acquisition module. The control structure diagram is shown in figure 4.

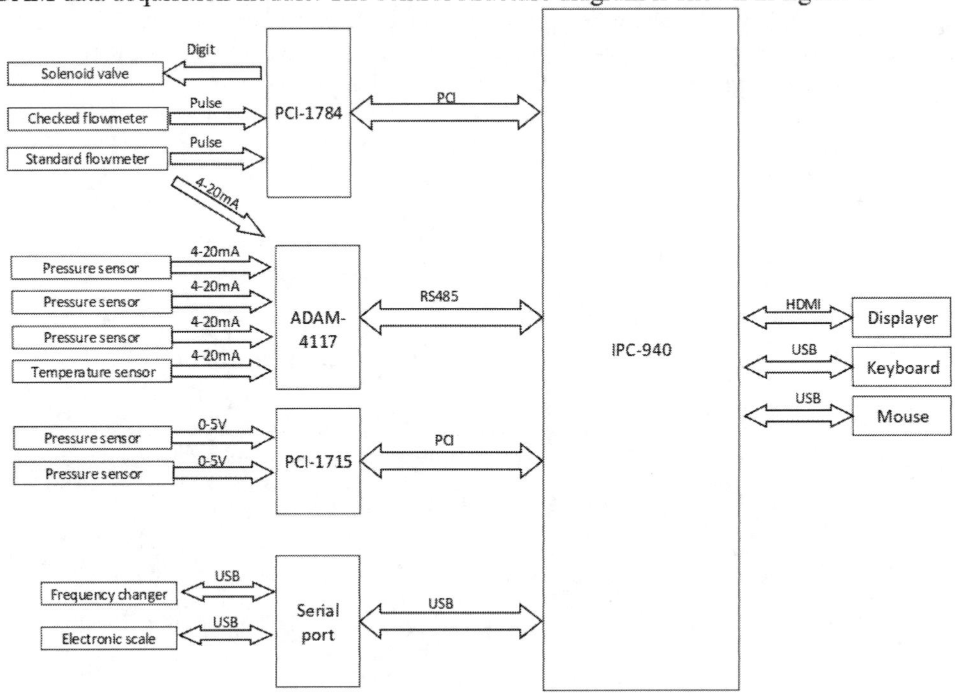

Figure 4. Schematic diagram of system control structure.

IPC-940, as the upper computer of the whole system, is externally connected with Advantech's adam-4117 current data acquisition card, PCI-1715 high-speed voltage data acquisition card, and PCI-1780 high-speed counter board card. Among them, ADAM - 4117 modules realize the standard flowmeter, checked flowmeter and some of the current output of the sensor data acquisition function; PCI-1715 high speed voltage data acquisition board card to achieve the electromagnetic valve before and after the voltage output pressure sensor data collection function; PCI-1780 high speed counter board card can count the pulse output of the standard meter and the inspected meter as well as the function of starting and stopping the solenoid valve [4].

3. Innovative design of the device

Referring to the relevant requirements of this device in the documents of "JJG 643-2003 standard table method flow standard device" and "gbt34268-2017", the device shall meet the following indicators :(1) the overall measurement accuracy of the device shall reach 0.5%;(2) output stability of the device is 0.2%.Therefore, the device has high requirements on the stability of water circulation and the accuracy of filling performance.

3.1. Design of water cycle stability

The stability of water circulation determines the excellent performance of the whole device. Therefore, the following two aspects are mainly considered when designing the water circulation system of the device: optimizing the water circulation control algorithm; optimize the hardware pipeline of water circulation system.

3.1.1. Optimization of water cycle control algorithm. The water circulation control system of the device is mainly composed of ACS355 inverter and Helix V603 pump, as shown in figure 5 and figure 6.The upper computer sends frequency conversion control instructions to ACS355 inverter through RS485 serial port communication to adjust the frequency output of the inverter, thus controlling the flow output of Helix V603 stainless steel multi-stage centrifugal pump.

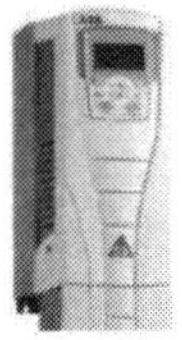

Figure 5. ABB ACS355 inverter

Figure 6. Helix V603 stainless steel multi-stage centrifugal pump

The water circulation control system of this device adopts PID control algorithm, the standard table is the measuring feedback element; ABB frequency conversion, water pump and regulating valve are executive components; The flow standard device is the controlled object; The PID control algorithm module of Labwindows/CVI is the controller, and the block diagram of the control system is shown in figure 7.

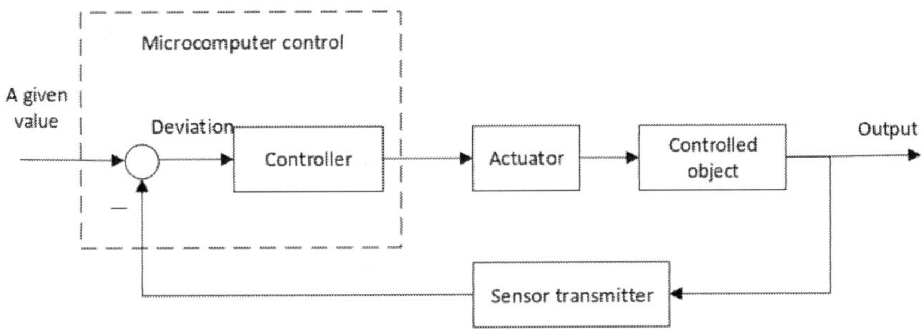

Figure 7. Basic block diagram of microcomputer process control system.

Due to the time delay of the feedback signal, namely the main loop electromagnetic flowmeter, in the process of signal acquisition and transmission, the PID control algorithm is easy to adjust the water source stability when there is a large fluctuation near the target value, the stability is not high. Therefore, the water source control algorithm is optimized in the design of this device. By reading the value of the feedback signal, it can judge whether the flow reaches the allowable error of the target value. When entering the error range, it can read the current output frequency of the inverter, close the PID control and open the output mode of frequency stabilization. Because in the acquisition of the signal delay of control system, frequency stabilization control on water stability of the control effect is better than the frequency conversion control, and so ended up in the device adopts the following algorithm: use PID algorithm to find within the set target error of inverter output power, find closed after frequency conversion control frequency stabilization control is used to improve the stability of the water [5].

3.1.2. Optimization of water circulation hardware pipeline. In the design of hardware pipeline, the device uses the method of large pipe diameter drainage of small pipe diameter to supply water to the test functional circuit. The working process of water source is as follows: firstly, the water pump is controlled by frequency converter to control the flow in the main loop of DN50 pipe diameter to a stable state; secondly, water is drawn from the main loop of DN50 pipe diameter for the test function loop of DN4 pipe diameter is shown in figure 8.

Figure 8. Large tube diameter drains small tube diameter

The pipeline design method of radial drainage by large pipe and small pipe diameter can effectively further reduce the influence of water source fluctuation on the measurement accuracy of equipment. Working principle is as follows: when the water wave of big diameter, as a result of fluctuations in water conductivity, when the fluctuation is conducted from DN50 pipe diameter to DN4 test function circuit, it will attenuate to some extent, so the function of the test circuit of the effect on the stability of the water will become smaller and tested to ensure that the water stability function circuit experiment.

3.2. Test function loop design

The test functional circuit mainly includes: calibration functional circuit and quick filling functional circuit, as shown in figure 9 and figure10. When the calibration function is performed, open the valve on the left side of figure10 and close the valve on the right side to introduce the calibration water flow into the water tank; In figure 9 the left side is the standard flowmeter, and the right side is the checked flowmeter. The ADAM4117 and PCI1780 boards are used to collect the output pulse and current signals of the checked flowmeter and the standard flowmeter, so as to calibrate the checked flowmeter. When carrying out the calibration function, open the valve on the right side and close the valve on the left side in figure.10, and introduce the filling water into the measuring cup. In figure 9, the checked flowmeter on the right is changed into E+H Dosimag 5BH04 high-precision electromagnetic filling flowmeter. The PCI1780 board card is used to collect the electromagnetic filling flowmeter for weighing and comparing with the electronic scale reading under the measuring cup.

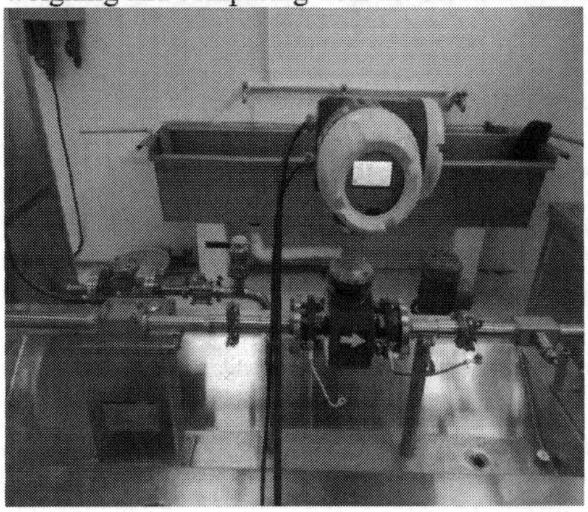

Figure 9. Calibration function part of pipeline

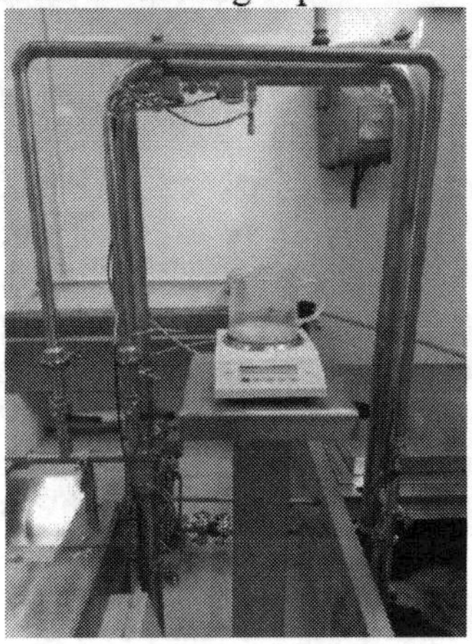

Figure 10. Quick filling function part of pipeline

The solenoid valve is used to switch the filling flow in the hardware design of the quick filling function. However,the solenoid valve working principle of the defect: When the solenoid valve on electricity, by the electromagnetic force to close the coil, action time is short; after the loss of electricity, by mechanical force to open the coil, the action time is slow. As a result, the speed of the switch controlling the filling water flow is not ideal [6]. Therefore, the design of this device adopts the mode of dual solenoid valve control, as shown in figure 10.Normally closed solenoid valve and normally open solenoid valve are installed successively before filling outlet. Normally closed solenoid valve is used to control water flow opening and normally open solenoid valve is used to control water flow closing. The purpose is to make the water flow in the opening and closing is driven by the electromagnetic force after the power, thus greatly improving the rapid filling performance.

4. Experimental analysis

According to the calculation method of flow stability verification in "JJG 643-2003 standard table method flow standard device", the stability analysis of the whole device was carried out. By weighing the same filling mass point through filling experiment, the repeatability of fast filling function is calculated. According to the above method, the device has passed the verification and inspection of Shanghai institute of metrology and testing technology. Experimental data are shown in table 1.

Table 1. Experimental results of flow stability and repeatability of filling.

Serial number	project	Calibration results		Extended uncertainty(k=2)
1	Flow stability	$0.12m^3/h$	0.06%	/
		$0.63m^3/h$	0.08%	/
2	Filling repeatability	50g	0.20%	$U_{rel} = 0.10\%$
		500g	0.03%	

5. Conclusion

The high-precision dynamic flow controller detection device designed in this paper can realize the functions of high-precision calibration and rapid filling of the tested flowmeter. Moreover, the stability of the flow rate of the device can reach 0.08%, and the repeatability of the filling function can reach 0.20%.The control precision of dynamic flow controller is improved, which can meet the control precision of pharmaceutical, food, beverage and other industries.

References

[1] John D Wright, Pedro Espina. Flow Calibration Services at NIST. Flow Measurement and Instrumentation, 2000,8(5):27-28.
[2] Pöschel W, Engel R. The Concept of a New Primary Standard for Liquid Flow Measurement at PTB Braunschweig [J]. Proceedings, 1998.
[3] Hiro shi S ato, Noriyuki Furuichi, Yo shiya Terao and Masaki Takamoto, Basic Design of Very Large Water Flow Calibration Facility for Nuclear Power Application. National Metrology Institute of Japan/AIST.
[4] Jin-Hai L I, Yan-Xun S U. Uncertainty of the Commutator and Stability Measurement in Liquid Flowmeter Standard System[J]. Acta Metrologica Sinica, 2008.
[5] Basicevic I , Kukolj D , Popovic M . On the application of fuzzy-based flow control approach to High Altitude Platform communications[M]. Kluwer Academic Publishers, 2011.
[6] Cheng C, Li B, Qin S. Design and Research of Water Flow Standard Facilities Based on Field Service [M]// Advanced Computational Methods in Energy, Power, Electric Vehicles, and Their Integration. 2017.

Importance Measure Analysis Method for Maintenance based on Mutil-Function Testability States

Zhiyuan Cao[1,*], Haiyong Dong[2], Qingfan Gu[3] and Guoqing Wang[4]

[1]University of Michigan – Shanghai Jiao Tong University Joint Institute, Shanghai, 200240, China
[2]School of Computer Science and Engineering, Northwestern Polytechnical University, Xi' an, 710072, China
[3]Shanghai Avionics Corporation, Shanghai, 200233, China
[4]School of Aeronautics and Astronautics, Shanghai Jiao Tong University, Shanghai, 200240, China

[*]E-mail: caozhiyuan@sjtu.edu.cn

Abstract. Maintenance order plays an important role in industrial field, which influences multiple factors such as costs. Current method is based on importance measure using the fault tree analysis. Simply based on the fault tree of the failure condition, such method ignore some vital information hidden in other testability states, and may results in unsuitable maintenance order. This paper puts forward a novel method to evaluate importance measure based on multi-function testability states, which can be used to optimize the maintenance order. To illustrate the utility of this method, the article uses a vehicle model, and compares the different results to validate the effect of the proposed importance measure. This method can be utilized to evaluate the maintenance order containing multiple functions.

1. Introduction

Maintenance for a complicated system is important in industrial field. Usually, when a testability state is in its failure condition, maintenance engineer replaces the possibly broken units in a particular order based on their experience. Although the traditional method is aimed at maintaining the system with the least replacements of units, in practice it is inefficient and may ignore some other vital factors, resulting in high cost and an unnecessary waste of energy. 5.5% of the companies' turnover in Finland is spent on maintenance on average, and this ratio may rise to 25%. Hence some companies try to improve the equipment reliability by increasing their Mean Time Between Failure (MTBF), etc. [1]. Besides, in fields of vehicle, some maintenance activities are cost much time and money, which make it necessary to carefully determine the order of maintenance [2].

To solve the above problem, researchers have tried a variety of modifications to optimize the process and take into consideration more factors to make the process of maintenance more efficient. Kang used Fault Tree Analysis (FTA) to qualitatively and quantitatively evaluate the failure characteristics of semi-submersible floating offshore wind turbine and gave some suggestions to improve the system performance [3]. Zhang analysed the heave compensation systems using Importance Measure (IM) and assessed the importance of each component to the whole system [4]. Cui used probability density evolution method to optimize the evaluation of importance measure [5]. Gao considered the cost factor during the process of the importance measure evaluation [6]. And

Almugrin emphasized the vital status of testability in the field of maintenance [7]. However, for a system with multi-function testability, the working states of other functions are often ignored, which in fact may be an advantage in guiding the process of maintenance.

In this paper, a novel analysis method based on multi-function testability state is put forward, which provides a different choice of maintenance order to better meet the requirements in industrial practice. The structure of the paper is listed as follows: Section 2 introduces the method to analyse importance measure based on multi-function testability. Section 3 gives a model which deeply illustrates the method as well as its applicability to explain its practical usage. Section 4 concludes the work done in this paper, and proposes some suggestions of future improvements.

2. Importance Measure Analysis Based on Multi-Function Testability States

Importance measure analysis contains quantitative assessment of the components in a defined system or function. By choosing the failure condition which can be analyzed as the top event, the corresponding intermediate or basic events, connected with logic gates "and" or "or", can be written below. By analysing the event thoroughly, all the events could be divided into basic ones and a fault tree is completed. By comparing the values of importance measure in the fault tree, a preliminary and a final maintenance order can be determined accordingly. However, when considering multi-function testability, the method should be modified as the figure followed.

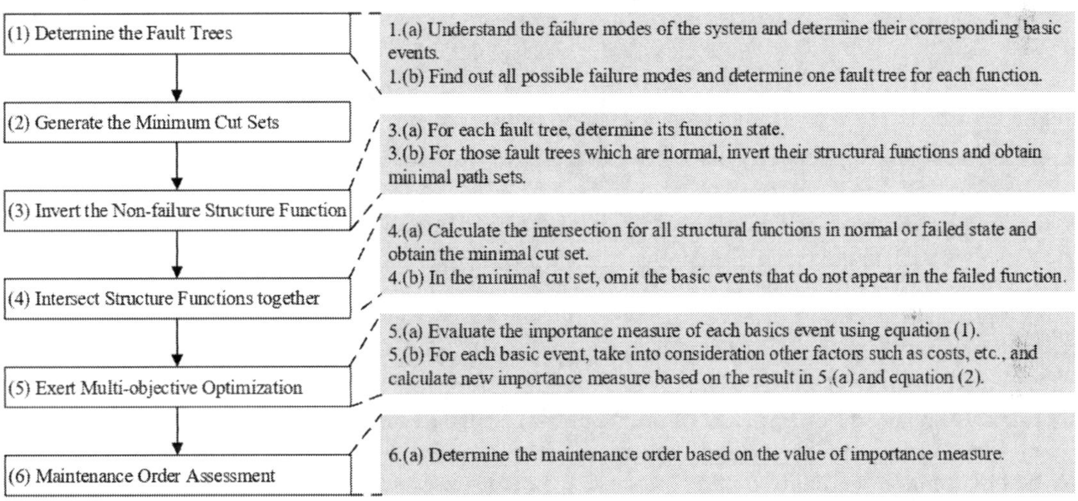

Figure 1. Importance measure analysis process based on multi-function testability.

As shown in figure 1, first, the fault tree for each possible failure condition should be modeled and their minimal cut set and structural function could be generated. Then, according to what happens in practice, determine whether a testability state is in its normal or failed state. For those in normal state, invert their structural function by using De Mogan's Law, i.e. $\overline{\prod_{i=1}^{n} x_i} = \sum_{i=1}^{n} \overline{x_i}$, and $\overline{\sum_{i=1}^{n} x_i} = \prod_{i=1}^{n} \overline{x_i}$. After that, multiply the structural functions of all top events together and use Boolean algebra to simplify them. Specifically, when a basic event intersects with its inverse form, the whole term is canceled. For example, $x_1x_2 \times \overline{x_1}$ should be canceled. Furthermore, if one term is the subset of another term, the subset ought to be canceled. For instance, given a structural function $x_1x_2x_3 + x_2x_3$ the former term should be canceled, and the structural function will turn into x_2x_3. As a result, the minimal cut set are generated, and equation (1) can be used to evaluate their importance measure IM_{mf} (multiple factor optimization):

$$IM_{mf} = \frac{1}{k}\sum_{j=1}^{n}\frac{(-1)^{\lambda}}{n_j}\prod_{p=1}^{P}\Big(1-\frac{1}{k_p}\sum_{j=1}^{m}\frac{1}{m_j}\Big)\prod_{q=1}^{Q}\Big(1+\frac{1}{k_q}\sum_{j=1}^{m}\frac{1}{m_j}\Big) \tag{1}$$

In equation (1), let x_i denote the element being evaluated, then k refers to the number of minimal cut set, n_j refers to the number of homogenous elements (non-inverse basic event and inverse basic event) in a cut set containing x_i, λ refers to sign-judging coefficient, k_p and k_q refer to the number of cut set in other normal function or failure condition respectively, and m_j refers to n_j in the specific failure condition. Here λ equals to 1 when the basic event is non-inverse (i.e., x_i), and -1 when inverse (i.e., $\overline{x_i}$). In this equation, the basic events and their inverse forms are considered separately, and the final result is the difference between the importance measure of the basic events and their inverse forms. Since during the process of subset cancel, some terms disappear which actually contribute to the importance measure, the terms $\prod_{p=1}^{P}(1-\frac{1}{k_p}\sum_{j=1}^{m}\frac{1}{m_j})$ and $\prod_{q=1}^{Q}(1+\frac{1}{k_q}\sum_{j=1}^{m}\frac{1}{m_j})$ are added. To be more specific, although some events in other failure conditions do not influence the final minimal cut set, the failure possibility of a basic event is decreased/increased due to its appearance in another normal/failure condition.

To better determine the importance measure to obtain an optimal maintenance order, other vital factors, such as cost or risks, should be added, and the importance measure IM_{mo} (with multiple factor optimization) can be evaluated as shown in equation (2):

$$IM_{mo} = IM_{mf}\prod_{i=1}^{A}\alpha_i\prod_{j=1}^{B}\frac{1}{\beta_j} \tag{2}$$

In equation (2), α_i refers to the value of factors which has positive correlations (i.e. higher value of α_i is expected), and β_i refers to the value of factors with negative correlations (i.e. lower value of β_i is expected).

Eventually, the importance measure is generated, and the maintenance order can be determined accordingly based on the value of importance measure IM_{mo}.

3. Case Study

3.1. Vehicle failure Model

A vehicle is comprised of a variety of units possessing different functions. Common malfunctions of vehicles include failure in steering, transmission, electric control, or braking systems, etc. Ovidiu verified that during the entire running process, maintenance is vital for vehicles of all types [8]. And Atkinson mentioned that adequate attention on vehicle maintenance will result in economic benefits[9]. Often the function units are made in terms of Field Replace Unit (FRU) which enables quick and easy maintenance replacements. Due to its complexity, a simple FRU may serve as the basic event of different failure conditions. Hence even a normally working state provides information about the failure conditions, and has influence on the maintenance order. To better and clearly illustrate our model, examples in this paper have been simplified. There are three failure conditions in the model, including engine failure, lighting system failure and black exhaust failure, which are shown in figure 2 (a), (b), (c) respectively. Among them there exist 13 different basic events corresponding to 13 FRUs. When maintaining, an optimized order is generated with lower costs and less time, of replacing the possibly broken FRUs with new ones to successfully fix the vehicle with malfunctions.

3.2. Assumptions

In order that the system model is more accurate and easier to analyze, in this paper several common assumptions are taken [10].

- Failure will only occur on the devices and external systems will not have effect on the system;
- The state of a failure mode can only be classified into two categories, one is *Normal* and the other is *Failure*;

IOP Conf. Series: Materials Science and Engineering **825** (2020) 012042 doi:10.1088/1757-899X/825/1/012042

- The devices do not interact with each other.

3.3. Safety Model Construction

Based on the assumptions above, the fault tree analysis can be done as the follow figure.

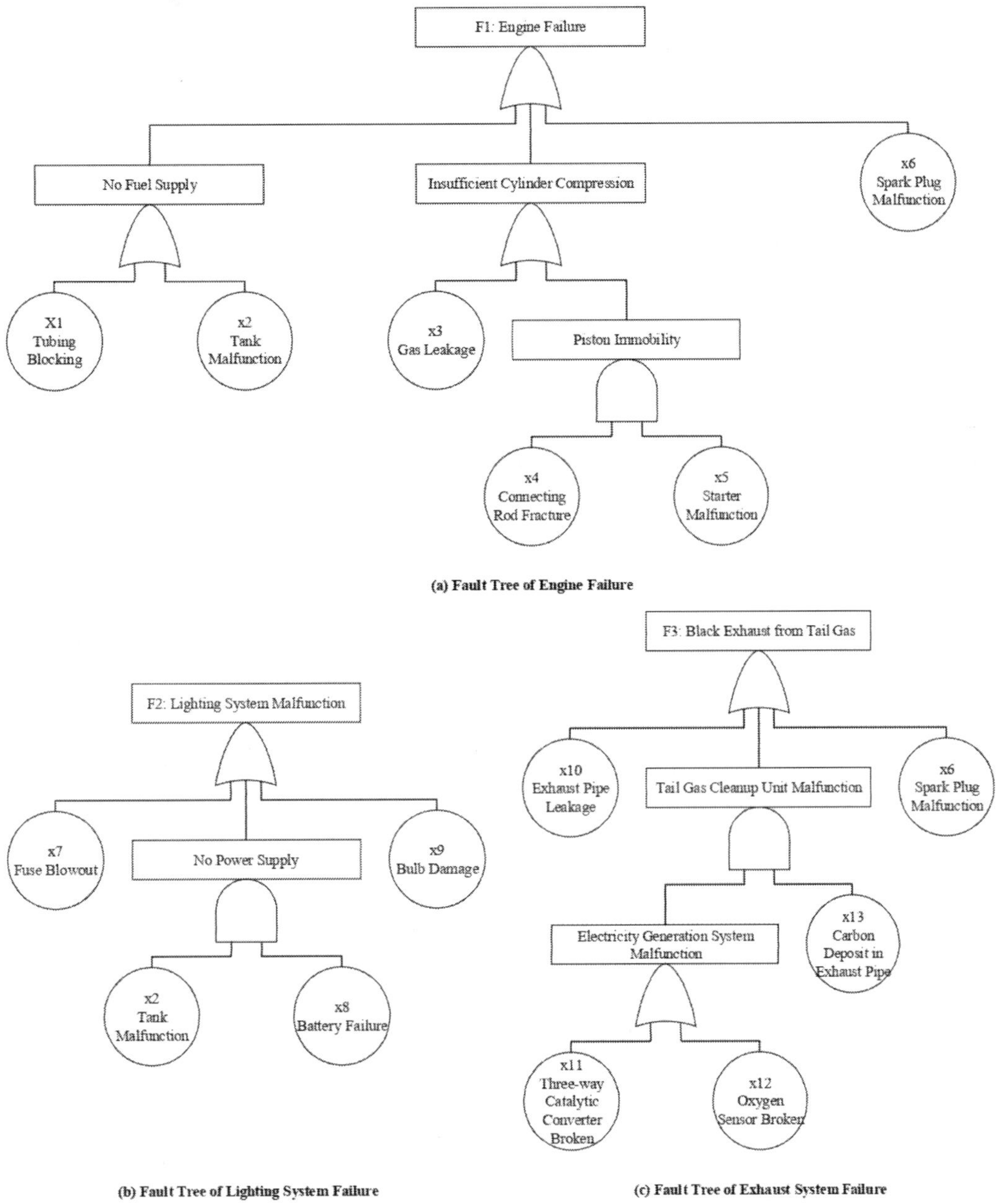

(a) Fault Tree of Engine Failure

(b) Fault Tree of Lighting System Failure

(c) Fault Tree of Exhaust System Failure

Figure 2. Fault Tree of System Function Failure.

As shown in figure 2(a), if the engine cannot be started, it may be caused by lack of fuel supply, insufficient cylinder compression or spark plug malfunction. Among them, lack of fuel supply happens when the oil tank is broken and the tube is blocked. Insufficient cylinder compression takes place when there is a gas leakage or piston immobility, which is caused by the joint effect of connecting rod fracture and starter malfunction. In figure 2(b), similarly, if the lighting system is in its failed state, it may be caused by fuse blowout, no power supply or bulb damage. And malfunctions of tank and battery are responsible for power supply shortage. In figure 2(c), exhaust system may emit black tail gas, which is a common failure condition. Three factors may result in this failure: exhaust pipe leakage, tail gas cleanup unit malfunction, or spark plug malfunction. Among them, the failure of cleanup system is caused by electricity generation system malfunction and carbon deposit in exhaust pipe together. And either three-way catalytic converter or oxygen sensor failure will lead to the malfunction in electricity generation system. Since each event have been divided into basic event, three fault trees are constructed. It can be observed that the three fault trees share some basic events, and thus multi-function testability is applicable.

To illustrate the usage of proposed method, further assume that such condition is satisfied: The engine mode is in its failure condition, while the lighting system and the exhaust system are in their normal state. According to the process explained in section 2, the structural functions of these fault trees can be written as follows:

$$\Phi_1 = x_1 + x_2 + x_3 + x_4 x_5 + x_6$$

$$\Phi_2 = x_2 x_8 + x_7 + x_9$$

$$\Phi_3 = x_6 + x_{10} + x_{11} x_{13} + x_{12} x_{13}$$

3.4. Importance measure

To find out the importance measure, since F_2 and F_3 are in normal state, they ought to be inverted. The results $\overline{\Phi_2}$ and $\overline{\Phi_3}$ are:

$$\overline{\Phi_2} = \overline{x_2} \cdot \overline{x_7} \cdot \overline{x_9} + \overline{x_7} \cdot \overline{x_8} \cdot \overline{x_9}$$

$$\overline{\Phi_3} = \overline{x_6} \cdot \overline{x_{10}} \cdot \overline{x_{11}} \cdot \overline{x_{12}} + \overline{x_6} \cdot \overline{x_{10}} \cdot \overline{x_{13}}$$

Hence, by intersecting $\Phi_1, \overline{\Phi_2}$ and $\overline{\Phi_3}$, the multi-function testability based structural function Φ_{mft} is calculated as followed equation.

$$\begin{aligned}
\Phi_{mft} &= \Phi_1 \cdot \overline{\Phi_2} \cdot \overline{\Phi_3} \\
&= x_1 \cdot \overline{x_2} \cdot \overline{x_6} \cdot \overline{x_7} \cdot \overline{x_9} \cdot \overline{x_{10}} \cdot \overline{x_{11}} \cdot \overline{x_{12}} + x_1 \cdot \overline{x_2} \cdot \overline{x_6} \cdot \overline{x_7} \cdot \overline{x_9} \cdot \overline{x_{10}} \cdot \overline{x_{13}} + x_1 \cdot \overline{x_6} \cdot \overline{x_7} \cdot \overline{x_8} \cdot \overline{x_9} \cdot \overline{x_{10}} \cdot \overline{x_{11}} \cdot \overline{x_{12}} \\
&+ x_1 \cdot \overline{x_6} \cdot \overline{x_7} \cdot \overline{x_8} \cdot \overline{x_9} \cdot \overline{x_{10}} \cdot \overline{x_{13}} + x_2 \cdot \overline{x_6} \cdot \overline{x_7} \cdot \overline{x_8} \cdot \overline{x_9} \cdot \overline{x_{10}} \cdot \overline{x_{11}} \cdot \overline{x_{12}} + x_2 \cdot \overline{x_6} \cdot \overline{x_7} \cdot \overline{x_8} \cdot \overline{x_9} \cdot \overline{x_{10}} \cdot \overline{x_{13}} \\
&+ x_3 \cdot \overline{x_2} \cdot \overline{x_6} \cdot \overline{x_7} \cdot \overline{x_9} \cdot \overline{x_{10}} \cdot \overline{x_{11}} \cdot \overline{x_{12}} + x_3 \cdot \overline{x_2} \cdot \overline{x_6} \cdot \overline{x_7} \cdot \overline{x_9} \cdot \overline{x_{10}} \cdot \overline{x_{13}} + x_3 \cdot \overline{x_6} \cdot \overline{x_7} \cdot \overline{x_8} \cdot \overline{x_9} \cdot \overline{x_{10}} \cdot \overline{x_{11}} \cdot \overline{x_{12}} \\
&+ x_3 \cdot \overline{x_6} \cdot \overline{x_7} \cdot \overline{x_8} \cdot \overline{x_9} \cdot \overline{x_{10}} \cdot \overline{x_{13}} + x_4 \cdot x_5 \cdot \overline{x_2} \cdot \overline{x_6} \cdot \overline{x_7} \cdot \overline{x_9} \cdot \overline{x_{10}} \cdot \overline{x_{11}} \cdot \overline{x_{12}} + x_4 \cdot x_5 \cdot \overline{x_2} \cdot \overline{x_6} \cdot \overline{x_7} \cdot \overline{x_9} \cdot \overline{x_{10}} \cdot \overline{x_{13}} \\
&+ x_4 \cdot x_5 \cdot \overline{x_6} \cdot \overline{x_7} \cdot \overline{x_8} \cdot \overline{x_9} \cdot \overline{x_{10}} \cdot \overline{x_{11}} \cdot \overline{x_{12}} + x_4 \cdot x_5 \cdot \overline{x_6} \cdot \overline{x_7} \cdot \overline{x_8} \cdot \overline{x_9} \cdot \overline{x_{10}} \cdot \overline{x_{13}}
\end{aligned}$$

Then, $x_7 \sim x_{13}$ are omitted because they are irrelevant to the failure condition F_1.

$$\begin{aligned}
\Phi_{mft} &= x_1 \cdot \overline{x_2} \cdot \overline{x_6} + x_1 \cdot \overline{x_2} \cdot \overline{x_6} + x_1 \cdot \overline{x_6} + x_1 \cdot \overline{x_6} + x_2 \cdot \overline{x_6} + x_2 \cdot \overline{x_6} + x_3 \cdot \overline{x_2} \cdot \overline{x_6} + x_3 \cdot \overline{x_2} \cdot \overline{x_6} + x_3 \cdot \overline{x_6} + x_3 \cdot \overline{x_6} \\
&+ x_4 \cdot \underline{x_5} \cdot \overline{x_2} \cdot \underline{\overline{x_6}} + x_4 \cdot \underline{x_5} \cdot \overline{x_2} \cdot \overline{x_6} + \underline{x_4} \cdot x_5 \cdot \overline{x_6} + x_4 \cdot x_5 \cdot \overline{x_6} \\
&= x_1 \cdot x_6 + x_2 \cdot x_6 + x_3 \cdot x_6 + x_4 \cdot x_5 \cdot x_6
\end{aligned}$$

Therefore the four minimal cut sets $x_1 \cdot \overline{x_6}, x_2 \cdot \overline{x_6}, x_3 \cdot \overline{x_6}, x_4 \cdot x_5 \cdot \overline{x_6}$ are generated. According to equation (1), the importance measure IM_{mf} (without multiple factor optimization) for $x_1 \sim x_6$ can be evaluated. Due to the fact that the bigger the value of IM_{mf}, the more possible the unit is broken, the preliminary maintenance order is determined.

However, since the importance measure of x_1 equal to x_3, and x_4 equal to x_5, maintenance order of their corresponding FRU U_{x_1} and U_{x_3} cannot be decided. Thus multiple factors regarding maintenance are introduced, which help optimize the result further.

The information about the costs and MTBF are listed in Table 1. According to equation (2), since both MTBF and costs have negative correlations, IM_{mo} can be calculated from multiplying IM_{mf} by the influence factor of MTBF and costs. Based on the optimized importance measure, final results can be determined, which are shown in Table 1.

Table 1. Importance measure with multiple factor optimization and final maintenance order

	Information		Traditional Method		Optimized Method			
Basic Event	MTBF (h)	Average Cost ($)	IM_{tr}	Maintenance Order	IM_{mf}	Preliminary Maintenance Order	IM_{mo}	Final Maintenance Order
x_1	1200	180	0.2	1	0.250	1	1.157E-6	2
x_2	1000	370	0.2	1	0.208	3	5.622E-7	5
x_3	800	120	0.2	1	0.250	1	2.604E-6	1
x_4	900	220	0.1	5	0.125	4	6.313E-7	4
x_5	1000	100	0.1	5	0.125	4	1.150E-6	3
x_6	800	30	0.2	1	-0.750	$+\infty$	-3.142E-5	$+\infty$

3.5. Results Analysis

As shown in Table 1, by using the traditional method which simply based on importance measure of the failure condition, it may cost more money and time to carry out maintenance. However, by considering the influence of other normal states, maintenance will cost less money and time as x_6 is confirmed to be well. Finally, if the influence of other factors such as costs and MTBF are included, maintenance order can be further optimized, resulting in a better maintenance order.

4. Conclusions

This paper introduces a novel importance measure analysis method for maintenance based on mutil-function testability state. Different from traditional methods, the proposed method considers the influence of testability of other function state on the studied one, which ensures a more reasonable maintenance order. Taking the vehicle model as an example, there exists 3 function corresponding to 3 failure conditions. Besides the failed function, the importance measure considers the influence of the normal ones, and presents an optimized maintenance order, which is more reasonable comprehensively.

However, when dealing with systems with high complexity, the method may be too complicated to evaluate the value of importance measure quickly.

References

[1] Komonen K 2002 *Int. J. Prod. Econ.* **79** 15-31
[2] López-Arquillos A, Rubio-Romero, J C, Súarez-Cebador, M and del Carmen Pardo-Ferreira, M 2015 Int. J. Ind. Ergonom. **47** 53-60
[3] Jichuan K, Liping S and Guedes S C 2018 *Renew. Energ.* **133** 1455-1467
[4] Zhang C, Qian Y, Dui H, Wang S and Shi J 2020 *J. Loss Prevent Proc.* **63** 103996
[5] Cui L, Lü Z and Zhao X 2010 *Sci. China Technol. Sc.* **53**(4) 1138-1145
[6] Gao X, Barabady J and Markeset T 2010 *Int. J. Syst. Assur. Eng. Manag.* **1** 17-23
[7] Almugrin S, Albattah W and Melton A 2016 J. Syst. Software **121** 298-310
[8] Vulcu O I and Arghir M 2016 *Mechatronics: Ideas, Challenges, Solutions and Applications* **414** 219-238
[9] Atkinson J and Postle O 1977 The Effect of Vehicle Maintenance on Fuel Economy *Fuel Economy of the Gasoline Engine* ed Blackmore D R and Thomas A (London: Springer)chapter 7 pp 119-132
[10] Dong H, Gu Q, Wang G, Zhai Z, Lu Y and Wang M 2019 *Processes* **7** 117

IOP Conf. Series: Materials Science and Engineering **825** (2020) 012043 doi:10.1088/1757-899X/825/1/012043

Neural Network Based Modelling and Steering Control of an Intelligent Vehicle Under Dynamic Sensitive Conditions

K Zhang[1,*], S J Wang[2], L Ji[3] and C Wang[1]

[1]E/E department, Brilliance Automobile Engineering Research Institute, shen yang 110141, China
[2]School of Mechanical Engineering, Shenyang University of Technology, shen yang, 110023, China
[3]School of Mechatronics Engineering, Shenyang Aerospace University, shen yang, 110136, China

[*]hit_zhangkun@126.com

Abstract. A neural network based modelling and steering control method of an intelligent vehicle under dynamic sensitive conditions is proposed in this paper. Four radius function neural networks are used to compensate the modelling error of a traditional dynamic vehicle steering model. The designed control law is consisted of four terms: a neural network term, a proportion term, an integral term and a robust term. Stability is analysed through Lyapunov function and the training law is generated at the same time. Test results show that the designed steering controller works quite well and is of great value in path tracking of an intelligent vehicle when the vehicle dynamics getting significant.

1. Introduction

With the development of the automobile, the problems of traffic safety, traffic congestion and traffic pollution have become more and more significant. Intelligent Transportation System (ITS) has proven to be an effective solution of these problems. As the core component of ITS, research on the connected intelligent vehicle has been highly valued and widely concerned by countries all over the world, including all major automobile manufacturers, automobile component suppliers, internet enterprises, universities and research institutes, which was also highly fuelled by the development in computer science and artificial intelligence recent years.

The DARPA Grand Challenge started in 2004 and DARPA Urban Challenge in 2007 is a major milestone of modern intelligent driving development. The raised abstracted system plan, layered software architecture by those successful entries are still well used by many most popular intelligent driving prototypes all around the world. To limit the scope of this paper, the perception, localization, route planning, behavioural decision and motion planner are omitted, only the motion controller is studied, whose role is to track the reference path or trajectory that was provided by the motion planner, with insurance of vehicle stability and a small tracking error.

There have been plenty of studies regrading motion control of intelligent driving vehicles. Some of these control methods exploit the geometric relationship between vehicle and the reference path, which are simple, robust, and can be easily implemented but can still achieve accurate path tracking in a broad set of driving scenarios[1, 2]. As the vehicle kinematics become significant, simplifying the vehicle to a kinematic bicycle model is a common approximation. Some well-known control methods are implemented on the kinematic model, and are widely used in platooning and motion control of vehicles with trailers[3-5]. When the vehicle speed, acceleration and side slip of the vehicle become significant, control methods based on the dynamic vehicle steering model and more advanced control methods, like Model Predictive Control[6], Sliding Model Control[7] and fuzzy control[8] are widely used, which are more complicate but can achieve desirable performance in these dynamic sensitive scenarios.

The commonly used dynamic model is built under some assumptions, which may fail in dynamic sensitive steering scenarios. Thus, a neural network compensated dynamic vehicle steering model is built, and the corresponding control method is proposed in this paper.

2. Dynamic vehicle steering model

The most commonly used dynamic vehicle steering model is called bicycle model, which in fact is two wheels that were rigidly linked. The same dynamic vehicle steering model built in [9] is used in this paper, detailed modelling process is not listed here due to limited space. The model in state space is

$$\boldsymbol{A}\ddot{\boldsymbol{q}} + \boldsymbol{B}\dot{\boldsymbol{q}} + \boldsymbol{C}\boldsymbol{q} - \boldsymbol{H} = \boldsymbol{\delta} \tag{1}$$

Where $\boldsymbol{q} = \begin{bmatrix} e_{cg} & \theta_e \end{bmatrix}^T$

$$\boldsymbol{q} = \begin{bmatrix} e_{cg} \\ \theta_e \end{bmatrix}, \quad \boldsymbol{A} = \begin{bmatrix} \frac{m}{c_f} & 0 \\ 0 & \frac{I_z}{l_f c_f} \end{bmatrix}, \quad \boldsymbol{B} = \begin{bmatrix} -\frac{c_f + c_r}{c_f v_x} & -\frac{l_f c_f - l_r c_r}{c_f v_x} \\ -\frac{l_f c_f - l_r c_r}{l_f c_f v_x} & \frac{l_f^2 c_f + l_r^2 c_r}{l_f c_f v_x} \end{bmatrix}, \quad \boldsymbol{C} = \begin{bmatrix} 0 & \frac{c_f + c_r}{c_f} \\ 0 & \frac{l_f c_f - l_r c_r}{l_f c_f v_x} \end{bmatrix},$$

$$\boldsymbol{H} = \begin{bmatrix} \frac{m}{c_f}\left(\frac{l_f c_f - l_r c_r}{m v_x} + v_x\right)\omega(s) \\ \frac{l_f^2 c_f + l_r^2 c_r}{l_f c_f v_x}\omega(s) - \frac{I_z}{l_f c_f}\dot{\omega}(s) \end{bmatrix}$$

As is shown above, e_{cg} is the cross track error, θ_e is the heading error, m is the mass of the vehicle, ω is the yaw rate, I_z is the inertial moment, l_f and l_r are the length of front and rear half axis, c_f and c_r is the corner stiffness of the front and rear wheels.

This model is built under some assumptions, so it is supposed to fail under some dynamic sensitive conditions such as when the vehicle speed and acceleration getting larger, the lateral slip getting significant. Thus, neural network is used to compensate these modelling errors in this paper.

3. Control algorithm design

3.1. Control law design

During the modeling process of the dynamic vehicle steering model, there exist many uncertainties and assumptions, thus, four radial basis function network are used to model $\boldsymbol{A}$, $\boldsymbol{B}$, $\boldsymbol{C}$ and $\boldsymbol{H}$:

$$\begin{cases} \boldsymbol{A} = \boldsymbol{A}_{NN} + \boldsymbol{E}_A \\ \boldsymbol{B} = \boldsymbol{B}_{NN} + \boldsymbol{E}_B \\ \boldsymbol{C} = \boldsymbol{C}_{NN} + \boldsymbol{E}_C \\ \boldsymbol{H} = \boldsymbol{H}_{NN} + \boldsymbol{E}_H \end{cases} \tag{2}$$

Where $\boldsymbol{A}_{NN}$, $\boldsymbol{B}_{NN}$, $\boldsymbol{C}_{NN}$ and $\boldsymbol{H}_{NN}$ are the output of the corresponding networks, $\boldsymbol{E}_A$, $\boldsymbol{E}_B$, $\boldsymbol{E}_C$

and $\boldsymbol{E}_H$ are the corresponding network modelling errors.

Thus, $\boldsymbol{A}\ddot{\boldsymbol{q}}_r + \boldsymbol{B}\dot{\boldsymbol{q}}_r + \boldsymbol{C}\boldsymbol{q}_r - \boldsymbol{H}$ can be represented as

$$\begin{aligned}\boldsymbol{A}\ddot{\boldsymbol{q}}_r + \boldsymbol{B}\dot{\boldsymbol{q}}_r + \boldsymbol{C}\boldsymbol{q}_r - \boldsymbol{H} &= \boldsymbol{A}_{NN}\ddot{\boldsymbol{q}}_r + \boldsymbol{B}_{NN}\dot{\boldsymbol{q}}_r + \boldsymbol{C}_{NN}\boldsymbol{q}_r - \boldsymbol{H}_{NN} + \boldsymbol{E} \\ &= \boldsymbol{\omega}_A^T\boldsymbol{h}_A\ddot{\boldsymbol{q}}_r + \boldsymbol{\omega}_B^T\boldsymbol{h}_B\dot{\boldsymbol{q}}_r + \boldsymbol{\omega}_C^T\boldsymbol{h}_C\boldsymbol{q}_r - \boldsymbol{\omega}_H^T\boldsymbol{h}_H + \boldsymbol{E}\end{aligned} \tag{3}$$

Where, $\boldsymbol{\omega}_A$, $\boldsymbol{\omega}_B$, $\boldsymbol{\omega}_C$ and $\boldsymbol{\omega}_H$ are the corresponding network weight vector, $\boldsymbol{h}_A$, $\boldsymbol{h}_B$, $\boldsymbol{h}_C$ and $\boldsymbol{h}_H$ are the corresponding network hidden layer outputs, $\boldsymbol{E}$ is the error term, which is represented as $\boldsymbol{E} = \boldsymbol{E}_A\ddot{\boldsymbol{q}}_r + \boldsymbol{E}_B\dot{\boldsymbol{q}}_r + \boldsymbol{E}_C\boldsymbol{q}_r - \boldsymbol{E}_H$, $\ddot{\boldsymbol{q}}_r$ and $\dot{\boldsymbol{q}}_r$ are the inputs with error terms, which is represented as

$$\begin{cases}\dot{\boldsymbol{q}}_r = \boldsymbol{r} + \dot{\boldsymbol{q}} = \dot{\boldsymbol{e}} + \boldsymbol{\Lambda}\boldsymbol{e} + \dot{\boldsymbol{q}}_d - \dot{\boldsymbol{e}} = \dot{\boldsymbol{q}}_d + \boldsymbol{\Lambda}\boldsymbol{e} \\ \ddot{\boldsymbol{q}}_r = \dot{\boldsymbol{r}} + \ddot{\boldsymbol{q}} = \ddot{\boldsymbol{e}} + \boldsymbol{\Lambda}\dot{\boldsymbol{e}} + \ddot{\boldsymbol{q}}_d - \ddot{\boldsymbol{e}} = \ddot{\boldsymbol{q}}_d + \boldsymbol{\Lambda}\dot{\boldsymbol{e}}\end{cases} \tag{4}$$

Where, $\boldsymbol{\Lambda}$ is a diagonal matrix, designed as $\boldsymbol{\Lambda} = \text{eye}(5)$.

Represent the estimated value of $\boldsymbol{A}_{NN}$, $\boldsymbol{B}_{NN}$, $\boldsymbol{C}_{NN}$ and $\boldsymbol{H}_{NN}$ as $\hat{\boldsymbol{A}}_{NN} = \hat{\boldsymbol{\omega}}_A^T\boldsymbol{h}_A$, $\hat{\boldsymbol{B}}_{NN} = \hat{\boldsymbol{\omega}}_B^T\boldsymbol{h}_B$, $\hat{\boldsymbol{C}}_{NN} = \hat{\boldsymbol{\omega}}_C^T\boldsymbol{h}_C$ and $\hat{\boldsymbol{H}}_{NN} = \hat{\boldsymbol{\omega}}_H^T\boldsymbol{h}_H$. Where, $\hat{\boldsymbol{\omega}}_A^T$, $\hat{\boldsymbol{\omega}}_B^T$, $\hat{\boldsymbol{\omega}}_C^T$ and $\hat{\boldsymbol{\omega}}_H^T$ are the estimated weight vector of $\boldsymbol{A}_{NN}$, $\boldsymbol{B}_{NN}$, $\boldsymbol{C}_{NN}$ and $\boldsymbol{H}_{NN}$.

Thus, the original dynamic model can be represented as

$$\begin{aligned}\boldsymbol{\delta} &= \boldsymbol{A}\ddot{\boldsymbol{q}} + \boldsymbol{B}\dot{\boldsymbol{q}} + \boldsymbol{C}\boldsymbol{q} - \boldsymbol{H} = \boldsymbol{A}\ddot{\boldsymbol{q}}_r + \boldsymbol{B}\dot{\boldsymbol{q}}_r + \boldsymbol{C}\boldsymbol{q}_r - \boldsymbol{H} - \boldsymbol{A}\dot{\boldsymbol{r}} - \boldsymbol{B}\boldsymbol{r} \\ &= \boldsymbol{\omega}_A^T\boldsymbol{h}_A\ddot{\boldsymbol{q}}_r + \boldsymbol{\omega}_B^T\boldsymbol{h}_B\dot{\boldsymbol{q}}_r + \boldsymbol{\omega}_C^T\boldsymbol{h}_C\boldsymbol{q}_r - \boldsymbol{\omega}_H^T\boldsymbol{h}_H - \boldsymbol{A}\dot{\boldsymbol{r}} - \boldsymbol{B}\boldsymbol{r} - \boldsymbol{C}\int_0^t \boldsymbol{r}\,\mathrm{d}t + \boldsymbol{E}\end{aligned} \tag{5}$$

The steering angle is designed as

$$\boldsymbol{\delta} = \boldsymbol{\delta}_m + \boldsymbol{K}_p\boldsymbol{r} + \boldsymbol{K}_i\int_0^t \boldsymbol{r}\,\mathrm{d}t + \boldsymbol{\delta}_r \tag{6}$$

$\boldsymbol{\delta}_m$ is the estimated steering angle of the network, which is $\boldsymbol{\delta}_m = \hat{\boldsymbol{A}}_{NN}\ddot{\boldsymbol{q}}_r + \hat{\boldsymbol{B}}_{NN}\dot{\boldsymbol{q}}_r + \hat{\boldsymbol{C}}_{NN}\boldsymbol{q}_r - \hat{\boldsymbol{H}}_{NN}$, $\boldsymbol{K}_p = \text{eye}(100)$ is the proportion term; $\boldsymbol{K}_i = \text{eye}(100)$ is the integral term, $\boldsymbol{\delta}_r$ is the robust term considering the network modelling error, designed as $\boldsymbol{\delta}_r = \boldsymbol{K}_r\,\text{sgn}(\boldsymbol{r})$, $\boldsymbol{K}_r = \text{diag}\left[k_{rii}\right], k_{rii} \geq \left|\boldsymbol{E}_i\right|$.

3.2. Neural network design

Substitute the estimated steering angle and the robust term into the control law (6), yields

$$\boldsymbol{\delta} = \hat{\boldsymbol{\omega}}_A^T\boldsymbol{h}_A\ddot{\boldsymbol{q}}_r + \hat{\boldsymbol{\omega}}_B^T\boldsymbol{h}_B\dot{\boldsymbol{q}}_r + \hat{\boldsymbol{\omega}}_C^T\boldsymbol{h}_C\boldsymbol{q}_r - \hat{\boldsymbol{\omega}}_H^T\boldsymbol{h}_H + \boldsymbol{K}_p\boldsymbol{r} + \boldsymbol{K}_i\int_0^t \boldsymbol{r}\,\mathrm{d}t + \boldsymbol{K}_r\,\text{sgn}(\boldsymbol{r}) \tag{7}$$

Substitute this control law (7) into the dynamic model (5), yields

$$\begin{aligned}&\boldsymbol{A}\dot{\boldsymbol{r}} + \boldsymbol{B}\boldsymbol{r} + \boldsymbol{C}\int_0^t \boldsymbol{r}\,\mathrm{d}t + \boldsymbol{K}_p\boldsymbol{r} + \boldsymbol{K}_i\int_0^t \boldsymbol{r}\,\mathrm{d}t + \boldsymbol{K}_r\,\text{sgn}(\boldsymbol{r}) \\ &= (\boldsymbol{\omega}_A^T - \hat{\boldsymbol{\omega}}_A^T)\boldsymbol{h}_A\ddot{\boldsymbol{q}}_r + (\boldsymbol{\omega}_B^T - \hat{\boldsymbol{\omega}}_B^T)\boldsymbol{h}_B\dot{\boldsymbol{q}}_r + (\boldsymbol{\omega}_C^T - \hat{\boldsymbol{\omega}}_C^T)\boldsymbol{h}_C\boldsymbol{q}_r - (\boldsymbol{\omega}_H^T - \hat{\boldsymbol{\omega}}_H^T)\boldsymbol{h}_H + \boldsymbol{E} \\ &= \tilde{\boldsymbol{\omega}}_A^T\boldsymbol{h}_A\ddot{\boldsymbol{q}}_r + \tilde{\boldsymbol{\omega}}_B^T\boldsymbol{h}_B\dot{\boldsymbol{q}}_r + \tilde{\boldsymbol{\omega}}_C^T\boldsymbol{h}_C\boldsymbol{q}_r - \tilde{\boldsymbol{\omega}}_H^T\boldsymbol{h}_H + \boldsymbol{E}\end{aligned} \tag{8}$$

Training law is designed as $\dot{\hat{\boldsymbol{\omega}}}_{Ak} = \boldsymbol{\Gamma}_{Ak}\boldsymbol{h}_{Ak}\ddot{\boldsymbol{q}}_r\boldsymbol{r}_k$, $\dot{\hat{\boldsymbol{\omega}}}_{Bk} = \boldsymbol{\Gamma}_{Bk}\boldsymbol{h}_{Bk}\dot{\boldsymbol{q}}_r\boldsymbol{r}_k$, $\dot{\hat{\boldsymbol{\omega}}}_{Ck} = \boldsymbol{\Gamma}_{Ck}\boldsymbol{h}_{Ck}\boldsymbol{q}_r\boldsymbol{r}_k$ and $\dot{\hat{\boldsymbol{\omega}}}_{Hk} = \boldsymbol{\Gamma}_{Hk}\boldsymbol{h}_{Hk}\boldsymbol{r}_k$, where $\boldsymbol{\Gamma}_{Ak}$, $\boldsymbol{\Gamma}_{Bk}$, $\boldsymbol{\Gamma}_{Ck}$ and $\boldsymbol{\Gamma}_{Hk}$ are symmetric definite matrix, $\boldsymbol{h}_{Ak}$, $\boldsymbol{h}_{Bk}$, $\boldsymbol{h}_{Ck}$ and $\boldsymbol{h}_{Hk}$ are the corresponding network hidden layer outputs, and k is number of inputs.

IOP Conf. Series: Materials Science and Engineering **825** (2020) 012043 doi:10.1088/1757-899X/825/1/012043

3.3. Stability analysis

Design Lyapunov function as

$$V=\frac{1}{2}\boldsymbol{r}^T\boldsymbol{A}\boldsymbol{r}+\frac{1}{2}\left(\int_0^t\boldsymbol{r}\,\mathrm{d}t\right)^T\boldsymbol{K}_i\int_0^t\boldsymbol{r}\,\mathrm{d}t+\frac{1}{2}\left(\int_0^t\boldsymbol{r}\,\mathrm{d}t\right)^T\boldsymbol{C}\int_0^t\boldsymbol{r}\,\mathrm{d}t+\frac{1}{2}\sum_{k=1}^{n}\tilde{\boldsymbol{\omega}}_{Ak}^T\boldsymbol{\Gamma}_{Ak}^{-1}\tilde{\boldsymbol{\omega}}_{Ak}$$
$$+\frac{1}{2}\sum_{k=1}^{n}\tilde{\boldsymbol{\omega}}_{Bk}^T\boldsymbol{\Gamma}_{Bk}^{-1}\tilde{\boldsymbol{\omega}}_{Bk}+\frac{1}{2}\sum_{k=1}^{n}\tilde{\boldsymbol{\omega}}_{Ck}^T\boldsymbol{\Gamma}_{Ck}^{-1}\tilde{\boldsymbol{\omega}}_{Ck}-\frac{1}{2}\sum_{k=1}^{n}\tilde{\boldsymbol{\omega}}_{Hk}^T\boldsymbol{\Gamma}_{Hk}^{-1}\tilde{\boldsymbol{\omega}}_{Hk} \tag{9}$$

The derivation of Lyapunov function (9) is

$$\dot{V}=\boldsymbol{r}^T\left(\boldsymbol{A}\dot{\boldsymbol{r}}+\frac{1}{2}\dot{\boldsymbol{A}}\boldsymbol{r}+\boldsymbol{K}_i\int_0^t\boldsymbol{r}\,\mathrm{d}t+\boldsymbol{C}\int_0^t\boldsymbol{r}\,\mathrm{d}t\right)+\sum_{k=1}^{n}\tilde{\boldsymbol{\omega}}_{Ak}^T\boldsymbol{\Gamma}_{Ak}^{-1}\dot{\tilde{\boldsymbol{\omega}}}_{Ak}$$
$$+\sum_{k=1}^{n}\tilde{\boldsymbol{\omega}}_{Bk}^T\boldsymbol{\Gamma}_{Bk}^{-1}\dot{\tilde{\boldsymbol{\omega}}}_{Bk}+\sum_{k=1}^{n}\tilde{\boldsymbol{\omega}}_{Ck}^T\boldsymbol{\Gamma}_{Ck}^{-1}\dot{\tilde{\boldsymbol{\omega}}}_{Ck}-\sum_{k=1}^{n}\tilde{\boldsymbol{\omega}}_{Hk}^T\boldsymbol{\Gamma}_{Hk}^{-1}\dot{\tilde{\boldsymbol{\omega}}}_{Hk} \tag{10}$$

Considering the symmetric of $\dot{\boldsymbol{A}}-2\boldsymbol{B}$, yields

$$\dot{V}=\boldsymbol{r}^T\left(\boldsymbol{A}\dot{\boldsymbol{r}}+\boldsymbol{B}\boldsymbol{r}+\boldsymbol{C}\int_0^t\boldsymbol{r}\,\mathrm{d}t+\boldsymbol{K}_i\int_0^t\boldsymbol{r}\,\mathrm{d}t\right)+\sum_{k=1}^{n}\tilde{\boldsymbol{\omega}}_{Ak}^T\boldsymbol{\Gamma}_{Ak}^{-1}\dot{\tilde{\boldsymbol{\omega}}}_{Ak}$$
$$+\sum_{k=1}^{n}\tilde{\boldsymbol{\omega}}_{Bk}^T\boldsymbol{\Gamma}_{Bk}^{-1}\dot{\tilde{\boldsymbol{\omega}}}_{Bk}+\sum_{k=1}^{n}\tilde{\boldsymbol{\omega}}_{Ck}^T\boldsymbol{\Gamma}_{Ck}^{-1}\dot{\tilde{\boldsymbol{\omega}}}_{Ck}-\sum_{k=1}^{n}\tilde{\boldsymbol{\omega}}_{Hk}^T\boldsymbol{\Gamma}_{Hk}^{-1}\dot{\tilde{\boldsymbol{\omega}}}_{Hk} \tag{11}$$

Substitute (8) into (11), yields

$$\dot{V}=-\boldsymbol{r}^T\boldsymbol{K}_p\boldsymbol{r}+\boldsymbol{r}^T\boldsymbol{E}-\boldsymbol{r}^T\boldsymbol{K}_r\,\mathrm{sgn}(\boldsymbol{r})+\boldsymbol{r}^T\tilde{\boldsymbol{\omega}}_A^T\boldsymbol{h}_A\ddot{\boldsymbol{q}}_r+\boldsymbol{r}^T\tilde{\boldsymbol{\omega}}_B^T\boldsymbol{h}_B\dot{\boldsymbol{q}}_r+\boldsymbol{r}^T\tilde{\boldsymbol{\omega}}_C^T\boldsymbol{h}_C\boldsymbol{q}_r-\boldsymbol{r}^T\tilde{\boldsymbol{\omega}}_H^T\boldsymbol{h}_H$$
$$+\sum_{k=1}^{n}\tilde{\boldsymbol{\omega}}_{Ak}^T\boldsymbol{\Gamma}_{Ak}^{-1}\dot{\tilde{\boldsymbol{\omega}}}_{Ak}+\sum_{k=1}^{n}\tilde{\boldsymbol{\omega}}_{Bk}^T\boldsymbol{\Gamma}_{Bk}^{-1}\dot{\tilde{\boldsymbol{\omega}}}_{Bk}+\sum_{k=1}^{n}\tilde{\boldsymbol{\omega}}_{Ck}^T\boldsymbol{\Gamma}_{Ck}^{-1}\dot{\tilde{\boldsymbol{\omega}}}_{Ck}-\sum_{k=1}^{n}\tilde{\boldsymbol{\omega}}_{Hk}^T\boldsymbol{\Gamma}_{Hk}^{-1}\dot{\tilde{\boldsymbol{\omega}}}_{Hk}$$
$$=-\boldsymbol{r}^T\boldsymbol{K}_p\boldsymbol{r}+\boldsymbol{r}^T\boldsymbol{E}-\boldsymbol{r}^T\boldsymbol{K}_r\,\mathrm{sgn}(\boldsymbol{r})+\sum_{k=1}^{n}\tilde{\boldsymbol{\omega}}_{Ak}^T\boldsymbol{h}_{Ak}\ddot{\boldsymbol{q}}_r\boldsymbol{r}_k+\sum_{k=1}^{n}\tilde{\boldsymbol{\omega}}_{Bk}^T\boldsymbol{h}_{Bk}\dot{\boldsymbol{q}}\boldsymbol{r}_k+\sum_{k=1}^{n}\tilde{\boldsymbol{\omega}}_{Ck}^T\boldsymbol{h}_{Ck}\boldsymbol{q}_r\boldsymbol{r}_k \tag{12}$$
$$-\sum_{k=1}^{n}\tilde{\boldsymbol{\omega}}_{Hk}^T\boldsymbol{h}_{Hk}\boldsymbol{r}_k+\sum_{k=1}^{n}\tilde{\boldsymbol{\omega}}_{Ak}^T\boldsymbol{\Gamma}_{Ak}^{-1}\dot{\tilde{\boldsymbol{\omega}}}_{Ak}+\sum_{k=1}^{n}\tilde{\boldsymbol{\omega}}_{Bk}^T\boldsymbol{\Gamma}_{Bk}^{-1}\dot{\tilde{\boldsymbol{\omega}}}_{Bk}+\sum_{k=1}^{n}\tilde{\boldsymbol{\omega}}_{Ck}^T\boldsymbol{\Gamma}_{Ck}^{-1}\dot{\tilde{\boldsymbol{\omega}}}_{Ck}-\sum_{k=1}^{n}\tilde{\boldsymbol{\omega}}_{Hk}^T\boldsymbol{\Gamma}_{Hk}^{-1}\dot{\tilde{\boldsymbol{\omega}}}_{Hk}$$
$$=-\boldsymbol{r}^T\boldsymbol{K}_p\boldsymbol{r}+\boldsymbol{r}^T\boldsymbol{E}-\boldsymbol{r}^T\boldsymbol{K}_r\,\mathrm{sgn}(\boldsymbol{r})$$

Let $k_{rii}\geq|\boldsymbol{E}_i|$, it can induced that

$$\boldsymbol{r}^T\boldsymbol{E}-\boldsymbol{r}^T\boldsymbol{K}_r\,\mathrm{sgn}(\boldsymbol{r})=\boldsymbol{r}^T(\boldsymbol{E}-\boldsymbol{K}_r\,\mathrm{sgn}(\boldsymbol{r}))\leq 0 \tag{13}$$

Substitute (13) into (12), yields $\dot{V}=-\boldsymbol{r}^T\boldsymbol{K}_p\boldsymbol{r}+\boldsymbol{r}^T(\boldsymbol{E}-\boldsymbol{K}_r\,\mathrm{sgn}(\boldsymbol{r}))\leq 0$

So this designed system is stable.

4. Simulation and test results

4.1. The Test Vehicle

The test vehicle is developed by brilliance automobile research institute, as is shown in figure 1. The software is mainly hierarchically structured into 4 layers: sensor interface layer, perception & localization layer, decision-making layer and vehicle interface layer, as is shown in figure 2. The decision-making layer is further abstracted into mission plan, behavioral decision, motion plan and motion control sub-layers.

Figure 1. Test vehicle

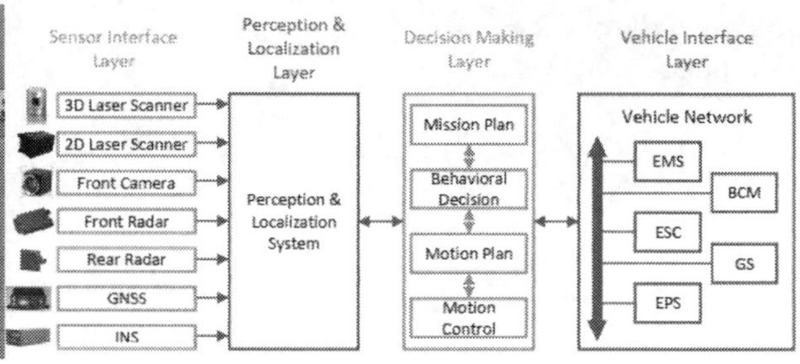

Figure 2. Test vehicle software architecture

Sensor interface layer is the hardware driver of different kinds of sensors, including laser scanner, camera, radar, global navigation satellite system (GNSS) and inertial navigation system (INS), which can provide the raw sensor data.

Perception & localization layer is responsible for the process of the raw sensor data and prior knowledges like high resolution map data, which can provide the observation of the surroundings and the movement state estimation of the host vehicle.

Mission planner is responsible for the producing of the global route that is a sequence of waypoints across the road network according to the prior map data and user specified information. The behavioral decision layer is responsible for selecting a specific and appropriate behavioral based on the perception information and ego vehicle movement information. The selected driving behavioral is then translated into a dynamically feasible, safe and comfort path or trajectory, which is accomplished by the motion plan module.

The planned local reference path or trajectory is then tracked by the motion planner, and the corresponding control signal is sent to vehicle interface through CAN bus.

4.2. Test Result

The steering controller proposed in this paper is implemented as a part of the motion controller. The output front road wheel steer angle is transformed to the steering wheel angle according a pre-calibrated steer ratio table. During the test, the vehicle speed is separately controlled by a self-adaptive longitudinal motion controller, which is out the scope of this paper.

The test was conducted on a field test road of brilliance automobile research institute. The reference path and the footprint of the test vehicle is illustrated in figure 3, and a more clear and intuitive cross tracking error is shown in figure 4.

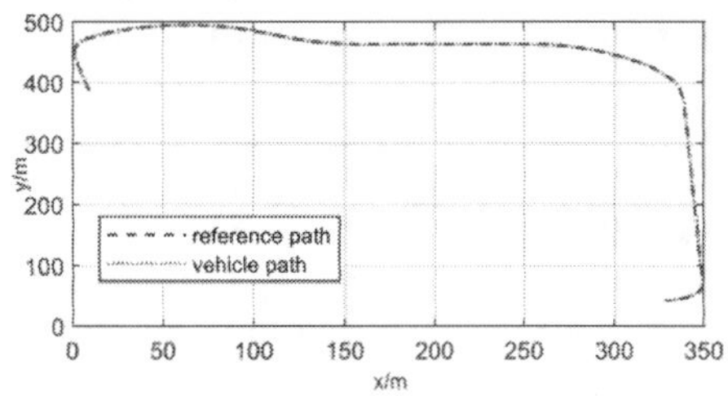

Figure 3. field test reference path and vehicle path

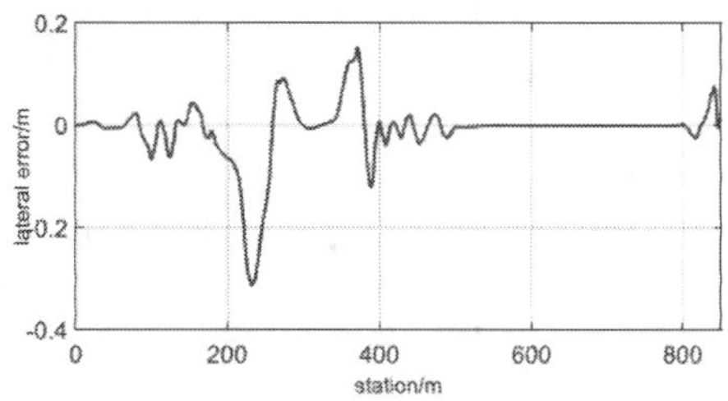

Figure 4. field test lateral tracking error

As is shown in figure 3 and figure 4, the designed steering controller tracked the reference path quite well, with a maximum lateral tracking error of -0.3m, this was acceptable but still some improvement is needed in developing of level 3+ intelligent vehicles.

Double Lane Change (DLC) tests under 10m/s, 15m/s, 20m/s and 25m/s are conducted to test the dynamic sensitive scenario tracking effect of the designed controller, the reference path and the footprint of the test vehicle is illustrated in figure 5, and the cross tracking error is shown in figure 6.

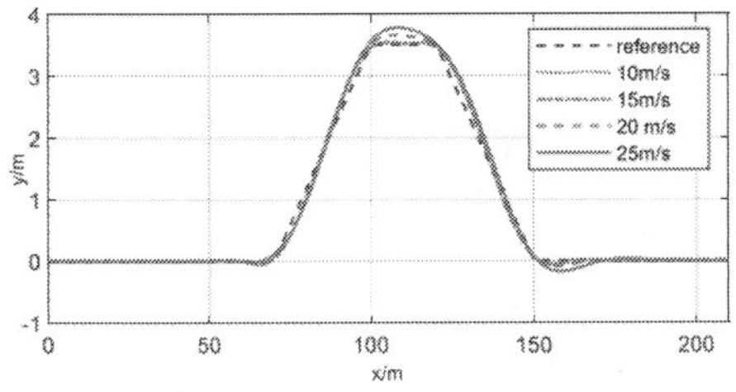

Figure 5. DLC reference path and vehicle path

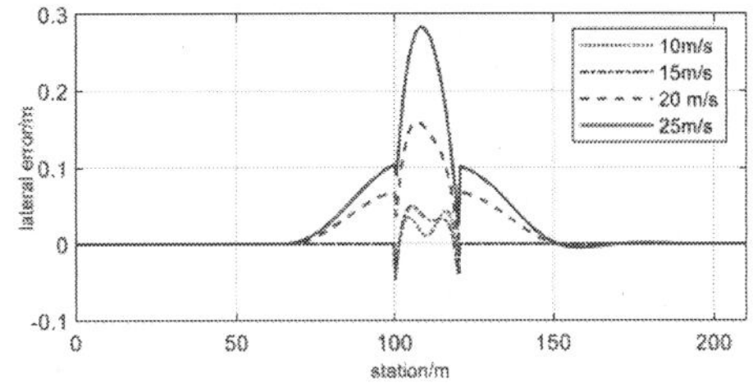

Figure 6. DLC lateral tracking error

The maximum lateral tracking error of DLC test is 0.27m, prove the effectiveness of the designed controller under dynamic sensitive conditions.

5. Conclusion

To handle dynamic sensitive steering scenarios, a neural network based modelling and steering control method of an intelligent vehicle is proposed in this paper. The modelling error of a traditional dynamic vehicle steering model is compensated by four radius function neural networks. A corresponding four term control law is designed, and the stability is analysed through Lyapunov function. Test results show that the designed steering controller works quite well especially when the vehicle dynamics getting significant. A coupled lateral and longitudinal motion controller is expected to achieve better results.

Acknowledgments

This work is supported by the China National key research and development plan key special project of intergovernmental international science and technology innovation cooperation development project "automotive passive and active collaborative protection technology for pedestrian safety" (No. 2018YFE0192900).

References

[1] Sebastian T, Montemerlo M and Palatucci M 2009 Stanley: The Robot that Won the DARPA Grand Challenge *Journal of Field Robotics* vol 23(9) pp 661-92.

[2] Chirs U, Joshua A, Drew B, et al 2008 Autonomous driving in urban environments : Boss and the urban challenge *Journal of Field Robotics* vol 25(8) pp 425-66.

[3] Luca A D, Oriolo G and Samson C 1998 Feedback control of a nonholonomic car-like robot *In Robot Motion Planning and Control* pp 171–249.

[4] Samson C 1993 Mobile robot control Part 2 : control of chained systems and application to path following and time-varying point-stabilization of wheeled vehicles *HAL - INRIA*.

[5] Campbell S F 2007 Steering control of an autonomous ground vehicle with application to the DARPA Urban Challenge *Massachusetts Institute of Technology*.

[6] Falcone P, Borrelli F, Asgari J, Tseng H E and Hrovat D 2007 Predictive active steering control for autonomous vehicle systems *Transactions on Control Systems Technology* vol 15 pp 566–80.

[7] Guo J, Hu P, Li L, et al 2012 Design of Automatic Steering Controller for Trajectory Tracking of Unmanned Vehicles Using Genetic Algorithms *IEEE Transactions on Vehicular Technology* vol 61(7) pp 2913-24.

[8] Nguyen A T, Sentouh C and Popieul J C 2017 Fuzzy Steering Control for Autonomous Vehicles under Actuator Saturation: Design and Experiments *Journal of the Franklin Institute* S0016003217306130.

[9] Zhang K, Cui S M and Wang J F 2014 Intelligent vehicle's path tracking control based on self-adaptive RBF network compensation *Control and Decision* vol 29(4) pp 627-31.

IOP Conf. Series: Materials Science and Engineering **825** (2020) 012044 doi:10.1088/1757-899X/825/1/012044

Adaptive Sliding Mode Control of Artillery Coordinator Based on Disturbance Observer

Jinjie Gan* **and Longmiao Chen**

School of Mechanical Engineering, Nanjing University of Science and Technology, Nanjing 210094, Jiangsu, China

*E-mail:569862472@qq.com

Abstract: In order to improve the control accuracy of the artillery coordinator composed of hydraulic servo system under various uncertain conditions such as extreme environment, unknown disturbances, and system parameter changes, an adaptive sliding mode control strategy based on disturbance observer is designed. Sliding mode control can make the system follow the designed sliding mode dynamic surface state trajectory, and realize the control of non-linear system. Increasing adaptive control can prevent system failure due to excessive sliding mode switching gain. Based on this, disturbance observers are added, feed-forward compensation is performed, and the influence of system parameters and load torque changes on the system is reduced to reduce chattering and effectively improve control performance. The electro-hydraulic joint simulations show that the algorithm can make the system meet the control requirements under complex working conditions under different coordination angles, ammunition and no-load conditions.

1. Introduction

The characteristics of sliding mode variable structure control are suitable for non-linear control, which has little influence on interference and parameter changes, and is easy to implement. Suitable for electro-hydraulic servo system control. However, the actual system is non-ideal. The state trajectory will pass back and forth on the sliding mode surface, causing jitter, which will affect the state and control accuracy of the system. Chattering is an inevitable problem for sliding mode variable structure control [1-2]. Adaptive control: It can identify the dynamic characteristics of the object and take decisions to change the system action [3-4]. The disturbance observer does not need to establish an accurate mathematical model, and has a simple structure and strong engineering usability. Feed-forward compensation for unknown parameters and unmodeled states of the system can effectively reduce the chattering of the sliding mode controller, improve the stability and robustness of the system, and enhance the anti-interference ability of the system [5].

Based on the study of sliding mode control, adaptive control and perturbation observer algorithm, based on the study and inspiration of the above literature, this paper designs a perturbation observer based on the problem of artillery coordinator control with slow time-varying perturbations. Adaptive sliding mode control. DOB is used to feed forward the system disturbance to compensate for changes in system parameters and load torque. The unknown parameter is not prevented from being over-estimated. By introducing adaptive rate estimation parameters, the chattering of sliding mode control is greatly reduced, and the feasibility of the control scheme is verified by the joint simulation of AMEsim and MATLAB [6].

2. System description and modeling

The actuator of the coordinator is a single-rod cylinder, which is controlled by a servo valve and an electrical system. It is a typical electromechanical-hydraulic structure. As shown in Figure 1, the coordinator is fixed at point B, the coordination cylinder is connected to the coordinator through a hinge C, and the lower fulcrum of the cylinder is point A. The angle is the rotation angle of the coordinator. The angle between the oil cylinder and the coordinator is. Is the angle between AB and vertical = 62 °

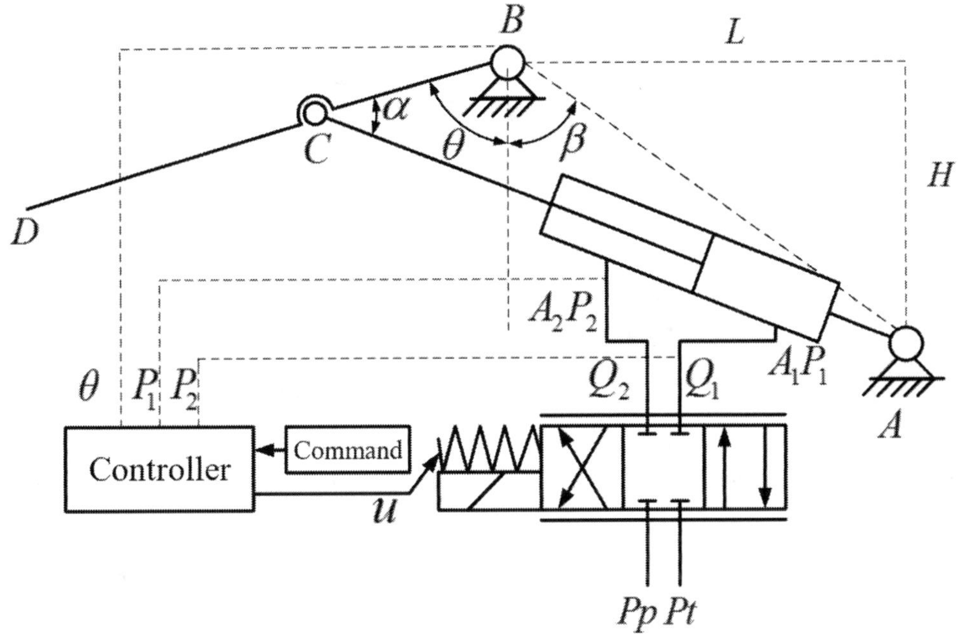

Figure 1. Coordinator system diagram.

The inlet pressure of the hydraulic system is P_P, the return pressure is P_t, the flow rate of the rodless cavity is Q_1, the area is A_1, the pressure is P_1, the flow rate of the rod cavity is Q_2, the area is A_2, and the pressure is P_2. The controller receives a command and sends a signal to the servo valve based on the system's motion angle and the pressure in the two chambers of the cylinder u, and the core displacement is X_t,

$$x_t = k_t u \tag{1}$$

Assuming 1: no leakage, the continuous flow equation of the coordinator cylinder is:

$$\begin{cases} Q_1 = A_1\dot{x} + \dfrac{V_1}{\beta_e}\dot{P}_1 \\ Q_2 = A_2\dot{x} - \dfrac{V_2}{\beta_e}\dot{P}_2 \end{cases} \tag{2}$$

where A_1 is the area of the cavity without a rod, A_2 is the area of the cavity with a rod, β_e is the volume elastic modulus of the hydraulic oil, and x is the displacement of the piston rod.

As shown in Figure (1), assuming that the driving force of the coordinator cylinder is F, it can be obtained from Newton's second law:

$$Fl_{BC}\sin\alpha - mgl_{BD}\sin\theta = J_{eq}\ddot{\theta} + B_{eq}\dot{\theta} + d \tag{3}$$

where $m=100kg$ is the mass of the coordinator , $g=9.8m/s^2$, J_{eq} is the equivalent moment of inertia, B_{eq} is the equivalent viscous friction coefficient, and d is the unmodeled disturbance torque.

According to the force balance equation of the hydraulic cylinder:

$$F = P_1A_1 - P_2A_2 \tag{4}$$

define the state variables $X=\left[\theta_1,\theta_2,\theta_3\right]^T=\left[\theta,\dot{\theta},\ddot{\theta}\right]^T$ .with (1)~(4),the system state can be expressed as follows:

$$\begin{cases}\dot{\theta}_1=\theta_2\\ \dot{\theta}_2=\theta_3\\ a_1\theta_3=-a_2\theta_2-a_3\theta_3+u-d\end{cases} \tag{5}$$

where a_1,a_2,a_3,d are uncertain parameters, so there are **assumptions 2**: Unknown parameters a_1,a_2,a_3,d bounded, $a_{1\max}>a_{1\min}>0$, $a_{2\max}>a_{2\min}>0$, $a_{3\max}>a_{3\min}>0$, $\eta>0$ make $\theta_{1\min}\le\theta_1\le\theta_{1\max},\theta_{2\min}\le\theta_2\le\theta_{2\max}$, $\theta_{3\min}\le\theta_3\le\theta_{3\max}$,$|d|<\eta$.

3. Controller design

The structure of the coordinator controller algorithm is shown in Figure (2). θ_d is the target angle, $\theta,\dot{\theta},\ddot{\theta}$ is the actual angle, angular velocity, and angular acceleration. $\hat{d}$ is the disturbance observer output.

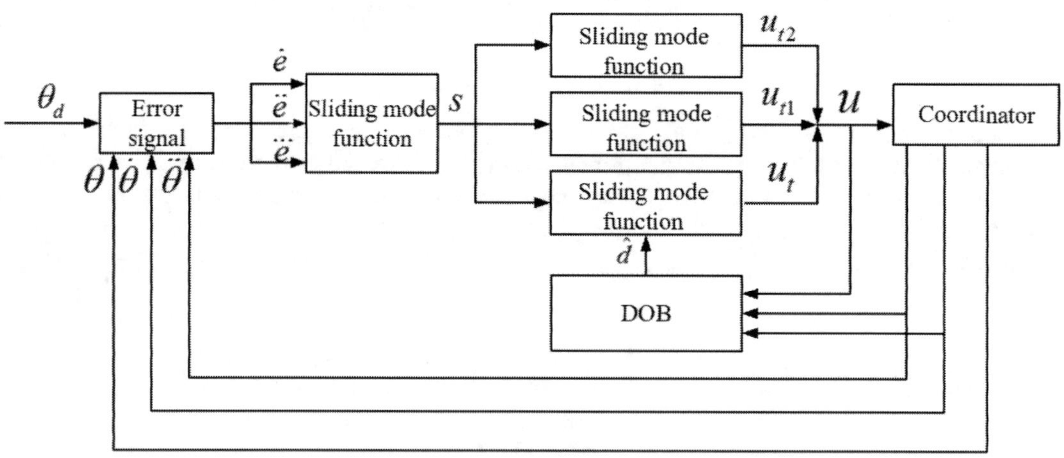

Figure 2. Control structure diagram.

3.1. Design of Sliding mode controller

The angle tracking error and the sliding mode function are defined separately

$$e=\theta_d-\theta \tag{6}$$

$$s=\ddot{e}+y_1\dot{e}+y_2e \tag{7}$$

where y_1,y_2 is the sliding surface parameter to be designed, which can be determined according to the actual control situation.

Define the Lyapunov function

$$V_1 = \frac{1}{2}a_1 s \tag{8}$$

$$\begin{aligned}\dot{V}_1 &= s a_1 \dot{s} \\ &= s\left[-a_1(\dddot{\theta}_d - y_1\ddot{e} - y_2\dot{e}) - a_2\theta_2 - a_3\theta_3 + u - d\right]\end{aligned} \tag{9}$$

design the robust controller u as follows

$$u = a_1(\dddot{\theta}_d - y_1\ddot{e} - y_2\dot{e}) + a_2\theta_2 + a_3\theta_3 - k_s s - \eta \cdot sat(s) \tag{10}$$

where k_s is the robustness coefficient, η is the switching gain, $sat(s)$ is a saturation function, the purpose is to reduce chattering

$$sat(s) = \begin{cases} 1, s => \Delta \\ \frac{s}{\Delta}, |s| \le \Delta \\ -1.s < -\Delta \end{cases} \tag{11}$$

where Δ is the thickness of the boundary layer.

With (8)~(11)

$$\begin{aligned}\dot{V}_1 &= s\left[-a_1(\dddot{\theta}_d - y_1\ddot{e} - y_2\dot{e}) - a_2\theta_2 - a_3\theta_3 + u - d\right] \\ &= s\left[-k_s s - \eta sat(s) - d\right] \\ &= -k_s s^2 - \eta|s| - ds \\ &\le -k_s s^2\end{aligned} \tag{12}$$

the system can stable by these functions.

3.2. Design of DOB

Define[7]

$$\dot{\hat{d}} = K(d - \hat{d}) \tag{13}$$

$$z = \hat{d} + K\ddot{a}_1\theta \tag{14}$$

$$\begin{aligned}\dot{z} &= \dot{\hat{d}} - Ka_1\dddot{\theta} \\ &= K(-a_1\ddot{\theta} - a\dot{\theta} - a\ddot{\theta} + u) - K\hat{d} + Ka\ddot{\theta}_1\end{aligned} \tag{15}$$

therefore, DOB could be designed as follows

$$\begin{cases} \dot{z} = K(-a_2\dot{\theta} - a_3\ddot{\theta} + u) - K\hat{d} \\ \hat{d} = z - Ka_1\ddot{\theta} \end{cases} \tag{16}$$

Define the error of DOB as

$$\tilde{d} = d - \hat{d} \tag{17}$$

IOP Conf. Series: Materials Science and Engineering **825** (2020) 012044 doi:10.1088/1757-899X/825/1/012044

assumption 3: $\dot{d} = 0$, derivative calculation for Equation (17)

$$\dot{\tilde{d}} = \dot{d} - \dot{\hat{d}} = -\dot{\hat{d}} = -\dot{z} + Ka_1\ddot{\theta} \tag{18}$$

substitute equation (15) into (18)

$$\begin{aligned} \dot{\tilde{d}} &= -K(-a_2\dot{\theta} - a_3\ddot{\theta} + u + Ka_1\ddot{\theta}) + Kz + Ka_1\ddot{\theta} \\ &= K(z - Ka_1\ddot{\theta}) - K(-a_2\dot{\theta} - a_3\ddot{\theta} + u + a_1\ddot{\theta}) \\ &= K(\hat{d} - d) \\ &= -K\tilde{d} \end{aligned} \tag{19}$$

therefore, the error equation of the DOB is $\dot{\tilde{d}} + K\tilde{d} = 0$, and the solution is $\tilde{d}(t) = \tilde{d}(t_0)e^{-Kt}$,the value of $\tilde{d}(t_0)$ is determined, and the convergence accuracy of the observer depends on $\hat{d}$, so that the estimated value e approaches the interference d exponentially.

With the introduction of the DOB equation (16),the robust control law equation (10) becomes

$$u = a_1(\ddot{\theta}_d - y_1\ddot{e} - y_2\dot{e}) + a_2\theta_2 + a_3\theta_3 + \hat{d} - k_s s - \eta \cdot sat(s) \tag{20}$$

Define the Lyapunov function

$$V_2 = \frac{1}{2}a_1 s^2 + \frac{1}{2}\tilde{d}^2 \tag{21}$$

$$\begin{aligned} \dot{V}_2 &= s a_1 \dot{s} + \tilde{d}\dot{\tilde{d}} \\ &= -k_s s^2 - \eta|s| - ds + (d - \hat{d})\dot{\tilde{d}} \\ &= -k_s s^2 - \eta|s| - ds - K\tilde{d}^2 \\ &\le -k_s s^2 - K\tilde{d}^2 \le -k_1(\frac{1}{2}a_1 s^2 + \frac{1}{2}\tilde{d}^2) \\ &= -k_1 V_2 \end{aligned} \tag{22}$$

where $k_2 = 2\min(a_1 k_s, K)$,the solution of the equation $\dot{V}_2 \le -k_1 V_2$ as follows[8]

$$V_2(t) \le e^{-k_1(t-t_0)}V(t_0) \tag{23}$$

it can be seen that the control system converges exponentially, and the convergence accuracy depends on the value k_1.

3.3. Design of adaptive control

In the actual control process, the value of a_1, a_2, a_3 is constantly changing and uncertain. Therefore, an adaptive algorithm is added on the basis of control law equation (20) to estimate parameter changes. Define the error of estimation as

$$\tilde{a}_i = a_i - \hat{a}_i \tag{24}$$

where $\tilde{a}_i$ is the estimation error, $\hat{a}_i$ is the estimated value, $i = 1, 2, 3$.

Then equation (20) becomes

$$u=\hat{a}_1(\ddot{\theta}_d-y_1\ddot{e}-y_2\dot{e})+\hat{a}_2\theta_2+\hat{a}_3\theta_3+\hat{d}-k_s s-\eta\cdot sat(s) \tag{25}$$

define the Lyapunov function

$$V_3=V_2+\frac{1}{2}c_1\tilde{a}_1^{\,2}+\frac{1}{2}c_2\tilde{a}_2^{\,2}+\frac{1}{2}c_3\tilde{a}_3^{\,2} \tag{26}$$

$$\begin{aligned}\dot{V}_3&=\dot{V}_2+c_1\tilde{a}_1\dot{\tilde{a}}_1+c_2\tilde{a}_2\dot{\tilde{a}}_2+c_3\tilde{a}_3\dot{\tilde{a}}_3\\&=s\left[-\tilde{a}_1(\ddot{\theta}_d-y_1\ddot{e}-y_2\dot{e})-\tilde{a}_2\theta_2-\tilde{a}_3\theta_3-k_s s-\eta sat(s)-d\right]-K\tilde{d}^2+c_1\tilde{a}_1\dot{\tilde{a}}_1+c_2\tilde{a}_2\dot{\tilde{a}}_2+c_3\tilde{a}_3\dot{\tilde{a}}_3\\&=\tilde{a}_1\left[c_1\dot{\tilde{a}}_1-s(\ddot{\theta}_d-y_1\ddot{e}-y_2\dot{e})\right]+\tilde{a}_2(c_2\dot{\tilde{a}}_2-s\theta_2)+\tilde{a}_3(c_3\dot{\tilde{a}}_3-s\theta_3)-k_s s^2-\eta|s|-ds-K\tilde{d}^2\end{aligned} \tag{27}$$

where c_1, c_2, c_3 are positive numbers.

Assumption 4: $\dot{a}_1=0$, $\dot{a}_2=0$, $\dot{a}_3=0$.

Design the adaptive law as follows

$$\begin{cases}\dot{\hat{a}}_1=-\dfrac{1}{c_1}s(\ddot{\theta}_d-y_1\ddot{e}-y_2\dot{e})\\ \dot{\hat{a}}_2=-\dfrac{1}{c_2}s\theta_2\\ \dot{\hat{a}}_3=-\dfrac{1}{c_3}s\theta_3\end{cases} \tag{28}$$

substitute equation (28) into quation (27)

$$\begin{aligned}\dot{V}_3&=\tilde{a}_1\left[c_1\dot{\tilde{a}}_1-s(\ddot{\theta}_d-y_1\ddot{e}-y_2\dot{e})\right]+\tilde{a}_2(c_2\dot{\tilde{a}}_2-s\theta_2)+\tilde{a}_3(c_3\dot{\tilde{a}}_3-s\theta_3)-k_s s^2-\eta|s|-ds-K\tilde{d}^2\\&=-k_s s^2-\eta|s|-ds-K\tilde{d}^2\\&\le -k_s s^s-K\tilde{d}^2\end{aligned} \tag{29}$$

the system can stable by these functions.

In order to prevent the overestimation of a_1, a_2, a_3 from affecting the system and ensuring its bounds, a discontinuous projection algorithm is used [9]

$$\dot{\hat{a}}_i=\mathrm{Proj}_{a_i}(\frac{1}{c_i}\beta_i)=\begin{cases}0, \hat{a}_i\ge a_{i\max} and \dfrac{1}{c_i}\beta_i>0 and \dfrac{1}{c_i}\beta_i>0\\ 0, \hat{a}_i\le a_{i\min} and \dfrac{1}{c_i}\beta_i>0 and \dfrac{1}{c_i}\beta_i<0\\ \dfrac{1}{c}\beta_i, \text{otherwise}\end{cases} \tag{30}$$

where $i=1,2,3$, $\beta_1=-s(\theta_d-y_1\ddot{e}-y_2\dot{e})$, $\beta_2=-s\theta_2$, $\beta_3=-s\theta_3$.**Assumption 5:** equation $a_{i\min}\le\hat{a}_i(0)\le a_{i\max}$ can satisfy equation $a_{i\min}\le\hat{a}_i(\mathrm{t})\le a_{i\max}$ in any case, therefore, the modified algorithm can guarantee the boundedness of the estimated parameters.

4. The Co-simulation and Results

The electro-hydraulic servo simulation model is established by the AMESim and Simulink co-simulation method to improve modelling efficiency and accuracy.

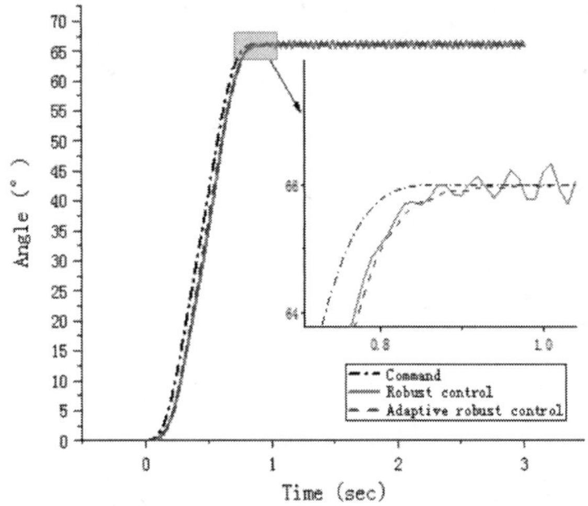

Figure 3. The Angle tracking trajectory

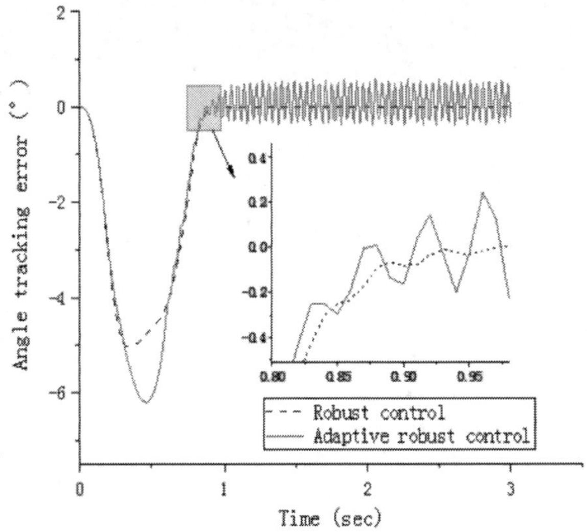

Figure 4. The angle tracking error

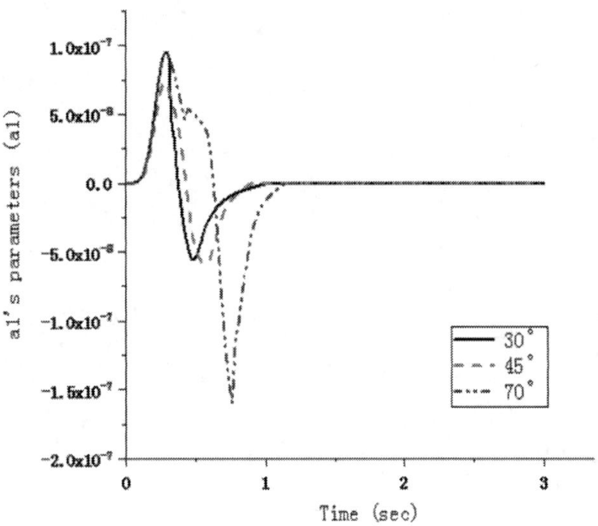

Figure 5. The *a1's* parameters

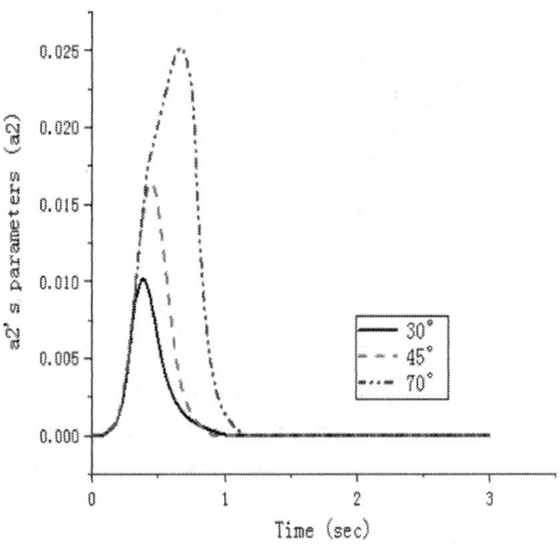

Figure 6. The *a2's* parameters

The parameters of the electro-hydraulic servo system:DOB gain K=50,sliding surface parameters $y_2 = 150$, $y_2 = 12000$, adaptive parameters $c_1 = 1\times10^{20}$, $c_2 = 1\times10^{13}$, $c_3 = 1\times10^{17}$,robust coefficient $k_s = abs(s)\times10^{-8}$, switching gain $\eta = 15$, boundary layer thickness $\Delta = 2\times10^4$. $a_{1\max} = 2\times10^{-7}$, $a_{1\min} = -2\times10^{-7}$, $a_{2\max} = 4\times10^{-3}$, $a_{2\min} = -2\times10^{-3}$, $a_{3\max} = 9\times10^{-5}$, $a_{3\min} = -6\times10^{-5}$.Set the target angle to 66 degrees, the simulation step size is 0.001 seconds, and the simulation time is 3 seconds. The angle tracking trajectory is shown in Figure(3).The angle tracking error is shown in Figure(4).

The adaptive parameters at three different target angles are shown in Figure(5)~Figure(7).The DOB parameters at three target angles is shown in Figure(8).

It can be seen from Figures 3 and 4 that the traditional sliding mode control error has good tracking performance in the early stage, with a maximum dynamic error of -6 °. When it reaches the target angle, the tracking error hovers around ± 0.4 °. Chatter, unable to position smoothly. The maximum dynamic tracking error of the adaptive sliding mode control based on the DOB is -5 °, and the dynamic error when reaching the target angle is ± 0.15 °. The in-position time is about 0.8 seconds, and the chatter phenomenon is significantly weakened, which improves the positioning accuracy and enables The coordinator smoothly and quickly locates to the target angle.

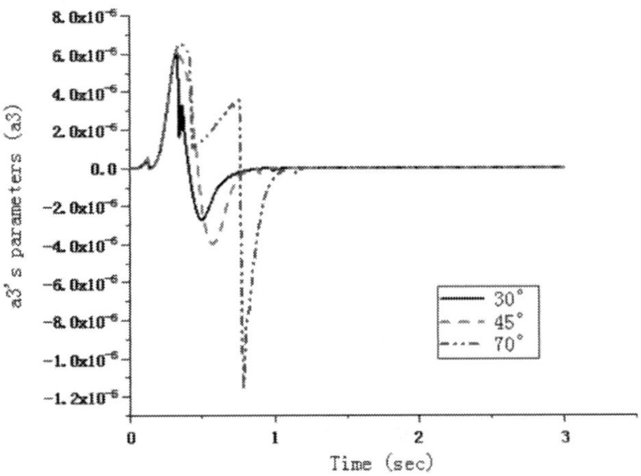

Figure 7. The *a3's* parameters

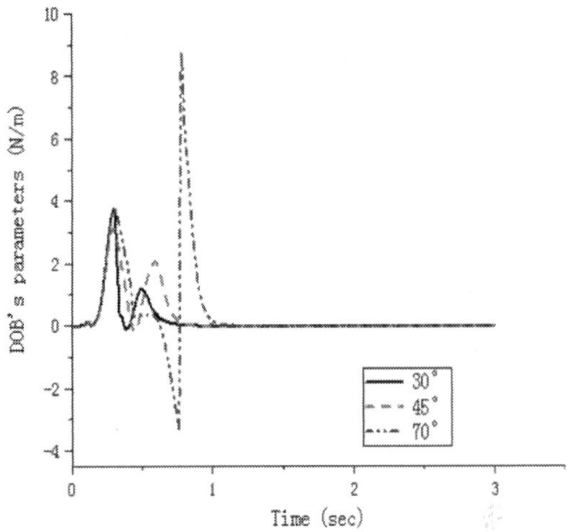

Figure 8. The *d's* parameters

From Figure 5 to Figure 7, it can be seen that the adaptive parameters are within the set range. The parameter curve of the high-angle motion of the coordinator is shaken than the parameter curve of the low-angle motion, and the overall impact on the system is more obvious at high angle. It can be seen from Fig. 8 that the DOB can quickly respond to changes in the speed of the system in a short time to changes in disturbance torque and system parameters, and compensate for external disturbances, so that the coordinator reduces severe chattering and improves control accuracy. When the target angle is 30 ° and 45 °, the perturbation torque of the perturbation observer is relatively small, but at 70 °, the system impact is large, and the perturbation torque generated is significantly increased.

5. Conclusion

In this paper, an adaptive sliding mode control strategy based on DOB is applied to control research of artillery coordinator, and it is applied to hydraulic servo system. The AMEsim and MATLAB electro-hydraulic joint simulations show that the adaptive algorithm can well estimate the parameter values and guarantee fluctuations within the boundary. The disturbance observer can feed forward compensate the system, which greatly reduces the unmodeled disturbance pairs and unknown parameters. It affects the system control accuracy, and greatly reduces the chattering phenomenon existing in traditional sliding mode control, which makes the positioning process more stable and improves the positioning accuracy.

References

[1] Cai Q D. Optimal design and control technology of an artillery coordinator [D]. Nanjing University of Science and Technology, 2016.

[2] Zheng Y, Ma D W, Yao J Y and Hu J. Active Disturbance Rejection Control of Rocket Servo System [J]. *Acta Armamentarii*,**2014, 35 (5)**: 597-603

[3] Ji J J. Analysis of hydraulic control system of self-propelled artillery bomber lifting mechanism [D]. Harbin Institute of Technology, 2014.

[4] Dong S Y, Chen L M and Zhang T. Adaptive dynamic surface sliding mode control of artillery automatic loading manipulator [J]. *Journal of Ordnance and Equipment Engineering*, **2018, 39 (1)**: 97-101.

IOP Conf. Series: Materials Science and Engineering **825** (2020) 012044 doi:10.1088/1757-899X/825/1/012044

[5] Zhou Y L and Chen M.The input of disturbance observer based on robust control of nonlinear systems [J]. *Journal of Nanjing University of Science and Technology*, **2014, 38 (1)**: 40-47.
[6] Lu J S, Feng G B, Sun H G and Zhang Y F. Electromechanical-hydraulic joint simulation of automatic ammunition supply and delivery system based on AMESim / RecurDyn / Simulink [J]. *Fire Power and Command Control*. **2016, 41 (7)**: 188-192.
[7] Liu J K and Sun F C.A novel dynamic terminal sliding mode control of uncertain nonlinear system[J]. *Journal of Control Theory and Applications*,**2007,5(2)**:189-193.
[8] Liu J K and Wang X H.Advanced Sliding Mode Control for Mechanical System :Design,Analysis and MATLAB Simulation[M]. Beijing:Tsinghua&Springer Press,2011
[9] Yao B.Advanced Motion Control:From Classical PID to Nonlinear Adaptive Robust Control[C]. The 11th IEEE International Workshop on Advanced Motion Control,2010,Nagaoka,Japan.

IOP Conf. Series: Materials Science and Engineering **825** (2020) 012045 doi:10.1088/1757-899X/825/1/012045

Temperature Variation of Wet Clutches in the DSG Vehicle during a 10km Driving Cycle

Jikai Liu[1,2] and Jie Chen[1]

[1] Institute of Microelectronics, Chinese Academy of Sciences, Beijing, 100029, China
[2] Beijing AI Force Technology Co., Ltd. Beijing, 100086, China

Email: jackylauno1@gmail.com

Abstract. The temperature of wet clutches has a vital influence on the vehicle in terms of dynamic performance, service life, wear characteristics, etc. To investigate the temperature variations during the operation process, a detailed dynamic powertrain model of the DSG vehicle is established firstly to calculate the dynamic load of different clutches. Then, a thermal resistance model for the gearbox cooling system is built to predict the temperature of wet clutches. Integrating the dynamic model and the heat transfer model, a co-simulation model is finally obtained. The simulation results illustrate that the temperature of the clutch used for the launching process has a sharp increase from 100℃ to 123℃; however, within a 10km driving cycle, the temperature of five clutches are all lower than 125℃, which meet the requirements of the vehicle.

1. Introduction

Combining advantages of automatic transmission and manual transmission, the direct shift gearbox (DSG), which also called dual clutch transmission (DCT), not only has the advantages of high efficiency, compactness and low weight, but also can achieve power shift process without power interruption and circulation [1]. Recently, DSG gets extensive and growing attention from the industry and is becoming widely used in vehicles. Similar to other transmissions, a lot of researches have been done to achieve excellent vehicle performance, including optimizing the clutch engagement process during the launching process and the torque transfer process during the gearshift process.

For the launching process, a sliding mode variable structure control strategy is proposed to improve the dynamic performance of DSG vehicle [2]. From the simulation results, the designed control strategy can not only reflect driving intention adequately, but also has strong robustness. However, launching the vehicle with only one clutch is not sufficient to utilize the structure advantage of DSG. A control method based on twin-clutch torque distribution is provided for the DSG vehicle [3]. The simulation results indicate that the proposed method could greatly shorten the launching time, share the friction work by two clutches and avoid possible power coupling and power cycling.

As for the DSG shift process, it can be divided into the torque phase and the inertia phase. During the torque phase, the engine torque is transferred from the leaving clutch to the coming clutch. To obtain a smooth transfer process, a detailed dynamic model of DSG vehicle is established and a control strategy is proposed in [4]. The results show that the vehicle can achieve good performance during the torque phase. Furthermore, to investigate the effects of clutch application timing, three different clutch pressure profiles are studied, and the optimized profile can obtain the best possible shift quality [5]. When it comes to the torque relationship between two clutches, a torque based control strategy is proposed with only oncoming clutch slipping [6]. In terms of the inertia phase, the power-on upshift and power-off downshift process are optimized by the feedback control and torque based logic control [7]. In addition, an integral linear quadratic regulator is introduced by Guoqiang to regulate the relative speed between the engine and the slipping clutch during the inertia phase [8]. From the simulation results, the proposed controller can reduce jerk level and improve shift quality significantly, compared with conventional controllers.

Owning to these considerable researches, the performance of DSG vehicle has been improved significantly. However, when it comes to the thermal problems during the vehicle operation process, it is seldom considered, especially for the temperature variation of clutches during a continues driving cycle. To fill that gap in knowledge, the thermal status of the clutches in the DSG vehicle is investigated.

In the second part, a powertrain model for the DSG vehicle is proposed to calculate the dynamic load during the vehicle operating process. In the third part, a thermal resistance model for the gearbox cooling system is built. After that, the simulation results are presented in the forth part. Finally, the conclusions are presented.

2. Dynamic model of the DSG vehicle

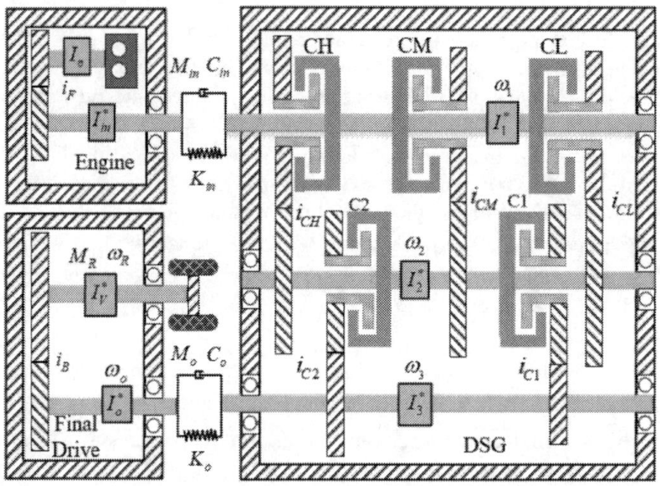

Figure 1. Powertrain system of the DSG vehicle.

The structure of the DSG vehicle is presented in Figure 1. As there are no synchronizers in this DSG, the launch and the shift processes are completed with different slipping clutches. By controlling clutches according to Table 1, various gear ratios can be achieved. The sign "•" presents that the corresponding clutch is engaged.

Table 1. Clutch schedule for each gear.

Gear	CL	CM	CH	C1	C2
1	•			•	
2		•		•	
3			•	•	
4	•				•

2.1. Launching process description

During the launching process, C1 is pre-engaged, and only CL is slipping to start the vehicle. Then, the engine power is transferred from the input shaft to the output shaft through DSG. According to the Newton second law, the system dynamic can be deduced as follows.

$$(I_e + \frac{I_{in}^*}{i_F})\frac{d\omega_e}{dt} = M_e - \frac{M_{in}}{i_F} \tag{1}$$

$$I_1^* \frac{d\omega_1}{dt} = M_{in} - M_{CL} \tag{2}$$

$$I_2^* \frac{d\omega_2}{dt} = M_{CL}i_{CL} - M_{C1} \tag{3}$$

$$I_3^* \frac{d\omega_3}{dt} = M_{C1}i_{C1} - M_o \tag{4}$$

$$(I_o^* + \frac{I_V^*}{i_B^{\ 2}})\frac{d\omega_o}{dt} = M_o - \frac{M_R}{i_B} \tag{5}$$

$$M_{in} = S_{in}(\theta_{in} - \theta_1) + C_{in}(\omega_{in} - \omega_1) \tag{6}$$

$$M_o = S_o(\theta_3 - \theta_o) + C_o(\omega_3 - \omega_o) \tag{7}$$

$$M_R = \left(\underbrace{\frac{1}{2}\rho_{air}C_D A v_e^2}_{\text{air resistance}} + \underbrace{mgf\cos\beta}_{\text{wheel resistance}} + \underbrace{mg\sin\beta}_{\text{gradient resistance}} + \underbrace{\delta m \frac{dv_e}{dt}}_{\text{acceleration resistance}} \right) r_w \tag{8}$$

2.2. Upshift with two clutches

As the downshift process is the opposite process of the upshift process, only the upshift process is studied. Taking the 1st gear to the 2nd gear as an example, the clutch CL is released, and the clutch CM is applied. Then, equations (2,3) should be changed into equations (9,10).

$$I_1^* \frac{d\omega_1}{dt} = M_{in} - M_{CL} - M_{CM} \tag{9}$$

$$I_2^* \frac{d\omega_2}{dt} = M_{CL}i_{CL} + M_{CM}i_{CM} - M_{C1} \tag{10}$$

Before the torque phase and after the inertia phase, the system torque transferred by CL and CM are illustrated in equations (11, 12), respectively. During the torque phase, the relationship between these clutches is described by the equation (13).

$$M_{CL} = \frac{M_{in}I_2^* + M_{C1}I_1^* i_{CL}}{I_1^* i_{CL}^2 + I_2^*} \tag{11}$$

$$M_{CM} = \frac{M_{in}I_2^* + M_{C1}I_1^* i_{CM}}{I_1^* i_{CM}^2 + I_2^*} \tag{12}$$

$$M_{CL} = \frac{M_{in}I_2^* - M_{CM}(I_2^* + I_1^* i_{CL} i_{CM}) + M_{C1}I_1^* i_{CL}}{I_1^* i_{CL}^2 + I_2^*} \tag{13}$$

2.3. Upshift with four clutches

For the gearshifts between 1st and 3rd gear, only two clutches are involved; while for the gearshift between 3rd and 4th gear, four clutches are involved. Then, equations (2-4) should be changed into equations (14-16).

$$I_1^* \frac{d\omega_1}{dt} = M_{in} - M_{CH} - M_{CL} \tag{14}$$

$$I_2^* \frac{d\omega_2}{dt} = M_{CH}i_{CH} + M_{CL}i_{CL} - M_{C1} - M_{C2} \tag{15}$$

$$I_3^* \frac{d\omega_3}{dt} = M_{C1} i_{C1} + M_{C2} i_{C2} - M_o \tag{16}$$

Similarly, the torque relationship between four clutches are optimized, shown in equations (17,18).

$$M_{CH} = \frac{M_{in}(I_2^* i_{C1}^2 + I_3^*) + M_o I_1^* i_{CH} i_{C1} - M_{CL}(I_1^* i_{CH} i_{CL} i_{C1}^2 + I_2^* i_{C1}^2 + I_3^*) + M_{C2} I_1^* (i_{C1} - i_{C2}) i_{C1} i_{CH}}{I_1^* i_{C1}^2 i_{CH}^2 + I_2^* i_{C1}^2 + I_3^*} \tag{17}$$

$$M_{C1} = \frac{M_{in} I_3^* i_{CH} + M_o (I_1^* i_{CH}^2 i_{C1} + I_2^* i_{C1}) + M_{CL} I_3^* (i_{CL} - i_{CH}) - M_{C2}[(I_1^* i_{CH}^2 + I_2^*) i_{C1} i_{C2} + I_3^*]}{I_1^* i_{C1}^2 i_{CH}^2 + I_2^* i_{C1}^2 + I_3^*} \tag{18}$$

For the engine torque, it is modeled as a mean value torque generator [5]. As for the friction torque generated by clutches, it is expressed as follows.

$$M_{Clutch} = \mu N P A_f \frac{2}{3}\left(\frac{R_o^3 - R_i^3}{R_o^2 - R_i^2}\right) \tag{19}$$

3. Lumped thermal resistance model

At present, the Finite Element Method (FEM) is the most widely used numerical calculation method for engineering analysis of thermal problems. However, if this method is used to predict and evaluate the dynamic temperature of the friction element in real time, there are shortcomings such as too complicated theory, huge computational workload, long calculation time and difficulty in programming. Therefore, this section builds a thermal resistance model based on the lumped parameter method to predict the average temperature rise of the wet clutches [9]. As shown in Figure 2, the entire cooling system is divided into different lumped elements, including the operating environment (E), the radiator (R), the hydraulic fluid reservoir (F), the relief safety valve (RV), the pump (P), the pressure regulating valve (V) and clutches.

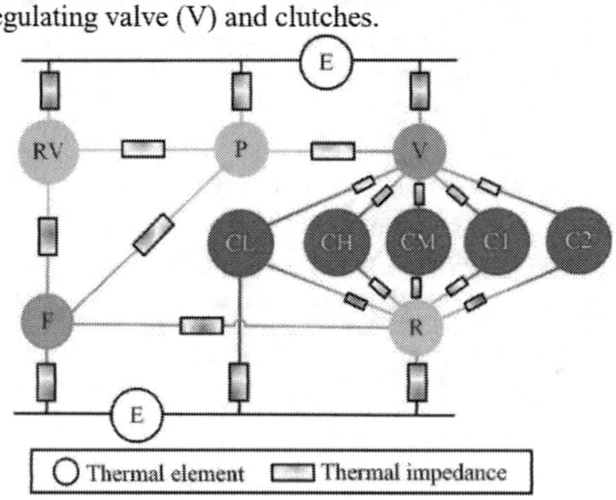

Figure 2. Lumped thermal resistance heat transfer model.

Considering the heat transfer of the components between elements, the heat transfer between the lubricants in each element, the heat exchange between the components and the lubricating oil, and the heat generation of the components in each element, etc., the temperature variations of the component and the lubrication oil in each element can be deduced by the following equation.

$$\begin{cases} C_i \dfrac{\mathrm{d}T_i}{\mathrm{d}t} = \Phi_i - K_{i,io}(T_i - T_{io}) - \sum\limits_j K_{i,j}(T_i - T_E) \\ C_{io} \dfrac{\mathrm{d}T_{io}}{\mathrm{d}t} = \Phi_o - K_{io,i}(T_{io} - T_i) - \sum\limits_j K_{io,jo}(T_{io} - T_{jo}) \end{cases} \tag{20}$$

with $C_{i(o)} = \rho_{i(o)} \cdot c_{i(o)} \cdot V_{i(o)}$ and K is the equivalent heat transfer coefficient [10].

IOP Conf. Series: Materials Science and Engineering **825** (2020) 012045 doi:10.1088/1757-899X/825/1/012045

Taking the clutch CL as an example to investigate its temperature variation, the heat state equation of clutch CL can be further expressed as follows according to the equation (20).

$$\begin{cases} C_{CL}\dfrac{dT_{CL}}{dt}=\dot{Q}_{CL}-K_{CL,E}(T_{CL}-T_E)-K_{CL,CLo}(T_{CL}-T_{CLo}) \\ C_{CLo}\dfrac{dT_{CLo}}{dt}=-K_{CLo,Vo}(T_{CLo}-T_{Vo})-K_{CLo,Ro}(T_{CLo}-T_{Ro}) \\ \qquad -K_{CLo,CL}(T_{CLo}-T_{CL})-K_{CLo,E}(T_{CLo}-T_E) \end{cases} \tag{21}$$

It is amused that all the generated friction heat is absorbed by the lubrication oil and the heat flux can be described as follows.

$$\dot{Q}_{CL}=M_{Clutch}\Delta\omega_{Clutch} \tag{22}$$

4. Thermal condition simulation and analysis

With the established dynamic model to calculate the friction torque during the launch and shift process and the lumped thermal model to predict the temperature of wet clutches, the thermal conditions of different clutches can be simulated. In this paper, a 10km driving cycle, shown in Figure 3, is taken as an example to investigate the temperature variations of clutches.

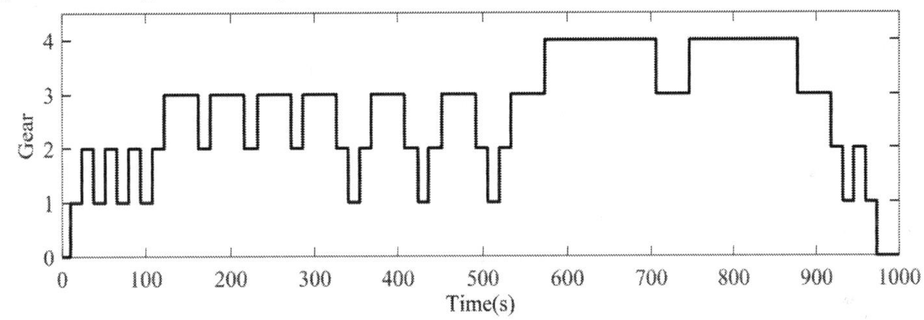

Figure 3. 10km driving cycle.

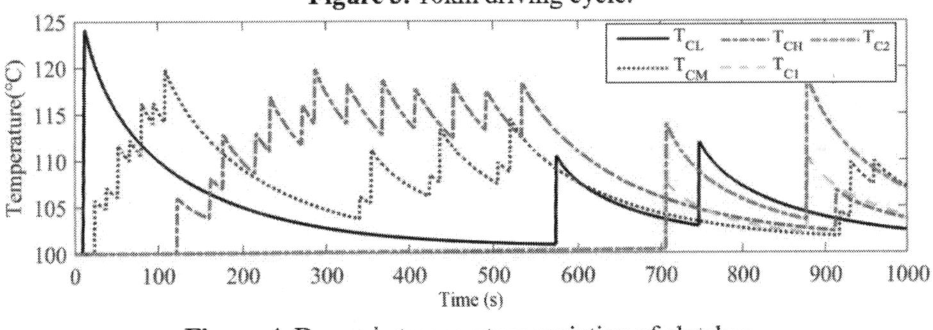

Figure 4. Dynamic temperature variation of clutches.

The simulation results are shown in Figure 4. After the clutch slips, its temperature rises, then decreases due to the cooling of the lubricating oil. In comparison, CM and CH are used frequently, and each time there is a certain temperature rise during the friction process. However, since the time interval between shifts is relatively long and the cooling is enough, the temperature always varies at a low level. As for the CL, C1, and C2, although the slipping frequency is low, a large amount of friction work is generated, resulting into the high temperature rise. Particularly, during the launching process, the temperature of the clutch CL increases from 100℃ to 123℃ directly, suffering a great dynamic load. However, in general, when operating in accordance with the shift cycle, the temperature of each clutch changes between 100℃ and 125℃, within the controllable range [9].

5. Conclusions

In this paper, the dynamic temperature variations of wet clutches are investigated for the DSG vehicle. To achieve this outcome, a detailed dynamic model is established to calculate the dynamic load of different clutches within a 10km driving cycle. Then, to predict the temperature of clutches, a lumped thermal model is developed. From simulation results, the clutch used for the launching process suffers a large heat flux and its temperature increase significantly. As for the clutches used for the shift process, the temperature varies wavily. While, on the whole, the temperature of clutches is under 125℃, satisfying the requirements of vehicle.

References

[1] Matthes B 2005 Dual clutch transmissions—lessons learned and future potential *SAE Trans* **114** 941–952
[2] Zhao Z, Chen H, Wang Q 2014 Sliding mode variable structure control and real-time optimization of dry dual clutch transmission during the vehicle's launch *Mathematical Problems in Engineering* **2014**
[3] Jiang Z, Liu Q, Dong S, Chen H 2018 Launch Coordination Control Based on Twin-Clutch Torque Distribution for DCT Vehicle *IFAC-PapersOnLine* **31** 904-909
[4] Goetz M, Levesley M C, Crolla D A 2005 Dynamics and control of gearshifts on twin-clutch transmissions *Proceedings of the institution of mechanical engineers, Part D: Journal of Automobile Engineering* **8** 951-963
[5] Kulkarni M, Shim T, Zhang Y 2007 Shift dynamics and control of dual-clutch transmissions *Mech Mach Theory* **842** 168–182
[6] Liu J, Ma B, Li H, Chen M, Li G 2018 Control strategy optimization for a dual-clutch transmission downshift with a single slipping clutch during the torque phase *Proceedings of the Institution of Mechanical Engineers, Part D: Journal of Automobile Engineering* **232** 651-666
[7] Liu Y, Qin D, Jiang H, Zhang Y 2014 Shift control strategy and experimental validation for dry dual clutch transmissions *Mechanism and Machine Theory* **75** 41-53
[8] Li G, Görges D 2018 Optimal control of the gear shifting process for shift smoothness in dual-clutch transmissions *Mechanical Systems and Signal Processing* **103** 23-38
[9] Wu J, Ma B, Li H, Liu J 2019 Creeping control strategy for Direct Shift Gearbox based on the investigation of temperature variation of the wet multi-plate clutch *Proceedings of the Institution of Mechanical Engineers, Part D: Journal of Automobile Engineering* 0954407019836313
[10] Holman J P 1986 *Heat transfer* (New York: McGraw-hill) p 521-565

Appendix

Parameters	Meanings	Parameters	Meanings
A	front area of vehicle	θ	rotation angle
A_f	surface area of friction pairs	f	friction coefficient of road
C_D	drag coefficient	g	acceleration of gravity
I	moment of inertia of components	i	gear ratio
M	torque transferred by components	m	vehicle weight
N	number of friction pairs	v_e	vehicle speed
P	pressure applied on clutch	r_w	radius of wheel
R	radius of friction disc	β	road grade angle
$\dot{Q}_{CL}$	generated heat flux in CL	μ	friction coefficient
S	stiffness coefficient of the shaft	ω	angular speed of components
T	temperature of components	$c_{i(o)}$	specific heat capacity
$V_{i(o)}$	volume	$\rho_{i(o)}$	density
$\dot{\Phi}_{i(o)}$	heat source	δ	increasing coefficient of mass

IOP Conf. Series: Materials Science and Engineering **825** (2020) 012046 doi:10.1088/1757-899X/825/1/012046

Sound Quality Prediction of Vehicle Door Closing Based on Experiment and Boundary Element Method

Gao Yunkai and Liu Zhe

School of Automobile, Tongji University, Shanghai 201804, China

Email:350762275@qq.com

Abstract. The sound quality of the vehicle door can only be improved after the testing vehicle has been manufactured. In this paper, a simulation model of the door system closing collision is established, and the vehicle test and the bench test are designed. In the vehicle test, according to the design characteristics of the door structure and the situation that the operating load cannot be obtained directly. The impact load on the door are simplified as the concentrated loads when the door is closed. The reference points are arranged near the location of the concentrated loads, and the number is twice of the concentrated loads. The acceleration response at the reference points are measured. In the bench test, the frequency response functions (FRFs) from the concentrated load points to the reference points are measured. Using the inverse matrix method, the value of the concentrated loads are obtained. A boundary element model is established and the concentrated load are applied to the model to collect sound pressure signals at the field point. The results show that the sound pressure values obtained by experiment and simulation are highly consistent. It shows that this method is feasible in predicting the sound quality.

1. Introduction

With the continuous escalation of consumption in the automotive industry, consumers are increasingly paying attention to the Noise, Vibration, Harshness (NVH) of the passenger cars. The first thing consumers encounter when buying a car is the door NVH, which has become a problem faced by Original Equipment Manufacturer (OEM). The subjective and objective evaluation of the sound quality of door closing is only after the prototype is produced, and the engineer is not fundamentally instructed to predict the sound quality. Therefore, it is necessary to predict the sound quality of the door in advance to fundamentally improve the sound quality.

David Hamilton [1] analyzed the main factors affecting the sound quality of door-closing collision, including: duration, amplitude, frequency distribution, etc. Based on the analysis of these influencing factors, the main influencing factors of door sound quality are analyzed using 3D drawing technology, and the subjective evaluation performance value of door sound quality is proposed, which provides a theoretical basis for door sound quality improvement. Ford Motor Technology Research Office [2] proposed that loudness and sharpness have a greater impact on the door impact sound quality, and proposed a design scheme for the door system sheet metal, door lock system and sealing system. Zhidong Zhang [3] established the finite element model and boundary element model of the door. First, the finite element model was used to perform the door closing collision analysis to extract the response speed of the set measurement points. The measured results were used as the boundary condition of the boundary element model. The analysis and calculation of the sound pressure signal at the set point of the door panel system. Yang Chuan [4] established an optimization model with the goal of reducing the collision energy of door closing, and compared and analyzed the sound pressure level of sound radiation from door closing before and after optimization. The effectiveness of the optimization model for reducing noise sound pressure level is verified. Ouyang Lian-ge [5] used the cab finite element model to perform a free modal calculation and compared the results with the experimental modal. Confirm the accuracy of modeling, use LMS Virtual.Lab software to perform modal superposition, and then get the frequency response of the cab. Fang Yuan [6] estimated the electromagnetic vibration and noise characteristics of the electric vehicle powertrain during the design phase. According to the structural characteristics of centralized drive electric vehicle power assembly, a joint simulation method of electric vehicle power assembly noise analysis under multi-source dynamic excitation was established. Chen Jian [7] took an excavator cab as an example, and established a structural finite element and acoustic boundary element model. Based on the frequency domain inverse matrix method to solve the excitation load under working conditions. The modal-based forced response method is used to obtain the vibration speed response of the cab under this excitation. Luo Mingjun [8] established a finite element model of the automotive cabin cavity based on Hypermesh finite element software. Then it is imported into the SYSNOISE environment in NASTRAN format to build a single-region boundary element calculation model. Feng Lei [9] used a forklift cab as a research object, and applied the PML method and boundary element method to analyze and predict the external radiation sound field caused by the vibration of the cab structure. Deng Yongcheng [10] taked the oil pan of a minicar engine as an object to study the control of the radiated noise of the oil pan. Zhao Yuwei [11] used the body-in-white model and the finite element model of the door to obtain structural modal information. The vibration velocity characteristics of the structure from 20 to 200 Hz were obtained and the boundary element model of the interior cavity was established. Zhao Jing [12] selected the engine mount position as the excitation force on the structural finite element model, and obtained the structural vibration response by modal superposition method. Using the vibration response as the boundary condition of the acoustic boundary element model, the acoustic sensitivity curve of a point in the vehicle is calculated.

In this paper, the sound pressure response of the target point and the acceleration response of the reference point are measured through the vehicle test, and the frequency response function from the excitation point to the reference point is measured through the bench test. Use the inverse matrix method to find the work force of the excitation point. According to the door structure of a new energy vehicle, its finite element model is established using finite element software. Boundary element model of the door was established using acoustic software. The obtained working force is taken as a boundary condition. Acoustic radiation analysis is performed on the established boundary element model to extract the sound pressure data of the field points. Comparative analysis shows that the measured values of the target point are in good agreement with the predicted values.

2. Load identification

When the system is subjected to an external excitation load $f(t)$, the differential equation of the system is:

$$M\ddot{x} + C\dot{x} + Kx = f(t) \tag{1}$$

Where M, C, and K respectively represent the mass matrix, damping matrix, and stiffness matrix of the system.

Perform fast Fourier transform on equation (2) and convert the time domain to the frequency domain to obtain:

$$H(\omega) = \frac{X(\omega)}{F(\omega)} \tag{2}$$

Where $H(\omega)$ is the frequency response function. It is an inherent property of the system.

The system is assumed to be linear and time invariant. When the system receives excitation loads $F_1, F_2, \ldots, F_n$, the system has responses $X_1, X_2, \ldots, X_m$, so the system response can be written as:

IOP Conf. Series: Materials Science and Engineering **825** (2020) 012046 doi:10.1088/1757-899X/825/1/012046

$$\begin{bmatrix} X_1 \\ X_2 \\ \cdots \\ X_m \end{bmatrix} = \begin{bmatrix} H_{11} & H_{12} & \cdots & H_{1n} \\ H_{21} & H_{22} & \cdots & H_{2n} \\ \cdots & \cdots & \ddots & \cdots \\ H_{m1} & H_{m2} & \cdots & H_{mn} \end{bmatrix} \begin{bmatrix} F_1 \\ F_2 \\ \cdots \\ F_n \end{bmatrix} \tag{3}$$

Where H_{mn} is the frequency response function from the excitation point load F_n to the reference point X_m.

The excitation point load can be obtained by multiplying both sides of equation (3) by the inverse matrix of the frequency response function:

$$\begin{bmatrix} F_1 \\ F_2 \\ \cdots \\ F_n \end{bmatrix} = \begin{bmatrix} H_{11} & H_{12} & \cdots & H_{1n} \\ H_{21} & H_{22} & \cdots & H_{2n} \\ \cdots & \cdots & \ddots & \cdots \\ H_{m1} & H_{m2} & \cdots & H_{mn} \end{bmatrix}^{-1} \begin{bmatrix} X_1 \\ X_2 \\ \cdots \\ X_m \end{bmatrix} \tag{4}$$

3. Data collection

3.1 Vehicle test

The test object is the left front door of a new energy vehicle, and the test is carried out on an equipped vehicle. The test system mainly includes sensors, data acquisition and recording systems, and working condition control devices. The microphone is a built-in ICP omnidirectional microphone produced by B & K Company, and the acceleration sensor is a structural test ICP three-direction acceleration sensor manufactured by PCB Company. LMS SCADAS III SC316-UTP data acquisition system and LMS. Test. Lab test data processing and analysis system are used. The door closing speed is controlled by a door closing speed meter and a rubber rope energy storage device.

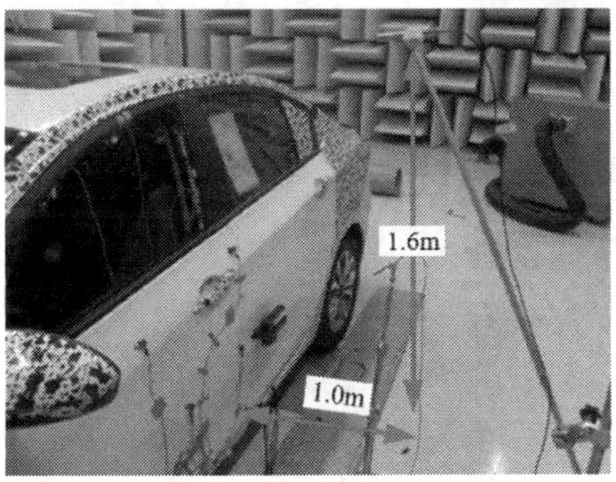

Figure 1. Microphone position

Figure 2. Door closing speed meter and rubber rope energy storage device

In the vehicle test, the microphone was arranged at the user's right ear position at the door opening outside the vehicle to collect sound pressure signals. The position of the microphone is 1.0m outward from the left front door and 1.6m above the ground, pointing to the body side, as shown in Figure 1. Door closing speed meter and rubber rope energy storage device are used to control the door closing speed, as shown in Figure 2.

In this section, the working load when the door is closed is discretely concentrated into 11 degrees of freedom. The load on the seal is discretely concentrated into 8 degrees of freedom (S1~S8). The load on the lock is dispersed into a concentrated load (L1) with 3 degrees of freedom, and the direction is the X Y Z direction, as shown in Figure 3. When using the inverse matrix method to obtain the working point load, the degree of freedom of the reference point is required to be more than twice that of the working point, the degree of freedom of the reference point is 48 (X1 ~ X16), and the direction is X Y Z-direction, as shown in Figure 4. Meet the calculation requirements of the inverse matrix

IOP Conf. Series: Materials Science and Engineering **825** (2020) 012046 doi:10.1088/1757-899X/825/1/012046

method. Paste the three-direction acceleration sensor to the position of the working point and the reference point.

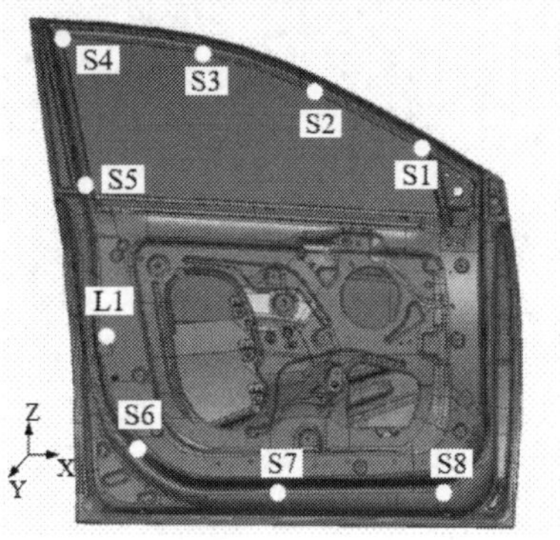

Figure 3. The position of working points

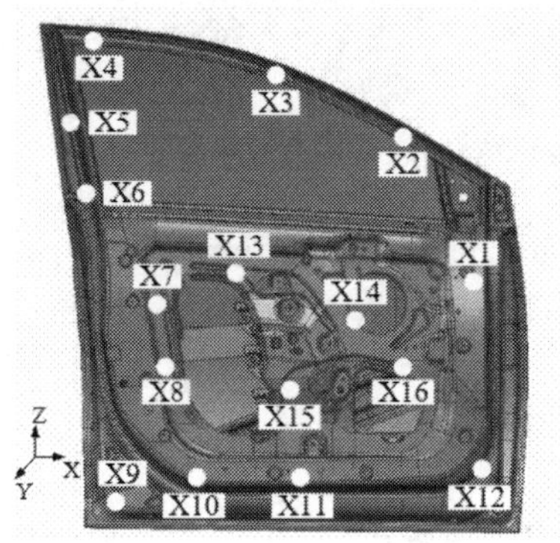

Figure 4. The position of reference points

3.2 Bench test

The bench test system is mainly composed of two parts: an excitation and measurement system, a data acquisition system. The excitation system is mainly a force hammer. Modally Tuned ICP impact hammer manufactured by PCB company is used as the exciter. The measurement system and data acquisition system are the same as the vehicle test.

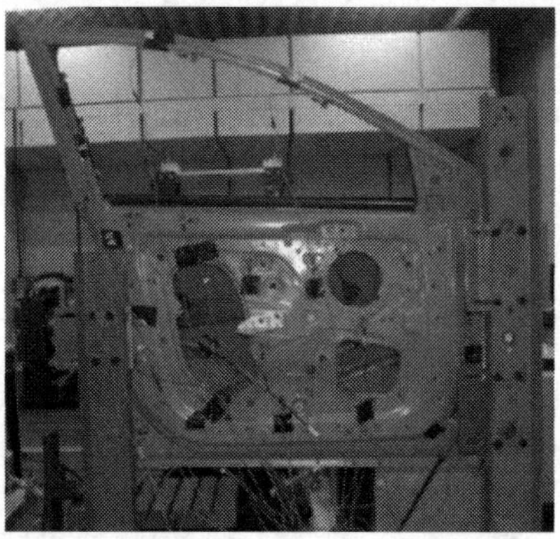

Figure 5. Bench test

In order to approximate the boundary conditions required by the actual structure of the door, the door fixture is designed, the door is fixed on the fixture, and its frequency response function is measured, as shown in Figure 5. The natural frequency of the tested fixture is much higher than that of the door, and the fixture meets the experimental requirements. The analysis frequency range of this paper is 0~200Hz. Hit the working point as shown in Figure 3 with a hammer, paste a three-direction acceleration sensor at the reference point as shown in Figure 5, and the frequency response function from each working point to the reference point can be measured, as shown in Figure 6. The working force of the working point can be obtained by formula 4, as shown in Figure 7.

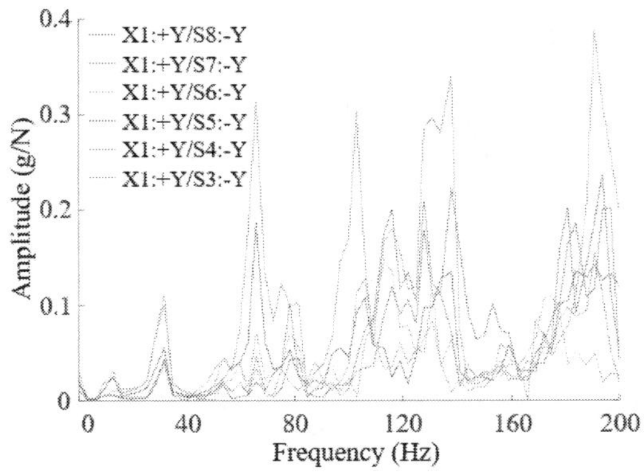

Figure 6. Partial frequency response function

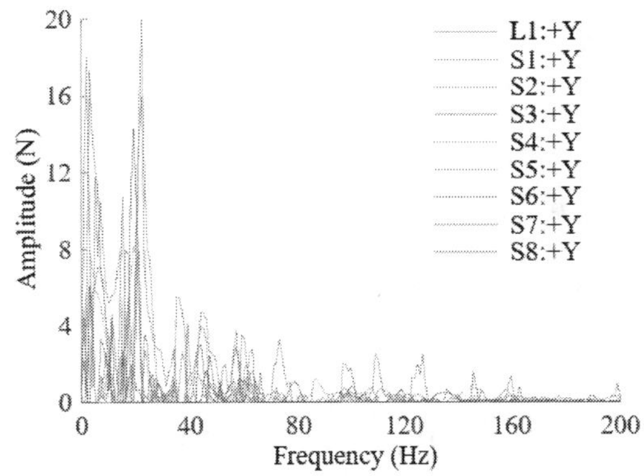

Figure 7. Identification of working force

4. Simulation analysis

This section briefly introduces the process of establishing the finite element model of the left front door. Sheet metal parts and glass use shell elements, sealing strips, structural glue and hinges use hexahedral elements. The number of finite element model elements is more than 130 000. The door structural materials are mainly stamped steel plates, and its main mechanical performance indicators are yield strength and tensile strength. The names, materials, and performance parameters of the main door structures are shown in Table 1.

Table 1. Model materials and parameters

Component	material	Yield Strength (MPa)	tensile strength (MPa)
Outer panel	HC180B	205	330
Inner panel	DC04	210	270
Reinforcing plate	DC01	260	270
Hinge	Q345	345	490

Boundary element method is a calculation method for discrete problems proposed after finite element. The finite element method is a global discretization. The boundary element method only discretizes the boundary, which can more accurately simulate the original model than the finite element method. It retains the information of the original model more completely and reduces the dispersion error.

The door structure in contact with the air is discretized into a finite number of element, and the mesh size of the boundary element model is 5 mm. Y-direction excitation load is applied at the discrete points (S1~S8) of the sealing strip, and X, Y, Z-direction load excitation is applied at the lock (L1). The excitation load value is shown in Figure 7. The fluid medium is defined as air, the speed of sound is 340m/s, the density is 1.225kg/m^2, and the coordinates of the response point are X=2500mm, Y=-1000mm, Z=1600mm, as shown in Figure 8.

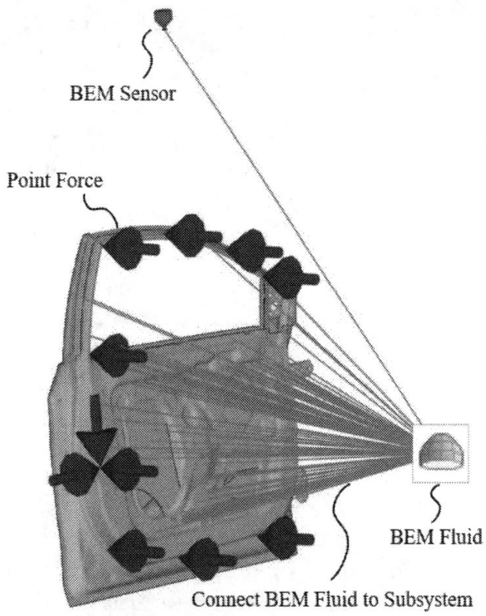

Figure 8. Boundary Element Model

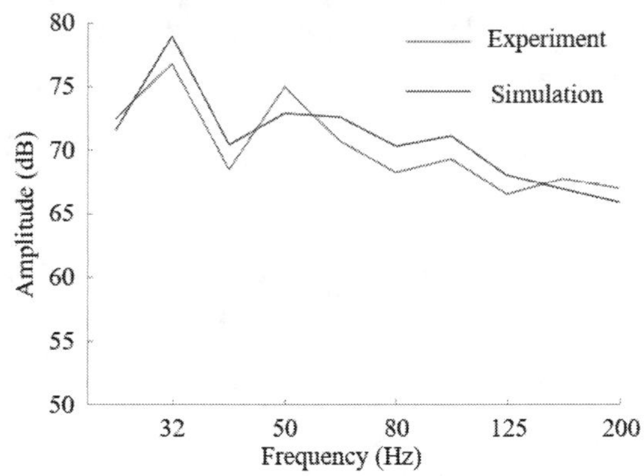

Figure 9. Frequency domain curve of sound pressure level

The modal information of the door is imported into the boundary element and calculated. Then calculate the sound pressure level at the field point according to the sound pressure level calculation formula.

$$P_{dB} = 20\log_{10}(p / p_0) \quad (5)$$

Where P_0 is the standard sound pressure, and the value is 2×10^{-5}Pa.

The frequency curve of the sound pressure level of the microphone at the set field point is shown in Figure 9. It can be seen that the simulation curve agrees well with the experimental curve, and the effectiveness of the prediction of the door closing sound is valid.

5. Conclusion

This paper designs vehicle tests and bench tests. In the vehicle test, the sound pressure level curve at the target field point and the vibration acceleration response at the reference point are obtained when the door of a passenger car is closed. The frequency response function from the discrete excitation point to the reference point was obtained in the bench test. Using the inverse matrix method, the vibration acceleration response at the reference point, and the frequency response function from the discrete excitation point to the reference point, the working load at the discrete excitation point is obtained. This paper also establishes a finite element model of the vehicle door closing collision simulation analysis to obtain the modal information of the door, and imports the modal information into the boundary element model. The calculated working load at the discrete points is input to the boundary element model, and a field point is set at the target point to collect the sound pressure signal. Finally, the data obtained from the simulation are compared with the data measured by the test. The validity and feasibility of the method are verified.

References

[1] Hamilton D . Sound Quality of Impulsive Noises: An Applied Study of Automotive Door Closing Sounds[C]// Noise & Vibration Conference & Exposition. 1999.
[2] Petniunas A , Otto N C , Amman S , et al. Door System Design for Improved Closure Sound Quality[J]. SAE Technical Papers, 1999.

IOP Conf. Series: Materials Science and Engineering **825** (2020) 012046 doi:10.1088/1757-899X/825/1/012046

[3] Zhang Z , Young S . SAE Technical Paper Series [SAE International SAE 2005 Noise and Vibration Conference and Exhibition -, (MAY. 16, 2005)] SAE Technical Paper Series - Low Frequency Transient CAE Analysis for Vehicle Door Closure Sound Quality[J]. 2005, 1.
[4] Yang Chuan, Yu Dejie. The Study on Optimization Methods of the Door-Slamming Noise[J]. Automotive Engineering, 2013,35(3): 243-248.
[5] Ouyang Lian-ge, Yi Buhui, Ge Xiaohong. Analysis and Evaluation of Cab's Interior Acoustic Characteristics of Crawler Beach Cleaners[J]. Noise and Vibration Control, 2015, 35(01): 114-118.
[6] Fang Yuan, Zhang Tong, Yu Peng, et al. Joint simulation of radiated noise of electric powertrain based on finite element modeling and boundary element method[J]. Transactions of the Chinese Society of Agricultural Engineering, 2014, 30(21): 40－46.
[7] Chen Jian, Jiang Fengxin, Xiao Yue. Low Frequency Noise Analysis and Control of Excavator Cab[J]. China Mechanical Engineering, 2014, 25(15): 2124-2130.
[8] Luo Mingjuan, Zhou Fan, Zhao Yongling, et al. Application of boundary element method in the interior noise modeling[J]. Journal of Nanchang University (Engineering & Technology), 2014, 36(02): 157-160.
[9] Feng Lei, Fang Yuying, Zuo Yanyan. Prediction Analysis on Exterior Acoustic Radiation of a Forklift Cab[J]. Science Technology and Engineering, 2013, 13(25): 7422-7426.
[10] Deng Yongcheng, Yu Dejie. A Research on the Radiation Noise Control of Oil Pan Based on Structural-acoustic Sensitivity Analysis[J]. Automotive Engineering, 2012, 34(12): 1125-1130.
[11] Zhao Yuwei, Zhou Hong. Bus Body Panel Acoustic Contribution Analysis[J]. Automobile Technology, 2012(12): 30-33.
[12] Zhao Jing, Zhou Hong, Liang Yingzhen, et al. Vehicle Body Panel Acoustic Contribution Analysis and Optimization[J]. Journal of Mechanical Engineering, 2010, 46(24): 96-100.

IOP Conf. Series: Materials Science and Engineering **825** (2020) 012047 doi:10.1088/1757-899X/825/1/012047

Hybrid Impedance Control of Robot Manipulators based on Generalized Momentum

Kaining Li, Xianmin Zhang and Yanjiang Huang*

School of Mechanical and Automotive Engineering, South China University of Technology, Guangzhou 510640, China

* Email: mehuangyj@scut.edu.cn

Abstract. In the hybrid impedance control strategy, the task space is divided into position control subspace and force control subspace, and the constraint direction in the manipulator task space is controlled by torque-based impedance control, while the free movement direction is controlled by position-based impedance control. The hybrid impedance control has strong force tracking characteristics in the direction of force control and strong flexibility in the direction of position control. In this paper, we use the generalized momentum observer instead of the torque sensor to estimate the torque exerted by the external environment on the robot. Simulation results are shown for a three-link SCARA robot.

1. Introduction

The manipulator impedance control realizes the interaction between manipulator and environment through the dynamic relationship between position and environmental forces [1]. By specifying the relationship between contact forces and position, it ensures that the robot operates in a restricted environment while maintaining the appropriate contact forces. Impedance control is divided into two different control structures based on torque and position, torque-based impedance control adjusts the contact force and displacement of the end-effector by controlling the joint driving torque [2], while position-based impedance control adjusts the position/velocity of the end-effector according to the contact force deviation between the robot and the environment [3]. Hybrid impedance control [4] combines the above two control strategies, achieving accurate force control and position control with good robustness. In [5] and [6], hybrid impedance control strategy is designed and proved accurately and robustly. There are still some main issues being studied on the impedance control, such as the capability of position tracking and force tracking [7]. Moreover, the external contact force of the manipulator in contact with the environment is usually measured by a six-dimensional force sensor [8]. If without the sensor, external contact torque can also be obtained by means of a generalized momentum observer [9-10].

In this paper, we provide a hybrid impedance control strategy which has precise force tracking characteristics in the direction of torque-based impedance control and strong position tracking in the direction of position-based impedance control. Generalized momentum observer is used instead of torque sensor to estimate external contact torque.

2. Robot dynamics and properties

In this section, we summarize the dynamics model of joint space and the relevant properties of robot. The torque of the external environment on the manipulator is taken into account and the torque is estimated by using a generalized momentum observer.

2.1. Robot dynamics model

The robot dynamic model is expressed by the following equation:

$$M(\theta)\ddot{\theta} + C(\theta,\dot{\theta})\dot{\theta} + G(\theta) + \tau_f(\dot{\theta}) = \tau + \tau_e \tag{1}$$

where θ , $\dot{\theta}$ and $\ddot{\theta}$ are the joint angle, joint angular velocity and joint angular acceleration, respectively. $M(\theta)$ is the inertia matrix, $C(\theta,\dot{\theta})\dot{\theta}$ is the centrifugal and Coriolis vector, $G(\theta)$ is the gravity vector, $\tau_f(\dot{\theta})$ is torque of joint friction vector, τ is the torque applied to the robot by the actuators, and τ_e is the torque of the external environment on the manipulation.

The robot dynamic equation satisfies the following two characteristics:

1) $M(\theta)$ is the symmetric and positive-definite matrix;

2) $\dot{M}(\theta) - 2C(\theta,\dot{\theta})$ is the skew-symmetry matrix, which is also equivalent to the identity

$$\dot{M}(\theta) = C(\theta,\dot{\theta}) + C^T(\theta,\dot{\theta}). \tag{2}$$

2.2. Generalized momentum observer

The generalized momentum of the robot is defined as

$$p = M(\theta)\dot{\theta}. \tag{3}$$

Combining equations (1) and (2), the derivative of equation (3) is

$$\dot{p} = C^T(\theta,\dot{\theta})\dot{\theta} - G(\theta) - \tau_f(\dot{\theta}) + \tau + \tau_e. \tag{4}$$

Define the residual vector r as

$$r = K\left[p(t) - p(0) - \int_0^t (C^T\dot{\theta} - G - \tau_f + \tau + r)ds\right] \tag{5}$$

with $r(0)=0$, where K is a diagonal gain matrix and satisfies $K>0$, $p(t)$ is the robot generalized momentum at time $t \geq 0$, as defined in equation (3). The vector r can be calculated from the measured $(\theta,\dot{\theta})$ and the commanded joint torque τ. The advantage of this method is that the inverse of the inertial matrix and the joint acceleration are not required [11].

Combining equations (4) and (5), the resulting dynamics of r is

$$\dot{r} = K(\tau_e - r). \tag{6}$$

And then, a transfer function for the residual dynamics can be written as

$$\frac{r_i(s)}{\tau_{e,i}(s)} = \frac{K_i}{K_i + s}, \qquad i = 1,\ldots,n, \tag{7}$$

where i is the serial number of links. It can be seen that r tracks τ_e in the form of a first-order system, and $r(0) = \tau_e(0)$ at steady state. So we can use r as an estimate of the external torque, meaning

$$K \to \infty \qquad \Rightarrow \qquad r \approx \tau_e. \tag{8}$$

For the compactness of the formula, we denote $\gamma(\theta,\dot{\theta}) = C^T(\theta,\dot{\theta})\dot{\theta} - G(\theta) - \tau_f(\dot{\theta})$ in Figure 1 that shows the block diagram of the observer. It's worth noting that equation (8) is the ideal result if the model is correct. In fact, due to the inaccuracy of the robot dynamic model and parameters, there is an error between the contact force and the actual external force. For the knowledge of model and parameter identification to improve model accuracy, please refer to the literature [12], we will not repeat them here.

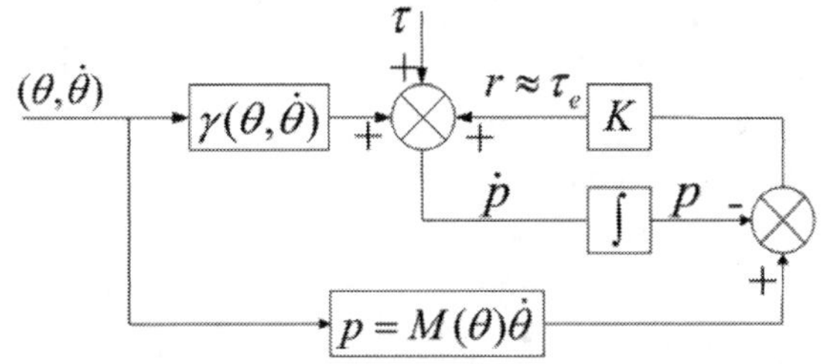

Figure 1. The block diagram of the generalized momentum observer.

3. Robot impedance control strategy

In this section, we will show how to use the external contact force information provided by the residual vector (5) to design an impedance control scheme to ensure that the movement of the contact point along the obstacle surface and the approximate adjustment of the contact force. The performance of the proposed control scheme will indeed depend on the accuracy of the "measurement" of joint contact torque provided by the residual r, that is, on the accuracy of the model and on the system affecting the input and measurement noise.

3.1. Joint space position-based impedance control

For simplicity, let us denote $h(\theta,\dot{\theta}) = C(\theta,\dot{\theta})\dot{\theta} + G(\theta) + \tau_f(\dot{\theta})$, so that (1) can be rewritten as

$$M(\theta)\ddot{\theta} + h(\theta,\dot{\theta}) = \tau + \tau_e. \tag{9}$$

In the joint space, the desired dynamic behavior of the robot after applying position-based impedance control can be expressed as

$$M_d(\ddot{\theta}_d - \ddot{\theta}) + B_d(\dot{\theta}_d - \dot{\theta}) + K_d(\theta_d - \theta) = \tau_e \tag{10}$$

where M_d, B_d, and K_d are symmetrical matrices that denote the desired inertia, damping, and stiffness matrices, respectively. The vector θ_d denotes the desired joint angle. The equation (9) is arranged as

$$\ddot{\theta} = \ddot{\theta}_d - M_d^{-1}[\tau_e - B_d(\dot{\theta}_d - \dot{\theta}) - K_d(\theta_d - \theta)]. \tag{11}$$

The resulting control law after combining the equations (9) and (11) becomes

$$\tau = M(\theta)(\ddot{\theta}_d - M_d^{-1}[\tau_e - B_d(\dot{\theta}_d - \dot{\theta}) - K_d(\theta_d - \theta)]) + h(\theta,\dot{\theta}) - \tau_e. \tag{12}$$

3.2. Joint space torque-based impedance control

In terms of force control direction of the mechanical arm, the mechanical arm is equivalent to a mass-damping system, and its stiffness is set to 0, thus achieving accurate force control. For the torque-based impedance control case, the desired relationship between θ and τ_e can be expressed as the impedance function

$$M_d(\ddot{\theta}_d - \ddot{\theta}) + B_d(\dot{\theta}_d - \dot{\theta}) = \tau_d - \tau_e \tag{13}$$

IOP Conf. Series: Materials Science and Engineering **825** (2020) 012047 doi:10.1088/1757-899X/825/1/012047

where τ_d is the desired force. Form (9) and (13), the resulting control law can be deduced as

$$\tau = M(\theta)(\ddot{\theta}_d - M_d^{-1}[\tau_d - \tau_e - B_d(\dot{\theta}_d - \dot{\theta})]) + h(\theta,\dot{\theta}) - \tau_e. \tag{14}$$

3.3. Hybrid impedance control strategy based on r-observer

Based on the above ideas, a switching matrix S is introduced into the hybrid impedance control. This matrix, a diagonal matrix with elements either 1 or 0, is used to separate the torque-based and position-based impedance control. Combining (12) and (14) with the switching matrix, the desired dynamics behavior of the system can be described as

$$M_d(\ddot{\theta}_d - \ddot{\theta}) + B_d(\dot{\theta}_d - \dot{\theta}) + SK_d(\theta_d - \theta) = (I - S)\tau_d - \tau_e \tag{15}$$

where I is the unit matrix. Then we can deduce the following control law:

$$\tau = M(\theta)(\ddot{\theta}_d - M_d^{-1}[(I - S)\tau_d - \tau_e - B_d(\dot{\theta}_d - \dot{\theta}) - SK_d(\theta_d - \theta)]) + h(\theta,\dot{\theta}) - \tau_e. \tag{16}$$

The method of generalized momentum observer is used to obtain the external contact torque instead of a torque sensor. The symbol $\hat{\tau}_e$ represents an estimate of the external torque τ_e. As shown in Figure 2, the hybrid impedance controller receives torque error between the desire torque and the contact torque of the environment on the robot manipulation and then sends the position θ to the robot servo controller to modify the joint position in order to maintain a constant torque.

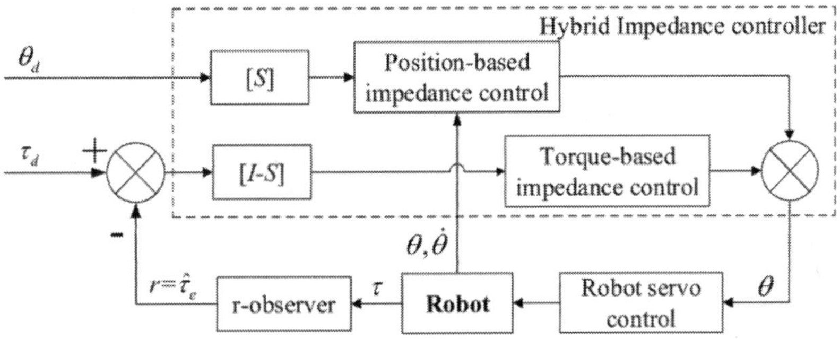

Figure 2. The implementation of the hybrid impedance controller based on r-observer.

4. Simulation

To verify the proposed hybrid impedance control strategy, a series of simulation studies are conducted and presented in this section. The simulation platform is based on Matlab/Simulink which mainly consists of the following parts: the hybrid impedance controller, robot dynamics model, and the generalized momentum observer.

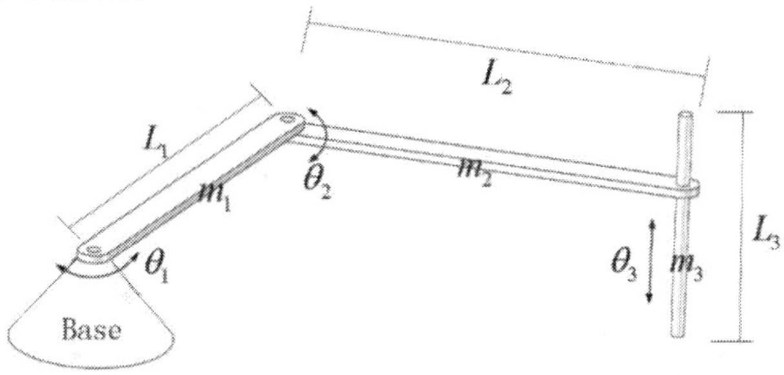

Figure 3. The model of SCARA robot.

4.1. Simulation setup

In the simulation, a SCARA robot [13], as shown Figure 3, is used as an object to be controlled. This robot has three rotational degrees of freedom in the horizontal direction and one in the vertical direction in general. Since the 3rd-axis rotating joint and the 4th-axis prismatic joint are coupled, and the 4th-axis prismatic joint can be independently realized without accompanying the rotation of the 3rd joint, the 3rd joint and the 4th joint are merged as a combined link 3 in modelling, holding the joint type as a prismatic joint and keeping the 3rd joint locked during simulation experiment. Table 1 shows the SCARA robot model parameters which including the mass, length and inertia of each manipulation.

Table 1. SCARA robot model parameters.

Axle	m (kg)	L (m)	$I\,(\mathrm{kg\cdot m^2})$
1	2.4312	0.25	32.1907
2	3.7860	0.25	33.7615
3	0.5552	0.14	—

In the hybrid impedance controller, the first two planar rotating joints of the SCARA robot execute the position-based impedance control strategy while the 3rd vertical prismatic joint executes the torque-based impedance control strategy by choosing the elements of the matrix S, namely, $S = \mathrm{diag}(1,1,0)$.

To verify the tracking error performance of the hybrid impedance control, the control parameters need to be set reasonably. The value of diagonal gain matrix K needs to be large enough, so set it to K=diag(2000,2000,1600). The hybrid impedance controller parameters $M_d = \mathrm{diag}(5,5,2)$, $B_d = \mathrm{diag}(50,50,30)$, $K_d = \mathrm{diag}(500,500,0)$ are selected when the two rotating joints execute position tracking and the 3rd joint executes force tracking. The sample time t in simulation is set to 5 s. The input of position tracking is $\theta_{d1} = \theta_{d2} = 1-\cos(\pi t)$ rad while the desired tracking force is $\tau_d = 10$ N at steady state. Moreover, we take the viscous and coulomb friction into consideration:

$$\tau_f = f_v\dot{\theta} + f_c\mathrm{sign}(\dot{\theta}) \tag{17}$$

where $f_v = \mathrm{diag}(0.5,0.2,0.1)$ and $f_c = \mathrm{diag}(1.5,0.5,0.3)$.

4.2. Simulation results

Figure 4 and Figure 5 show the joint driving torque of joint 1 and 2 when executing the above position tracking in the hybrid impedance controller. As shown in the simulation results, the torque of the robot will fluctuate slightly during start-up because of the friction and so on. This is a major source of tracking error.

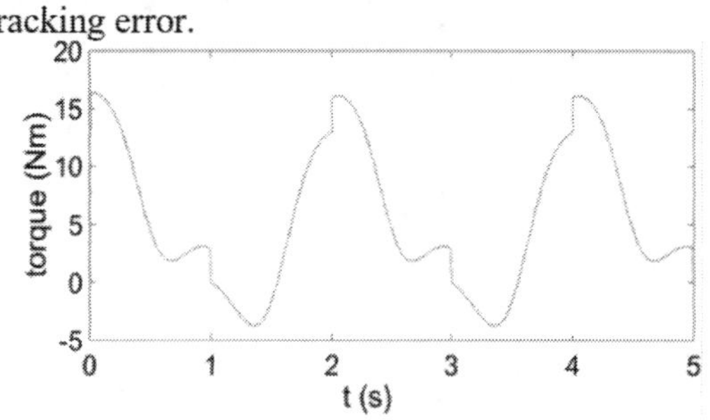

Figure 4. The driving torque of joint 1.

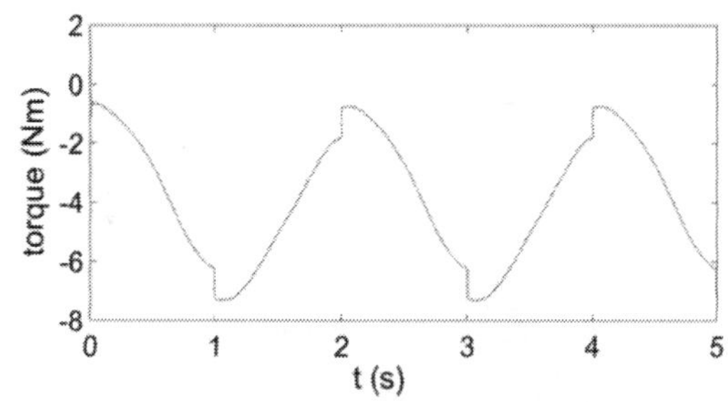

Figure 5. The driving torque of joint 2.

The position tracking trajectories of the first two joints are shown in Figure 6 (a) and Figure 7 (a), respectively. Meanwhile, the corresponding performance of the position tracking error is drawn in Figure 6 (b) and Figure 7 (b). It is observed that the position tracking errors are under one degree. Thus, the hybrid impedance controller performs well at the first two rotating joints.

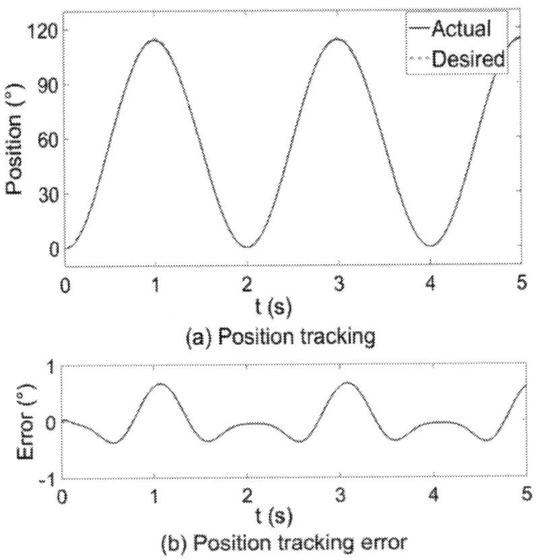

Figure 6. The position tracking performance of joint 1.

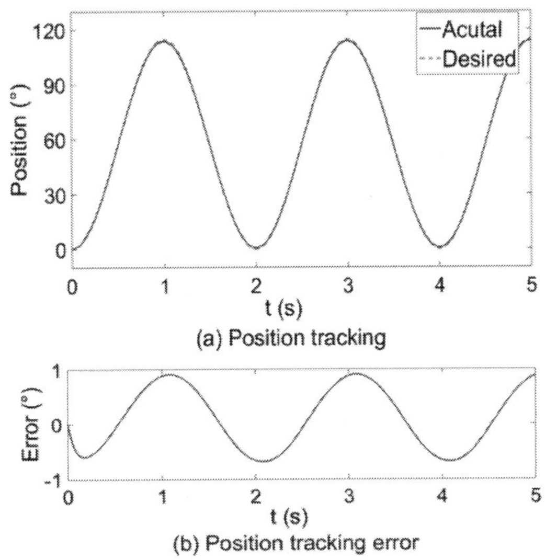

Figure 7. The position tracking performance of joint 2.

The force tracking performance on a plane trajectory, a slope trajectory and a sine trajectory of joint 3 is shown Figure 8, Figure 9 and Figure 10, respectively. Namely, execute force tracking control under different command trajectory $\theta_d = 0.05$, $\theta_d = 0.02t$, and $\theta_d = 0.05\sin(0.4\pi t)$. Figures 8 (a) - 10 (a) show that the actual trajectory tracks the command trajectory on a plane, on a slope and in sine working conditions, respectively, while Figures 8 (b) - 10 (b) show that the actual contact force tracks the desired contact force on a plane, on a slope and in sine working conditions, severally. In the simulation, the desired tracking force is set to 10 N at steady state, but we change the direction of the desired force at 2.5 seconds to better demonstrate the force tracking performance of the hybrid impedance controller.

It can be seen that the force tracking performance is brilliant for the third joint when given the corresponding position command. At the moment of starting and intermediate desired force change direction, the actual force will reach steady state within 0.3 seconds.

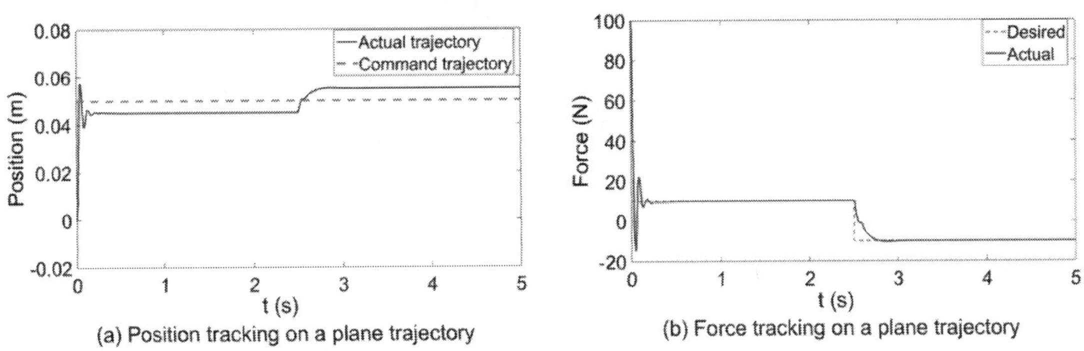

Figure 8. The force tracking performance on a plane trajectory of joint 3.

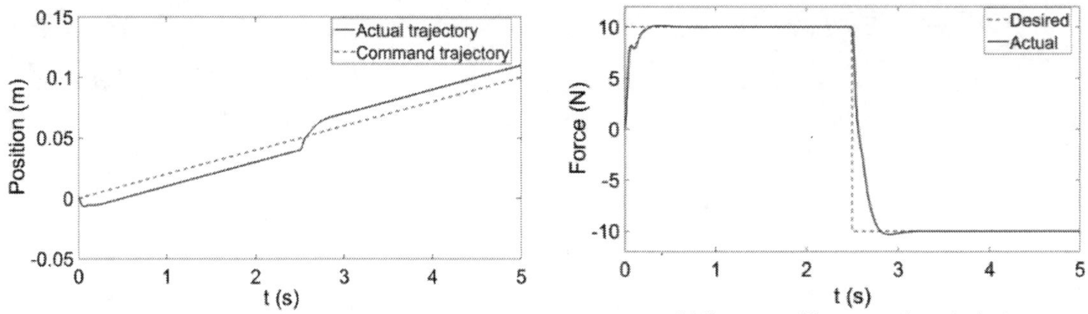

(a) Position tracking on a slope trajectory (b) Force tracking on a slope trajectory

Figure 9. The force tracking performance on a slope trajectory of joint 3.

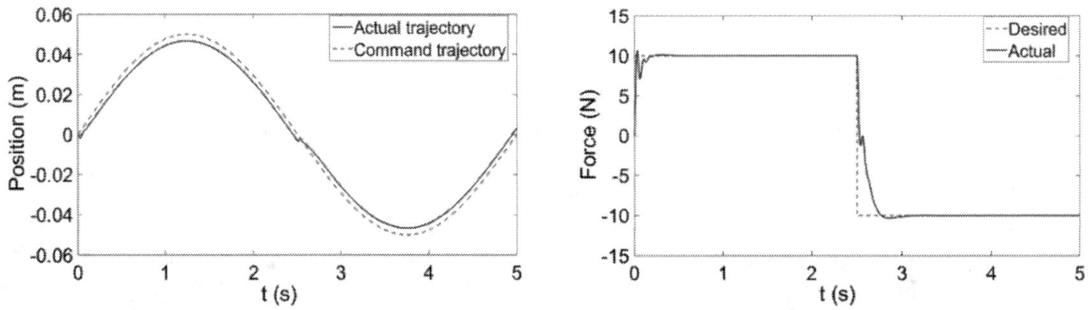

(a) Position tracking on a sine trajectory (b) Force tracking on a sine trajectory

Figure 10. The force tracking performance on a sine trajectory of joint 3.

5. Conclusion

With the emergence of interactive robot, the importance of robot contact operation becomes more and more prominent. A hybrid impedance controller based on generalized momentum is proposed for dynamic contact force tracking and non-contact axis position tracking. Thus, it can be determined that position and force can be controlled simultaneously by impedance control.

The above simulation shows that the hybrid impedance control strategy has good tracking performance. Besides, by estimating the external force on the manipulation, we don't need to know the environmental stiffness and also need no force sensors.

Acknowledgments

This work was supported by the National Nature Science Foundation of China (Grant Nos. 51820105007 and U1501247).

References

[1] Hogan N 1985 Impedance control: An approach to manipulation: Part I—Theory *J. Dyn. Sys., Meas., Control* **107** 1–7
[2] Bonitz R C and Hsia T C 1996 Internal force-based impedance control for cooperating manipulators *IEEE Trans. on Robotics and Automation* **12** 78–89
[3] Yuan J, Qian Y, Gao L, Yuan Z and Wan W 2019 Position-based impedance force controller with sensorless force estimation *Assembly Automation* **39** 489–496
[4] Cao H, Chen X, He Y and Zhao X 2019 Dynamic adaptive hybrid impedance control for dynamic contact force tracking in uncertain environments *IEEE Access* **7** 83162–74
[5] Anderson R J and Spong M W 1988 Hybrid impedance control of robotic manipulators *IEEE Journal on Robotics and Automation* **4** 549–556

[6] Liu G J and Goldenberg A A 1991 *Proc. IEEE Int. Conf. on Robotics and Automation* (Sacramento: IEEE) pp 287–292
[7] Duan J, Gan Y, Chen M and Dai X 2018 Adaptive variable impedance control for dynamic contact force tracking in uncertain environment *Robotics and Autonomous Systems* **102** 54–65
[8] Keemink A Q, van der Kooij H and Stienen A H 2018 Admittance control for physical human–robot interaction *The International Journal of Robotics Research* **37** 1421–44
[9] De Luca A and Mattone R 2005 *Proc. IEEE Int. Conf. on Robotics and Automation* (Barcelona: IEEE) pp 999–1004
[10] Wahrburg A, Morara E, Cesari G, Matthias B and Ding H 2015 *IEEE Int. Conf. on Automation Science and Engineering* (Gothenburg: IEEE)
[11] De Luca A, Albu-Schaffer A, Haddadin S and Hirzinger G 2006 *Proc. IEEE Int. Conf. on Intelligent Robots and Systems* (Beijing: IEEE) pp 1623–30
[12] Afrough M and Hanieh A A 2019 Identification of dynamic parameters and friction coefficients *Journal of Intelligent & Robotic Systems* **94** 3–13
[13] Zhang Y, Qiu Z and Zhang X 2019 Calibration method for hand-eye system with rotation and translation couplings *Applied optics* **58** 5375–87

IOP Conf. Series: Materials Science and Engineering **825** (2020) 012048 doi:10.1088/1757-899X/825/1/012048

Research on Process Reliability of RF Coaxial Cable in Spacecraft Assembly

Yougao Fan[1,2,*], Peirong Zhao[1,2], Guangtong Liu[1,2] and Shuang Liu[1]

[1]Beijing Institute of Spacecraft Environment Engineering, Beijing 100094, China
[2]Beijing Engineering Research Center of the Intelligent Assembly Technology and Equipment for Aerospace Product, Beijing 100094, China

[*]Corresponding author E-mail: fanygao@163.com

Abstract. This paper is concerned with the process reliability of RF coaxial cable in the spacecraft assembly. By carrying out some process tests, the effects of coaxially of the inner and outer conductors of the RF coaxial cable assembly, the number of cable winding coils, the bending radius of the RF coaxial cable, and the extreme conditions of the assembly on the performance of the RF coaxial cable are studied. According to the test results, a more comprehensive process method and specific process parameters are obtained, thereby further improving the reliability and safety of the RF coaxial cable in the spacecraft assembly.

1. Introduction

Radio frequency, abbreviated as RF, is an abbreviation for high-frequency alternating-current electromagnetic waves, with a frequency range from 300KHz to 30GHz [1]. Radio frequency technology is widely used in the fields of wireless communications. RF cable is a cable that transmits electromagnetic energy in the radio frequency range. It is an indispensable component in various radio communication systems and electronic equipment. It is widely used in wireless communication and broadcasting, television, navigation and other aspects [2].

The RF cable has the characteristics of transmission frequency bandwidth, strong resistance to external interference, small radiation loss, simple structure, convenient installation, and economical characteristics. The structure of RF cables can be classified according to different forms and uses. For instance, it is classified as RF coaxial cable, symmetrical RF cable, and spiral RF cable according to the structure; solid insulated cable, air insulated cable, and semi-air insulated cable according to the insulation form; flexible cables, semi-rigid cables, and rigid cables according to their flexibility [3].

With the continuous improvement of the spacecraft technology level and capabilities, the use of coaxial radio frequency cable assemblies in spacecraft has become more extensive and deeper. The assembly of the cables and the connectors runs through the various stages of spacecraft assembly, tests, and large-scale tests [4]. RF coaxial cables and RF connectors are mainly used for the connection between high-frequency equipment of spacecraft. It is an important work of the spacecraft during the assembly process. The quality of the assembly is directly related to the spacecraft measurement and control, antennas and transponders whether the sub-system can work normally [5, 6]. Due to the lack of in-depth understanding of the structure and performance of RF coaxial cable assemblies, quality problems caused by the operation of RF coaxial cables and their electrical connectors have occurred

IOP Conf. Series: Materials Science and Engineering **825** (2020) 012048 doi:10.1088/1757-899X/825/1/012048

frequently in recent years, so it is urgent to study their assembly process methods in order to improve the reliability and safety of the assembly of RF coaxial cables and electrical connectors.

The purpose of this paper is to carry out some process tests to study the effects of assembly extreme conditions on the performance of RF coaxial cable assemblies, and to determine the operation methods and specific technical indicators of various types of RF coaxial cables and connectors, so as to further improve the reliability and safety in spacecraft assembly.

2. Overview of RF coaxial cable assemblies installation

A RF coaxial cable assembly is composed of a RF coaxial cable and two RF electrical connectors, and is used to connect radio frequency electronic instruments and related parts (groups). RF coaxial cable assembly is classified as flexible cables, semi-rigid cables, hard cables, etc. according to bending conditions, or general cables, phase-stable cables, low-loss cables, etc. according to performance indicators, and is usually classified according to bendability.

2.1. RF coaxial cable

At present, there are two types of RF coaxial cables used in spacecraft assembly, including flexible cables and semi-rigid cables. The working frequency of RF coaxial cables is about 2GHZ to 35GHZ, and the characteristic impedance is 50Ω. The cable is composed of four parts, including inner conductor, insulation medium, outer conductor or additional shielding and sheath, as shown in Figure 1 and Table 1.

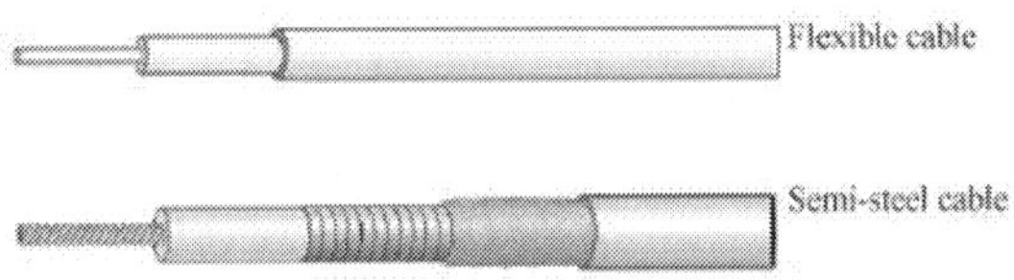

Figure 1. Schematic diagram of RF coaxial cable

Table 1. RF coaxial cables commonly used in spacecraft assembly

Name	Type	Outer diameter (mm)	Features
Flexible cable	UFB205A/UFB311A MCJ205A/MCC-311A	5/8 5/8	Low loss, small attenuation, easy installation
Semi-rigid cable	SFT-50-3	3.6	Superior shielding performance, good stability, easy to form
Semi-rigid threaded cable	SFC-50-7-3	10	Good radiation resistance, easy to bend

2.2. RF coaxial electrical connector

RF coaxial electrical connector is mainly composed of contact, insulator, housing, and accessories. Except for accessories, other components should be made of non-magnetic materials, and incompatible metals should not be placed in contact with each other, as shown in Figure 2.

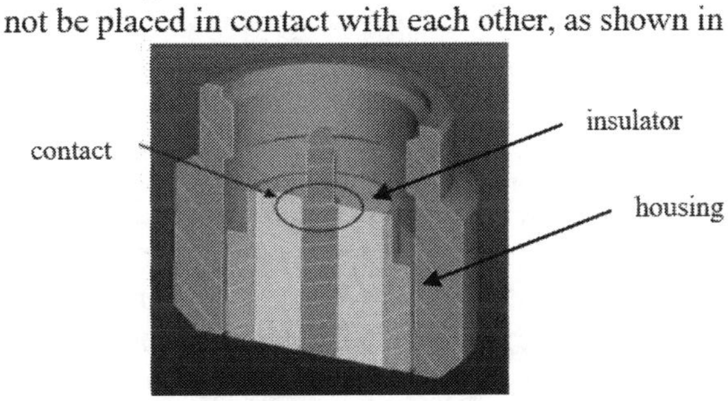

Figure 2. Schematic diagram of RF coaxial electrical connector

IOP Conf. Series: Materials Science and Engineering **825** (2020) 012048 doi:10.1088/1757-899X/825/1/012048

At present, the types of RF electrical connectors used in spacecraft assembly include SMA, TNC, SSMA, and so on. The connection method of the electrical connector and the cable is soldered, and the connection method of the plug and the socket is screwed. See Table 2 for common models of spacecraft RF electrical connectors.

Table 2. RF coaxial electrical connector commonly used in spacecraft assembly

Type	Frequency range	Features	Tightening torque of electrical connector
SMA	0～18GHz	Small size, frequency bandwidth, superior electromechanical performance, high reliability, the most widely used RF connector	0.8Nm~1.2Nm
TNC	DC～11GHz	Small and medium power connectors with strong vibration resistance, high reliability and excellent electromechanical performance	2.3Nm~2.65Nm
SSMA	0～30GHz	Small size, high frequency bandwidth, superior electromechanical performance, high reliability, etc.	0.8Nm~1.2Nm

3. Reliability scheme of RF coaxial cable assembly process

3.1. Research content

At present, the assembly of flexible RF coaxial cable has strong implementation capabilities in terms of technology, operation, inspection and other aspects, but there are still some problems. For instance, the process standards cannot provide strong guidance for the installation of flexible RF coaxial cable assemblies, the level of quantification and refinement are not enough; operators do not have a deep understanding of the performance and installation impact factors of flexible RF coaxial cable, which are mainly reflected in two aspects, (1) the effects of bending radius of flexible RF coaxial cable on installation, and (2) the effects of the number of cable winding coils on installation.

In view of the above-mentioned influencing factors, some process tests were carried out to obtain some refinement and quantitative indicators of process parameters. The specific research content is as follows: (1) verify the reliability and safety of flexible RF coaxial cables commonly used in satellite assembly under various working conditions (bending radius, number of coils, etc.), (2) determine the process parameters of flexible RF coaxial cables and electrical connectors during satellite assembly, and (3) determine the operation methods and quality control requirements of flexible RF coaxial cables and electrical connectors during satellite assembly.

3.2. Process reliability scheme of RF coaxial cable assembly

According to the current installation of flexible RF coaxial cables in satellite assembly, the values of the key performance parameters of the cable assembly (voltage standing wave ratio (VSWR), phase) before and after temperature shock test, shock test and vibration test are compared with the standard requirements, to obtain the reliable process methods and process parameters.

RF coaxial cable assemblies use temperature-stabilized, low-loss aerospace-grade products, operating temperature is -55℃ to +125℃, operating frequency is DC to 18GHz, and VSWR is no greater than 1.3.

3.3. Process test scheme for flexible RF coaxial cable

Select two types of flexible RF coaxial cables, UFB205A and UFB311A, and SMA and TNC connectors used with them to simulate their typical conditions on a spacecraft, design different comparison test items, and analyze the test data to obtain correct and reliable process methods and process parameters. It mainly includes related technical parameters such as cable laying and fixing requirements, minimum bending radius and the number of cable winding coils. The installation state of the flexible RF cable test pieces on the vibration test bench are shown in Figure 3.

The temperature shock test conditions are carried out in accordance with article 4.5.5 of the standard GJB1215A-2005 (test temperature -55℃~+100℃, heating and cooling rate no greater than 10℃/min, number of cycles 5 times, ultimate holding time 2 hours). After the test, the flexible RF cable assemblies are left for more than 24 hours for electrical performance testing.

The vibration test conditions are carried out in accordance with the provisions of article 4.5.18 of GJB1215A-2005, and no electrical interruption longer than 1μs is allowed. The test conditions are shown in Table 3.

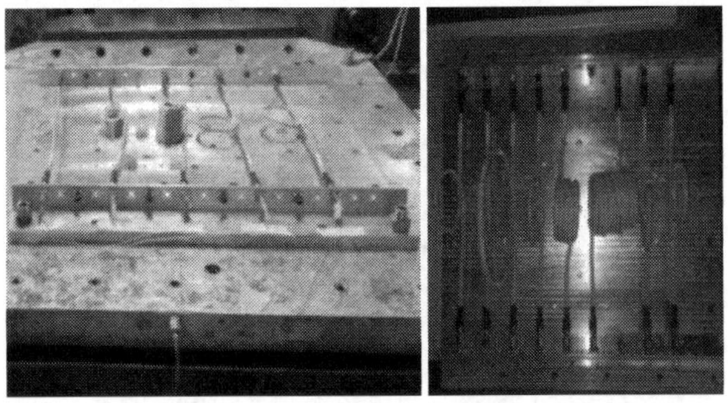

Figure 3. Flexible RF cable process test

Table 3. Conditions for vibration test

Sine vibration	Frequency Range	10Hz~2000Hz
	Acceleration	15g
	Vibration amplitude	1.5mm
	Scan speed	1 oct/min
	Test duration	one cycle, 10Hz~2000Hz~10Hz
Random vibration	Frequency Range	20Hz~2000Hz, 1.5g
	Acceleration	15g
	Power spectral density	$0.1g^2/Hz$
	Test duration	10min, each axis

4. Test data and analysis

4.1. Effects of bending radius on RF parameters of cable assemblies

4.1.1. Effects of bending radius on RF parameters of cable assemblies under temperature shock test. 1) Effects of bending radius on VSWR of cable assemblies under temperature shock test

It can be seen from Table 4 that there is no regular relationship between the change rate of the VSWR and the cable bending radius for the two types of cable samples (whether equal length or not).

Table 4. Test data of cable assemblies before and after temperature shock test Ⅰ

Cable model Connector model (Test frequency)	Sample No. (length mm)	Bending radius (mm)	Before the test	After the test	Rate of change	Whether to exceed 20%
UFB205A SMA-J210G (DC-16GHz)	No. 1 (520)	12.7	1.279	1.407	10.01	No
	No. 2 (580)	25.4	1.254	1.529	21.9	Yes
	No. 3 (660)	38.1	1.207	1.44	19.3	No
	Additional No. 1 (420)	12.7	1.215	1.213	0.16	No
	Additional No. 2 (420)	25.4	1.158	1.151	0.6	No
	Additional No. 3 (420)	38.1	1.168	1.162	0.5	No
UFB311A TNC-J311AG (DC-7GHz)	No. 9 (650)	32	1.318	1.335	1.29	No
	No. 10 (700)	40	1.159	1.257	8.46	No
	No. 11 (950)	80	1.25	1.266	1.28	No

2) Effects of cable bending radius on phase of cable assemblies under temperature shock test

From Table 5, if we only look at the data of sample No. 1, No. 2, No. 3 and No. 9, No. 10, No. 11, it seems that as the cable bending radius increases, the phase change rate of the two cables gradually decreases. However, referring to the data of the additional samples, the phase change rate is irregular. Therefore, it cannot be said that the cable bending radius and the phase change rate have a regular relationship.

Table 5. Test data of cable assemblies before and after temperature shock test II

Cable model Connector model (Test frequency)	Sample No. (length mm)	Bending radius (mm)	Before the test (°)	After the test (°)	Change (°)	Rate of change (°/m)
UFB205A SMA-J210G (DC-16GHz)	No. 1 (520)	12.7	-125.16	-14.16	111	213.46
	No. 2 (580)	25.4	-71.86	44.79	116.65	201.12
	No. 3 (660)	38.1	153.86	-93.2	112.94	171.12
	Additional No. 1 (420)	12.7	121.44	-138.61	99.95	237.98
	Additional No. 2 (420)	25.4	130.76	-39.82	189.42	451
	Additional No. 3 (420)	38.1	126.04	-109.71	124.25	295.83
UFB311A TNC-J311AG (DC-7GHz)	No. 9 (650)	32	-91.7	76.56	168.26	258.86
	No. 10 (700)	40	151.32	-43.44	165.24	236.06
	No. 11 (950)	80	-60.65	106.72	167.37	176.18

4.1.2. Effects of bending radius on RF parameters of cable assemblies under sine vibration test. 1) Effects of cable bending radius on VSWR of cable assemblies under sine vibration test

The data in Table 6 shows that the changes of VSWR of the two types of cable assemblies are very small, and the changes of the bending radius and the VSWR are irregular.

Table 6. Test data of cable assemblies before and after sine vibration test I

Cable model Connector model (Test frequency)	Sample No. (length mm)	Bending radius (mm)	Before the test	After the test	Change
UFB205A SMA-J210G (DC-16GHz)	No. 1 (520)	12.7	1.405	1.381	-0.024
	No. 2 (580)	25.4	1.43	1.389	-0.041
	No. 3 (660)	38.1	1.433	1.424	-0.009
UFB311A TNC-J311AG (DC-7GHz)	No. 9 (650)	32	1.354	1.422	0.068
	No. 10 (700)	40	1.213	1.275	0.062
	No. 11 (950)	80	1.215	1.273	0.058

2) Effects of bending radius on phase of cable assemblies under sine vibration test

It can be seen from Table 7 that the phase changes of the two types of cables are not large, and the phase changes of the UFB205A cable are irregular. Combined with the phase change of UFB205A and UFB311A, the bending radius of the cable and the change rate of the phase of the cable assemblies should have no regular relationship.

Table 7. Test data of cable assemblies before and after sine vibration test II

Cable model Connector model (Test frequency)	Sample No. (length mm)	Bending radius (mm)	Before the test (°)	After the test (°)	Change (°)
UFB205A SMA-J210G (DC-16GHz)	No. 1 (520)	12.7	-14.58	-14.36	0.22
	No. 2 (580)	25.4	37.78	37.32	0.46
	No. 3 (660)	38.1	-91.27	-90.98	0.29
UFB311A TNC-J311AG (DC-7GHz)	No. 9 (650)	32	73.75	74.03	0.28
	No. 10 (700)	40	-43.99	-43.61	0.38
	No. 11 (950)	80	106.45	106.96	0.51

IOP Conf. Series: Materials Science and Engineering **825** (2020) 012048 doi:10.1088/1757-899X/825/1/012048

4.1.3. Effects of bending radius on RF parameters of cable assemblies under random vibration test. 1) Effects of bending radius on VSWR of cable assemblies under random vibration test

For the UFB205A cable, it can be seen that the variation of VSWR of samples No. 1 and No. 4 is 0.006. The bending radius of sample No. 2 is twice as large as that of sample No. 1. The variation of VSWR is also larger than that of sample No. 1. It cannot be considered that there is a regular relationship between the bending radius and the change of the VSWR. For the UFB311A cable, it can be seen from Table 8 that the variation of VSWR of the three cable assemblies of samples No. 9, No. 10 and No. 11 is also irregular.

Table 8. Test data of cable samples before and after random vibration test I

Cable model Connector model (Test frequency)	Sample No. (length mm)	Bending radius (mm)	Before the test	After the test	Change
UFB205A SMA-J210G (DC-16GHz)	No. 1 (520)	12.7	1.381	1.394	0.013
	No. 2 (580)	25.4	1.389	1.41	0.021
	No. 3 (660)	38.1	1.424	1.456	0.032
	No. 4 (520)	12.7	1.439	1.446	0.007
UFB311A TNC-J311AG (DC-7GHz)	No. 9 (650)	32	1.422	1.479	0.057
	No. 10 (700)	40	1.275	1.283	0.008
	No. 11 (950)	80	1.273	1.225	-0.048

2) Effects of cable bending radius on phase of cable assemblies under random vibration test

From Table 9, the amount of phase change is small, and the effects of the bending radius on the phase is irregular.

Table 9. Test data of cable samples before and after random vibration test II

Cable model Connector model (Test frequency)	Sample No. (length mm)	Bending radius (mm)	Before the test (°)	After the test (°)	Change (°)
UFB205A	No. 1 (520)	12.7	-14.36	-14.46	-0.13
SMA-J210G	No. 2 (580)	25.4	37.32	38.07	0.75
(DC-16GHz)	No. 3 (660)	38.1	-90.98	-90.85	-0.13
UFB311A	No. 9 (650)	32	74.03	74.65	0.62
TNC-J311AG	No. 10 (700)	40	-43.61	-43.38	0.23
(DC-7GHz)	No. 11 (950)	80	106.96	106.95	-0.01

4.2. Effects of cable winding number on RF parameters of cable assemblies

4.2.1. Effects of cable winding number on RF parameters of cable assemblies under temperature shock test. 1) Effects of cable winding number on VSWR of cable assemblies under temperature shock test

From the data in Table 10, there is no regular change rate of VSWR. Moreover, the data of additional samples also prove that there is no regular relationship between the number of cable winding coils and the change rate of VSWR.

Table 10. Test data of cable samples before and after temperature shock test III

Cable model Connector model (Test frequency)	Sample No. (length mm)	Cable winding number	Before the test	After the test	Rate of change	Whether to exceed 20%
UFB205A SMA-J210G (DC-16GHz)	No. 4 (520)	1	1.219	1.41	15.7	No
	No. 5 (950)	5	1.21	1.384	14.4	No
	No. 6 (1530)	10	1.23	1.413	14.9	No
	Additional No. 4 (1250)	1	1.219	1.213	0.5	No
	Additional No. 5 (1250)	5	1.186	1.178	0.7	No
	Additional No. 6 (1250)	10	1.282	1.284	0.2	No
UFB311A	No. 12 (650)	1	1.22	1.27	4.1	No

TNC-J311AG	No. 13 (1650)	5	1.183	1.26	6.51	No
(DC-7GHz)	No. 14 (2910)	10	1.195	1.243	4.02	No

2) Effects of cable winding number on phase of cable assemblies under temperature shock test

From Table 11, the change rate of the phase of the formal sample decreases with the increase of the cable winding number. However, the change rate of the phase of the additional sample does not support this rule. Therefore, it cannot be explained that there is a regular relationship between the phase change rate and the number of cable winding coils under the conditions of the temperature shock test.

Table 11. Test data of cable samples before and after temperature shock test Ⅳ

Cable model Connector model (Test frequency)	Sample No. (length mm)	Cable winding number	Before the test (°)	After the test (°)	Change (°)	Rate of change (°/m)
UFB205A SMA-J210G (DC-16GHz)	No. 4 (520)	1	-96.86	9.24	106.1	204
	No. 5 (950)	5	53.81	166.21	112.4	118.3
	No. 6 (1530)	10	71.51	179.66	108.15	70.68
	Additional No. 4 (1250)	1	-4.59	-85.94	81.35	65.08
	Additional No. 5 (1250)	5	46.12	-30.9	77.02	61.62
	Additional No. 6 (1250)	10	-151.82	17.28	190.9	152.72
UFB311A TNC-J311AG (DC-7GHz)	No. 12 (650)	1	-132.81	31.1	163.91	252.2
	No. 13 (1650)	5	-144.86	14.91	159.71	96.8
	No. 14 (2910)	10	-100.95	62.2	163.15	56.1

4.2.2. Effects of cable winding number on RF parameters of cable assemblies under sine vibration test. 1) Effects of cable winding number on VSWR of cable assemblies under sine vibration test

From the data in Table 12, there is no regular change rate of VSWR.

Table 12. Test data of cable samples before and after sine vibration test Ⅲ

Cable model Connector model (Test frequency)	Sample No. (length mm)	Bending radius (mm)	Cable winding number	Before the test	After the test	Change
UFB205A	No. 4 (520)	12.7	1	1.443	1.439	-0.004
SMA-J210G	No. 5 (950)	12.7	5	1.422	1.423	0.001
(DC-16GHz)	No. 6 (1530)	12.7	10	1.414	1.455	0.041
UFB311A	No. 12 (650)	32	1	1.244	1.242	-0.002
TNC-J311AG	No. 13 (1650)	32	5	1.264	1.248	-0.016
(DC-7GHz)	No. 14 (2910)	32	10	1.248	1.251	0.003

2) Effects of cable winding number on phase of cable assemblies under sine vibration test

It can be seen from Table 13 that although the phase change rate is not large, the phase change rate decreases as the number of cable winding coils increases.

Table 13. Test data of cable samples before and after sine vibration test Ⅳ

Cable model Connector model (Test frequency)	Sample No. (length mm)	Bending radius (mm)	cable winding number	Before the test (°)	After the test (°)	Change (°)	Rate of change (°/m)
UFB205A	No. 4 (520)	12.7	1	100.35	99.64	0.71	1.36
SMA-J210G	No. 5 (950)	12.7	5	163.77	163.3	0.47	0.49
(DC-16GHz)	No. 6 (1530)	12.7	10	178.03	177.42	0.61	0.40
UFB311A	No. 12 (650)	32	1	30.17	30.54	0.37	0.57
TNC-J311AG	No. 13 (1650)	32	5	13.93	13.61	0.32	0.19
(DC-7GHz)	No. 14 (2910)	32	10	62.85	62.76	0.21	0.09

4.2.3. Effects of cable winding number on RF parameters of cable assemblies under random vibration test. 1)Effects of cable winding number on VSWR of cable assemblies under random vibration test

It can be seen from Table 14 that, whether it is a UFB205A cable or a UFB311A cable, the change in the voltage standing wave ratio is very small, and the change and the number of coil windings are irregular.

Table 14. Test data of cable assemblies samples before and after random vibration test III

Cable model Connector model (Test frequency)	Sample No. (length mm)	Bending radius (mm)	Cable winding number	Before the test	After the test	Change
UFB205A	No. 4 (520)	12.7	1	1.439	1.446	0.007
SMA-J210G	No. 5 (950)	12.7	5	1.423	1.41	-0.013
(DC-16GHz)	No. 6 (1530)	12.7	10	1.455	1.456	0.001
UFB311A	No. 12 (650)	32	1	1.242	1.263	0.023
TNC-J311AG	No. 13 (1650)	32	5	1.248	1.242	-0.006
(DC-7GHz)	No. 14 (2910)	32	10	1.251	1.239	-0.012

2) Effects of cable winding number on phase of cable assemblies under random vibration test

From Table 15, the rate of change of the phases of the two types of cables is not large and irregular.

Table 15. Test data of cable assemblies samples before and after random vibration test IV

Cable model Connector model (Test frequency)	Sample No. (length mm)	Bending radius (mm)	Cable winding number	Before the test (°)	After the test (°)	Change (°)	Rate of change (°/m)
UFB205A	No. 4 (520)	12.7	1	99.24	99.18	-0.06	-0.115
SMA-J210G	No. 5 (950)	12.7	5	163.3	163.61	0.31	0.326
(DC-16GHz)	No. 6 (1530)	12.7	10	178.42	178.71	0.29	0.190
UFB311A	No. 12 (650)	32	1	30.54	30.67	0.13	0.2
TNC-J311AG	No. 13 (1650)	32	5	13.61	13.83	0.22	0.133
(DC-7GHz)	No. 14 (2910)	32	10	62.76	62.35	-0.41	-0.187

5. Conclusion

This paper studies the process reliability of flexible RF cable assemblies in spacecraft assembly. Due to the small number of test samples, the test data has a certain degree of dispersion, but it does not affect the conclusion of the overall test and achieves the expected effect and purpose. Through the study of this subject, it is concluded that (1) the static bending radius and dynamic bending radius of the RF cable during installation should not exceed the ranges given by the cable manufacturer, and (2) the number of cable winding coils of the RF cable does not affect the performance index of the cable, but in the case of high cable phase requirements, the fewer the number of coils the better.

References

[1] David and Vye. 2012 RF connector and cable assembly outlook [J]. Microwave journal, 2012, 55: 6-18.

[2] Richard M. and Pat H. RF cable and connector outlook [J]. Microwave Journal, 2014, 57: 6-10.

[3] Slack D. Microwave and RF cable assemblies: the neglected system component. Part 1: Electrical and mechanical aspects of the coaxial assembly [J]. Applied Microwave and Wireless, 1997, 9(6): 36-45.

[4] Zhang W., Chen C., Chen H., Sun G., and Zhao S. Application of Augmented Reality in Satellite Cable Network Assembly [C]. IOP Conference Series: Materials Science and Engineering, 2019, 542: 012034.

[5] Guo T., Hu R., Xiao Z., Zhao J., and Fang Z. Research to Assembly Scheme for Satellite Deck Based on Robot Flexibility Control Principle[C]. IOP Conference Series: Materials Science and Engineering, 2018, 324: 012086.

[6] Jiao C., Jiang X., Li X., and Liu Y. Vision based cable assembly in constrained environment [C]. IEEE International Conference on Robotics and Biomimetics (ROBIO), 2018, 8-13.

IOP Conf. Series: Materials Science and Engineering **825** (2020) 012049 doi:10.1088/1757-899X/825/1/012049

Experimental Study on Friction Torque of Wet Clutch under Three Kind of Contact Areas

YAN Xu[1], CHEN Man[1,2], LI He-yan[3], ZHANG Bin[1] and ZHANG Xiao-peng[1]

[1.]School of Mechnical Engineering, Beijing Institute of Technology, Beijing 100081, China;

[2.]Collaborative Innovation Center of Electric Vehicles in Beijing, Beijing Institute of Technology, Beijing 100081, China;

[3.]College of Urban Transportation and Logistics, Shenzhen Technology University, Shenzhen 518118, China

[4.]Jianglu Machinery Electronics Group Co. Ltd, Xiangtan 411100, China

Email: 18651058502@163.com

Abstract. It is of great significance to study the friction characteristics in the process of slippage and friction of the wet clutch for optimizing vehicle start, improving vehicle driving comfort and prolonging the service life of the clutch. Based on the test data of wet multi-plate clutch under braking condition in SAE#2 test bench friction torque test, we analyze the influence of rotate speed and average pressure on the friction torque through the comparison of friction torque under two friction pairs. The results show that in the case of sufficient lubricating oil flow, the attenuation of multiple pairs of friction plates has nothing to do with the proportion of the groove area of friction plates. Friction torque attenuation will occur under any groove area, but different groove areas will affect the rotate speed and average pressure of this phenomenon. In general, the frictional torque attenuation of wet multi-plate clutch gradually increases with the increase of relative speed and decreases with the increase of average surface pressure.

1. Introduction

As an indispensable mode of transportation in modern society, automobile has been developing towards the direction of high life span, high power density and high reliability[1]. As an important part of automobile transmission system, the working performance of the wet clutch directly determines the working performance of the transmission system, thus affecting the safety, comfort and shifting smoothness of the whole vehicle. The friction torque transferred by the clutch in the process of sliding and friction has a significant impact on the wet clutch. The study on the friction torque characteristics in the process of sliding and friction of the clutch is of great significance for improving the comfort of the vehicle in the driving process, extending the service life of the clutch and improving the working performance of the transmission system.

At present, domestic and foreign scholars have done a lot of theoretical analysis and experimental research on the friction torque in the process of sliding friction of wet clutch. Cao fei and Zhu hongqing studied the friction performance and thermal load performance of wet clutch[2-3]. Yu liang, Li heyan et al. established a torque calculation model considering the axial average specific pressure attenuation of the friction pair with key friction, and simulated the influence of key friction on the axial average specific pressure attenuation of the friction pair with clutch[4]. Zhang heng et al. proposed the friction torque attenuation coefficient of wet multi-plate clutch based on the friction torque attenuation phenomenon in SAE#2 friction torque test[5]. Hashimoto and Kato first studied the oil film distribution between the friction sheet and the steel sheet under the action of centrifugal force, which laid a framework for establishing an accurate model of the towing torque[6-7]. Holgerson et al. conducted the bench test of the wet clutch with the SAE#2 test bed, and concluded that appropriate reduction of the joint pressure and temporary closure of the driving torque could effectively reduce the instability of the transfer torque caused by the maximum temperature rise of the friction pair during the engagement of the wet clutch[8]. Chinar et al. studied the influence of grooves on gas-liquid flow in the separation state through experiments and numerical calculations[9].

In order to study the torque transfer capacity of the clutch multi-friction pair, this paper studies the friction torque attenuation phenomenon according to the friction torque test data of two and six friction pairs measured on the SAE#2 test bench. We find that this phenomenon will occur under any groove area, but different groove areas affect the relative speed and average surface pressure when this phenomenon occurs. Finally, we analyze the reason of friction torque attenuation of wet multi-plate clutch.

2. Study on friction torque mechanism of clutch

In the process of clutch sliding and friction, the friction torque is calculated by the formula[10]

$$T = \frac{2\pi}{3} \cdot N \cdot \mu \cdot p_0 \cdot \left(R_O^3 - R_I^3\right) \tag{1}$$

In formula (1), N is the number of friction pairs; p_0 is the average surface pressure of friction pair; R_0 and R_i are the outer diameter and inner diameter of the friction plate respectively; μ is the average friction coefficient of two pairs of clutches.

2.1 Pressure p_0

In this test, five working conditions of clutch pressure were selected, namely 25kPa, 50kpa, 75kpa, 100kpa and 200kpa. The average surface pressure of the piston on the friction pair was 0.2Mpa, 0.4Mpa, 0.6Mpa, 0.8Mpa and 1.6Mpa respectively.

2.2 Friction characteristic μ

Based on the data obtained from the pin-disk friction and wear test, the empirical formula of the comprehensive friction coefficient of the wet friction pair (65Mn-copper-base powder metallurgy) with velocity, temperature and pressure as independent variables was fitted as

$$\mu = 23e^{\left(\frac{-3v}{(\ln T - 3.2)((28.3p)^{0.44} - 0.87)} - 5.16\right)} + 0.08\left(e^{-0.005T} - 1\right)\left(e^{-0.2v} - 1\right) + \frac{0.008\ln(4v+1)}{e^{0.005T}} - 0.005\ln(28.3p) + 0.020 \tag{2}$$

In formula (2), v is the relative sliding linear velocity of the median diameter of the friction pair, $v = (n \cdot r_e)/60$, its unit is *m/s*; *T* is the temperature, the unit is ℃.

Therefore, the formula for calculating the friction torque is

$$T = \frac{2\pi}{3} \cdot N \cdot \left[23e^{\left(\frac{-3v}{(\ln T - 3.2)((28.3p)^{0.44} - 0.87)} - 5.16\right)} + 0.08\left(e^{-0.005T} - 1\right)\left(e^{-0.2v} - 1\right) + \frac{0.008\ln(4v+1)}{e^{0.005T}} - 0.005\ln(28.3p) + 0.020\right] \cdot p_0 \cdot \left(R_O^3 - R_I^3\right) \tag{3}$$

By the reference 3, it can be seen that due to the existence of friction at the spline, the total friction torque has a non-linear relationship with the number of friction pairs. The more the number of friction pairs, the smaller the average single pair of friction torque, and the more obvious the role of friction at the key. So the formula can only be used for single pair ideally. In many cases, wet clutch friction damping torque, in order to study the differences, this paper analyzes SAE# 2 brake friction testing machine working conditions of the test data. Combined with formula (3), we studied the influence of groove area and the number of friction pair on friction torque.

3. Test method

The SAE#2 test bed is a test device used to study the friction characteristics of the clutch. The whole test system is shown in figure 1. The motor provides power to drive the inertia and the active end of the clutch to rotate. The compressed air of the air pump ACTS directly on the piston to complete the combination of the clutch friction pair. The clutch oil by the hydraulic pump through the filter and cooling switch, into the lubricating oil road.

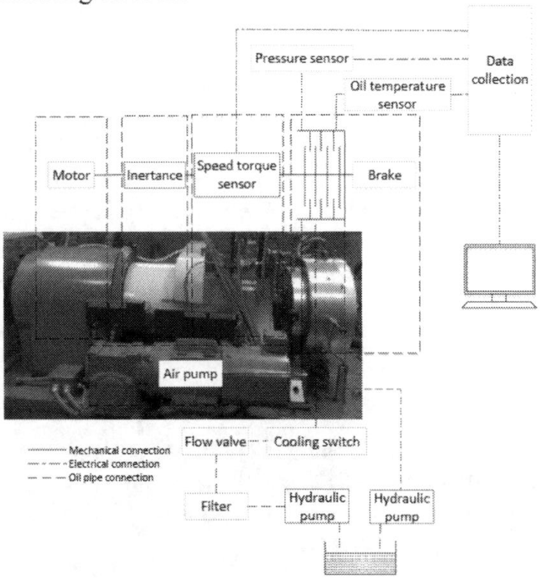

Figure 1. SAE#2 clutch test system

The dual steel sheet material used in the test clutch is 65Mn. the friction sheet material is copper based powder metallurgy material and nano-modifier is added. And the lubrication groove is double circular groove, with groove area accounting for 25%, 35% and 50%, respectively, as shown in figure 2.The lubricating oil model is 10W/40-CF, the lubrication flow is 4mL/min/cm^2. The lubricating oil temperature is controlled by the cooling switch.

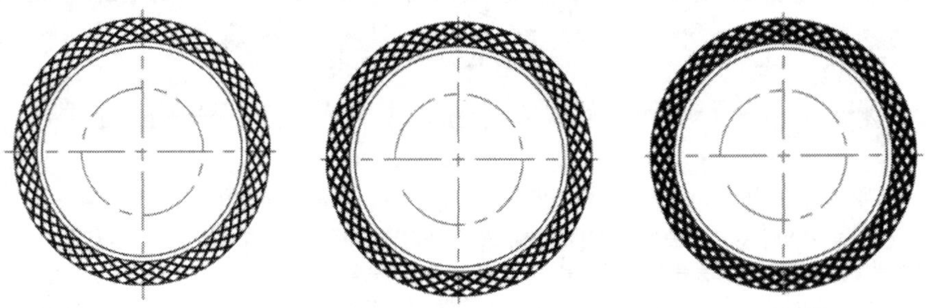

(a)25% groove area (b)35% groove area (c)50% groove area

Figure 2. The friction plate form of three lubricating oil groove

4. Friction torque

It can be obtained from formula (3)

$$T_6 = 6 \times r_e \cdot p \cdot A_c \cdot \xi_{2-6} \cdot \mu = 3 \times (2 \times r_e \cdot p \cdot A_c \cdot \xi_{2-6} \cdot \mu) = 3T_2 \quad (4)$$

In the formula (4), T_2 represents the measured friction torque of two friction pairs, and T_6 represents the calculated friction torque of six friction pairs relative to the friction torque of two friction pairs. Therefore, under the same working condition, the calculated friction torque of six friction pairs should be three times of the benchmark friction torque of two friction pairs.

According to the test data of friction torque of two and six friction pairs under the three grooves of 25%, 35% and 50% measured by SAE#2 friction testing machine, the friction torque of wet multi-plate clutch under braking conditions is drawn as shown in Figure 3, 4 and 5.

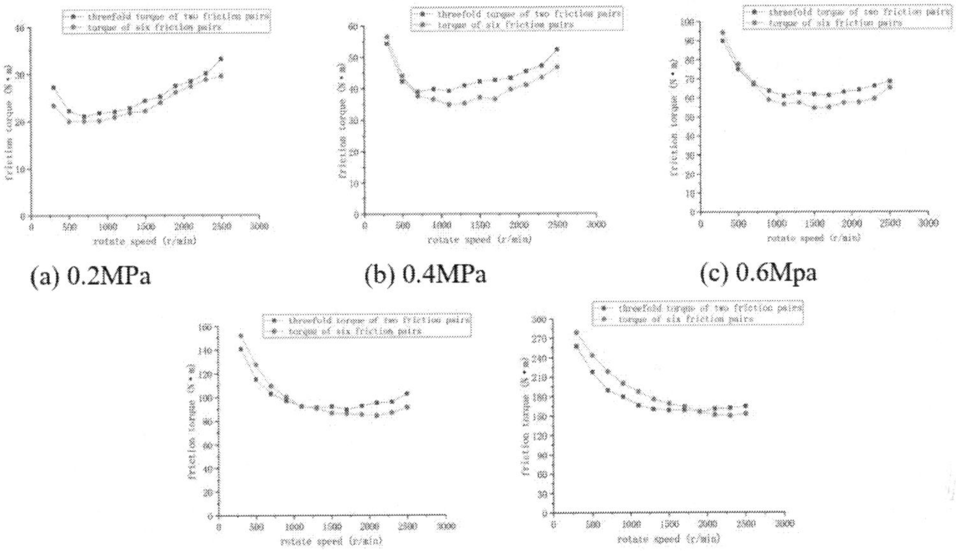

(a) 0.2MPa (b) 0.4MPa (c) 0.6Mpa

(d) 0.8MPa (e) 1.6MPa

Figure 3. Comparison of friction torque under 25% groove area

Figure 3 shows the comparison between the threefold friction torque of two friction pairs and the friction torque of six friction pairs under different surface-pressure when the groove area accounts for 25%. As figure 3 (a) shows, the friction torque of 3 times two friction pairs is always greater than that of six friction pairs. As figure 3 (b) shows, when the speed is lower than 600r/min, the friction torque of six friction pairs is greater than 3 times that of two friction pairs. When the speed is higher than 600r/min, the friction torque of the six friction pairs is less than 3 times that of the two friction pairs. As figure 3 (c) shows, when the speed is less than 700r/min, the friction torque of the six friction pairs is greater than 3 times of the friction torque of the two friction pairs; when the speed is higher than 700r/min, the friction torque of the six friction pairs is less than 3 times that of the two friction pairs. As figure 3 (d) shows, when the speed is less than 1100r/min, the friction torque of the six friction pairs is more than 3 times that of the two friction pairs. When the speed is higher than 1100r/min, the friction torque of the six friction pairs is less than 3 times that of the two friction pairs. As figure 3 (e) shows, when the speed is lower than 1900r/min, the friction torque of six friction pairs is greater than 3 times of the friction torque of two friction pairs; when the speed is higher than 1900r/min, the friction torque of the six friction pairs is less than 3 times that of the two friction pairs. The speed at which the frictional torque is equal is called the critical speed.

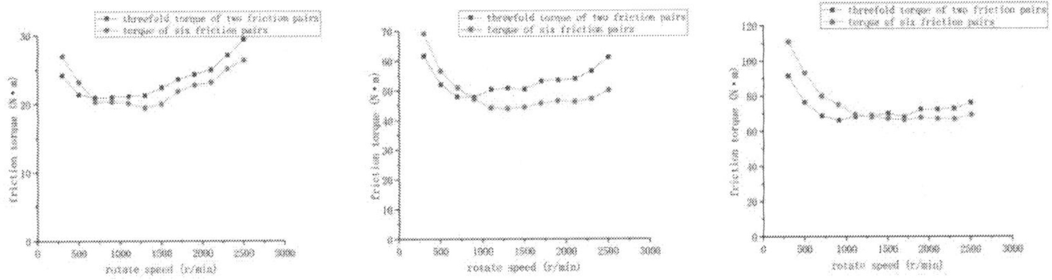

(a) 0.2MPa (b) 0.4MPa (c) 0.6Mpa

IOP Conf. Series: Materials Science and Engineering **825** (2020) 012049 doi:10.1088/1757-899X/825/1/012049

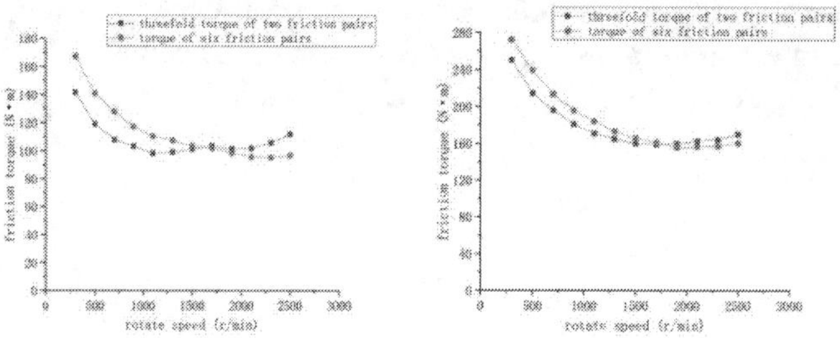

(d) 0.8MPa (e) 1.6MPa

Figure 4. Comparison of friction torque under 35% groove area

Figure 4 shows the comparison between the friction torque of two friction pairs and the friction torque of six friction pairs under three times the pressure on different surfaces when the proportion of groove area is 35%. As figure 4 (a) shows, when the speed is lower than 600r/min, the friction torque of six friction pairs is greater than 3 times that of two friction pairs; when the speed is higher than 600r/min, the friction torque of the six friction pairs is less than 3 times that of the two friction pairs. As figure 4 (b) shows, when the speed is lower than 900r/min, the friction torque of six friction pairs is greater than 3 times of the friction torque of two friction pairs. When the speed is higher than 900r/min, the friction torque of the six friction pairs is less than 3 times that of the two friction pairs. As figure 4 (c) shows, when the speed is less than 1200r/min, the friction torque of the six friction pairs is more than 3 times that of the two friction pairs; when the speed is higher than 1200r/min, the friction torque of the six friction pairs is less than 3 times that of the two friction pairs. As figure 4 (d) shows, when the speed is lower than 1600r/min, the friction torque of six friction pairs is greater than 3 times of the friction torque of two friction pairs. When the speed is higher than 1600r/min, the friction torque of the six friction pairs is less than 3 times that of the two friction pairs. As figure 4 (e) shows, when the speed is lower than 1800r/min, the friction torque of six friction pairs is greater than 3 times of the friction torque of two friction pairs; when the speed is higher than 1800r/min, the friction torque of the six friction pairs is less than 3 times that of the two friction pairs.

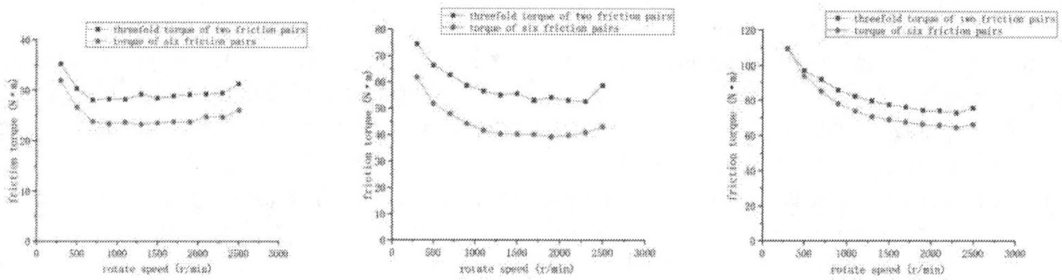

(a) 0.2MPa (b) 0.4MPa (c) 0.6Mpa

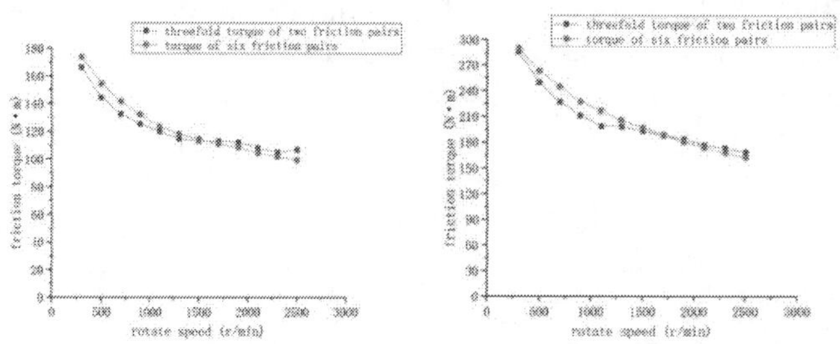

(d) 0.8MPa (e) 1.6MPa

Figure 5. Comparison of friction torque under 50% groove area

Figure 5 shows the comparison between the friction torque of two friction pairs and the friction torque of six friction pairs under three times the pressure on different surfaces when the groove area accounts for 25%. As figure 5(a) and (b) show, the friction torque of three times of two friction pairs is always greater than that of six friction pairs. As figure 5 (c) shows when the speed is higher than 300r/min, the friction torque of the two friction pairs 3 times is greater than that of the six friction pairs; when the speed is higher than 1700r/min, the friction torque of the six friction pairs is less than 3 times that of the two friction pairs. As figure 5 (d) shows, when the speed is lower than 1500r/min, the friction torque of six friction pairs is greater than 3 times that of two friction pairs; when the speed is higher than 1500r/min, the friction torque of the six friction pairs is less than 3 times that of the two friction pairs. As figure 5 (e) shows, when the speed is less than 1700r/min, the friction torque of six friction pairs is greater than 3 times of the friction torque of two friction pairs; when the speed is higher than 1700r/min, the friction torque of the six friction pairs is less than 3 times that of the two friction pairs.

5. Analysis of influence factors

As can be seen from literature 3, the theoretical required lubrication flow rate of a single friction pair covered by the whole oil film decreases with the increase of the average surface pressure of the friction pair and increases with the increase of the relative speed, and the required maximum lubrication flow rate is lower than 0.03L/min. In this test, the lubrication flow per unit area of the clutch friction pair was 4mL/min/cm2, and that of the single friction pair was 0.217L/min, which was much higher than the theoretical lubrication flow required by the single friction pair under the full oil film. Lubricating oil film dynamic pressure effect. Therefore, the remaining lubricating oil covered by all-oil film will lead to an increase in the ratio of viscous torque of multiple friction pairs and a decrease in the contact area of micro-convex peak, resulting in a smaller friction coefficient of six pairs under the same working condition. And a smaller friction torque of multiple friction pairs under the same working condition is equivalent to a smaller friction torque of two friction pairs.

Take the friction torque when the proportion of groove area is 35% as an example. When the pressure is constant, when the speed is lower than the critical speed, the friction torque of the six friction pairs is greater than 3 times the friction torque of the two friction pairs. When the speed is higher than the critical speed, the friction torque of the six friction pairs is gradually less than 3 times the friction torque of the two friction pairs. This is because the friction pair mean surface pressure is constant, with the increase of relative speed, full oil film lubrication theory demand under the cover of friction pair traffic increases, the oil membrane covering the rest of lubricating oil flow is reduced, oil film thickness increases, the rough peak contact gradually reduce, declining real contact area, six friction pair friction coefficient decreases, and six vice friction torque decreases. With the increase of the average surface pressure, the critical speed increases gradually. This is because with the increase of the average surface pressure, the flow rate of the remaining lubricating oil covered by the whole oil film increases, the oil film thickness gradually decreases, and the rough peak contact gradually increases, the friction coefficient of the six friction pairs increases accordingly, resulting in the increase of the friction torque of the six friction pairs, which requires a higher speed to reduce the pressure on the flow rate of the lubricating oil.

6. Conclusion

Based on the test data on SAE#2 test bench on the sliding and friction process of wet multi-plate clutch under braking conditions, the variation trend of friction torque of two and six friction pairs under different groove areas was measured, and the causes of the trend were analyzed. The research results show that:

1) The test results show that, the friction torque of the six friction pairs is less than 3 times that of the two friction pairs when the relative speed is larger the critical speed. This is because under the condition of sufficient flow of lubricating oil of friction pair, the ratio of rough contact torque of multi-pair clutch is small, and the ratio of viscous torque is large. With the same relative speed and average surface pressure, the friction coefficient of multiple friction pairs is less than that of two friction pairs.

2) The frictional torque attenuation phenomenon occurred under the three kinds of groove areas of 25%, 35% and 50%, and the groove area only affected the relative speed and average surface pressure when the phenomenon occurred. The larger the groove area, the greater the average surface pressure.

3) When the critical speed is reached, the friction torque of the six pairs is equal to that of the three pairs, and then the friction torque of the six pairs is gradually less than that of the three pairs. The critical speed increases as the pressure increases.

Reference

[1] WALKER P D, ZHANG N, TAMBA R. Control of gear shifts in dual clutch transmission powertrains[J]. Mechanical Systems & Signal Processing, 2011, 25(6):1923-1936.

[2] Zhang Heng, Li Heyan, et al. Experimental Study on Attenuation Characteristics of Friction Torque Transferred by the Wet Multi-disc Clutch[J]. Journal of Harbin Institute of Technology, ,2018,(07):94-102(in Chinese).

[3] Yu Liang, Li Heyan, Ma Biao, et al. An analysis model for axial pressure decay of Multi-disc clutch with the effect of spline friction[J]. Journal of Jilin University(Engineering and Technology Edition),2018,(04):990-997(in Chinese).

[4] Cao Fei. Study on the Characteristic of Wet Multi-disk Clutch Engagement Process[D]. Central South University,2013(in Chinese).

[5] Zhu Hongqing. Study on slipping and thermal characteristics of wet clutch[D]. Zhe Jiang University,2012(in Chinese).

[6] Kato Y, Murasugi T, Hirano H, et al. Fuel economy improvement through tribological analysis of the wet clutches and brakes of an automatic transmission[J]. Soc. Automot. Eng.[Spec. Publ.], 1993, 16: 57-60.

[7] Hashimoto H, Wada S, Murayama Y. The performance of a turbulent-lubricated sliding bearing subject to centrifugal effect[J]. Trans. Jpn. Soc. Mech. Eng. Ser. C, 1984, 49(446): 1753-1761.

[8] Mansouri M , Holgerson M , Khonsari M M , et al. Thermal and dynamic characterization of wet clutch engagement with provision for drive torque[J]. Journal of Tribology, 2001, 123(2):313-323.

[9] Aphale C R , Schultz W W , Ceccio S L . The Influence of Grooves on the Fully Wetted and Aerated Flow Between Open Clutch Plates[J]. Journal of Tribology, 2010, 132(1):011104.

[10] Qu Zaiying, Huang Yuechu. Friction materials for powder metallurgy[M]. Metallurgical industry press, 2005.

AUTHOR INDEX